Monitoring and Assessment of Environmental Quality in Coastal Ecosystems, 4th Edition

Monitoring and Assessment of Environmental Quality in Coastal Ecosystems, 4th Edition

Guest Editor

Sílvia C. Gonçalves

MDPI

Basel • Beijing • Wuhan • Barcelona • Belgrade • Novi Sad • Cluj • Manchester

Guest Editor
Sílvia C. Gonçalves
Marine and Environmental
Sciences Center
Polytechnic of Leiria
Peniche
Portugal

Editorial Office
MDPI AG
Grosspeteranlage 5
4052 Basel, Switzerland

This is a reprint of the Special Issue, published open access by the journal *Environments* (ISSN 2076-3298), freely accessible at: https://www.mdpi.com/journal/environments/special_issues/10R4S3J1J4.

For citation purposes, cite each article independently as indicated on the article page online and as indicated below:

Lastname, A.A.; Lastname, B.B. Article Title. *Journal Name* **Year**, *Volume Number*, Page Range.

ISBN 978-3-7258-7889-5 (Hbk)
ISBN 978-3-7258-7890-1 (PDF)
https://doi.org/10.3390/books978-3-7258-7890-1

Cover image courtesy of Sílvia C. Gonçalves

Contents

About the Editor . . . . **vii**

Sílvia C. Gonçalves
Monitoring and Assessment of Environmental Quality in Coastal Ecosystems, 4th Edition
Reprinted from: *Environments* **2026**, *13*, 157, https://doi.org/10.3390/environments13030157 . . **1**

Chiara Fratini, Serena Anselmi and Monia Renzi
Dredge Sediment as an Opportunity: A Comprehensive and Updated Review of Beneficial Uses in Marine, River, and Lagoon Eco-Systems
Reprinted from: *Environments* **2025**, *12*, 200, https://doi.org/10.3390/environments12060200 . . **6**

Roberto Zonta, Giuliano Lorenzetti, Giorgia Manfè, Simone Leoni, Gian Marco Scarpa, Luca Zaggia, et al.
Water Discharge and Sediment Transport from the Dese River Estuary to the Venice Lagoon (Italy): Annual Dynamics and the Influence of Floods
Reprinted from: *Environments* **2024**, *11*, 294, https://doi.org/10.3390/environments11120294 . . **24**

Adriano Sfriso, Yari Tomio and Andrea Augusto Sfriso
Climatic Changes Shift Macroalgal Assemblages from Cold- to Warm-Adapted Species: The Venice Lagoon as a Study Case
Reprinted from: *Environments* **2025**, *12*, 149, https://doi.org/10.3390/environments12050149 . . **50**

Gerald Schernewski, Maria Schneider, Thomas Neumann and Mario von Weber
Phytoplankton Dynamics in a Large Lagoon: Nutrient Load Reductions, Climate Change, and Cold- and Heatwaves
Reprinted from: *Environments* **2025**, *12*, 370, https://doi.org/10.3390/environments12100370 . . **70**

Gerald Schernewski, Thomas Neumann, Sarah Piehl and Nicole M. Swer
Ecosystem-Model-Based Valuation of Ecosystem Services in a Baltic Lagoon: Long-Term Human Technical Interventions and Short-Term Variability
Reprinted from: *Environments* **2025**, *12*, 35, https://doi.org/10.3390/environments12020035 . . . **94**

Jessica Sullivan, Michael Foster, James Chassereau and Robert Sullivan, Jr.
Geomorphic Comparison of Three Globally Significant Wetland Landscapes
Reprinted from: *Environments* **2025**, *12*, 458, https://doi.org/10.3390/environments12120458 . . **122**

José Luis Yanes, Alejandro Jiménez-Bonilla, Marina Martínez-Caro, Ana Fernández-Ayuso and Miguel Rodríguez-Rodríguez
Groundwater/Surface Water Temperature Variations and Hydrogeological Implications in Doñana National Park
Reprinted from: *Environments* **2025**, *12*, 83, https://doi.org/10.3390/environments12030083 . . . **137**

Lisa Tondelli, Nicola Bettoso, Oriana Blasutto, Massimo Celio and Alessandro Acquavita
Toxic Phytoplankton in Mussel Farms in the Gulf of Trieste, Adriatic Sea (Italy): A Preliminary Analysis of Long-Term Data (2001–2022) in Relation to Environmental Conditions
Reprinted from: *Environments* **2025**, *12*, 152, https://doi.org/10.3390/environments12050152 . . **157**

James Murphy, Jonathan E. Higham, Andrew J. Plater, Kasey E. Clark and Rachel Collin
Integration of Earth Observation and Field-Based Monitoring for Morphodynamic Characterisation of Tropical Beach Ecosystems
Reprinted from: *Environments* **2025**, *12*, 205, https://doi.org/10.3390/environments12060205 . . **179**

Björn Kjerfve, Hazel A. Oxenford, Rachel Collin, Inácio Abreu Pestana, Jimena Samper-Villarreal, Israel Medina-Gómez, et al.
Two Decades of CARICOMP Mangrove Monitoring (1992–2013) Reveal Variability in Tree Structure and Productivity of *Rhizophora mangle* Across the Wider Caribbean
Reprinted from: *Environments* **2025**, *12*, 463, https://doi.org/10.3390/environments12120463 . . **211**

Sidnei Aranha, Felipe Rakauskas, Leonardo Ferreira da Silva, Caio Fernando Fontana and Maurício Lamano Ferreira
Threat of Alien Species to Native Biodiversity in Mangroves near Latin America's Largest Port: Pathways for Technological Innovation and Strengthening of Regulations
Reprinted from: *Environments* **2025**, *12*, 483, https://doi.org/10.3390/environments12120483 . . **236**

Francisca Morales, Pamela T. Muñoz, Agustina Undabarrena, Paula S. M. Celis-Plá, Consuelo Rámila, Polette Aguilar-Muñoz, et al.
Climate Change-Related Salinity Fluctuations and Warming Induce Physiological Stress and Cellular Alterations in an Antarctic Intertidal Brown Alga
Reprinted from: *Environments* **2025**, *12*, 390, https://doi.org/10.3390/environments12100390 . . **260**

About the Editor

Sílvia C. Gonçalves

Sílvia C. Gonçalves is a Senior Researcher at MARE–Marine and Environmental Sciences Centre and holds a PhD in Biology (specialization in Ecology) from the University of Coimbra, Portugal. She is presently a Coordinating Professor at the School of Tourism and Maritime Technology—Polytechnic of Leiria (Portugal), lecturing mainly in the scientific areas of biology, ecology, physiology, and marine environment. She focuses her research on marine ecology, in the following domains: (1) the bio-ecology of crustacean populations on sandy beaches and their potential as ecological indicators of human impacts, (2) structure and function of macrobenthic communities; (3) monitoring and evaluation of environmental quality in coastal ecosystems, especially on sandy beaches and coastal lagoons. Her scientific research activities also include environmental biotechnology (e.g., bioremediation of trace metals by halophyte plants) and aquaculture (e.g., maintenance and reproduction of polychaetes, sea urchins, and fish).

environments

MDPI

Editorial

Monitoring and Assessment of Environmental Quality in Coastal Ecosystems, 4th Edition

Sílvia C. Gonçalves

MARE—Marine and Environmental Sciences Centre/ARNET, ESTM, Polytechnic Institute of Leiria, 2520-641 Peniche, Portugal; silvia.goncalves@ipleiria.pt

Marine environmental quality is a complex and multifactorial concept that encompasses the assessment of the physical, chemical and biological status of a given marine environment as opposed to its natural state, as well as the measurement of the levels of contaminants present in water, sediments and organisms. To preserve the structure and function of marine ecosystems and their processes it is therefore pivotal to ensure their good environmental status. Coastal environments, in particular, are especially prone to degradation, fragmentation, pollution and destruction since these are worldwide human densely populated areas, where the anthropogenic pressure has been increasing for centuries. Also, as areas of intense exchange of energy and materials between the land and the oceans, they are especially sensitive and vulnerable to natural and anthropogenic disturbances [1]. Global climate changes and their multiple effects, such as ocean warming and acidification, rising sea-levels, and extreme storms and floods, impose additional threats to coastal environments. These several types of pressures act synergistically, increasing the detrimental effects beyond what would be anticipated from the addition of independent pressures [2].

Monitoring and assessment are fundamental components for effective marine management [3]. In fact, across the world there are several marine conventions and legislations (e.g., Cartagena Convention—Convention for the Protection of the Caribbean Sea; Water Framework Directive and Marine Framework Directive in the European Union; Clean Water Act in the USA) with monitoring and assessment programs aimed at defining ecological quality of marine ecosystems in an integrative way, by making use of several biological, physico-chemical and pollution elements [4]. Scientists have been developing distinct approaches to address those issues, with hydrological and sediment dynamics modeling, water quality monitoring and assessment, presence of pollutants and the development of ecological indicators, being some of the most common ones.

Continuous monitoring and management strategies driven by data collections are essential in the process of decision-making regarding coastal areas vulnerable to erosion [5]. Also, the development of structured frameworks that allow researchers to steer through the available technologies and datasets, according to their suitability to distinct monitoring scenarios and needs, is very important to support coastal management and improve the resilience of coastlines [6]. As such, developing new methodologies and technologies, or improving and integrating already existing ones, for monitoring and managing coastal erosion and other coastal hazards is crucial.

Multidisciplinary approaches that combine water quality, chemical contamination, and biological or ecological elements (e.g., 7, 8), are very useful for environmental monitoring and assessment, and become especially powerful tools when long-term datasets are available. The synergies obtained from the involvement of various expertise can help to deepen the understanding of environmental quality, which is fundamental for the protection of coastal ecosystems [7]. Furthermore, the combination of biological and chemical monitoring methods offers the comprehensive assessment currently needed, given the increasing anthropogenic impacts [8]. Finally, the use of integrated methods, such as multiple criteria analysis, that combine environmental, ecological, social and economic priorities, can help all parties involved to rationalize their views and create policies dependent on sound science [9]. As previously stated by [10], integration is the key to safeguarding beaches and their unique biodiversity, and to maintaining their benefits for the following generations through sustainable development: this statement was originally applied by those authors in reference to sandy beaches, but it can be further extended to all coastal ecosystems.

The fourth edition of the Special Issue on "*Monitoring and Assessment of Environmental Quality in Coastal Environments*" aimed to bring together contributions from several international experts focusing on the application of distinct tools and methods, in relation to diverse coastal environments with a varied geographical origin. After the rigorous peer review processes of *Environments*, twelve out of seventeen initial submissions were accepted for publication and inclusion in this Special Issue (eleven research articles and one review).

Contribution 1 reviewed the innovative remediation techniques available for dredged sediments in order to promote their multiple sustainable beneficial reuses, by turning them into valuable resources. The authors emphasize the integrated strategies and multidisciplinary approaches combining technological innovations that further improve sediment reuse to foster economic growth and environmental protection.

Several contributions explored the highly complex transition environments. Contributions 2 and 3 were focused on the iconic Venice Lagoon (Italy). Due to the relevance of the sediment load in sustaining tidal and subtidal habitats of the lagoon, contribution 2 analyzed the freshwater discharge and the sediment transport in two estuarine sections of the Dese River, the main tributary of this large wetland. The study highlighted that high-resolution monitoring of sediment delivery is critical for the protection of the estuarine areas of the Venice Lagoon, which has been facing an erosive trend in recent decades [11]. Major changes in the macroalgal assemblages of the Venice Lagoon in the last twenty years, caused by temperature increases, have been addressed in contribution 3. The authors concluded that the native cold-water taxa, mainly Phaeophyceae, are in sharp decline, while thermophilic species, namely non-indigenous Rhodophyceae species, are increasing and replacing the once dominant native taxa.

One of the largest lagoons in Europe, the Oder/Szczecin Lagoon in the Baltic Sea region, was analyzed in contributions 4 and 5. Using a combination of monitoring data from the last 40 years and 3D ecosystem modeling, contribution 4 evaluated changes in phytoplankton abundance and diversity on distinct temporal scales, from short-term effects to long-term trends. The authors concluded that, while phytoplankton biomass and composition presented pronounced interannual and short-term variability, the average annual phytoplankton biomass and the long-term species composition did not change significantly. However, heat- and cold waves did not show consistent immediate effects. In contribution 5, 3D ecosystem model simulations over a 10-year period, complemented with socio-economic data, to assess the ecosystem services of the Oder/Szczecin Lagoon were performed and analyzed. Model simulations revealed that nitrogen retention was the lagoon's most important ecosystem service, with the highest economic income, and that commercial fisheries had been diminishing its economic significance in the lagoon. According to their conclusions, the authors highlighted that ecosystem services and po-

 https://doi.org/10.3390/environments13030157

tential changes derived from the implementation of human interventions (e.g., channel deepening) should be considered in environmental impact assessments.

The morphology, the drainage density and drainage efficiency of three distinct marsh landscapes of the Ashepoo, Combahee, and Edisto (ACE) Basin, of the southeastern North American coastal plain in South Carolina (USA), were explored in contribution 6, using high-resolution digital elevation models. The findings of those authors suggest that drainage density and drainage efficiency should be considered separately, with drainage efficiency being mostly influenced by the structure and spatial arrangement of creek networks within a marsh.

Contribution 7 analyzed the sand dune ponds of the wetlands of Doñana National Park (in southwestern Spain), which is one of the most important wetlands in Europe, recognized as a UNESCO World Heritage Site and a Ramsar Wetland of International Importance. This study aimed to characterize the thermal regime of surface water and groundwater on those environments to identify potential trends related to the rising temperatures and the decreasing precipitation over an eight-year time series, as well as to study patterns in groundwater flow dynamics. The findings of this contribution warrant the analyses of the sensitivity of those wetlands to global environmental change and grant the characterization of recharge and discharge flows in this coastal aquifer.

In the coastline of the Gulf of Trieste (Adriatic Sea, Italy), contribution 8 analyzed the spatial distribution, the seasonal patterns and the temporal trends of toxic phytoplankton, together with physico-chemical parameters, nutrient concentrations and precipitation, along four selected shellfish farming sites on a long-term dataset. The results demonstrated that the spatial distribution of toxic phytoplankton presented significant gradients and that nutrients and the physico-chemical parameters varied seasonally. Therefore, the study provides a valuable basis for further approaches with the aim of analyzing the impact of climate variability in the dynamics of toxic phytoplankton at the Gulf of Trieste.

Within the context of coastal erosion monitoring, contribution 9 proposes a novel multi-platform methodology, integrating field-based measurements with online Earth Observation data, to create a comprehensive and accessible toolkit for quantifying beach morphodynamical characteristics. The methodology was tested with real conditions using beaches of the Bocas del Toro Archipelago, on the Caribbean coast of Panama, as case study. The use of the multi-platform toolkit revealed to be advantageous in capturing multiscale dynamics, deepening the understanding of coastal morphodynamics, which is especially relevant for remote tropical beaches that often lack essential baseline data.

A mangrove monitoring program across 18 countries over the wider Caribbean, on a long-term temporal series and focused on the dominant mangrove species *Rhizophora mangle*, was performed under the Caribbean Coastal Marine Productivity (CARICOMP) program and two decades of those results were presented by contribution 10 of this Special Issue. The main objective of this contribution was to reveal the spatial and temporal variability of the monitored mangroves across the region and investigate potential drivers of change on these coastal ecosystems. The results revealed high variability in mangrove structure and productivity across the study region and a lack of consistency among sites in long-term trends of mangrove productivity, suggesting that local-level drivers of forest structure, productivity and change are important. Also, according to the authors, conservation and management strategies should be site-specific and reinforced by the monitoring of local environmental events.

Aimed at analyzing the risks of biological invasion in mangrove ecosystems caused by port activities in their vicinity, contribution 11 presents the opportunities and challenges faced by Latin America's largest and most important port complex, the Port of Santos (Brazil), regarding its activities and transport flows and the potential for technological use

 https://doi.org/10.3390/environments13030157

to mitigate environmental conflicts. The results revealed that the intense and constant flow of international vessels into the Port of Santos leads to a persistent threat of biological invasion, reinforcing the role of ships as vectors for the transport of exotic species. Also, the authors highlight that control measures, mediated by effective policies, regulations and technologies, should be implemented in the short term to avoid destabilizing the adjacent mangrove ecosystems and causing socio-economic losses.

Finally, in light of the rapidly warming rates faced by Antarctica and their effects on seawater temperature and salinity at intertidal environments, contribution 12 investigated the combined effects of salinity fluctuations and temperature rise on the health status and physiology of the Antarctic intertidal macroalga *Adenocystis utricularis*. The findings demonstrated that warming temperature had a negative effect on *A. utricularis* oxidative response at a control salinity (33 psu), but that an altered stress response to salinity fluctuations arises under thermal stress, revealing an interactive effect between salinity and temperature.

Conflicts of Interest: The author declares no conflicts of interest.

List of Contributions:

1. Fratini, C.; Anselmi, S.; Renzi, M. Dredge Sediment as an Opportunity: A Comprehensive and Updated Review of Beneficial Uses in Marine, River, and Lagoon Eco-Systems. *Environments* **2025**, *12*, 200. https://doi.org/10.3390/environments12060200.
2. Zonta, R.; Lorenzetti, G.; Manfè, G.; Leoni, S.; Scarpa, G.M.; Zaggia, L.; Bendoricchio, C.; Cerni, M.; Dominik, J. Water Discharge and Sediment Transport from the Dese River Estuary to the Venice Lagoon (Italy): Annual Dynamics and the Influence of Floods. *Environments* **2024**, *11*, 294. https://doi.org/10.3390/environments11120294.
3. Sfriso, A.; Tomio, Y.; Sfriso, A.A. Climatic Changes Shift Macroalgal Assemblages from Cold-to-Warm-Adapted Species: The Venice Lagoon as a Study Case. *Environments* **2025**, *12*, 149. https://doi.org/10.3390/environments12050149.
4. Schernewski, G.; Schneider, M.; Neumann, T.; von Weber, M. Phytoplankton Dynamics in a Large Lagoon: Nutrient Load Reductions, Climate Change, and Cold- and Heatwaves. *Environments* **2025**, *12*, 370. https://doi.org/10.3390/environments12100370.
5. Schernewski, G.; Neumann, T.; Piehl, S.; Swer, N.M. Ecosystem-Model-Based Valuation of Ecosystem Services in a Baltic Lagoon: Long-Term Human Technical Interventions and Short-Term Variability. *Environments* **2025**, *12*, 35. https://doi.org/10.3390/environments12020035.
6. Sullivan, J.; Foster, M.; Chassereau, J.; Sullivan, R., Jr. Geomorphic Comparison of Three Globally Significant Wetland Landscapes. *Environments* **2025**, *12*, 458. https://doi.org/10.3390/environments12120458.
7. Yanes, J.L.; Jiménez-Bonilla, A.; Martínez-Caro, M.; Fernández-Ayuso, A.; Rodríguez-Rodríguez, M. Groundwater/Surface Water Temperature Variations and Hydrogeological Implications in Doñana National Park. *Environments* **2025**, *12*, 83. https://doi.org/10.3390/environments12030083.
8. Tondelli, L.; Bettoso, N.; Blasutto, O.; Celio, M.; Acquavita, A. Toxic Phytoplankton in Mussel Farms in the Gulf of Trieste, Adriatic Sea (Italy): A Preliminary Analysis of Long-Term Data (2001–2022) in Relation to Environmental Conditions. *Environments* **2025**, *12*, 152. https://doi.org/10.3390/environments12050152.
9. Murphy, J.; Higham, J.E.; Plater, A.J.; Clark, K.E.; Collin, R. Integration of Earth Observation and Field-Based Monitoring for Morphodynamic Characterisation of Tropical Beach Ecosystems. *Environments* **2025**, *12*, 205. https://doi.org/10.3390/environments12060205.
10. Kjerfve, B.; Oxenford, H.A.; Collin, R.; Pestana, I.A.; Samper-Villarreal, J.; Medina-Gómez, I.; Cortés, J.; Smith, S.R.; Koltes, K.; Feller, I.C.; et al. Two Decades of CARICOMP Mangrove Monitoring (1992–2013) Reveal Variability in Tree Structure and Productivity of Rhizophora mangle Across the Wider Caribbean. *Environments* **2025**, *12*, 463. https://doi.org/10.3390/environments12120463.

11. Aranha, S.; Rakauskas, F.; da Silva, L.F.; Fontana, C.F.; Ferreira, M.L. Threat of Alien Species to Native Biodiversity in Mangroves near Latin America's Largest Port: Pathways for Technological Innovation and Strengthening of Regulations. *Environments* **2025**, *12*, 483. https://doi.org/10.3390/environments12120483.
12. Morales, F.; Muñoz, P.T.; Undabarrena, A.; Celis-Plá, P.S.M.; Rámila, C.; Aguilar-Muñoz, P.; Molina, V.; Navarro, N.P.; Sáez, C.A.; Lavergne, C.; et al. Climate Change-Related Salinity Fluctuations and Warming Induce Physiological Stress and Cellular Alterations in an Antarctic Intertidal Brown Alga. *Environments* **2025**, *12*, 390. https://doi.org/10.3390/environments12100390.

References

1. Lu, Y.; Yuan, J.; Lu, X.; Su, C.; Zhang, Y.; Wang, C.; Cao, X.; Li, Q.; Su, J.; Ittekkot, V.; et al. Major threats of pollution and climate change to global coastal ecosystems and enhanced management for sustainability. *Environ. Pollut.* **2018**, *239*, 670–680. [CrossRef]
2. Solan, M.; Whiteley, N.M. *Stressors in the Marine Environment: Physiological and Ecological Responses; Societal Implications*; Oxford University Press: Oxford, UK, 2016.
3. Katsanevakis, S.; Stelzenmüller, V.; South, A.; Kirk Sørensen, T.; Jones, P.J.S.; Kerr, S.; Badalamenti, F.; Anagnostou, C.; Breen, P.; Chust, G.; et al. Ecosystem-based marine spatial management: Review of concepts, policies, tools, and critical issues. *Ocean Coast. Manag.* **2011**, *54*, 807–820. [CrossRef]
4. Borja, Á.; Dauer, D.M.; Grémare, A. The importance of setting targets and reference conditions in assessing marine ecosystem quality. *Ecol. Indic.* **2012**, *12*, 1–7. [CrossRef]
5. Andrade, F.A.A.; Brasil, P.; Lima, M. Portugal coastal erosion monitoring system using unmanned aerial vehicles and aerial photogrammetry. *Sustain. Geosci. People Planet Prosper.* **2025**, *1*, 100007. [CrossRef]
6. McLaughlin, D.; Doyle, T.B.; Asbridge, E.F.; Rogers, K. A framework for using coastal monitoring data: A foredune case study. *Ocean Coast. Manag.* **2026**, *272*, 107989. [CrossRef]
7. Ferraro, A.; Siciliano, A.; Spampinato, M.; Morello, R.; Trancone, G.; Race, M.; Guida, M.; Fabbricino, M.; Spasiano, D.; Fratino, U. A multi-disciplinary approach based on chemical characterization of foreshore sediments, ecotoxicity assessment and statistical analyses for environmental monitoring of marine-coastal areas. *Mar. Environ. Res.* **2024**, *202*, 106780. [CrossRef] [PubMed]
8. Chaikasem, S.; Na Roi-et, V. Marine algae as environmental health bio-sentinels: Benzene genotoxicity assessment using the comet assay for coastal biomonitoring. *Water Air Soil Pollut.* **2026**, *237*, 491. [CrossRef]
9. Karydis, M.; Kitsiou, D. Marine water quality monitoring: A review. *Mar. Pollut. Bull.* **2013**, *77*, 23–36. [CrossRef] [PubMed]
10. Rodil, I.F.; Harris, L.R.; Lucrezi, S.; Cerrano, C. Sandy beach management and conservation: The integration of economic, social and ecological values. In *Sandy Beaches as Endangered Ecosystems: Environmental Problems and Possible Assessment and Management Solutions*; Gonçalves, S.C., Ferreira, S.M.F., Eds.; CRC Press, Taylor and Francis Group: Boca Raton, FL, USA, 2022; pp. 251–294.
11. Suman, D.; Guerzoni, S.; Molinaroli, E. Integrated coastal management in the Venice lagoon and its watershed. *Hydrobiologia* **2005**, *550*, 251–269. [CrossRef]

environments

MDPI

Review

Dredge Sediment as an Opportunity: A Comprehensive and Updated Review of Beneficial Uses in Marine, River, and Lagoon Eco-Systems

Chiara Fratini [1,2], Serena Anselmi [1] and Monia Renzi [3,*]

[1] Bioscience Research Center, via A. Vecchia, 32, 58015 Orbetello, Italy; chiara.fratini@unipa.it (C.F.); serena.anselmi@bsrc.it (S.A.)
[2] Department of Earth and Sea Science, University of Palermo, via Archirafi, 22, 90123 Palermo, Italy
[3] Department of Life Science, University of Trieste, via L. Giorgieri, 10, 34127 Trieste, Italy
* Correspondence: author: mrenzi@units.it

Abstract: Dredging is essential for the maintenance of ports, waterways, lakes, and lagoons to ensure their operability and economic value. Over the last few decades, scientists have focused on the significant environmental challenges associated with dredging, including habitat destruction, loss of biodiversity, sediment suspension, and contamination with heavy metals and organic pollutants. The huge loss of sediment in coastal areas and the associated erosion processes are now forcing stakeholders to look ahead and turn potential problems into an opportunity to develop new sediment management strategies, beyond environmental protection, toward ecosystem restoration and coastal resilience. Moreover, the European and Italian strategies, such as the European Green Deal (EGD) and the Italian Ecological Transition Plan (PTE), highlight the need to reuse dredge sediment in circular economy strategies, transforming them into valuable resources for construction, agriculture, and environmental restoration projects. European legislation on dredging is fundamental to the issue of management and priorities of dredged materials, but the implementation rules are deferred to individual member states. In Italy, the Ministerial Decree 173/2016 covers the main aspects of dredge activities and dredge sediment management. Moreover, it encourages the remediation and reuse of the dredge sediment. This study starts with a comprehensive analysis of the innovative remediation techniques that minimize impacts and promote sustainable, beneficial sediment management. Different remediation methods, such as electrochemical treatments, chemical stabilization, emerging nanotechnologies, bioremediation, and phytoremediation, will be evaluated for their effectiveness in reducing pollution. Finally, we highlight new perspectives, integrated strategies, and multidisciplinary approaches that combine various technological innovations, including artificial intelligence, to enhance sediment reuse with the aim of promoting economic growth and environmental protection.

Keywords: dredging impact; restoration; reclamation; dredge sediment; biodiversity; beneficial use; circular economy; artificial intelligence

1. Introduction

Dredging is the removal of sediment from the bottom of water bodies; it is a necessary practice to ensure navigation in harbors and waterways and to manage coastal areas. The results of this practice are dredged sediment. Sediment is made up of sand, silt, clay, and other materials eroded from rock and soil and plays a vital role in aquatic ecosystems:

providing habitats and food sources for various organisms [1]. While dredging is crucial for human activities, sediment removal can have environmental impacts. It can disrupt benthic habitats, exacerbate coastal erosion, and increase water turbidity, affecting water quality and light penetration. Furthermore, it can lead to the loss of biodiversity by altering the sediment composition and habitat availability [2]. However, the appropriate management of the dredging project can reduce these consequences or even reverse some of these trends. Dredging can be an effective tool in environmental restoration projects, such as remediating contaminated sites and mitigating the effects of eutrophication [3,4]. In addition, the cleaned dredge sediment can be used for ecosystem restoration and coastal resilience (e.g., nourishment or development of coastal wetlands) [5]. Sediment can naturally contain trace contaminants due to their geological origins or anthropogenic activities, including industrial, agricultural, and domestic sources. Common pollutants include heavy metals (e.g., lead, mercury, cadmium, arsenic), organic pollutants (e.g., polycyclic aromatic hydrocarbons, polychlorinated biphenyls, persistent organic pollutants), and plastic materials (including microplastics) [6]. Contaminated sediment poses significant challenges to its management and reuse. Traditionally, these sediments are disposed of in landfills, and this represents an unacceptable waste of resources. Nowadays, globally, there has been a growing emphasis on sustainable sediment management practices, aligned with the principles of the circular economy [7]. In this respect, in Europe, the European Green Deal (EGD), and in Italy, the Ecological Transition Plan (PTE) emphasize the need for sustainable sediment management, promoting resource efficiency and minimizing environmental impacts [8,9]. A number of multiple directives have an indirect influence on dredging practices, such as the Water Framework Directive (2000/60/EC) and the Waste Framework Directive (75/442/EEC) [10,11]. These regulations also encourage practices like sediment recycling, recovery, and reuse, aiming to improve sustainability and reduce waste generation. Furthermore, the singular member states adopt laws to ensure the correct management of dredging practices. In Italy, Ministerial Decree 173/2016 states on dredge sediment management, basing the process on the level of contamination of the sediments. The characterized sediments can be used for various proposals (e.g., marine landfill and depositing site disposal). To promote sustainable reuse of dredge sediment, the decree promotes the changing of sediment class using remediation technologies [12]. Promising remediation techniques, such as electrochemical methods, chemical treatments, and biological remediation, offer ways to neutralize contaminants and restore sediment safety [13]; these strategies are discussed in detail in the following paragraphs. Sediment reuse presents a variety of possibilities, including applications in construction (e.g., as aggregates, road base materials, or landfill covers) and the development of green infrastructure such as wetlands or coastal restoration projects [14]. These solutions open new avenues for circular economy principles in sediment management and are also discussed.

2. Methodology

This review is divided into the following sections: (1) an introduction section focuses on the impacts of dredging, sediment remediation techniques, and a brief overview of the current regulatory framework for management in Italy; (2) the description of the methodology followed in this manuscript; (3) bibliometric analyses; (4) summary of principal remediation techniques for dredge sediment applied to freshwater and marine sediment and case studies; (5) summary of potential uses of dredge sediment and case studies; (6) the use of artificial intelligence toward the dredge sediment reclamation; and (7) the conclusions. To develop this study, a deep search on the Scopus, Google Scholar database, and Google search engine was carried out to identify relevant studies related to the environmental impacts of dredging, dredge sediment impact, and remediation or reuse strategies.

The search queries on both databases included combinations of the following keywords: "dredging impact", "dredge sediment impact", "dredge sediment remediation", "beneficial use of dredge sediment", "dredging management", "dredge sediment regulatory framework", and "artificial intelligence and dredging". The retrieved articles were manually reviewed and organized into a conceptual framework according to dredge sediment remediation technique or beneficial reuse, both divided by type of aquatic environment (marine, freshwater, lagoon). Publications were selected based on the following inclusion criteria: relevance to the main research themes and availability of sufficient methodological and thematic information. Exclusion criteria included duplicate entries, articles not directly related to dredging or sediment management, and publications lacking methodological transparency. The selected studies were compiled into a worksheet, including author, title, abstract, and relevant information, which provided the basis for this manuscript. To explore research trends and keyword patterns in the selected literature, a bibliometric analysis was performed using the Bibliometrix package in R ('bibliometrix' version 4.3.5). This analysis enabled the visualization of thematic networks and the identification of emerging research directions within the field. The combined use of Scopus, Google Scholar, and Google database allowed for a more comprehensive literature search by leveraging the strengths of each platform. Anyhow, we are aware of potential biases associated with the literature review process, such as algorithmic and quality variability bias and geographical bias due to the use of these databases.

3. Bibliometric Analysis

To understand the worldwide impact of dredging, we conducted a bibliometric analysis using data from the scientific search engine Scopus (Elselvier, Amsterdam, The Netherlands). The search was performed by entering the search string ("dredg*" AND "impact*") in the search document menu with a filter applied to select only scientific articles in English, and no temporal restrictions were imposed. Data were exported in bibText format and were analyzed using Bibliometrix, an open-source software package developed in R for bibliometric analysis and data visualization [15]. The search yielded 2482 documents from 766 different sources, covering a time span from 1971 to 2025. This dataset reflects a wide range of scientific output, with publications from multiple disciplines that have investigated various environmental, economic, and technological aspects of dredging. The annual growth rate was 5.01%, indicating a consistent increase in research activity related to dredging, especially in the last decade. The overall trend in scientific production from 1971 to 2024 is shown in Figure 1, indicating an increasing trend of published scientific articles relating to the impact of dredging over time.

The analysis of scientific production by country showed that the United States (frequency score 1125) is the most influential nation in terms of scientific production on dredging, followed by China (622), the United Kingdom (394), Australia (375), Brazil (285), France (276), Italy (258), Nether-lands (249), Canada (156), and Spain (137). A graphic visualization of the contribution of the major countries is shown in Figure 2.

To highlight keywords related to the impact of dredging, an analysis of the most frequently occurring terms was carried out. The central theme according to the occurrence was "dredging", followed by "environmental impact", "environmental monitoring", "sediment", "water quality", and "United States". As expected, "dredging" and "sediment" were among the most frequently recurring terms because they were the terms used for the search. The terms "environmental impact" and "environmental monitoring" reflect sensitivity towards environmental issues. The term "United States" confirms the centrality of the country with regard to the issue of dredging. The term "pollutant" comes up several times, referring to contamination of water or sediment. Also, some types of contaminants were highlighted, such as "heavy metal". The analyses underlined the terms "rivers", "estuary", and "marine" referring to these ecosystems. Notably, the term "biodiversity" emerges from research, emphasizing the term centrality. This refers to the variety and variability of life on Earth, encompassing the diversity within species, between species, and across ecosystems. In recent decades, biodiversity has become a central theme in global environmental discourse, not only for dredged activities but also in reference to species loss, habitat degradation, and the impacts of climate change. Preserving habitats and, consequently, biodiversity is essential to address future challenges. The search output is shown in Figure 3.

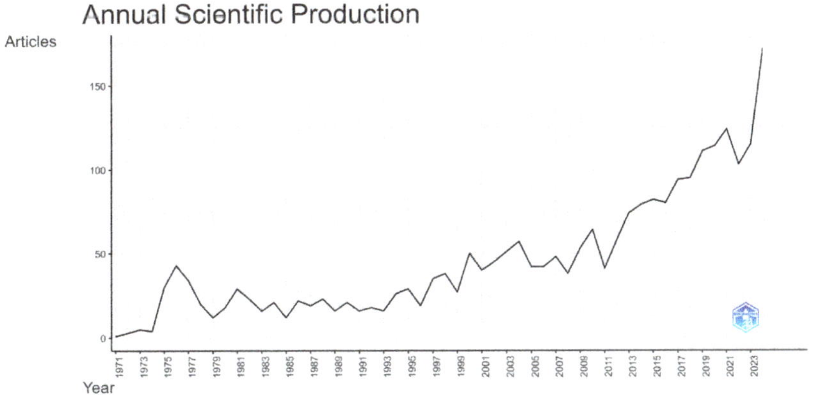

Figure 1. Annual scientific production, including the terms searching on the Scopus database, the string ("dredg*" AND "impact*"), selecting only scientific articles in English. The temporal range is from 1971 to 2024.

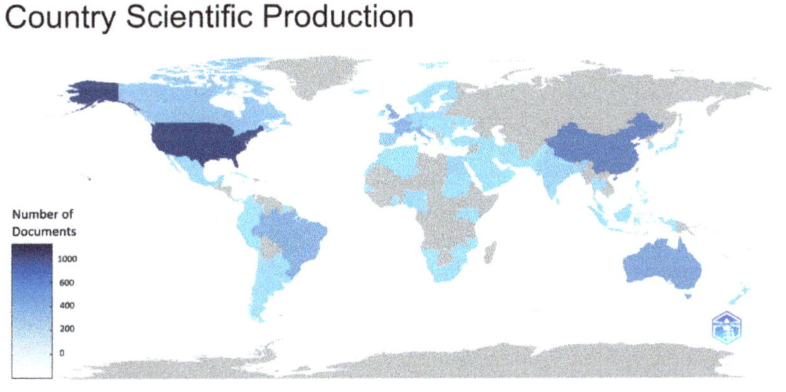

Figure 2. World map showing the scientific contribution of various countries for the topic "dredging impact". The color intensity indicates the number of scientific documents published per country. Darker blue corresponds to higher publication counts. Gray areas represent countries with no data available in the analyzed dataset.

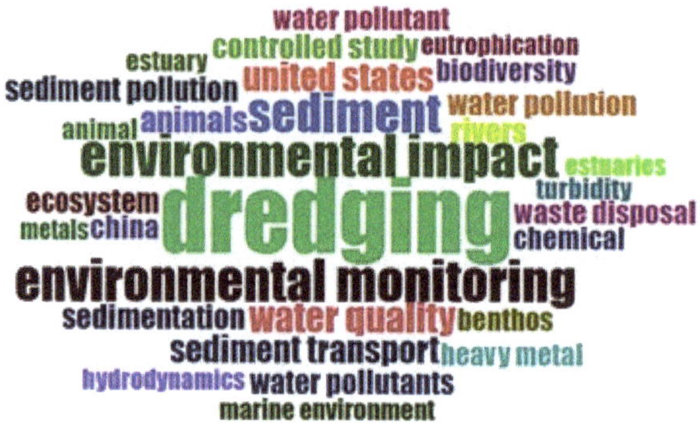

Figure 3. Word cloud map presenting keywords from the topic "dredging impact". The centrality, along with size, indicates the frequency value of the word.

4. Towards the Reuse of Dredge Sediment

4.1. Remediation Techniques on Dredge Sediment

Dredge sediment remediation techniques are increasingly recognized as effective solutions for managing the disposal of sediment, turning it from waste into valuable resources. These methods help reduce environmental risks by removing contaminants and making sediment suitable for reuse in construction, land reclamation, or habitat restoration. As the volume of dredge sediment continues to grow, implementing sustainable and efficient remediation strategies is essential for minimizing negative environmental impacts and supporting circular economic principles. The technologies reviewed offer a variety of approaches for the remediation and reuse of contaminated and non-contaminated dredge sediment. For further details on the technologies, see the technical appendix attached to this review (Supplementary Materials S1).

4.1.1. Electrochemical Remediation Technologies

Electrokinetic (EK) remediation is a promising electrochemical approach widely explored for the treatment of sediment contaminated with heavy metals. This technique involves the application of a direct electric field across the sediment matrix, promoting the mobilization of contaminants. Although EK remediation has been extensively studied under controlled conditions, its application has remained largely limited to laboratory-scale experiments. A growing body of research has examined the use of EK remediation in various types of sediments (marine, estuarine, lagoon, and lake sediments), each characterized by different physical and chemical properties. In a study conducted in the Darsena Pescherecci area of the Port of Piombino (Italy), a hexagonal EK configuration was employed to treat approximately 60 kg of marine sediment. The system demonstrated the effective removal of heavy metals under optimized conditions, offering a glimpse of the potential scalability of the approach [16]. Further investigations have explored EK treatment in both artificial and naturally contaminated marine sediment. For example, ref. [17] applied graphite electrodes directly to the sediment matrix and observed a significant removal of PAH, although the technique showed limited effectiveness for mercury (Hg) and arsenic (As). Other studies have focused on improving the efficiency of treatment by optimizing operational parameters, for example, in [18], where marine sediment co-contaminated with heavy metals and PAHs was dragged using a combination of periodic voltage gradients, citric acid (CA), and surfactants such as Tween 20 (TW20). This approach significantly enhanced contaminant desorption and mobility, resulting in improved remediation outcomes. The role of electrolyte composition has also been investigated in detail. In the case of lagoon sediment from the Camorim Lagoon in Rio de Janeiro, Brazil, ref. [19] demonstrated that the addition of complexing agents like ethylenediaminetetraacetic acid (EDTA) and humic substances markedly improved metal mobilization and removal. Similarly, ref. [20] examined the EK treatment of sediment stored for over 13 years in the Tancarville repository in the upper Seine estuary, France. A periodic EK regime (8 h ON, 16 h OFF) was applied over 21 days using 40 V (1 V/cm), with citric acid (0.1 M, pH 2.36) at the cathode and ultrapure water at the anode. The process effectively reduced contaminant levels in 64 L of estuarine sediment to meet criteria suitable for land reclamation. In a related study, ref. [21] investigated the use of EK remediation to mitigate ecological risks associated with metal-contaminated lake sediment. The treatment combined EK with chemical agents such as EDTA, nitric acid, and acetic acid. Despite being applied to relatively small sediment volumes (e.g., 0.54 L per 100 g), the treatment showed strong potential in reducing contaminant concentrations and associated risks.

4.1.2. Chemical Remediation

Chemical remediation, or sediment washing, is a technique that involves treating contaminated sediment with chemical solutions to desorb and transfer pollutants from the solid phase into a liquid washing medium. The efficiency of this process largely depends on the selection of chemical additives, which can include acids (e.g., nitric, hydrochloric, sulfuric), bases (e.g., sodium hydroxide), surfactants, and chelating agents (e.g., EDTA, EDDS). At the end of the process, these additives must be completely removed from the treated sediment to prevent secondary recontamination. Coarse-grained sediment typically responds better to chemical flushing, while fine-grained sediment presents greater challenges due to higher surface area and stronger binding of contaminants. For this reason, careful preliminary extraction tests are required to optimize key parameters such as chemical type and dosage, contact time, temperature, agitation, and the number of washing cycles necessary to meet environmental standards. Recent studies have demonstrated the versatility and effectiveness of chemical remediation in different types of sediments. For example, ref. [22] investigated a multi-stage washing process to treat two distinct contaminated sediments in Italy: mercury- and arsenic-rich sediment from the Bay of Augusta (Sicily) and the area of Bagnoli in the Gulf of Pozzuoli. The treatment protocol consisted of the following: (1) alkaline washing with NaOH to remove metals bound to humic substances; (2) a Fenton-like reaction using α-cyclodextrin (aCD) to stabilize Fe(II) and oxidize As(III) and Hg(0/I); (3) complexation of metals with aCD; and (4) a final reaction with sodium sulfide (Na_2S) to convert mercury into soluble polysulfide complexes. This sequence achieved removal efficiencies ranging from 56–72% for arsenic and 76–95% for mercury, demonstrating the potential of integrated washing-oxidation-complexation strategies. A different approach was taken by [23], who conducted batch extraction experiments to assess the effectiveness of chelating agents (EDTA and EDDS) in removing arsenic, lead, zinc, copper, nickel, and chromium from sediment collected in Malmfjärden Bay (Sweden). The study analyzed the impact of chelating agent type, concentration (0.01 M and 0.05 M), and pH on extraction performance. Results indicated that EDTA, being more cost-effective than EDDS, was particularly efficient at leaching cationic metals bound to the reducible fraction of the sediment. Additionally, acidic conditions enhanced arsenic extraction, while higher chelating agent concentrations favored the removal of copper, nickel, and chromium. Similarly, ref. [24] conducted two sediment-washing campaigns to evaluate the efficacy of various chelating agents (EDTA, EDDS, citric acid, and acetic acid) in removing heavy metals (Ni, Cu, Zn, Cr, Pb, Hg) and total petroleum hydrocarbons (TPH) from marine sediments collected in Augusta Bay (Italy). The batch extraction tests considered the influence of pH (4 and 9), contact time (0.5 to 24 h), and chelating agent concentration (0.05 M to 1 M) using a 1:10 solid-liquid weight ratio.

Results indicated that citric acid, due to its high biodegradability and cost-effectiveness, achieved superior removal efficiencies for both inorganic and organic pollutants, especially at acidic pH and prolonged contact time. EDTA was found to be effective mainly for inorganic contaminants, particularly under alkaline conditions, while EDDS showed good performance for the removal of Pb. The study concluded by proposing a conceptual layout for a large-scale sediment-washing plant, highlighting the technical and economic feasibility of citric acid as the primary extractant.

4.1.3. Solidification/Stabilization Methods

Dredge sediments are typically characterized by poor geotechnical properties, high compressibility, high water content, and low shear strength, which make them unsuitable for direct use in construction without prior treatment. Consequently, they are often treated as waste and disposed of in designated containment areas, which may lead to environmental degradation and loss of potentially reusable material. Stabilization has emerged as an effective approach to enhance the engineering properties of dredge soils, allowing their conversion into functional fill or construction materials. Although this technique does not aim to remove contaminants, it plays a critical role in immobilizing pollutants, thereby reducing their mobility and bioavailability in the treated matrix. Ordinary Portland cement (OPC) remains the most widely used binder in stabilization processes; however, its high production cost and significant environmental footprint have driven increasing interest in alternative or supplementary binders. Several studies have explored the partial replacement of OPC with industrial by-products (IBPs), which not only improve sustainability but also contribute positively to the mechanical performance of the treated sediment. For example, ref. [25] examined the incorporation of fly ash, ground granulated blast furnace slag (GGBFS), red mud, and phosphogypsum (PG) into OPC-based mixtures for marine sediment. These IBPs were shown to enhance hydration reactions by promoting the formation of contentious products such as C-(A)-S-H, CAH, and AFt, which help densify the matrix and improve long-term strength. In particular, the addition of PG and calcium aluminate cement (CAC) significantly boosted compressive strength, allowing for reductions in OPC usage while maintaining or enhancing structural integrity. CAC proved especially effective in improving both early strength development and deformation resistance, supporting the feasibility of using treated dredge sediment as a reliable fill material. A novel approach was proposed by [26], who developed a synergistic solidification/oxidation system (SOS) for stabilizing dredge sediment collected from a freshwater lake. This method combined OPC with sodium persulfate (SP), an oxidizing agent that also contributed to structural improvement. The presence of SP was found to substantially increase the unconfined compressive strength (UCS) of the sediment, particularly in samples with varying fulvic acid content, suggesting its suitability for complex organic-rich matrices. In a pilot-scale experiment, ref. [27] investigated the stabilization/solidification of marine sediment from the Mar Piccolo area of Taranto (Italy), using a combination of lime, organoclay, and activated carbon. The process enabled the transformation of 974 kg out of every 1000 kg of dredge material into a usable product, effectively reducing landfill disposal volumes to just 0.65 m^3 per ton of sediment. This study illustrates the real-world applicability of S/S techniques not only for improving the mechanical performance of dredge sediment but also for facilitating large-scale recycling and environmental sustainability.

4.1.4. Adding Sorbent Amendments

Amendments have gained increasing attention in recent years as effective in reducing or eliminating the bioavailability of contaminants bound to soil or dredge sediment and have been used in both in situ and ex situ tests. Among these, activated carbon (AC)—both in granular (GAC) and powdered (PAC) form—and Biochar (BC) have proven to be effective in immobilizing a wide array of pollutants through adsorption processes. For example, ref. [28] used activated carbon (PAC) and granular activated carbon (GAC) for in-situ remediation of PCB-contaminated sediments from Oskarshamn Harbor, Sweden. The thin layer coverage effectively reduced contamination, but negative effects on benthic communities were found. However, sorbent amendments can be magnetized; these offer the opportunity to recover them after usage. Laboratory-scale studies have demonstrated the practical efficacy of these magnetized sorbents in sediment remediation. For instance, ref. [29] developed a new activated carbon (AC) material for in situ remediation of PCB-contaminated sediments. The new material did not have acute toxicity effects on *Chironomusriparius* and *Lumbriculus variegatus*, with a significantly higher remediation potential than GAC (up to 89% reductions in PCB bioaccumulation). Ref. [30] evaluated magnetic activated carbon (MAC) derived from AC and BC, for the *in-situ* remediation of PAH-contaminated sediments. An 8.1% MAC amendment (5% AC content) reduced aqueous PAHs by 98% in three months, matching the performance of 5% pure AC. MAC recovery after three months was 77%; incomplete recovery led to a temporary PAH rebound, followed by a decline after six months. However, 77% recovery was insufficient to fully mitigate MAC-related ecotoxic effects on *Lumbriculus variegatus*.

4.1.5. Thermal Desorption

Thermal remediation methods use heat to desorb, volatilize, or degrade contaminants, often offering advantages over chemical or electrochemical treatments, especially for complex or fine-grained matrices where diffusion limitations hinder other approaches. It has been used for the remediation of numerous types of contaminants and has been applied to dredge sediments. For instance, ref. [31] investigated TD for the treatment of hydrocarbon-contaminated marine sediment, which is often challenging to remediate due to its cohesive and low-permeability nature. Their results demonstrated that heating at 200–280 °C for just 10 min could achieve TPH removal efficiencies ranging from 75% to 85.2%. The lighter hydrocarbon fractions (e.g., C12–C30) were more easily volatilized, while the heavier fractions (C32–C40) remained more persistent, underscoring the need for temperature optimization according to the specific contamination profile. In parallel, microwave heating (MWH) has emerged as a novel and energy-efficient thermal approach for the remediation of contaminated solids. Unlike conventional TD, MWH relies on the dielectric heating properties of the material itself, resulting in volumetric and rapid heating. This allows for faster treatment and reduced energy consumption. Ref. [32] evaluated the application of MWH for the remediation of severe hydrocarbon-contaminated sediment from Augusta Bay, integrating citric acid (CA) as a chelating agent. The study demonstrated that MWH at 650 W could raise the sediment temperature to 305 °C, achieving TPH removal rates of approximately 95.6% after 10 min and 99.6% after 15 min. Notably, the addition of 0.1 M CA improved removal efficiency to 99.7% in just 15 min, and 0.2 M CA reduced the required treatment time to 10 min. Life cycle assessment (LCA) further supported the environmental sustainability of MWH, showing a 75.74% reduction in total environmental damage compared to electrokinetic remediation. Beyond hydrocarbons, MWH has also been applied to heavy metal remediation. Ref. [33] studied a microwave-assisted hydrogen peroxide ($MW{-}H_2O_2$) oxidation process for river sediment contaminated with metals such as Cu, Zn, Cd, and Pb. The treatment was particularly effective in acidic conditions (pH 0–4), where metal desorption was enhanced. Increasing the microwave temperature (25–85 °C) and irradiation time (up to 6 min) significantly improved metal release, while maintaining metal stability post-treatment, thus minimizing environmental risk.

4.2. Biological Remediation Techniques

Biological remediation techniques use living organisms to degrade or immobilize contaminants. It offers an environmentally sustainable and cost-effective approach to treating contaminated sediment. Among these, bioremediation, vermiremediation, and phytoremediation have gained significant attention for their potential to address a wide range of contaminants, including heavy metals, organic pollutants, and petroleum hydrocarbons. These techniques rely on the metabolic capabilities of micro- and macro-organisms and plants to remove or neutralize pollutants in sediment.

4.2.1. Bioremediation

Bioremediation involves the use of microorganisms to degrade or transform contaminants into less harmful substances under optimal environmental conditions with the addition of nutrients. This method has proven to be effective for large-scale remediation due to its low energy requirements and cost-effectiveness. Furthermore, it can be applied to a broad spectrum of contaminants, including hydrocarbons, heavy metals, and pesticides. For example, ref. [34] demonstrated that the combination of bioaugmentation (the addition of specific microorganisms) and biostimulation (enhancement of microbial activity through nutrient addition) significantly reduced the half-life of tributyltin (TBT) in contaminated sediment. In this study, biostimulation alone resulted in the degradation of approximately 50% of TBT within 20 weeks, showing the potential for microbial remediation of persistent contaminants. Sulfur-oxidizing bacteria, such as *Acidithiobacillus thiooxidans*, have been identified as key agents for the oxidation of sulfide minerals in dredge sediment, thereby promoting the release of heavy metals [35]. Fungi have also shown promise in enhancing the mobilization of sediment contaminants. For example, ref. [36] evaluated the effectiveness of different microorganisms for the bioremediation of marine sediments heavily contaminated by heavy metals (HMs) in Portman Bay (Northwestern Mediterranean Sea). At the laboratory scale, autotrophic bacteria, acidophilic heterotrophic bacteria, and marine filamentous fungi (*Aspergillus niger* and *Trichoderma* sp.), alone or in combination, were compared for the bioleaching of metals. The results showed that the addition of fungi led to the best removal yields, with arsenic solubilization eight times higher than chemical treatments and twice as high as bacteria.

4.2.2. Vermiremediation

Vermiremediation utilizes earthworms to remediate contaminated soils and sediment. Earthworms not only degrade organic contaminants but also enhance soil structure and nutrient cycling, making them valuable in remediating contaminated environments. While their effectiveness can be limited in highly toxic soils, and their activity is influenced by environmental conditions, studies have shown that *Eisenia fetida* and other species can significantly reduce heavy metal concentrations in contaminated soils [37]. In another study, the addition of *Eudrilus eugeniae* earthworms reduced the concentrations of heavy metals like copper, zinc, manganese, chromium, and nickel. In addition, earthworms improved soil enzyme activities, improving soil health and reducing metal toxicity, as confirmed by seed germination assays [38].

4.2.3. Phyto-Remediation

Phytoremediation, the use of plants to decontaminate sediment and soils, offers a highly versatile and visually appealing approach to environmental restoration. This technique is especially effective in removing metals, oils, pesticides, and even plastic pollutants. The major advantage of phytoremediation is its ability to work on large-scale projects, offering a natural and low-cost solution to contamination. Salt-tolerant plant species, such as *Phragmites australis* (common reed) and *Halimione portulacoides*, have been successfully employed to remediate sediment contaminated with tributyltin (TBT), a toxic compound commonly found in marine sediment. These plants can accumulate pollutants in their tissues, thus reducing the concentration of contaminants in the surrounding sediment and improving the overall health of the ecosystem [39]. Similarly, other studies have explored the remediation potential of *Spartina maritima* and *Sarcocornia fruticosa* in estuarine sediment, with varying degrees of success depending on the plant species and the level of contamination. These plants not only accumulate contaminants but also improve sediment structure and nutrient cycling by oxygenating the soil with their root systems [40]. The use of plants to remediate petroleum-contaminated sediment has also shown great promise. In this context, ref. [41] investigated the performance of *Zea mays* and *Festuca arundinacea* in highly contaminated, aged soil. Both species demonstrated strong resilience and contributed to a marked decrease in total petroleum hydrocarbons (TPHs) over time. Notably, tall fescue achieved the highest remediation efficiency, removing up to 96.3% of initial TPHs. The application of peat as a soil amendment further enhanced plant growth and soil quality. The concentration of contaminants can represent a major challenge for effective phyto extraction, as highlighted in [42], which evaluated six months of phytoremediation using *Phragmites australis* grown on dredged sediments from urban retention tanks. The comparison between uncontaminated (S_{uncont}) and contaminated (S_{cont}) seedlings demonstrated that high levels of potentially toxic elements (Zn, Cu, Cd, Ni, Cr, Pb) can inhibit plant growth and function.

5. Turning Dredge Sediment into a Resource: Recently Tested Strategies

5.1. Dredge Sediment Utilization as Inert Materials

The preservation of natural resources and the sustainable utilization of waste materials are key components of sustainable development. Dredged sediment traditionally considered waste, is increasingly being explored as a valuable resource for various applications, such as concrete production, construction materials, and environmental remediation. One promising application is the incorporation of dredge sediment into cement formulations. Research on uncontaminated sediment from Dunkirk Harbor (France) showed that replacing up to 20% of cement with dredge sediment reduced mortar strength but still surpassed the performance of reference mortar. This sediment replacement also achieved a C30/37 strength class in concrete, indicating its potential in construction [43]. Additionally, combining dredge sediment with plastic waste has been explored. One study created lightweight aggregates by combining 70% sediment-based mineral filler with 30% thermoplastic waste. These aggregates met the European Standard for lightweight aggregates and displayed favorable mechanical properties, such as low water absorption compared to natural aggregates [44]. The creation of composite materials using marine dredge sediment, overburden soil, and lime production waste has also been investigated. These composites exhibited good mechanical properties, such as an axial resistance strength of 18.9 MPa, and are suitable for use in tiles, bricks, and road bases [45]. Additionally, marine dredge sediment has been tested for cement production by treating it at high temperatures (650 °C and 850 °C) to eliminate organic compounds and activate clay minerals. Sediment treated at 650 °C showed improved compressive strength and proved to be the most cost-effective treatment [46]. Beyond cement, dredge sediment has been utilized in brick production. One study incorporated incinerated waste and river dredge sediment into lightweight, thermally insulating clay bricks, achieving satisfactory mechanical properties after firing at temperatures between 800 and 1000 °C [47]. Additionally, the potential of combining specific river sediments with hemp shiv and palm oil fiber (POFl) was explored. Bricks made with these mixtures met the minimum tensile strength standard, demonstrating the feasibility of using dredge sediment in sustainable brick production [48]. Dredge sediment also shows promise as an adsorbent for environmental remediation. One study developed a low-cost adsorbent from dredge sediment (DSD) for the removal of crystal violet dye from aqueous solutions. The DSD, primarily composed of clay minerals like illite and kaolinite, exhibited a high surface area and mesoporous structure, which enhanced its adsorption capacity for dyes [49]. Another study combined dredge sediment with zeolite and bentonite to create ceramsite, a lightweight material suitable for water treatment applications. The resulting ceramsite had a high specific surface area (51.68 m^2/g) and large porosity (53.93%) and met the standards for macroporous materials [50]. Dredge sediment can also be utilized in road construction. Research has shown that dredge sediment, when used as embankment or subgrade material, can offer environmental and economic benefits. A field test conducted in the Port of Dunkirk, France, demonstrated that dredge sediment used as subbase material met design specifications and posed no significant environmental risks in terms of contaminant leaching, including trace metals and chlorides [51]. Within its uses as an inert material, the dredged material has been used as material for the covering of sanitary landfills. In this context, ref. [52] reused dredge material from ports in Istanbul and Kocaeli (Turkey) as capping material for a sanitary landfill. A comprehensive analysis was conducted to assess the physical, chemical, mineralogical, toxicological, and leaching characteristics of the DM. Both raw and processed forms were tested to evaluate their geotechnical suitability for landfill components such as cover layers, base or cap liners, and fill materials. The study concluded that non-hazardous dredge sediment could be effectively repurposed within different sections of landfills. This approach offers notable environmental and economic advantages over conventional disposal methods, such as marine dumping or upland storage.

5.2. *Agricultural Applications of Dredge Sediment*

Multiple studies have explored its suitability as a growing substrate, soil amendment, or input in soil regeneration processes. Ref. [53] investigated the use of dredge sediment from Malmfjärden Bay (Sweden), combining it with compost to grow lettuce. Although plant growth was achieved, the size of the lettuce was reduced, likely due to the soil's texture. The study emphasized the need for ecotoxicological assessments to evaluate potential risks to both plant development and human health. It concluded that while untreated sediment may not be appropriate for edible crops, it could be suitable for cultivating ornamental or bio-energy plants. Similarly, Ref. [54] evaluated the use of remediated marine dredge sediment as a substrate for strawberry and pomegranate cultivation. Their findings revealed no significant impact on fruit quality, and even when sediment with relatively high contamination levels was used, the product remained safe for consumption, supporting the feasibility of agricultural use following proper remediation. Further studies reinforced the importance of pre-treatment of dredge sediment. Ref. [55] showed that pre-washing dredge sediment to reduce salinity, followed by pH neutralization, enabled 100% barley germination, suggesting that proper treatment can significantly improve sediment viability as a plant-growing medium. Ref. [56] examined various solidification and stabilization (S/S) techniques, including the use of cement, clay, fly ash, and green-synthesized nanozero valent iron (nZVI) to reduce trace element concentrations in dredge sediment. Although most treatments lowered contaminants, some resulted in plant stress, underlining the importance of aligning the choice of treatment with the target plant species and intended agricultural use. In a pilot-scale experiment, ref. [57] utilized sediment from the Navicelli Canal (Pisa, Italy) mixed with green waste from Florence to cultivate *Photinia* × *fraseri*. The co-composted mixture increased microbial diversity (bacteria, fungi, and archaea) compared to untreated sediment, with no significant ecotoxicological effects and excellent plant growth results. However, risks remain, especially with contaminated sources. Ref. [58] assessed microcystin-contaminated sediment from Lake Taihu (China) used as a substrate for lettuce. While plant growth improved due to additional nutrients, lettuce grown in sediment from algal bloom periods contained detectable levels of microcystins, alongside reduced soluble sugar and Vitamin C content. These results highlight the critical need for monitoring and risk evaluation when repurposing contaminated sediment for agriculture. Instead, ref. [59] found positive outcomes in the United States, where sediment from the Illinois River enhanced the physical and chemical quality of sandy soils. Treated soils exhibited improved water-holding capacity, cation exchange capacity, and nutrient content, leading to increased biomass and grain yield in corn and soybeans. Metal concentrations in plant tissue and soils remained within safe limits. Dredge sediment also presents an opportunity for resource recovery, particularly as an alternative source of phosphorus for agricultural use. In this context, ref. [60] evaluated the fertilizing potential of lake dredge sediment, focusing on its phosphorus content and plant availability. The study found that freshly dredge sediment contains a form of phosphorus predominantly bound to amorphous iron (Fe-P), which can be bioavailable under appropriate conditions. This form was identified as a significant contributor to plant nutrition, suggesting that such sediment can act as a slow-release P fertilizer.

5.3. Sustainable Management and Large-Scale Reuse of Dredge Sediment

Large-scale applications include land reclamation, wetland and habitat restoration, beach nourishment, agricultural soil improvement, and use as construction fill for infrastructure such as roads, levees, and dikes. It is an effective strategy to manage dredge sediment, transforming it into a valuable resource. As an example, the Eems–Dollard project in the Netherlands aims to improve water quality and enhance natural habitats by reusing dredge material [61]. This initiative exemplifies the "beneficial use of dredge material" philosophy, contributing to estuarine restoration and climate resilience. Another flagship project is the Fehmarnbelt Tunnel, a submerged connection between Germany and Denmark. This infrastructure endeavor plans to reuse 100% of the dredge sediment generated during excavation for environmentally beneficial purposes, minimizing waste and supporting sustainability goals [62]. In addition to these initiatives, various studies demonstrate the technical feasibility and ecological value of using dredge sediment for habitat creation and coastal defense. In the Mississippi River Delta (United States), sediment deposition over 70 years has led to the formation of more than 800 hectares of new land in West Bay, as confirmed by remote sensing analyses [63]. Similarly, ref. [64] assessed the reuse of New York Harbor sediment for environmental enhancement, identifying viable applications such as artificial reef creation, oyster reef restoration, marsh and mudflat rehabilitation, and upland habitat construction. In the Westerschelde estuary (Netherlands), ref. [65] tested an alternative dredging strategy in which clean sand was placed seaward of an eroding intertidal flat to restore shallow-water habitats. Five years of monitoring revealed that sediment slowly migrated toward the flat, increasing the extent of subtidal and intertidal areas without negative ecological impacts, although no new habitats were formed. Ref. [66] introduced the "Mud Motor" concept to enhance salt marsh development, successfully increasing mud transport based on weather conditions and offering new insights into its feasibility for land reclamation. Moreover, ref. [67] demonstrated that fine-grained dredge sediment can support the recovery of tidal lagoons by increasing elevation, reducing erosion, and boosting system resilience. While nearshore disposal is common for dredging materials from maintenance operations, uncontrolled migration remains a risk. In this regard, ref. [68] developed a predictive tool to model sediment transport and sandbank movement, allowing for more strategic placement. Beach nourishment also represents a practical reuse strategy. Ref. [69] successfully applied sand from nearby port dredging operations to restore coastal areas, achieving outcomes such as dune field expansion, enhanced habitats for nesting shorebirds, and increased recreational space.

6. Artificial Intelligence (AI) and Its Potential Role in Dredge Sediment Remediation

Artificial Intelligence (AI) refers to the ability of machines to process data and make decisions autonomously, adapting their behavior without explicit programming. In recent years, AI has been increasingly applied across various environmental fields due to its capacity to analyze complex systems and optimize decision-making processes. However, to date, there is a lack of documented studies that specifically apply AI to the remediation or treatment of dredge sediment. Despite this, the potential for integrating AI into dredge sediment management is significant. AI could support multiple aspects of the remediation process, including the prediction of contaminant behavior, the optimization of treatment strategies, and the selection of the most sustainable reuse options based on site-specific conditions. For instance, machine learning models could assist in identifying suitable stabilization techniques or in forecasting long-term environmental impacts under different reuse scenarios. The application of AI has already demonstrated its value, particularly in the biodegradation of petroleum hydrocarbons. For example, an artificial neural network (ANN) applied an updated ANN to enhance fluoranthene degradation by *Mycobacterium litorale* [70]. Additionally, ref. [71] conducted an optimization study of total petroleum hydrocarbon (TPH) degradation in diesel-contaminated soils using *Paspalum scrobiculatum*, a tropical plant, with the support of ANN modeling. These works highlight how AI can be used to optimize biological remediation processes, guiding microbial or phytoremediation-based strategies more efficiently.

7. Conclusions

Dredging continues to be an essential activity for maintaining the operability of ports, navigable waterways, and lagoon systems. In recent years, there has been a significant shift in perspective from viewing dredge sediment as a waste product to recognizing it as a potential resource. This change aligns with broader efforts to embrace principles of circular economy and sustainable environmental practices. Dredge sediment is now considered a valuable input for environmental recovery, particularly through innovative remediation and reuse strategies. Recent advancements in sediment management have introduced a variety of technologies, such as electrochemical remediation, chemical washing, thermal desorption, biological treatment, and solidification/stabilization. These approaches not only help reduce the concentration of contaminants but also improve the physical and chemical characteristics of the sediment, making it suitable for beneficial applications such as land reclamation, habitat creation, and construction material production [72]. The integration of emerging technologies like artificial intelligence has further opened new avenues for optimizing treatment processes, improving environmental monitoring, and increasing remediation efficiency. This shift reflects a broader move toward sustainable and circular practices that aim not only to mitigate environmental harm but also to promote ecosystem restoration and resource efficiency. However, despite these promising developments, there are significant concerns regarding the economic feasibility and practical implementation of these technologies. High energy costs, complex logistics, and the uncertain scalability of many remediation methods pose major barriers to their widespread adoption [73]. Legislative frameworks also play a crucial role in determining how dredge sediment is managed and whether reuse projects can proceed. Italy's Ministerial Decree No. 173/2016 promotes sediment reuse, encouraging sediment reuse. But, in the UE complex permitting procedures, lack of harmonized standards, and insufficient public awareness contribute to the regulatory and institutional challenges facing sediment valorization efforts [74]. Despite that, one particularly promising area is the beneficial reuse of sediment, especially in high-volume dredging contexts where traditional disposal is no longer viable. Reuse options such as land reclamation, coastal defense, and wetland restoration offer tangible environmental and economic benefits and have already been successfully implemented in countries like the Netherlands. There, sediment has been repurposed to create green spaces and public parks, demonstrating its potential for both ecological enhancement and human use [75]. Dredging can be a useful tool in environmental restoration projects [76], for example, in cases of contamination, by removing contaminated sediment, and in cases of eutrophication. Nonetheless, care must be taken when intervening in restoration projects. The human impact on restoration projects is still unclear, and for this reason, these measures must be applied with care, planning, and execution of specific ecosystem restoration projects [77].

Supplementary Materials: The following supporting information can be downloaded at: https://www.mdpi.com/article/10.3390/environments12060200/s1.

Author Contributions: Conceptualization, C.F. and S.A.; methodology, C.F. and M.R.; software, C.F.; validation, C.F. and M.R.; formal analysis, C.F.; investigation, C.F.; resources, M.R.; data curation, C.F.; writing—original draft preparation, C.F., S.A. and M.R.; writing—review and editing, C.F. and M.R.; visualization, M.R.; supervision, M.R.; project administration, M.R.; funding acquisition, M.R. All authors have read and agreed to the published version of the manuscript.

Funding: This research was funded by Bioscience Research Center supporting industrial Ph.D. program at the university of Palermo [RG_61_2025_01].

Data Availability Statement: The original contributions presented in this study are included in the article and Supplementary Materials. Further inquiries can be directed at the corresponding authors.

Acknowledgments: During the preparation of this manuscript, language assistance tools were used to improve the clarity and fluency of the English text. All authors have reviewed and edited the final version and take full responsibility for the content. All contributors have consented to the use and acknowledgment of such tools. The authors are grateful to the Bioscience Research Center which supported and founded the Ph.D. research program linked to this review. Authors are, also, grateful to Luca Sittoni for his precious suggestions and critical review during the development of the manuscript.

Conflicts of Interest: The authors declare no conflicts of interest. The founders had no role in the design of the study; in the collection, analysis, or interpretation of data; in the writing of the manuscript; or in the decision to publish the results.

Abbreviations

The following abbreviations are used in this manuscript:

EGD	European Green Deal
PTE	Ecological Transition Plan
DO	Dissolved oxygen
GHG	Greenhouse gas
LCA	Life Cycle Assessment
EK	Electrochemical remediation
PAHs	Polycyclic aromatic hydrocarbons
CA	Citric acid
TW20	Tween 20

EDTA	Ethylenediaminetetraacetic acid
EDDS	Etilendiamminodisuccinic acid
S/S	Solidification/Stabilization
OPC	Portland cement
IBPs	Industrial by-products
PG	Phosphogypsum
CAC	Calcium aluminate cement
UCS	Unconfined compressive strength
GAC	Granular activated carbon
AC	Activated carbon
MAC	Magnetic activated carbon
BC	Biochar
PCBs	Polychlorinated biphenyls
DDT	Dichlorodiphenyltrichloroethane
TPHs	Total petroleum hydrocarbons
PCP	Pentachlorophenol
TBT	Tributyltin
DW	Dry weight
AI	Artificial intelligence
ANN	Artificial neural network
ESCAP	Soil petroleum pollution assessment prototype
HM	Heavy metal
POF	Palm oil fiber
DSD	Dredge sediment
nZVI	Nano-zero valent iron
TE	Trace element
GW	Green waste

References

1. Dorleon, G.; Rigaud, S.; Techer, I. Management of dredged marine sediments in Southern France: Main keys to large-scale beneficial reuse. *Environ. Sci. Pollut. Res.* **2024**. Available online: https://link.springer.com/article/10.1007/s11356-024-33129-9 (accessed on 21 March 2025). [CrossRef] [PubMed]
2. OSPAR Commission. Environmental Impacts to Marine Species and Habitats of Dredging for Navigational Purposes. In *OSPAR Biodiversity Series*; OSPAR Commission: London, UK, 2004. Available online: https://www.ospar.org/documents?v=6987 (accessed on 31 March 2025).
3. Bridges, T.S.; Gustavson, K.E.; Schroeder, P.; Ells, S.J.; Hayes, D.; Nadeau, S.C.; Palermo, M.R.; Patmontyy, C. Dredging processes and remedy effectiveness: Relationship to the 4 Rs of environmental dredging. *Integr. Environ. Assess. Manag.* **2010**, *6*, 619–630. [CrossRef] [PubMed]
4. Simonson, W.D.; Miller, E.; Jones, A.; García-Rangel, S.; Thornton, H.; McOwen, C. Enhancing climate change resilience of ecological restoration—A framework for action. *Perspect. Ecol. Conserv.* **2021**, *19*, 300–310. [CrossRef]
5. Costa-Pierce, B.A.; Weinstein, M.P. Use of dredge materials for coastal restoration. *Ecol. Eng.* **2002**, *19*, 181–186. [CrossRef]
6. Padhye, L.P.; Srivastava, P.; Jasemizad, T.; Bolan, S.; Hou, D.; Shaheen, S.M.; Rinklebe, J.; O'Connor, D.; Lamb, D.; Wang, H.; et al. Contaminant containment for sustainable remediation of persistent contaminants in soil and groundwater. *J. Hazard. Mater.* **2023**, *455*, 131575. [CrossRef]
7. Netherlands Enterprise Agency; Holland Circular Hotspot. Circular Economy & SDGs: How Circular Economy Practices Help to Achieve the Sustainable Development Goals. 2020. Available online: https://circulareconomy.europa.eu/platform/sites/default/files/3228_brochure_sdg_-_hch_cmyk_a4_portrait_-_0520-012.pdf (accessed on 31 March 2025).
8. European Commission. The European Green Deal (COM(2019) 640 Final). 2019. Available online: https://eur-lex.europa.eu/legal-content/EN/TXT/?uri=celex:52019DC0640 (accessed on 31 March 2025).
9. Ministry of the Environment and Energy Security. National Ecological Transition Plan (PTE). 2021. Available online: https://www.mase.gov.it/portale/piano-per-la-transizione-ecologica-pte- (accessed on 31 March 2025).
10. European Parliament and Council. Directive 2000/60/EC establishing a framework for Community action in the field of water policy (Water Framework Directive). *Off. J. Eur. Communities* **2000**, *L327*, 1–73. Available online: https://eur-lex.europa.eu/legal-content/EN/TXT/?uri=CELEX:32000L0060 (accessed on 31 March 2025).

11. Council of the European Communities. Council Directive 75/442/EEC on waste (Waste Framework Directive). *Off. J. Eur. Communities* **1975**, *L194*, 39–41. Available online: https://eur-lex.europa.eu/legal-content/EN/TXT/?uri=CELEX:31975L0442 (accessed on 31 March 2025).
12. Ministry for the Environment and Protection of Land and Sea. Ministerial Decree No. 173/2016: Procedures and technical criteria for the disposal at sea of dredged materials. *Off. Gaz. Ital. Repub.* **2016**. No. 208. Available online: https://www.fao.org/faolex/results/details/en/c/LEX-FAOC161852/ (accessed on 31 March 2025).
13. Liu, X.; Sathishkumar, K.; Zhang, H.; Saxena, K.K.; Zhang, F.; Naraginti, S.; Anbarasu, K.; Rajendiran, R.; Rajasekar, A.; Guo, X. Frontiers in environmental cleanup: Recent advances in remediation of emerging pollutants from soil and water. *J. Hazard. Mater. Adv.* **2024**, *16*, 100461. [CrossRef]
14. Solanki, P.; Jain, B.; Hu, X.; Sancheti, G. A Review of Beneficial Use and Management of Dredged Material. *Waste* **2023**, *1*, 815–840. [CrossRef]
15. Aria, M.; Cuccurullo, C. Bibliometrix: An R-tool for comprehensive science mapping analysis. *J. Informetr.* **2017**, *11*, 959–975. [CrossRef]
16. Pasciucco, E.; Pasciucco, F.; Castagnoli, A.; Iannelli, R.; Pecorini, I. Removal of heavy metals from dredging marine sediments via electrokinetic hexagonal system: A pilot study in Italy. *Heliyon* **2024**, *10*, e27616. [CrossRef] [PubMed]
17. Proietto, F.; D'Agostino, F.; Bonsignore, M.; Del Core, M.; Sprovieri, M.; Galia, A.; Scialdone, O. Electrochemical remediation of synthetic and real marine sediments contaminated by PAHs, Hg and As under low electric field values. *Chemosphere* **2024**, *350*, 141009. [CrossRef] [PubMed]
18. Ammami, M.T.; Portet-Koltalo, F.; Benamar, A.; Duclairoir-Poc, C.; Wang, H.; Le Derf, F. Application of biosurfactants and periodic voltage gradient for enhanced electrokinetic remediation of metals and PAHs in dredged marine sediments. *Chemosphere* **2015**, *125*, 1–8. [CrossRef] [PubMed]
19. Vallejo, A.D.D.; Araruna, J.T.; Gonçalves, R.A.; Magalhães de Souza, R.F.; Liu, S.; Wang, H. Enhance electrokinetic remediation of potentially toxic elements in the dredged sediment of Camorim Lagoon. *Eng. Sanit. Ambient.* **2024**, *29*, e2024025. [CrossRef]
20. Kanbar, H.J.; Ammami, M.T.; Benamar, A. Insights into processes and consequent metal (loid) behavior in dredged estuarine sediments upon electrokinetic treatment. *Environ. Chall.* **2024**, *15*, 100880. [CrossRef]
21. Ayyanar, A.; Thatikonda, S. Enhanced electrokinetic removal of heavy metals from a contaminated lake sediment for ecological risk reduction. *Soil Sediment Contam.* **2021**, *30*, 12–34. [CrossRef]
22. D'Agostino, F.; Bellante, A.; Bonsignore, M.; Del Core, M.; Clarizia, L.; Sabatino, N.; Giaramita, L.; Tranchida, G.; Chiavarini, S.; Sprovieri, M. A chemical remediation technique for a nearly-total removal of arsenic and mercury from contaminated marine sediments. *Heliyon* **2023**, *9*, e22633. [CrossRef]
23. Ferrans, L.; Jani, Y.; Hogland, W. Chemical extraction of trace elements from dredged sediments into a circular economy perspective: Case study on Malmfjärden Bay, south-eastern Sweden. *Resour. Environ. Sustain.* **2021**, *6*, 100039. [CrossRef]
24. Lumia, L.; Giustra, M.G.; Viviani, G.; Di Bella, G. Washing batch test of contaminated sediment: The case of augusta bay (SR, Italy). *Appl. Sci.* **2020**, *10*, 473. [CrossRef]
25. Zhang, S.; Liu, J.; Ding, J.; Gao, M.; Xue, C. Effect of recycled phosphogypsum and calcium aluminate cement on the strength behavior optimization of cement-treated dredged soil: A co-utilization of solid wastes. *J. Environ. Manag.* **2024**, *353*, 120020. [CrossRef]
26. Wang, S.; He, X.; Cai, G.; Li, J.; Qin, H.; Ma, Y.; Gong, S.; Lang, L. Strength characteristic and micro-mechanism of organic dredged sludge solidified by cement incorporating sodium persulfate. *Dev. Built Environ.* **2024**, *17*, 100323. [CrossRef]
27. De Gisi, S.; Todaro, F.; Mesto, E.; Schingaro, E.; Notarnicola, M. Recycling contaminated marine sediments as filling materials by pilot scale stabilization/solidification with lime, organoclay and activated carbon. *J. Clean. Prod.* **2020**, *269*, 122416. [CrossRef]
28. Rämö, R.; Bonaglia, S.; Nybom, I.; Kreutzer, A.; Witt, G.; Sobek, A.; Gunnarsson, J.S. Sediment Remediation Using Activated Carbon: Effects of Sorbent Particle Size and Resuspension on Sequestration of Metals and Organic Contaminants. *Environ. Toxicol. Chem.* **2022**, *41*, 1096–1110. [CrossRef]
29. Abel, S.; Akkanen, J. Novel, Activated Carbon-Based Material for in-Situ Remediation of Contaminated Sediments. *Environ. Sci. Technol.* **2019**, *53*, 3217–3224. [CrossRef]
30. Han, Z.; Sani, B.; Akkanen, J.; Abel, S.; Nybom, I.; Karapanagioti, H.K.; Werner, D. A critical evaluation of magnetic activated carbon's potential for the remediation of sediment impacted by polycyclic aromatic hydrocarbons. *J. Hazard. Mater.* **2015**, *286*, 41–47. [CrossRef]
31. Falciglia, P.P.; Lumia, L.; Giustra, M.G.; Gagliano, E.; Roccaro, P.; Vagliasindi, F.G.A.; Di Bella, G. Remediation of petrol hydrocarbon-contaminated marine sediments by thermal desorption. *Chemosphere* **2020**, *260*, 127576. [CrossRef]
32. Falciglia, P.P.; Ingrao, C.; De Guidi, G.; Catalfo, A.; Finocchiaro, G.; Farina, M.; Liali, M.; Lorenzano, G.; Valastro, G.; Vagliasindi, F.G.A. Environmental Life Cycle Assessment of marine sediment de-contamination by citric acid enhanced-microwave heating. *Sci. Total Environ.* **2018**, *619–620*, 72–82. [CrossRef]

33. Wang, L.; Yuan, X.; Zhong, H.; Wang, H.; Wu, Z.; Chen, X.; Zeng, G. Release behavior of heavy metals during treatment of dredged sediment by microwave-assisted hydrogen peroxide oxidation. *Chem. Eng. J.* **2014**, *258*, 334–340. [CrossRef]
34. Beolchini, F.; Rocchetti, L.; Dell'Anno, A. Degradation kinetics of butyltin compounds during the bioremediation of contaminated harbour sediments. *Chem. Ecol.* **2014**, *30*, 393–402. [CrossRef]
35. Lors, C.; Tiffreau, C.; Laboudigue, A. Effects of bacterial activities on the release of heavy metals from contaminated dredged sediments. *Chemosphere* **2004**, *56*, 619–630. [CrossRef]
36. Dell'anno, F.; Rastelli, E.; Buschi, E.; Barone, G.; Beolchini, F.; Dell'anno, A. Fungi Can Be More Effective Than Bacteria for the Bioremediation of Marine Sediments Highly Contaminated with Heavy Metals. *Microorganisms* **2022**, *10*, 993. [CrossRef] [PubMed]
37. Javed, F.; Hashmi, I. Vermiremediation-Remediation of Soil Contaminated with Oil Using Earthworm (*Eisenia fetida*). *Soil Sediment Contam.* **2021**, *30*, 639–662. [CrossRef]
38. Borah, G.; Deka, H. Vermiremediation of heavy metals (HMs)-contaminated agricultural land: Synergistic changes in soil enzyme activities and earthworm's growth parameters. *Environ. Sci. Pollut. Res. Int.* **2023**, *30*, 115266–115278. [CrossRef]
39. Carvalho, P.N.; Basto, M.C.P.; Silva, M.F.G.M.; Machado, A.; Bordalo, A.A.; Vasconcelos, M.T.S.D. Ability of salt marsh plants for TBT remediation in sediments. *Environ. Sci. Pollut. Res.* **2010**, *17*, 1279–1286. [CrossRef]
40. Moreira Da Silva, M.; Duarte, D.N.R.; Isidoro, J.M.G.P. The role of *Spartina maritima* and *Sarcocornia fruticosa* on trace metals retention in Ria Formosa, Portugal. 2013, *Unpublished manuscript*. Available online: https://www.researchgate.net/publication/258775532 (accessed on 31 March 2025).
41. Zand, A.D.; Bidhendi, G.N.; Mehrdadi, N. Phytoremediation of total petroleum hydrocarbons (TPHs) using plant species in Iran. *Turk. J. Agric. For.* **2010**, *34*, 429–438. [CrossRef]
42. Nawrot, N.; Wojciechowska, E.; Pazdro, K.; Szmagliński, J.; Pempkowiak, J. Uptake, accumulation, and translocation of Zn, Cu, Pb, Cd, Ni, and Cr by *P. australis* seedlings in an urban dredged sediment mesocosm: Impact of seedling origin and initial trace metal content. *Sci. Total Environ.* **2021**, *768*, 144983. [CrossRef]
43. Zhao, Z.; Benzerzour, M.; Abriak, N.E.; Damidot, D.; Courard, L.; Wang, D. Use of uncontaminated marine sediments in mortar and concrete by partial substitution of cement. *Cem. Concr. Compos.* **2018**, *93*, 155–162. [CrossRef]
44. Ennahal, I.; Maherzi, W.; Benzerzour, M.; Mamindy, Y.; Abriak, N.E. Performance of Lightweight Aggregates Comprised of Sediments and Thermoplastic Waste. *Waste Biomass Valor.* **2021**, *12*, 515–530. [CrossRef]
45. Mymrin, V.; Pan, R.C.Y.; Alekseev, K.; Avanci, M.A.; Stella, J.C.; Scremim, C.B.; Schiavini, D.N.; Pinto, L.S.; Berton, R.; Weber, S.L. Overburden soil and marine dredging sludge utilization for production of new composites as highly efficient environmental management. *J. Environ. Manag.* **2019**, *236*, 206–213. [CrossRef]
46. Dang, T.A.; Kamali-Bernard, S.; Prince, W.A. Design of new blended cement based on marine dredged sediment. *Constr. Build. Mater.* **2013**, *41*, 602–611. [CrossRef]
47. Adazabra, A.N.; Viruthagiri, G.; Atingabono, J. Developing fired clay bricks by incorporating scrap incinerated waste and river dredged sediment. *Process Saf. Environ. Prot.* **2023**, *179*, 108–123. [CrossRef]
48. Hussain, M.; Levacher, D.; Leblanc, N.; Zmamou, H.; Djeran-Maigre, I.; Razakamanantsoa, A.; Saouti, L. Reuse of harbour and river dredged sediments in adobe bricks. *Clean. Mater.* **2022**, *3*, 100046. [CrossRef]
49. Ouakouak, A.; Abdelhamid, M.; Thouraya, B.; Chahinez, H.O.; Hocine, G.; Hamdi, N.; Syafiuddin, A.; Boopathy, R. Development of a novel adsorbent prepared from dredging sediment for effective removal of dye in aqueous solutions. *Appl. Sci.* **2021**, *11*, 10722. [CrossRef]
50. Wang, L.; Shao, Y.; Zhao, Z.; Chen, S.; Shao, X. Optimized utilization studies of dredging sediment for making water treatment ceramsite based on an extreme vertex design. *J. Water Process Eng.* **2020**, *38*, 101603. [CrossRef]
51. Zentar, R.; Wang, H.; Wang, D. Comparative study of stabilization/solidification of dredged sediments with ordinary Portland cement and calcium sulfo-aluminate cement in the framework of valorization in road construction material. *Constr. Build. Mater.* **2021**, *279*, 122447. [CrossRef]
52. Çevikbilen, G.; Başar, H.M.; Karadoğan, Ü.; Teymur, B.; Dağlı, S.; Tolun, L. Assessment of the use of dredged marine materials in sanitary landfills: A case study from the Marmara sea. *Waste Manag.* **2020**, *113*, 70–79. [CrossRef]
53. Ferrans, L.; Schmieder, F.; Mugwira, R.; Marques, M.; Hogland, W. Dredged sediments as aplant-growing substrate: Estimation of health risk index. *Sci. Total Environ.* **2022**, *846*, 157463. [CrossRef]
54. Tozzi, F.; del Bubba, M.; Petrucci, W.A.; Pecchioli, S.; Macci, C.; Hernández García, F.; Martínez Nicolás, J.J.; Giordani, E. Use of a remediated dredged marine sediment as a substrate for food crop cultivation: Sediment characterization and assessment of fruit safety and quality using strawberry (*Fragaria* × *ananassa* Duch.) as model species of contamination transfer. *Chemosphere* **2020**, *238*, 124651. [CrossRef]
55. Kim, K.; Yoon, S.; Kwon, H.A.; Choi, Y. Effects of treatment agents during acid washing and pH neutralization on the fertility of heavy metal-impacted dredged marine sediment as plant-growing soil. *Environ. Pollut.* **2020**, *267*, 115466. [CrossRef]
56. Beljin, J.; Arsenov, D.; Slijepčević, N.; Maletić, S.; Đukanović, N.; Chalot, M.; Župunski, M.; Tomašević Pilipović, D. Recycling of polluted dredged sediment—Building new materials for plant growing. *Waste Manag.* **2023**, *166*, 13–24. [CrossRef]

57. Mattei, P.; Pastorelli, R.; Rami, G.; Mocali, S.; Giagnoni, L.; Gonnelli, C.; Renella, G. Evaluation ofdredged sediment co-composted with green waste as plant growing media assessed by eco-toxicological tests, plant growth and microbial community structure. *J. Hazard. Mater.* **2017**, *333*, 144–153. [CrossRef] [PubMed]
58. Cao, Q.; You, B.; Liu, W.; Xu, H.; Ma, S.; Wang, T. Using dredged sediments from Lake Taihu as a plant-growing substrate: Focusing on the impact of microcystins. *J. Environ. Manag.* **2024**, *370*, 122927. [CrossRef] [PubMed]
59. Darmody, R.G.; Ruiz Diaz, D. *Dredged Sediment: Application as an Agricultural Amendment on Sandy Soils;* TR Series (Illinois Sustainable Technology Center) TR-066; Illinois Sustainable Technology Center: Champaign, IL, USA, 2017. Available online: https://hdl.handle.net/2142/97824 (accessed on 31 March 2025).
60. Haasler, S.; Kragh, T.; Magid, J.; Gunnarsen, K.C.; Müller-Stöver, D.; Klamt, A.M.; Krogstrup, K.; Sorensen, H.; Nielsen, U.G.; Reitzel, K. Recycling of phosphorus from dredged lake sediment: Importance of iron-bound phosphates for plant growth. *Sustain. Environ.* **2024**, *10*, 693. [CrossRef]
61. Eems-Dollard 2050. Available online: https://eemsdollard2050.nl/ (accessed on 31 March 2025).
62. Femern A/S. Available online: https://femern.com/ (accessed on 31 March 2025).
63. Suedel, B.C.; McQueen, A.D.; Wilkens, J.L.; Saltus, C.L.; Bourne, S.G.; Gailani, J.Z.; King, J.K.; Corbino, J.M. Beneficial use of dredged sediment as a sustainable practice for restoring coastal marsh habitat. *Integr. Environ. Assess. Manag.* **2022**, *18*, 1162–1173. [CrossRef]
64. Yozzo, D.J.; Wilber, P.; Will, R.J. Beneficial use of dredged material for habitat creation, enhancement, and restoration in New York-New Jersey Harbor. *J. Environ. Manag.* **2004**, *73*, 39–52. [CrossRef]
65. van der Wal, D.; Forster, R.M.; Rossi, F.; Hummel, H.; Ysebaert, T.; Roose, F.; Herman, P.M.J. Ecological evaluation of an experimental beneficial use scheme for dredged sediment disposal in shallow tidal waters. *Mar. Pollut. Bull.* **2011**, *62*, 99–108. [CrossRef]
66. Baptist, M.J.; Gerkema, T.; van Prooijen, B.C.; van Maren, D.S.; van Regteren, M.; Schulz, K.; Colosimo, I.; Vroom, J.; van Kessel, T.; Grasmeijer, B.; et al. Beneficial use of dredged sediment to enhance salt marsh development by applying a 'Mud Motor'. *Ecol. Eng.* **2019**, *127*, 312–323. [CrossRef]
67. Staver, L.W.; Morris, J.T.; Cornwell, J.C.; Stevenson, J.C.; Nardin, W.; Hensel, P.; Owens, M.S.; Schwark, A. Elevation Changes in Restored Marshes at Poplar Island, Chesapeake Bay, MD: I. Trends and Drivers of Spatial Variability. *Estuaries Coasts* **2024**, *47*, 1784–1798. [CrossRef]
68. McFall, B.C.; Brutsché, K.E.; Priestas, A.M.; Krafft, D.R. Evaluation Techniques for the Beneficial Use of Dredged Sediment Placed in the Nearshore. *J. Waterw. Port Coast. Ocean Eng.* **2021**, *147*. Available online: https://ascelibrary.org/doi/10.1061/(ASCE)WW.1943-5460.0000648 (accessed on 25 March 2025). [CrossRef]
69. Silveira, T.M.; Santos, C.F.; Andrade, F. Beneficial use of dredged sand for beach nourishment and coastal landform enhancement—The case study of Tróia, Portugal. *J. Coast. Conserv.* **2013**, *17*, 825–832. [CrossRef]
70. Dudhagara, D.R.; Rajpara, R.K.; Bhatt, J.K.; Gosai, H.B.; Dave, B.P. Bioengineering for polycyclic aromatic hydrocarbon degradation by Mycobacterium litorale: Statistical and artificial neural network (ANN) approach. *Chemom. Intell. Lab. Syst.* **2016**, *159*, 155–163. [CrossRef]
71. Sanusi, S.N.A.; Halmi, M.I.E.; Abdullah, S.R.S.; Hassan, H.A.; Hamzah, F.M.; Idris, M. Comparative process optimization of pilot-scale total petroleum hydrocarbon (TPH) degradation by *Paspalum scrobiculatum* L. Hack using response surface methodology (RSM) and artificial neural networks (ANNs). *Ecol. Eng.* **2016**, *97*, 524–534. [CrossRef]
72. Cristóbal, J.; Foster, G.; Caro, D.; Yunta, F.; Manfredi, S.; Tonini, D. Management of excavated soil and dredging spoil waste from construction and demolition within the EU: Practices, impacts and perspectives. *Sci. Total Environ.* **2024**, *944*, 173859. [CrossRef] [PubMed]
73. Ulibarri, N.; Goodrich, K.A.; Wagle, P.; Brand, M.; Matthew, R.; Stein, E.D.; Sanders, B.F. Barriers and opportunities for beneficial reuse of sediment to support coastal resilience. *Ocean Coast. Manag.* **2020**, *195*, 105287. [CrossRef]
74. European Commission. Proposal for a Regulation on Soil Monitoring and Resilience (Soil Monitoring Law). Available online: https://eur-lex.europa.eu/legal-content/EN/TXT/HTML/?uri=CELEX:52023PC0160 (accessed on 30 March 2025).
75. Brils, J.; de Boer, P.; Mulder, J.; de Boer, E. Reuse of dredged material as a way to tackle societal challenges. *J. Soils Sediments* **2014**, *14*, 1638–1641. [CrossRef]
76. US Dredge. Top 6 Benefits of Dredging. Available online: https://www.usdredge.com/learn/top-6-benefits-of-dredging (accessed on 30 March 2025).
77. Moreno-Mateos, D.; Meli, P.; Vara-Rodríguez, M.I.; Aronson, J. Ecosystem response to interventions: Lessons from restored and created wetland ecosystems. *J. Appl. Ecol.* **2015**, *52*, 1528–1537. [CrossRef]

environments

MDPI

Article

Water Discharge and Sediment Transport from the Dese River Estuary to the Venice Lagoon (Italy): Annual Dynamics and the Influence of Floods

Roberto Zonta [1], Giuliano Lorenzetti [1], Giorgia Manfè [1], Simone Leoni [1], Gian Marco Scarpa [1], Luca Zaggia [2,*], Carlo Bendoricchio [3], Martino Cerni [3] and Janusz Dominik [1]

1 Istituto di Scienze Marine, Consiglio Nazionale delle Ricerche (ISMAR CNR), Castello 2737/F, 30122 Venezia, Italy; r.zonta@ismar.cnr.it (R.Z.); g.lorenzetti@ismar.cnr.it (G.L.); g.manfe@ismar.cnr.it (G.M.); simone.leoni@ve.ismar.cnr.it (S.L.); gianmarco.scarpa@ve.ismar.cnr.it (G.M.S.); janusz.dominik@ve.ismar.cnr.it (J.D.)

2 Istituto di Geoscienze e Georisorse, Consiglio Nazionale delle Ricerche (IGG CNR), Corso Stati Uniti 4, 35131 Padova, Italy

3 Acque Risorgive—Consorzio di Bonifica, via Rovereto 12, 30174 Venezia, Italy; c.bendoricchio@acquerisorgive.it (C.B.); m.cerni@acquerisorgive.it (M.C.)

* Correspondence: luca.zaggia@igg.cnr.it

Abstract: The sediment load from the drainage basin may play a key role in sustaining the fragile tidal and subtidal habitats of the Venice Lagoon, which underwent significant erosional losses in recent decades. Freshwater discharge and suspended sediment transport were studied in two estuarine sections of the main tributary of the lagoon, Dese River, employing acoustic instruments and analyzing sediment accumulation in the estuary through bathymetric measurements. The results for 2021 are discussed in terms of discharge, rainfall–runoff relationships, suspended sediment load, and the impact of floods. Compared to previous estimates from 1999, the mean water flow was similar in both years, but the annual suspended sediment load nearly doubled. A single flood event is described, which produced extreme sediment transport and delivered 21.8% of the annual load over a 22 h period. This study provides a framework for assessing whether the sediment load from the Dese River may offset degradation in the morphodiversity of its estuary and the habitats it supports. High-resolution monitoring of sediment delivery from the freshwater tributaries is emphasized as a critical approach for protecting the estuarine areas at the water–land interface of the Venice Lagoon.

Keywords: freshwater discharge; flood hysteresis; tidal estuary; suspended sediment load; Venice Lagoon; morphodiversity; salt marshes

1. Introduction

Human interventions can significantly modify hydrodynamics, sediment transport, and biological activities that shape the morphology of a lagoon over time. The typical activities that affect the evolution of a lagoon include canal dredging, land reclamation, altering boundaries, modifying the structure of the barrier islands, and introducing pollutants and nutrients that affect water quality. These interventions can lead to adverse processes, affecting the morphological and biological diversity and the natural development of lagoons [1–3]. Adding to these challenges, sea level rise further exacerbates the effects, accelerating the degradation of lagoon environments and hindering their natural development [2,4].

The Venice Lagoon, the largest wetland in the Mediterranean basin covering 550 km^2 (Figure 1), has historically experienced conflicts of interest between economic and recreational activities (navigation, tourism, fish and shellfish farming), the physical protection of the system, and the maintenance of morphological and biological diversities [5–8]. Balancing these components requires careful planning, sustainable management practices, and stakeholder engagement to find solutions that support long-term environmental conservation while permitting the development of human activities [9,10].

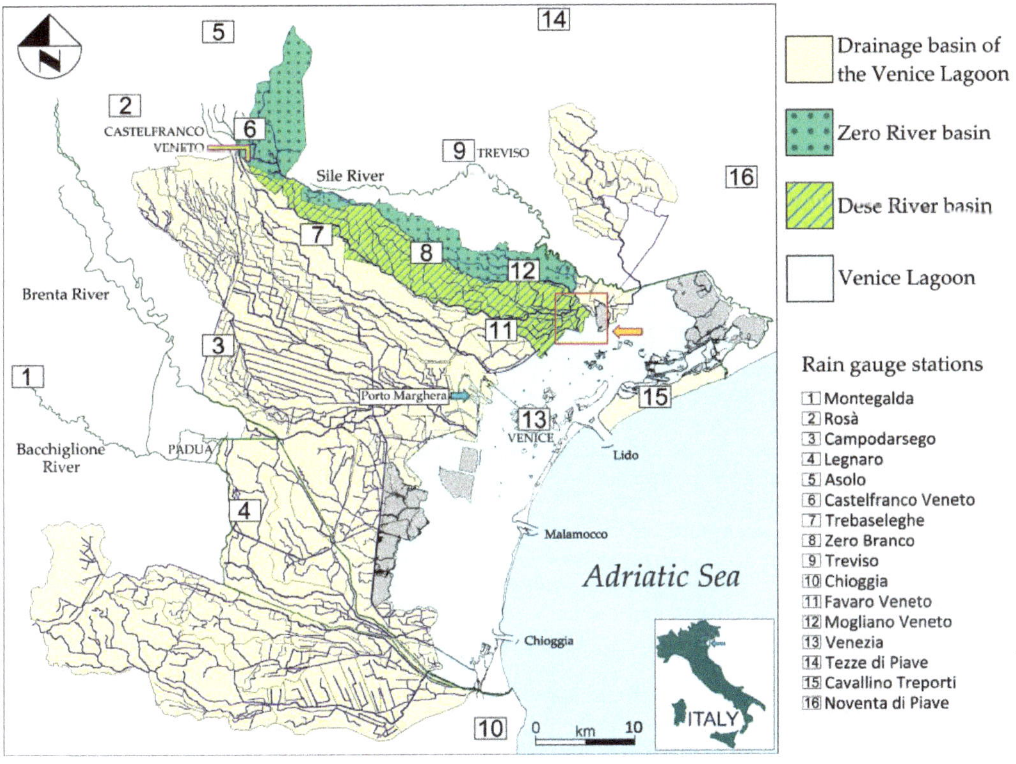

Figure 1. Map of the Venice Lagoon and its catchment area, with the Dese and Zero river basins highlighted. The area covered by the measurements is boxed. The locations of the 16 rainfall data collection stations are indicated.

The lagoon has exhibited an erosive trend in recent decades [11], with a progressive deepening of subtidal systems and flattening of tidal habitats, resulting in a significant loss of the morphodiversity of its transitional environments. From 1927 to 2002, salt marsh areas decreased from 68 to 32 km^2, while the area of subtidal flats, ranging between depths of -0.75 and -2.00 m, expanded from 88 to 206 km^2 [12]. The estimated net loss of sediment is approximately 110×10^6 m^3, 70% of which occurred in the period prior to 1970.

The MOSE (*MOdulo Sperimentale Elettromeccanico*, i.e., experimental electromechanical module) system [13], designed to protect Venice and its lagoon with mobile barriers during high sea level events, started to operate in October 2020 at the three inlets, marking the onset of a new era of "regulated flows" for the lagoon. While this project addressed immediate concerns of flood protection and the safeguarding of economic interests [14,15], it raised concerns for possible short- and long-term environmental issues such as reduced water renewal rates [16], increasing hypoxic events [17], alterations in sedimentological patterns [18,19], and negative impacts on the lagoon's morphological and biological diversity [20]. In this context, attention should be focused on the processes that govern the sediment balance of the lagoon, including the quantification and fate of suspended sediment load (SSL) delivered via freshwater inputs from drainage basins.

The importance of monitoring and studying SSL from land to coastal areas, estuaries, and lagoons as a means to protect the health and sustainability of marine and coastal ecosystems has been emphasized by the Clean Water Act (CWA, 1972), the Coastal Zone Management Act (CZMA, 1972), and the Water Framework Directive (WFD, 2000). Recent European initiatives such as the European Green Deal (2019) and the EU Biodiversity Strategy for 2030 (2020) further emphasize the importance of sediment flow monitoring for promoting sustainability, protecting biodiversity, and controlling pollution within an integrated watershed management framework.

An instrumental network is currently operating in the Venice Lagoon, providing continuous measurements of the discharge (Q) in the main freshwater tributaries of the drainage basin using acoustic instruments on behalf of the Provveditorato Interregionale per le Opere Pubbliche del Veneto—Trentino Alto Adige—Friuli Venezia Giulia. However, it does not record the suspended sediment concentration (SSC) and, as a result, the SSL cannot be directly determined.

In this study, we aimed to update the estimates of the annual Q and SSL delivered by the Dese River, the main tributary of the lagoon. The previous estimates, dating back to 1999 [21], were updated through an improved methodology involving high-resolution continuous monitoring of the water flow and SSC at two estuarine sections. The research also aimed to deepen the understanding of the estuary as a system and its role in the broader context of the lagoon, particularly in relation to sediment dynamics and their implications for the management and protection of tidal morphologies that are progressively decreasing in extent at the land–lagoon interface.

To this end, the hydrology and suspended sediment transport were analyzed with reference to the year 2021. The key findings include the identification of progressive siltation in the estuary's main channel, evidenced by bathymetric surveys, and the partial transfer of sediment deposited in the riverbed to the lagoon during significant flood events. A detailed analysis of a major flood event, using a hysteresis diagram, provides insight into the origins of extreme sediment transport. This study also highlights the importance of managing water volumes in the drainage basin, especially for irrigation during the warmer months, and emphasizes the potential of delivered sediment to mitigate the ongoing morphological evolution of the estuary area.

2. Materials and Methods

2.1. Study Sites

The Venice Lagoon, located in the northern part of the Adriatic Sea, is well described in terms of its ecosystem characteristics and functioning (e.g., [22]). It is connected to the sea through three inlets—Lido, Malamocco, and Chioggia—that allow water exchanges with the tide cycle. The mean tidal range is 55 cm, with 30 cm and 110 cm during neap and spring tides, respectively [23]. With an average depth of about 1 m, the lagoon encompasses islands, salt marsh areas, mudflats, and a complex network of tidal channels, some exceeding depths of 10 m. Shallow water areas cover 75% of the total surface area.

The loss of sediment that has occurred in the lagoon over the last hundred years has mostly been due to human interventions [12]. The deepening of the three seaward inlets of the lagoon and their modification through the construction of jetties (1840–1934) promoted erosional processes and increased the net sediment export to the sea [24], simultaneously reducing positive inputs from the sea during storms. The dredging of navigation channels in the lagoon (1927–1970) triggered erosion processes, which were further intensified by the excavation of a large channel (Malamocco-Marghera, 1964–1968) to facilitate the passage of large vessels from the central inlet of Malamocco to the industrial area of Porto Marghera [12].

The erosion problem was aggravated by both natural and anthropogenic subsidence [25,26]. Additionally, boat traffic [27] and unregulated, intensive clam harvesting (1992–2001 [28] further contributed to the increased erosion of sediment from shallow water areas. This mobilized sediment was subsequently transported by tides into the lagoon channels and eventually lost into the Adriatic Sea.

Between the 1930s and the 1970s, the lagoon underwent a consistent loss of morphological diversity and became a sediment-starved system where subsidence is not compensated by depositional processes [12]. Particularly in its central basin, it subsequently evolved into its current condition, characterized by greater energy and a more open structure, resembling that of a bay [12].

Based on hydrodynamic measurements conducted at the inlets, the sediment transfer from the lagoon to the sea was estimated to be 572×10^3 t y^{-1} during the period 2004–2007 [29]. In contrast, the estimated total sediment contribution from the drainage basin via freshwater tributaries was merely 33.5×10^3 t y^{-1} in 1999 [30], representing approximately 6% of the estimated losses through the inlets.

Although relatively small within the overall sediment budget, this supply may be important in maintaining morphodiversity and ecological habitats in the estuarine areas of the lagoon, where peculiar tidal and intertidal landforms such as salt marshes and tidal creeks still exist but are becoming increasingly vulnerable to relative sea level rise (RSLR) conditions [4].

Despite its potential importance for the morphodynamics of the marshland, no systematic high-resolution observations or measurements of SSL have been made in the lagoon's tributaries.

Since the start of its operation in October 2020, the MOSE floodgates have been activated 84 times to defend the lagoon and Venice from high sea level events [31]. The system consists of 78 mobile gates clustered into four sets of barriers installed at the bottom of the three inlets of the lagoon (two at Lido, one at Malamocco and Chioggia) that close the inlet channels from the time around the peak of the storm surge. The barriers rise when the predicted tide exceeds 110 cm above the local datum of Punta Salute, which is set at 26 cm below the current mean sea level [32].

The lagoon catchment area (Figure 1), covering 2068 km^2, is inhabited by approximately 1 million people and includes significant urban centres, livestock and agricultural activities extending on approximately 70% of the land, and small- to medium-sized industrial districts. It comprises a series of freshwater tributary sub-basins with diverse morphological characteristics and extensions [33]. Approximately 90% of the total catchment water flows into the Venice Lagoon through twelve main tributaries, each characterized by different hydraulic regimes—natural, mechanical, or alternating mechanical—and subjected to flow regulation.

While the hydrographic basin is primarily characterized by a low-sloping alluvial plain, its northern section extends into the Veneto foothills, where a resurgence spring alignment originates several rivers, including the Dese and Zero Rivers (49 and 41 km long, respectively). The Zero River joins the Dese River in the vicinity of the lagoon, in the locality of Zuccarello (Figure 2); the two rivers drain a total area of 291 km^2, which corresponds to 14.3% of the overall surface of the lagoon drainage basin. Approximately half a kilometre downstream from the confluence, the Santa Maria Channel branches off from the main riverbed, flowing eastward toward the shallow waters of the Cona tidal marsh.

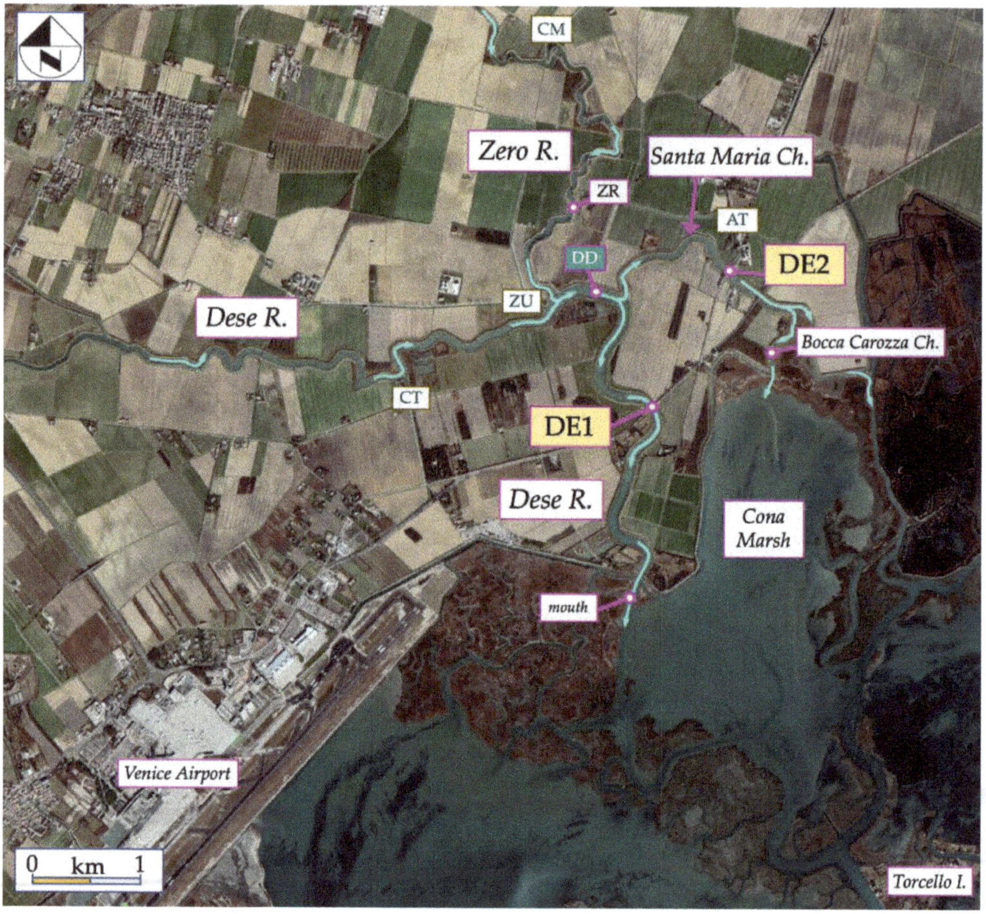

Figure 2. Details of the Dese estuary area covered by the measurements of this study. The main (DE1) and secondary (DE2) measurement sections of this study are indicated. CT (Cattal), ZU (Zuccarello), CM (Camarson), and AT (Altino) indicate the locations of the four pumping stations discharging freshwater into the estuary. ZR is the location where the flow rate is continuously monitored in the Zero River by the Consorzio Venezia Nuova (Venice, Italy). Flow rate was continuously measured in section DD in the period 1998–2000 in the framework of the DRAIN project [33]. Map background obtained from Google Earth (© 2024, Google).

In the estuary area, the river receives additional water inputs from four pumping stations serving territories that consist primarily of agricultural areas (for a supplementary drained surface of ca. 61 km^2), one of which discharges into the Santa Maria Channel. The Dese River also receives the occasional water inputs from an adjacent external territory (part of the Asolo Hills) to the north of the source area through the regulation of a hydraulic node in Castelfranco Veneto [34]. As a consequence of this complex system of inputs, the hydrological behaviour of the river basin is characterized by a large variability and its response cannot be uniquely defined within the basin, depending on the distribution of precipitations and regulation maneuvers.

In recent decades, the entire estuarine system—particularly the terminal stretch of the Dese River, from just before its confluence with the Zero River to its mouth in the lagoon, and the initial section of the Santa Maria Channel—has experienced progressive silting.

This phenomenon has caused significant navigation difficulties and adversely affected recreational activities while also creating problems for commercial transport and local hospitality businesses.

A comparison of bathymetric data for the river stretch corresponding to section DD (as shown in Figure 2) is presented in Figure A4 of Appendix A. The analysis includes data from four surveys conducted in 1982, 1998, 2011, and 2020, revealing that noticeable riverbed silting had occurred by 1998. However, the precise timing of the onset of this process following the 1982 survey remains indeterminate.

2.2. Measuring Stations

Self-recording measuring stations were set up in two tidal sections of the estuary to acquire continuous data series of discharge (Q) and suspended sediment concentration (SSC): the principal one on the Dese River (DE1) and a secondary station on the Santa Maria Channel (DE2).

In the DE1 section, an acoustic current meter SLD (Side Looking Doppler Sensor, 1 MHz/25 m, OTT HydroMet, Kempten, Germany) was utilized, along with a ceramic pressure sensor OTT PLS for measuring tide levels. DE2 section was equipped with a horizontal-oriented Acoustic Doppler Current Profiler (Channel Master 1200 kHz/25 m, Teledyne Technologies, Thousand Oaks, CA, USA). These instruments were mounted on the piers of two bridges. The widths of the riverbed are equal to 43.0 and 33.6 m, respectively, in DE1 and DE2. Details of instrument setup and calibration are provided in Appendix A. The instruments transmit acoustic pulses towards the opposite bank, crossing the river flow. When the sound wave reflects off suspended particles moving within the water column, it produces a Doppler effect, altering the frequency and intensity of the returning echo compared to the transmitted signal. This allows the measurement of the water speed. Conversely, the acoustic backscatter signal, which is a measure of sound attenuation due to suspended particles, indirectly permits an estimation of SSC [35,36].

Flow rate and backscatter data were acquired by the instruments at 1 min and 5 min sampling intervals, respectively. For the calculations, the mean values computed every 15 min were used. In section DE1, data were acquired continuously for 490 days (27 November 2020–31 March 2022), covering the calendar year 2021. In section DE2, the acquired data covered 308 days of 2021 due to an interruption caused by instrumentation malfunction (see Appendix B).

2.3. Instrument Calibration

To calibrate the time series acquired by the self-recording instruments, in situ measurements were carried out under various hydrodynamic and tide conditions, using a vessel-mounted acoustic Doppler current profiler (ADCP, StreamPro 2MHz Teledyne Techmologies, Thousand Oaks, CA, USA). In section DE1, flow rate was calculated using dedicated software from the self-recording instrument (OTT Prodis 2), as reported in [37]. In section DE2, a linear regression between Q_i values and the mean water velocities orthogonal to the section, measured by the self-recording instrument, was used to compute the time series of discharge (the index velocity method) [38]. Calibration diagrams are provided in Appendix A. The R^2 values of the regressions are 0.99 for section DE1 and 0.90 for section DE2.

For the instantaneous measurement of SSC, water samples were collected within the area sensed by the acoustic beam. Physico-chemical (temperature, turbidity, salinity) data were acquired along the vertical profile to assess the occurrence of water stratification, using a multiparametric probe (Ocean Seven 316 Plus, Idronaut, Brugherio, Italy).

SSC was determined in the laboratory by filtration of a 200 mL sample and referred to the dry weight after drying at 105 °C [39]. The grain size of the particulate was measured with a grain-size analyzer (LISST 100X, Sequoia Scientific Inc., Bellevue, DC, USA).

The acoustic backscatter signal of the acoustic Doppler profilers was converted into SSC values using the Surrogate Analysis and Index Developer (SAID) tool developed by the

U.S. Geological Survey (USGS) and written in the MATLAB® suite [40]. This tool enables the formulation of a linear regression model between the instantaneous SSC values (using $\log_{10}$SSC) measured in the water samples and the average backscatter values recorded by the instrument in the measurement cells corresponding to the sampled water. The backscatter values are corrected for the attenuation of the acoustic signal generated by the particle (meanSCB, sediment corrected backscatter), allowing the reconstruction of the SSC time series. Calibration diagrams are provided in Appendix A. The R^2 values of the regressions are 0.88 for section DE1 and 0.77 for section DE2. The time series of suspended sediment load (SSL) was then calculated by directly multiplying instantaneous values of Q and SSC.

2.4. Bathymetric Surveys

Bathymetric data were collected in DE1 and DE2 sections and at other sites of interest along the riverbeds using a single-beam echo sounder mounted on an autonomous surface vehicle (OpenSwap, ISMAR CNR and Proambiente Scrl, Bologna, Italy) [41].

2.5. Ancillary Data

To investigate the occurrence of significant (high-intensity) flood events in the river during the period 2015–2021, time series of hourly flow rates recorded in an upstream section of the Zero River were utilized (ZR, Figure 2). These data were collected by Consorzio Venezia Nuova (Venice, Italy) using acoustic instruments like those installed in sections DE1 and DE2.

The hourly data on the operation of the four pumping stations located in the estuary area, provided by the Consorzio Acque Risorgive (Venice, Italy), were used to estimate the water volumes introduced into the system during a significant flood event (see Section 3.5).

Precipitation data from 16 rainfall stations surrounding the sub-basin (Figure 1), provided by ARPAV (Veneto region), were processed using GIS (QGIS 3.26) to calculate the fortnightly and monthly rainfall inflows on the sub-basin and to analyze the hourly rainfall trend in the characterization of flood events. Isohyet maps of precipitations were obtained using the Module Thin Plate Spline (SAGA GIS 9.2).

Bathymetric data, acquired by our research group in 2011 using multibeam instrumentation, were used as a reference to observe changes in bed elevation in the estuary, which we related to ongoing siltation processes.

3. Results and Discussion

The data collected during the calendar year 2021 were used to characterize the hydrological behaviour of the Dese River. Specifically, Section 3.1 describes the annual Q and SSL, Section 3.2 analyses the rainfall-runoff relationship, and Section 3.3 examines the impact of flood events on these annual metrics. Section 3.4 describes the general siltation observed in the estuary, and the flood event that occurred on 13 April and resulted in an extreme transport of sediment towards the lagoon is analyzed in Section 3.5.

3.1. Annual Flow Rate and Suspended Sediment Load

Descriptive statistics of the hydrological behaviour of the two sections are reported in Table 1 for downstream ($Q > 0$) and upstream flow ($Q < 0$) conditions. The mean discharge (mQ) values in sections DE1 and DE2 (with downstream-to-upstream ratios of 2.46 and 2.07, respectively) indicate the greater importance of the main section DE1 in determining river discharge.

The mean SSC (mSSC) is also higher in DE1, and it is noteworthy that in both sections the values are similar in both flow directions, particularly in DE2 where they are nearly equivalent. The data reflect the cyclical movement of particulate matter with the tide in the estuary, with a net suspended sediment load (SSL) transported to the lagoon determined by the higher volumes during downstream flow compared to upstream flow in normal river conditions and by the contribution of flood events (see Section 3.3).

Table 1. Mean (m) and maximum (M) values of discharge (Q), suspended sediment concentration (SSC) and suspended sediment load (SSL) recorded in the principal (DE1) and secondary (DE2) sections during 2021 for downstream (Q > 0) and upstream flow (Q < 0) conditions. Data for section DE2 cover 308 days due to periods of instrumentation malfunction but can still be considered representative of the behaviour of the section.

Variable	Units	DE1 (Q > 0) 5592 h	DE1 (Q < 0) 3168 h	DE2 (Q > 0) 5108 h	DE2 (Q < 0) 2286 h
mQ *	$m^3\ s^{-1}$	16.0 ± 8.7	11.8 ± 6.7	6.5 ± 3.1	5.7 ± 3.0
MQ	$m^3\ s^{-1}$	64.7	32.7	32.5	18.2
mSSC *	$mg\ L^{-1}$	59.4 ± 74.6	49.4 ± 40.1	30.2 ± 21.9	33.0 ± 28.1
MSSC	$mg\ L^{-1}$	1762.3	794.5	523.0	635.6
mSSL *	$t\ 15\ m^{-1}$ °	1.0 ± 3.0	0.6 ± 0.7	0.2 ± 0.2	0.2 ± 0.2
MSSL	$t\ 15\ m^{-1}$ °	85.1	13.1	6.8	4.2

* ± standard deviation; ° expressed in the original unit to respect the data temporal resolution of the acquired data.

The generally lower SSC in the DE2 reflects the fact that the Santa Maria Channel becomes significantly shallower relative to the main riverbed within approximately one hundred metres downstream of its branching. This prevents the flow of water from the deep layer of the main river into the Santa Maria Channel (as it will be shown in Section 3.4, Figure 7).

The trend of the mean monthly discharge (^{mm}Q) in the two sections for 2021 is shown in Figure 3. For the DE2 section, due to periods of instrumentation malfunction, the values for November and December were calculated using a simple proportion, whereas the values for January and February were obtained through linear regression with the values recorded in section DE1. The details of the procedure are described in Appendix B.

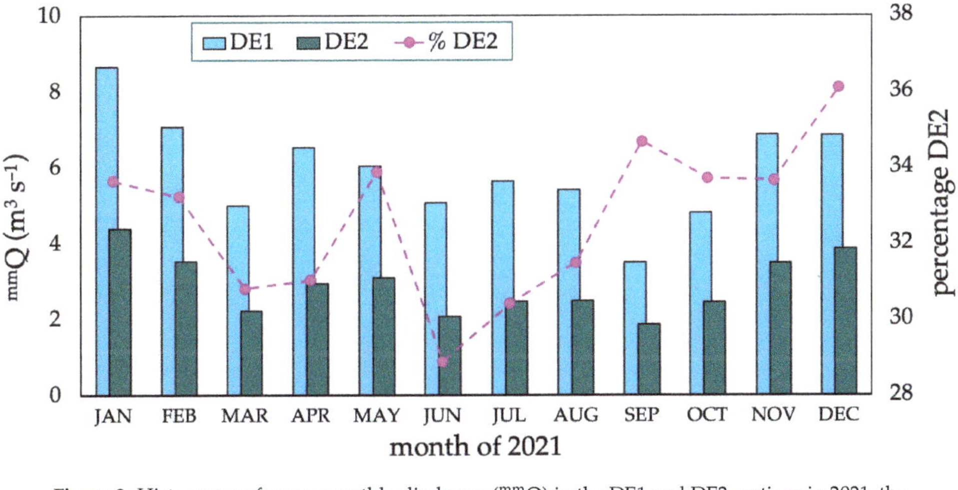

Figure 3. Histograms of mean monthly discharge (^{mm}Q) in the DE1 and DE2 sections in 2021; the dashed line shows the percentage of total discharge through DE2.

The calculated mean yearly discharge (^{my}Q) for DE1 and DE2 is 5.95 and 2.90 $m^3\ s^{-1}$, respectively. Therefore, the total discharge of the river for 2021 is 8.85 $m^3\ s^{-1}$. Approximately two-thirds (67 ± 2%, calculated from monthly values) of the river's total discharge flows through the main section DE1. The maximum monthly discharge of the river (combining the flow in the two measurement sections) was recorded in January (13.05 $m^3\ s^{-1}$), and the lowest was recorded in September (5.36 $m^3\ s^{-1}$).

The monthly values of SSL in the two measurement sections are shown in Figure 4 for the year 2021. To address the issue of missing data during the four months, a method similar to the one used for the Q was also used for the calculation of SSL (see Appendix B). On average, 82.1% ± 4.8 of the river sediment load transits through section DE1.

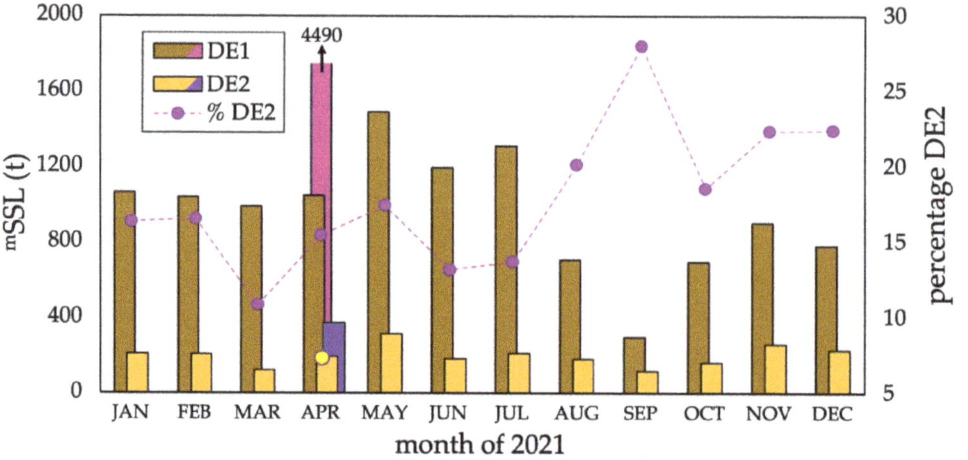

Figure 4. Monthly suspended sediment load (^{m}SSL) in the DE1 and DE2 sections in 2021. The April values do not include the contribution of the flood event on 13 April, the impact of which is shown with bars of a different colour (magenta for DE1 and purple for DE2). The dashed line shows the percentage of total load through DE2 (the yellow dot corresponds to accounting for the April flood).

In April, a double set of data is considered that excludes or includes the exceptional contribution to sediment transport from the 13 April flood event, which carried 3442 t through section DE1 (see Section 3.5). The total annual SSL (ySSL) value, through both sections DE1 and DE2, is 13,895 t (17,509 tons including the contribution of the 13 April flood).

A total of 52 and 33 water samples were collected at the measurement sections DE1 and DE2, respectively, and an additional amount of 128 samples were collected from other locations within the estuary. The analysis conducted on all 213 water samples showed that the particulate transported in the river estuary is predominantly in the silt size range (D50 = 24.2 ± 11.9 μm). By analyzing the surface sediments throughout the entire Dese River estuary, a strong correlation was observed between the content of finer particles (diameter < 7.8 μm) and the presence of heavy metals [42]. Finer particles are capable of attracting contaminants through different interactions, such as the adsorption of metal ions and ion exchange reactions [43,44]. The SSL can therefore be considered a proxy for the transport of some pollutants, primarily heavy metals, into the lagoon, as they are found in association with the fine particulate transported in suspension.

A previous investigation of the transfer of fresh water, suspended sediment, and pollutants from the drainage basin to the lagoon was carried out during the period 1998–2000 [21,33,45]. It included the Dese River among the set of investigated tributaries. The measurement section was located in DD (Figure 2), thus intercepting both flows towards DE1 and DE2. Table 2 summarizes the annual values of mean flow rate (^{my}Q) and suspended sediment load (^{y}SSL) obtained in the present study and compares them with measurements of 1999, which represent the only previous data of this kind. In 2021, the values are based on the sum of the data recorded in the DE1 and DE2 sections. For SSL, the data are presented both excluding and including the flood event on 13 April.

The yearly values of Q and SSL, and their variations over time, are not expected to remain consistent from year to year. Nonetheless, in the case of discharge, the two annual values are quite similar, even considering slightly higher precipitation in 2021 compared to 1999 (as it will be shown in Section 3.2, Table 3). In contrast, the difference in suspended sediment load is substantial (147% in 2021 compared to 1999) and becomes even greater when considering the contribution of the 13 April flood event (186%).

Table 2. Annual values for 2021 of mean flow rate (^{my}Q) and suspended sediment load (^{y}SSL) in sections DE1 and DE2, and the total values for the Dese River. For ^{y}SSL, values are shown both excluding and including (in brackets) the 13 April flood event. A comparison with estimates made in 1999 is provided.

Year	^{my}Q Total $m^3\ s^{-1}$	^{my}Q DE1 $m^3\ s^{-1}$	^{my}Q DE2 $m^3\ s^{-1}$	^{y}SSL total t	^{y}SSL DE1 t	^{y}SSL DE2 t
2021	8.85	5.95	2.90	13,895 (17,509)	11,522 (14,963)	2373 (2546)
1999	7.49			9428		

The amount of suspended sediment that is transferred from the sub-basin to the lagoon in a given year depends on various factors [46,47]. Beyond the inherent annual variations, our method for measuring the SSL ensures more accurate estimates compared to 1999, when SSL measurements were based on periodic water sampling and single-point sampling specifically during major flood events.

The availability of discrete instead of continuous SSC data introduced high uncertainty in the SSL estimation. The implementation of continuous SSC measurement using acoustic instrumentation in this study has enabled a higher temporal resolution and a more accurate estimate of the river SSL [48,49].

3.2. Rainfall-Runoff Analysis

The comparison between the volumes of precipitation falling on the catchment (rainfall) and of fresh water transferred to the lagoon from the tributary (runoff) provides an overview of the hydrology of the region and the effects of both land use and water management activities [50,51]. Monthly data for the year 2021 are shown in Table 3 and graphically represented in Figure 5.

The runoff is calculated as the sum of the flow in sections DE1 and DE2. The table also shows a simplified monthly runoff coefficient ($^{m}\delta$), defined as the ratio of runoff to rainfall volumes. For comparison, data for the year 1999 are also shown.

Table 3. Monthly rainfall and runoff volumes measured in 2021, along with monthly values of the runoff coefficient ($^{m}\delta$). For comparison, data from 1999 are also provided. Means of the monthly distributions and the total annual values are included.

Month	2021 Rainfall $10^6\ m^3$	Runoff $10^6\ m^3$	$^{m}\delta$	1999 Rainfall $10^6\ m^3$	Runoff $10^6\ m^3$	$^{m}\delta$
January	27.2	34.7	1.3	10.0	20.3	2.0
February	9.2	23.1	2.5	5.5	16.6	3.0
March	2.3	19.4	8.4	14.9	14.7	1.0
April	33.2	25.3	0.8	31.4	24.9	0.8
May	52.1	24.4	0.5	11.5	16.8	1.5

Table 3. *Cont.*

Month	2021 Rainfall 10^6 m^3	Runoff 10^6 m^3	$^m\delta$	1999 Rainfall 10^6 m^3	Runoff 10^6 m^3	$^m\delta$
June	15.3	19.1	1.3	38.0	18.4	0.5
July	28.3	21.7	0.8	29.0	15.6	0.5
August	15.4	21.2	1.4	14.4	15.2	1.1
September	10.6	14.4	1.4	10.8	14.7	1.4
October	9.8	19.4	2.0	27.3	17.3	0.6
November	39.1	28.2	0.7	31.3	34.7	1.1
December	13.5	29.3	2.2	16.4	26.2	1.6
Mean *	21.3 ± 14.7	23.4 ± 5.5	1.9 ± 2.1	20.0 ± 10.7	19.6 ± 6.1	1.3 ± 0.7
Mean * °			1.3 ± 0.6			
Yearly total	256.0	280.3	1.1 **	240.5	235.5	1.0 **

* ± standard deviation; ** ratio between the yearly totals of runoff and rainfall; ° excluding March 2021, considered as outlier.

A particularly high coefficient is observed in the month of March (8.4) due to a minimal volume of precipitation (2.3×10^6 m^3). The latter represents just 10.8% of the mean monthly distribution for the other 11 months. In contrast, runoff (19.4×10^6 m^3) remained close to the mean monthly values of 82.9%, partly due to contributions from the spring system of the high floodplain and partly due to external inputs to the drainage basin [34].

The March data resulted in a higher mean $^m\delta$ value with a much more widespread distribution recorded in 2021 (1.9 ± 2.1) compared to 1999 ($^m\delta$ = 1.3 ± 0.7). However, the yearly δ ($^y\delta$) values, calculated as the ratio between the yearly totals of runoff and rainfall, are very similar for both years. Excluding the March 2021 value, the distribution of $^m\delta$ over the remaining 11 months (1.3 ± 0.6) essentially matches that of 1999.

The distribution of monthly runoff values, particularly in 2021 (coefficient of variation CV = 23.5%), indicates that the regulation of flows in the catchment area of the Dese–Zero Rivers system helps to maintain the discharge relatively constant throughout the year. This is especially important during the spring–summer period, when infiltration and evapotranspiration losses are high, and there is a simultaneous need to store significant water volumes in the hydrographic network for irrigation. This regulation is also supported by contributions of water introduced from outside of the Dese River catchment; conversely, these contributions occasionally increase the river's runoff during periods of heavy rainfall in the areas north of the basin, where it is beneficial to divert some of the excess water away from the rivers in those territories.

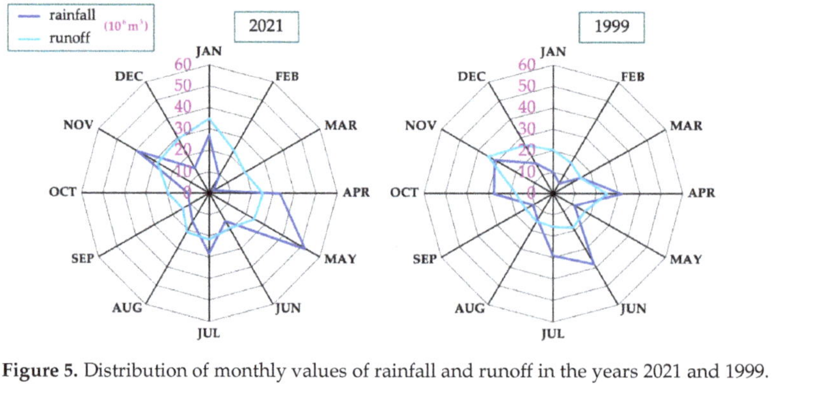

Figure 5. Distribution of monthly values of rainfall and runoff in the years 2021 and 1999.

3.3. Impact of Flood Events on Annual Q and SSL

To assess the contribution of flood events on the annual values of Q and SSL from the river basin, the data acquired in the main section DE1 were considered. Floods were identified as events in which the moving average discharge over a 25 h period (^{ma25}Q1) consistently exceeded 1.5 times the mean annual discharge measured at the section (^{my}Q = 5.92 m^3 s^{-1}). Ten flood events were recorded in 2021 (Table 4), with a total duration of 399.8 h (16.7 days), and contributed 7.8% of the annual discharge.

Nine of these events had an average flow rate of 13.6 m^3 s^{-1}, producing a total load of 1558 t. In comparison, the 13 April event (see Section 3.5), with a mean flow rate of 24.1 m^3 s^{-1} over 60.5 h, resulted in an extreme suspended sediment load (SSL), 2.2 times higher than the combined load of the other nine events. The flood runoff (FR) and the corresponding SSL for the 10 events are shown in Figure 6. The notable characteristics of the 13 April flood are highlighted by the mean SSC of the discharged water (calculated by dividing the SSL by the FR), distinguishing it as one of the significant events capable of producing a large percentage of the total annual SSL in the river [52,53].

Table 4. Descriptive parameters of the 10 flood events recorded in the Dese River in 2021; hydrodynamic and sediment load data refer to station DE1.

Date 2021	RD h	TR m^3 10^3	MR mm h^{-1}	mR mm h^{-1}	FD h	Qmax m^3 s^{-1}	FR m^3 10^3	SSL t	mSSL t h^{-1}
5 January	12	4075	11.4	3.6	50.0	45.4	2596	145	2.9
23 January	31	10,912	15.0	3.5	81.0	48.0	4331	212	2.6
31 January	12	5151	9.2	4.4	34.8	53.0	1391	72	2.1
8 February	23	2500	4.0	1.2	33.8	47.9	1121	96	2.8
12 May	14	10,726	38.0	6.9	19.5	40.4	633	49	2.5
24 May	16	13,611	19.6	9.3	40.8	50.1	2847	407	10.0
4 July	11	12,540	63.0	11.7	30.8	38.5	1453	345	11.2
27 July	5	9015	29.6	11.1	17.8	38.5	633	72	4.0
2 November	15	16,672	52.0	12.5	30.8	51.6	1576	162	5.3
mean	15	9467	26.9	7.1	37.7	45.9	1842	1.73	4.8
CV	49%	50%	76%	58%	50%	12%	65%	73%	71%
13 April	32	27,407	33.0	7.2	60.5	64.7	5246	3442	56.9
Total		112,607					21,824	5000	

RD = rainfall duration; TR = total rainfall; MR = maximum hourly rainfall; mR = mean hourly rainfall; FD = flood duration; Qmax = maximum flow rate; FR = total flood runoff; SSL = suspended sediment load; mSSL= mean hourly suspended sediment load.

Considering the parameters for nine floods listed in Table 4, no significant correlation (at $p < 0.05$) was observed between rainfall in the catchment and water flow variables, except for rainfall duration and flood duration (RD vs. FD: r = 0.77) or total flood runoff (RD vs. FR: r = 0.67). No significant correlation was found between sediment load (SLL, mSSL) and either rainfall in the catchment or water flow variables. These results indicate a high level of complexity in predicting SSL during floods in tidal-affected parts of the river and highlight the importance of high-resolution monitoring of both the Q and SSC. This is especially true for major flood events, which may be responsible for a significant proportion of the annual SSL. This issue is addressed in Section 3.5.

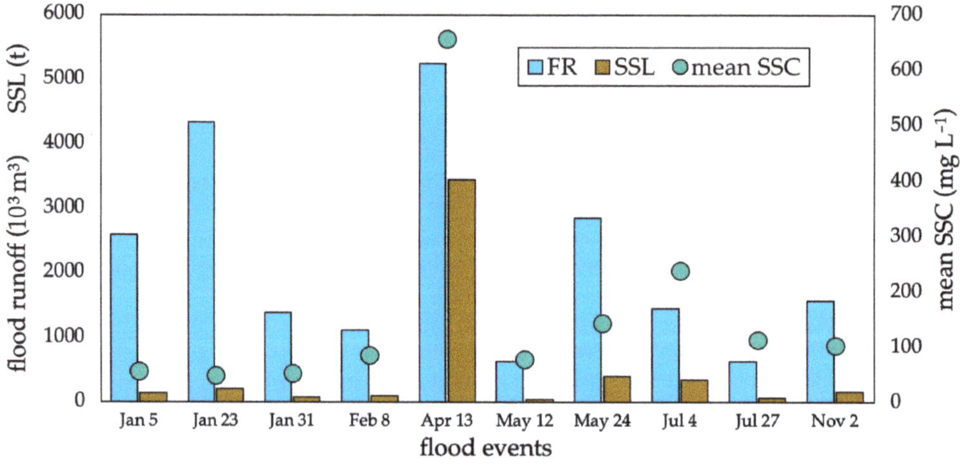

Figure 6. Flood runoff (FR) and SSL values for the 10 events, with the mean SSC of the discharged water (calculated as 10^3 SSL FR^{-1}) superimposed.

3.4. Bathymetry and Turbidity Maximum Zone

The bathymetric map shown in Figure 7 depicts the riverbed of the Dese River from its confluence with the Zero River to the measurement section DE1 (named E in Figure 7). The map was acquired in 2011 using multibeam instrumentation (data unpublished). It is observed that just downstream of the confluence, the Santa Maria Channel becomes significantly shallower over a distance of approximately one hundred metres. This bathymetric difference prevents the lower layer waters of the river from flowing into the channel. The figure also shows a comparison of bathymetries in five transects (A–E): the 2011 profiles were extracted from the bathymetric survey, while the 2020 profiles were acquired in this study with single-beam instrumentation. A general siltation of the sections is observed, resulting in a decrease in the wetted section over time, and an increase in the asymmetry of the sections, particularly in the main riverbed.

Studying water circulation and mixing between river fresh water and lagoon brackish water, Zonta and co-workers [54] identified the estuarine turbidity maximum zone (TMZ) [55,56] in the main river stretch approximately between stations C and DE1, as shown in Figure 7. In this area, resuspension phenomena triggered by tidal turbulence may occur, beginning at the start of the central phase of the tidal rise when the upstream current velocity rapidly increases in the lower layers of the water column. This process increases the concentration of suspended sediment in the water column, progressively affecting the entire system upstream of section DE1. A portion of the suspended material moves cyclically upstream and downstream during successive tidal phases, with some of this sediment eventually being transported to the shallow waters of Cona Marsh via the Santa Maria Channel [54].

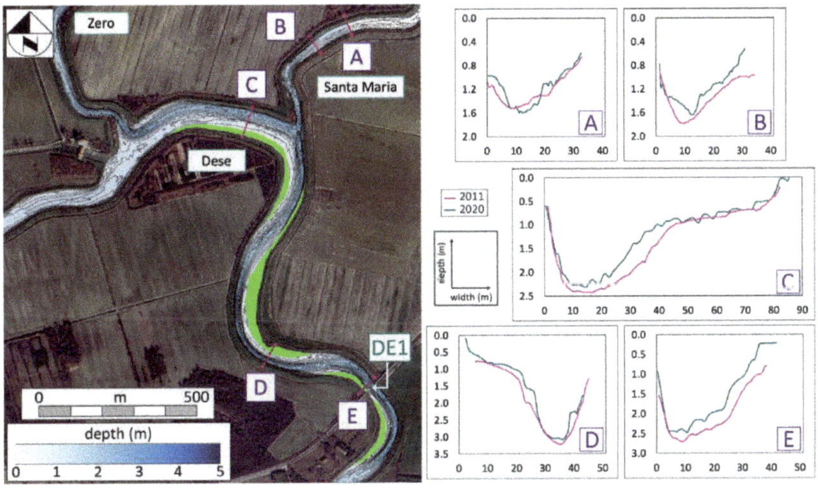

Figure 7. Bathymetric survey of the riverbed of the Dese River from the area of its confluence with the Zero River to downstream of the measurement section DE1, acquired in October 2011 using multibeam instrumentation (map background obtained from Google Earth—© 2024, Google). Silted areas in the riverbed are marked in green. The plots show the comparison in 5 transects (**A–E**), facing downstream) between the bathymetries extracted from the multibeam survey and measurements taken in December 2020 with single-beam instrumentation. All bathymetries are referenced to the zero-tide gauge for the lagoon.

3.5. The Flood Event of 13 April

A summary of the variables of interest for the 13 April flood is presented in Figure 8a–c, including the hyetograph (rainfall intensity), hydrograph (water discharge), and sedigraph (suspended solid concentration).

The main precipitation event (Figure 8a) that caused the flood (TR = 27,407 $\times$ 10^3 m^3, Table 4) was preceded by lighter rainfall, which started 25 h earlier (TR = 6436 $\times$ 10^3 m^3) and lasted for approximately 18 h (see Appendix C).

Based on the historical rainfall database [57], return periods for a 24 h precipitation [58] range between 2 and 3 years at all three rain gauge stations located within the catchment area (Castelfranco Veneto, Zero Branco, and Mogliano Veneto, Figure 1). Hourly time series of Q are continuously recorded at an upstream section of the Zero River (ZR Figure 2). With a peak discharge exceeding 35 m^3 s^{-1}, this flood ranks among the eight most significant events in the period from 2015 to 2021 (Appendix C, Figure A7), indicating a recurrence interval of approximately one year based on the Q scale.

Points B and E in Figure 8b, respectively, mark the start and end of the flood event, which has a duration of 60.5 h based on the moving average discharge over a 25 h period (^{ma25}Q1) for discharge in section DE1 (Q1).

The increase in Q1 (1) begins with the ebb of a spring tide excursion and exceeds 50 m^3 s^{-1} within a time of approximately 2 h (2). It then reaches a maximum of 63.7 m^3 s^{-1} after an additional 3 h (3). Subsequently, the discharge remains over 60 m^3 s^{-1} (mean value of 61.7 $\pm$ 1.4 m^3 s^{-1}), sustained by the inputs of waters drained by the four pumping stations. Although these stations are located at different points within the estuary, their combined throughput (ΣQps) is indicative of the overall forced contribution from the surrounding lowlands. The discharge remains elevated until around point (4), when the substantial volume of water discharged by the flood results in a relatively higher low tide and a deformation of the tidal wave shape [59].

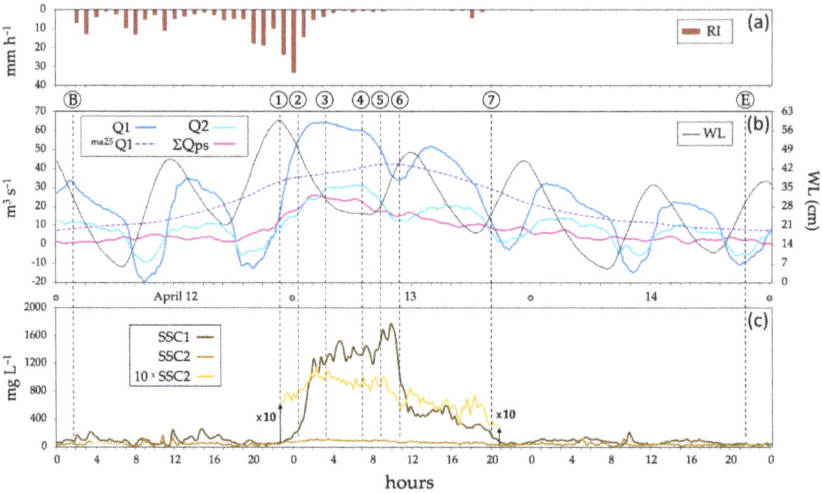

Figure 8. Flood event of 13 April 2021. Panel (**a**): Hyetograph represented by a rainfall intensity (RI) calculated as the mean of the hourly precipitation recorded at the gauging stations Castelfranco, Zero Branco, and Mogliano Veneto. Panel (**b**): Hydrograph: discharge recorded in the two sections (Q1 in DE1 and Q2 in DE2), moving average of Q1 based on a period of 25 h (^{ma25}Q1), and the instantaneous sum of discharge values from the four pumping stations (ΣQps); to reduce fluctuations in the trend, a moving average based on a 45 min period was applied to the Q2 and ΣQps data; water level recorded in DE1 (WL) displayed on the right axis. Panel (**c**): Sedigraph recorded in the two sections (SSC1 in DE1 and SSC2 in DE2); in the central part of the event, the SSC2 values are also shown multiplied by a factor of 10 for clarity. The key features of the event, described in the text, are indicated with markers (B, 1–7, E).

The intense river flow prevents the current reversal during the successive tidal flood phase, with Q1 decreasing to a largely positive minimum of 33.9 m^3 s^{-1} at the time of astronomical high tide (6). Based on the instantaneous Q1 measurements, the flood ends at point (7), with the flow returning to its typical tidal behaviour.

The trend of discharge in DE2 (Q2) closely resembles that of Q1, with a peak discharge of 32.5 m^3 s^{-1} and an instantaneous Q1 to Q2 ratio of 2.4 ± 0.5. The peak of the instantaneous sum of the Q1 and Q2 is 96.6 m^3 s^{-1}.

In section DE1, the relationship between the discharge (Q1) and suspended sediment concentration (SSC1) was analyzed during the event between points 1 and 7 of Figure 8. The analysis revealed a well-defined anticlockwise hysteresis loop (Figure 9), which occurs because the SSC1 peak lags behind the Q1 peak. The hysteresis reflects the non-linear behaviour and threshold effects in suspended sediment transport processes during the flood and provides insights into the sources of sediments and factors controlling their yield [50,60–62]. Similar hysteresis patterns can be the result of the interaction of different drivers and the interpretation may not be straightforward [47]. Anticlockwise loops may indicate that the sediment sources are located at a distance from the measurement section, and their mobilization is favoured by very high soil moisture and high antecedent rainfall precipitations [63]. Alternatively, they result from channel bed erosion, prolonged erosion processes during extended storm events, exhaustion of the available sediments due to previous events, and other reasons, as reviewed by Vercruysse and co-workers [47] and Malutta and co-workers [64].

As Q1 increases (1), SSC1 also begins to rise and, in less than two hours, exceeds values of 200 mg L^{-1} (2), which are only occasionally observed at this location under normal flow conditions. Subsequently, during the peak discharge period (3 and 4), SSC1 reaches and remains at unusually high levels (approximately 1300 mg L^{-1}).

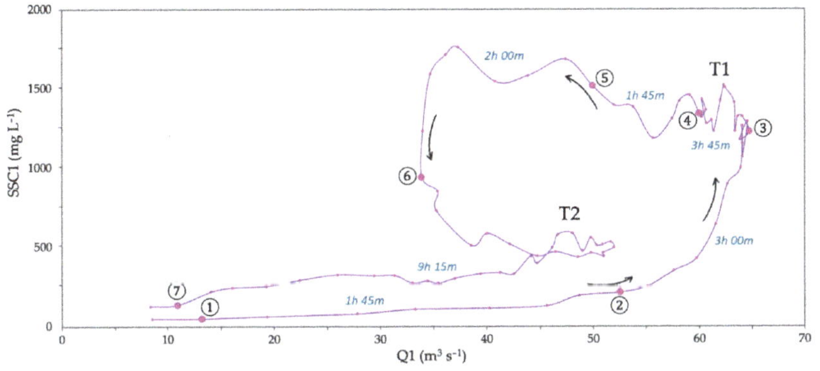

Figure 9. Anticlockwise hysteresis loop observed in section DE1 during the flood of 13 April 2021. The descriptive points (1–7) of the event correspond to those shown in Figure 8; the time intervals between points are indicated in italics. T1 and T2 identify the two transients as detailed in the text.

In our interpretation of the observed anticlockwise hysteresis, the sediment associated with the discharge peak is believed to be due to sediment resuspension [61,65] in the areas immediately upstream of section DE1. The lag time between the peaks in Q1 and SSC1 that determine the anticlockwise hysteresis is due to two distinct factors. Firstly, during the initial rise in Q1 (1 to 2 in Figures 8 and 9), the flow initially consists of water that was pushed upstream during the previous flood tide phase, which then flows back toward the lagoon. This water is characterized by the typical SSC found in the estuary. Secondly, the SSC begins to increase when Q1 exceeds the discharge usually observed in the section (2). When a critical threshold (as defined in [66]) of approximately Q1 = 50 m^3 s^{-1} is exceeded, the flow velocity can start to mobilize unconsolidated sediment from the riverbed, leading to its rapid entrainment and resulting in a rapid increase in SSC1.

The accumulation in the estuary, particularly in the riverbed, provides large amounts of sediment that are susceptible to resuspension and transport. Moreover, this extreme flood occurs after a long period of low to moderate flow (see Table 4 and Appendix C, Figure A7), which may have contributed to the sedimentation of particulate matter in the estuary.

The elevated SSC1, which remains well beyond the peak flow period (4), suggests that the antecedent riverbed storage of sediment is substantial and not rapidly exhausted. This concentration is also sustained by the input of turbid waters from the four pumping stations, which are activated once the storage capacity of their drainage network is reached, thereby transferring a significant wash load into the estuary.

The soil mobilization was presumably facilitated by both the antecedent rainfall on 11 April [63] and the extensive spring tilling of soils for crop planting.

During the period of maximum tidal excursion (mid-flood tide phase), a further increase in SSC1 to a maximum value of 1762 mg L^{-1} is observed (6). This increase is likely due to an additional contribution from processes typical of the TMZ occurring in the period from point (5) to (6). Although the flow remains largely seaward, the rising tide opposes the river flow, increasing bed shear stress and sediment resuspension, which leads to the observed rise in SSC1.

Additionally, approximately 10 h after the onset of the flood event, the increase in SSC1 may also be attributed to the transit of sediment from the nearby sections of the sub-basin at the measurement site and the possibility of final inputs of water from the pumping plants containing a higher concentration of material.

At the time corresponding to point (7), in accordance with the attenuation of Q1, the SSC1 returns to the typical values of the section. In the 22 h between points 1 and 7, the SSL transferred to the lagoon through section DE1 is 3261 t, corresponding to 21.8% of the annual sediment input.

In the hysteresis loop plot, two transients are observable. The first (from points 3 to 4, indicated as T1 in Figure 9) occurs during the period of maximum deformation of the tidal signal, when the Q1 stabilizes with a gradual decrease (from 63 to 53 $m^3 s^{-1}$ over approximately 6 h). During this time, the SSC1 remains consistently high, with values fluctuating around 1345 $\pm$ 91 mg L^{-1}. This shows that the ongoing sediment resuspension from the riverbed continues to play a key role in sustaining high sediment concentrations in the water column. The second transient, between points 6 and 7 (T2 in Figure 9), corresponds to the secondary discharge peak and highlights the stabilization of the sediment regime. In this interval of approximately 5 h, the distribution of the SSC has become uniform in the upstream areas of the estuary near the measurement section, resulting in nearly constant values (494 $\pm$ 53 mg L^{-1}) despite variations in Q1, which increases from 38.6 to 51.9 $m^3 s^{-1}$ and then reduces to 44.1 $m^3 s^{-1}$.

The SSC in DE2 (SSC2) follows a trend similar to that observed for SSC1 but with significantly lower values, averaging 16% of SSC1 from times (1) to (7)). Notably, section DE2 does not experience the additional increase in the SSC during the period from point (5) to (6), as the sediment resuspension caused by the opposing rising tide does not occur in the Santa Maria Channel because it is outside of the estuary TMZ.

Furthermore, the resuspension observed in DE1 likely affects primarily the lower part of the water column and therefore does not impact the Santa Maria Channel, which is mainly supplied by the upper half of the water column.

The first available satellite image, captured two days after the event (Figure 10, 15 April), shows the impact of solid transport generated by the flood. Water turbidity in the image shows a significant diffusion of suspended sediments throughout the estuary, reaching secondary channels, tidal creeks, and the tidal and subtidal flats of Cona Marsh and extending into the surrounding lagoon areas.

Figure 10. Satellite image captured two days after the event (15 April 2021, ESA Copernicus Sentinel-2 true-colour composite—20210415). The arrows delimit high turbidity zones caused by the 13 April flood. The main (DE1) and secondary (DE2) measurement sections of this study are indicated.

The occurrence of occasional transport pulses, driven by extreme floods, raises the question of whether this stepwise mechanism may positively contribute more to the preservation of salt marsh areas compared to a continuous but higher supply of suspended particulate matter. Furthermore, we must consider whether the long-term suspended sediment load delivered by the river is sufficient to balance or mitigate the erosional losses of sediment in the tidal and sub-tidal environments.

4. Conclusions

Understanding the hydrological behaviour of the main tributary of the Venice Lagoon provides the groundwork for investigating how the sediment load in the estuarine areas can reduce the loss of habitat in the adjacent tidal flats and marshlands, which underwent severe erosion over the past century and are still threatened by relative sea level rise.

The use of acoustic instrumentation for the continuous measurement of the discharge (Q) and suspended sediment concentration (SSC) in two tidal sections of the Dese River allowed the determination of the annual discharge and suspended sediment load (SSL) for the year 2021 and improved our knowledge on the dynamics of sediment exchanges with the lagoon at different timescales.

While the estimated mean annual flow (Q = 8.85 $m^3 s^{-1}$) was close to the previously reported data (7.49 $m^3 s^{-1}$ in 1999 [21]), the significant difference found for SSL (17,509 t in 2021 compared to 9428 t in 1999 [45]) could be partly attributed to the greater accuracy provided by the high-resolution measurements of the SSC in this study, as opposed to the periodical water sampling in the 1999 research.

Bathymetric measurements showed that the river estuary underwent a general silting, particularly in the main riverbed, leading to a reduction in the wetted cross-sectional area and an increase in its asymmetry. The streamflow energy is insufficient to transport the entire SSL produced in the catchment to the lagoon, which is then partially accumulated in the riverbed and adjacent estuarine areas. This storage can result from a combination of hydrodynamic changes caused by flow regulation, tidal influences, climatic trends, and possibly a recent variation in sediment yield. Understanding these processes is crucial to address interventions to improve the sedimentary balance of the estuary, including channel dredging in the more silted areas of the system.

Although the sediment accumulation in the estuary limits the transport of particulate matter to the adjacent salt marsh areas, a single flood event occurred during the investigations, resulting in an extreme pulse of SSL (3442 t, amounting to 3261 t within 22 h. This corresponds to 21.8% of the annual input to the lagoon through the main river section. This event mobilized part of the material accumulated in the riverbed of the estuary, favouring effective sediment transport to the salt marsh areas. The results of this study demonstrate that higher-resolution monitoring enhances the analysis of process dynamics at the event scale, enabling the detailed description of key sediment transport events that would otherwise remain undetected or inadequately described in coarser monitoring frameworks.

The characterization and quantification of suspended sediment load (SSL) from the river to the lagoon is indeed a key issue for the management and protection of the marshland habitats that still exist in its estuarine zone. Most freshwater tributaries of the Venice Lagoon are equipped with acoustic instrumentation for monitoring discharge. By extending the methodology applied in this study to the other measurement stations of the monitoring network, it would be possible to provide high-resolution time series of the SSC for the entire lagoon to investigate the effects of the climatic trend on the evolution of the habitat.

Author Contributions: Conceptualization, L.Z. and R.Z.; methodology, G.L., G.M., G.M.S. and R.Z.; validation, G.L., G.M. and R.Z.; formal analysis, G.L., G.M. and R.Z.; investigation, G.L., G.M. and G.M.S.; resources, G.L. and G.M.; data curation, G.L., G.M. and S.L.; writing—original draft preparation, J.D. and R.Z.; writing—review and editing, C.B., M.C., J.D., S.L., L.Z. and R.Z.; visualization, G.M.S., S.L. and R.Z.; supervision, R.Z.; project administration, R.Z.; funding acquisition, R.Z. All authors have read and agreed to the published version of the manuscript.

Funding: The research is part of the research programme Venezia 2021 (Scientific research program for a "regulated" lagoon), coordinated by CORILA (Consortium for coordination of research activities concerning the Venice Lagoon system) and funded by ministero delle Infrastrutture e dei Trasporti (Provveditorato Interregionale per le Opere Pubbliche del Veneto—Trentino Alto Adige—Friuli Venezia Giulia), through CORILA (Consorzio per il coordinamento delle ricerche inerenti al sistema lagunare di Venezia): 21/18/AC_AR02(04/12/2018).

Data Availability Statement: The datasets presented in this article are not readily available because dissemination requires authorization from the funding authority. Requests to access the datasets should be directed to the corresponding author.

Acknowledgments: We thank Loris Dametto (ISMAR CNR) and Daniele Curiel (SELC, Venice) for the support in field activities, Caterina Dabalà (CORILA) for the assistance provided in respecting the project plan, and Paola Focaccia (ISMAR CNR) for taking care of the administrative aspects of the project. Alessandro Zonta provided assistance in handling the large datasets. The authors also thank the staff of the consortium Acque Risorgive for the useful discussions on the results of the study. Roberto Chiarlo (Corr-Tek Idrometria, San Giovanni Lupatoto, VR, Italy) provided assistance in the installation and calibration of the measuring stations in the main river section DE1. Precipitation data were kindly provided by the Meteorological Center of Teolo (Padua, Italy) of the Agenzia Regionale per la Prevenzione e Protezione Ambientale del Veneto (ARPAV). The time series data on water discharge, recorded at station ZE on the Zero River and presented in Appendix C, were kindly provided by the Consorzio Venezia Nuova (Venice, Italy). We gratefully acknowledge Francesco Zonta for his artistic touch in crafting the graphical abstract.

Conflicts of Interest: The authors declare no conflicts of interest. The funders had no role in the design of the study; in the collection, analyses, or interpretation of data; in the writing of the manuscript; or in the decision to publish the results.

Appendix A. Measurement Sections, Instrumental Calibrations and Bathymetric Evolution of the Main River Stretch

This appendix describes the river sections where self-recording measurement stations were set up in the main course of the Dese River (DE1) and in the Santa Maria Channel (DE2) to acquire continuous data series on flow rate (Q) and suspended sediment concentration (SSC) using acoustic instrumentation. The plots of the linear regressions used to calibrate the instruments for measuring these two variables are presented. Additionally, the morphological evolution of a key section of the river is examined, highlighting progressive siltation trends over time.

The morphology of the two measurement sections is shown in Figure A1. The bathymetry is referenced to the zero-tide gauge for the lagoon (Punta della Salute), with its datum located 23 cm below the mean sea level [32]. The DE1 and DE2 sections have a maximum depth of 2.67 m and 1.55 m and a water cross-sectional area of 64 m^2 and 41 m^2, respectively.

The instruments measured in contiguous cells of equal width distributed along the section (nine cells of width 1.5 m in DE1, 22 cells of width 0.5 m in DE2). A blanking area (0.6 m) near the instrument corresponds to the length of the path travelled by the acoustic signal before the returning echo can be recorded. Both measurement stations were connected to a control unit, powered by a solar panel, for recording data on a data logger and transmitting via GSM to a workstation for storage and post-processing.

The regression lines used for calibrating the flow rate values obtained from the continuously self-recording instrumentation are shown in Figure A2 for the DE1 and DE2 sections, respectively. Figure A3 shows the calibration line used for the SSC values.

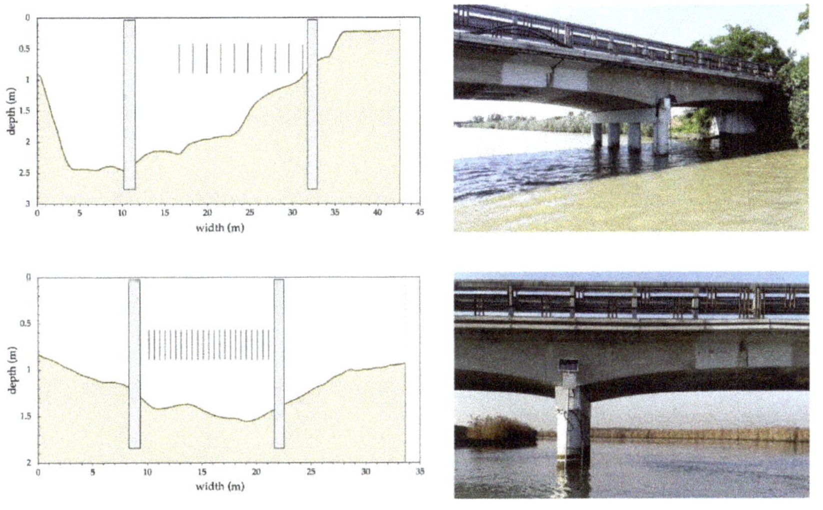

Figure A1. Morphology (**facing downstream**) of the measurement section DE1 on the Dese River (**upper left**) and DE2 on the Santa Maria Channel (**lower left**); the positions of the piers and the cells where the acoustic instrument takes measurements are indicated. Photographs show the installation of the instrumentation on the bridge pier (**facing upstream**).

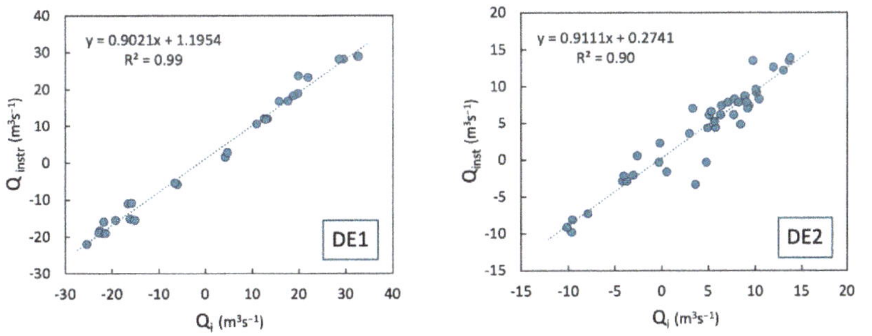

Figure A2. Regression diagram showing the relationship between the flow rate acquired by the self-recording instrument (Q_{instr}, instrumental) and those measured using a vessel-mounted instrument (Q_i, instantaneous) for the DE1 and DE2 sections, respectively.

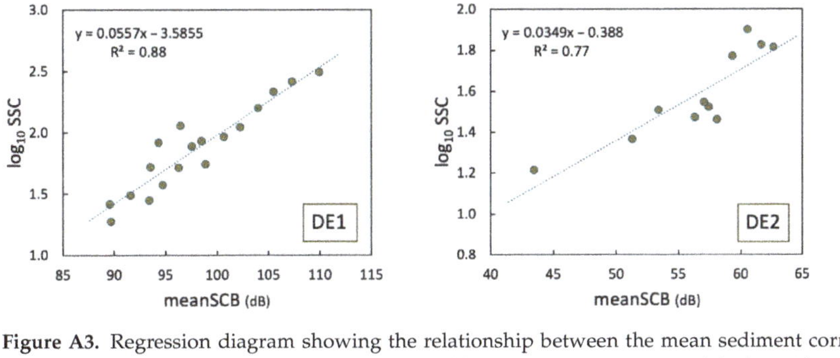

Figure A3. Regression diagram showing the relationship between the mean sediment corrected backscatter (meanSCB) values recorded by the self-recording instrument and the logarithm of the suspended sediment concentration (SSC) values determined in the laboratory from water samples collected from a vessel for the DE1 and DE2 sections, respectively.

Figure A4 presents four bathymetric surveys conducted along a short river stretch within the main section of the Dese River, with measurements taken in 1982, 1998, 2011, and 2020. The data illustrate the progressive siltation of the river stretch, reflecting ongoing depositional processes in the estuarine environment. A comparison of bathymetric data reveals that the wetted section in 1982 (142 m^2) was significantly reduced by 1998 (108 m^2). The 2011 survey indicates a similar wetted section (111 m^2) to that of 1998 but shows a marked asymmetry, with sediment accumulation on the right side and deepening on the left. The 2020 data reveal a further reduction in the wetted section (100 m^2), with additional siltation in the central area and a further decrease in depth compared to 2011.

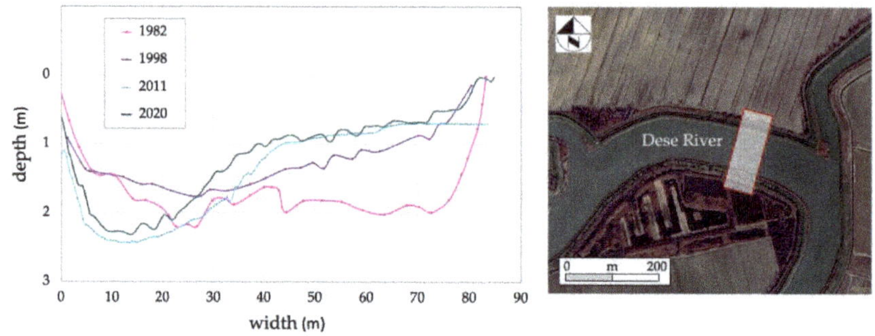

Figure A4. Bathymetric surveys of a short river stretch (indicated on the aerial map on the right, obtained from Google Earth—© 2024, Google) within the main section of the Dese River, acquired in four different years: 1982, 1998, 2011, and 2020. The bathymetric profiles from 2011 and 2020 are also presented in Figure 7 of the main text (Section C). The profiles are shown facing downstream and are referred to as the mean sea level.

Appendix B. Reconstruction of Monthly Q and SSL Data for Section DE2

This appendix describes the river sections where self-recording measurement stations were set up in the main course of the Dese River (DE1) and in the Santa Maria Channel (DE2) to acquire continuous data series on flow rate (Q) and suspended sediment concentration (SSC). The plots of the linear regressions used to calibrate the instruments for measuring these two variables are presented. Additionally, the morphological characteristics of the confluence between the Dese River and the Santa Maria Channel are highlighted.

In section DE1, the continuity of the acquired data enables the calculation of the annual flow rate for 2021. To extend the estimate to encompass the entire transfer of water and particulate matter from the Dese River to the lagoon, it is necessary to calculate the values in section DE2 as well. This allows for the rainfall/runoff analysis and comparison with previous estimates of yearly flow rate (${}^{y}Q$).

Due to malfunctions in the data recording system of the instrumentation installed in the DE2 section, 53 measurement days were lost in 2021, accounting for 14% of the year. The breakdown of lost days is as follows: 26 days in January, 11 days in February, 14 days in November, and 2 days in December.

For the four months, Table A1 displays the mean Q values calculated at section DE1 for both the entire month period (${}^{mm}Q$) and the period during which data are also available for DE2 (${}^{mp}Q$, mean Q in the period). The ${}^{mp}Q$ values recorded in the two sections for November and December showed good correspondence and were considered representative of the ${}^{mm}Q$ values. Moreover, in December, only two days of data were missing in DE2. Therefore, the ${}^{mm}Q$ in DE2 for both months was estimated using a simple proportion based on the data pair in DE1 and the available values in DE2 (Table A1).

Conversely, the pair of values for February shows a substantial difference, suggesting that the period during which DE2 collected data is not truly representative of the entire month. In the case of January, although the values reported in Table A1 are similar, only five days of data collection were considered insufficient to extrapolate the ^{mm}Q in DE2. The estimate of the ^{mm}Q values in DE2 for these two months was obtained using the equation resulting from the linear regression between the values in the two sections for the other 10 months (Figure A5, left). The values obtained are reported in Table A1.

Table A1. Comparison of mean Q and SSL values in the two sections for the months of January, February, November, and December 2021 (^{mm} = monthly mean; ^{mp} = mean in the period during which data are available for DE2). The ^{mm}Q and ^{mm}SSL values in DE2 obtained through proportion or regression line are reported.

Month 2021	DE1 ^{mm}Q $m^3\ s^{-1}$	DE1 ^{mp}Q $m^3\ s^{-1}$	DE2 ^{mp}Q $m^3\ s^{-1}$	DE2 ^{mm}Q $m^3\ s^{-1}$	DE1 ^{mm}SSL t	DE1 ^{mp}SSL t	DE2 ^{mp}SSL t	DE2 ^{mm}SSL t
January	8.66	8.30	4.22	4.39 [a]	1065	182	69	209
February	7.07	5.83	2.98	3.52 [a]	1041	418	111	207
November	6.86	7.04	3.57	3.48 [b]	903	508	146	260
December	6.85	7.08	4.00	3.87 [b]	780	717	208	226

Estimated [a] from the linear regression between the mean monthly values in the two sections for the period March to December, [b] using a proportion based on the data pair in DE1 and the available values for the month in DE2.

The measurement of SSL (suspended sediment load) through section DE2 is useful primarily for estimating the total solid transport of the Dese River on an annual basis and allows for comparison with previous estimates. To address the issue of missing data during the four months, a method similar to that used for ^{mm}Q was followed: the monthly values for November and December were estimated using a simple proportion based on the data pair in DE1 and the available values in DE2 (Table A1). The values for January and February were estimated using the linear regression equation derived from the data of the two sections for the other ten months (Figure A5, right).

For the regression calculation, the monthly SSL value for April in both sections was adjusted to exclude the sediment load from a flood event that produced extreme sediment transport (see Section 3.4 in the main text). Including this event's contribution would have introduced a bias in the regression. The monthly SSL values obtained are reported in Table A1.

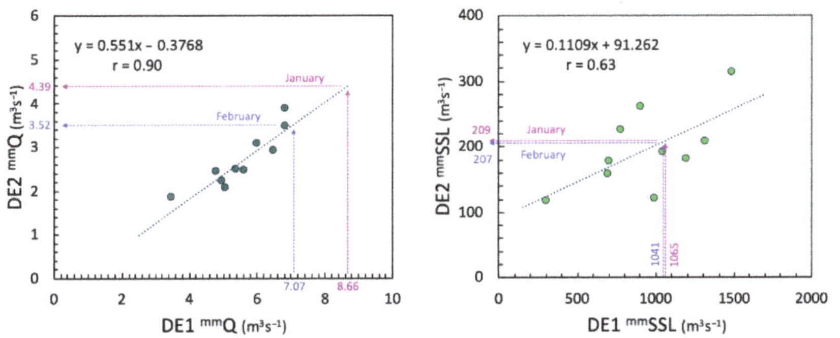

Figure A5. Regression diagrams showing the relationship between the mean monthly flow rate (^{mm}Q, **left**) and the mean monthly suspended sediment load (^{mm}SSL, **right**) values in sections DE1 and DE2 from March to December 2021, along with the estimated values for January and February in DE2. The relationship in the right diagram is more uncertain but significant at the $p = 0.05$ level.

Appendix C. Rainfall Event Generating the Flood of 13 April and Recurrence of Major Flood in the Zero River (2015–2021)

Figure A6 shows the isohyet map of the rainfall event that generated the flood of 13 April 2021 (TR = 27,407 × 10^3 m^3, Table 4 of the main text) and the hourly records from the three rain gauges located within the river basin (Castelfranco Veneto, Zero Branco, Mogliano Veneto); the latter includes the lighter rainfall (TR = 6436 × 10^3 m^3) approximately 20 h earlier.

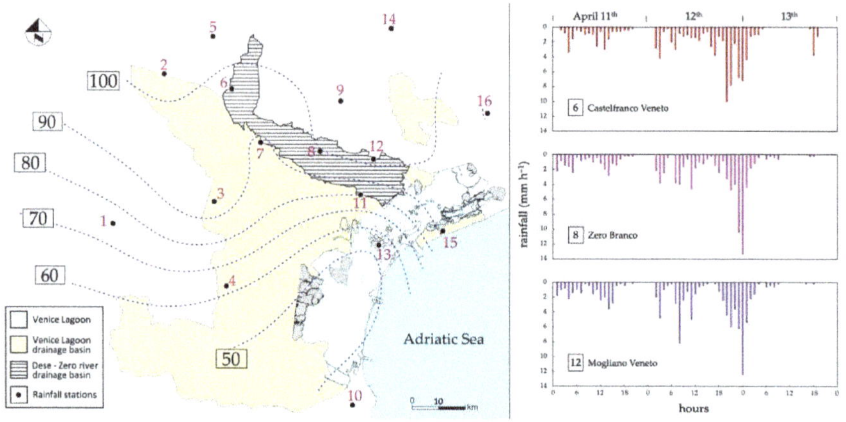

Figure A6. Isohyet map of precipitation associated with the flood event on 13 April 2021 (**left**). Precipitation values (in mm) are boxed within rectangles. The nomenclature for the 16 rainfall data collection stations (labelled in red) is provided in Figure 1 of the manuscript. Hourly rainfall depth recorded at the three rain gauge stations within the river drainage basin, including prior rainfall from 11 April (**right**).

Figure A7 shows the series of flow rate (Q) values recorded at the ZR station (see Figure 2 of the main text) on the Zero River during the period 2015–2021. Data provided by the Consorzio Venezia Nuova (Venice, Italy). The flood event of 13 April 2021 ranks among the eight most significant events from 2015 to 2021 (Figure A7), with a peak discharge exceeding 35 m^3 s^{-1}, indicating a recurrence interval of about one year based on the Q scale.

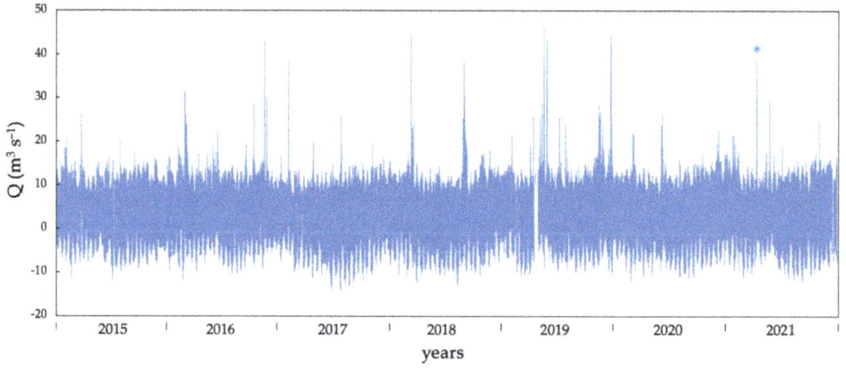

Figure A7. Time series of discharge (Q) values recorded by the ZR station (see Figure 2 of the main text) on the Zero River during the period 2015–2021. The flood event of 13 April 2021 is highlighted with an asterisk.

References

1. De Wit, R.; Leruste, A.; Le Fur, I.; Sy, M.M.; Bec, B.; Ouisse, V.; Derolez, V.; Rey-Valette, H. A Multidisciplinary Approach for Restoration Ecology of Shallow Coastal Lagoons, a Case Study in South France. *Front. Ecol. Evol.* **2020**, *8*, 108. [CrossRef]
2. Kuang, C.; Cong, X.; Dong, Z.; Zou, Q.; Zhan, H.; Zhao, W. Impact of Anthropogenic Activities and Sea Level Rise on a Lagoon System: Model and Field Observations. *J. Mar. Sci. Eng.* **2021**, *9*, 1393. [CrossRef]
3. Lacoste, É.; Jones, A.; Callier, M.; Klein, J.; Lagarde, F.; Derolez, V. A Review of Knowledge on the Impacts of Multiple Anthropogenic Pressures on the Soft-Bottom Benthic Ecosystem in Mediterranean Coastal Lagoons. *Estuar. Coast* **2023**, *46*, 2190–2207. [CrossRef]
4. Tosi, L.; Da Lio, C.; Cosma, M.; Donnici, S. Vulnerability of tidal morphologies to relative sea-level rise in the Venice Lagoon. *Sci. Total Environ.* **2024**, *931*, 173006. [CrossRef]
5. Bertocchi, D.; Camatti, N.; Giove, S.; van der Borg, J. Venice and Overtourism: Simulating Sustainable Development Scenarios through a Tourism Carrying Capacity Model. *Sustainability* **2020**, *12*, 512. [CrossRef]
6. Ravera, O. The Lagoon of Venice: The result of both natural factors and human influence. *J. Limnol.* **2000**, *59*, 19–30. [CrossRef]
7. Roggero, M. All in the same boat: Externalities, interdependence and the commons of Venice lagoon. *Marit. Stud.* **2017**, *16*, 15. [CrossRef]
8. Tagliapietra, D.; Umgiesser, G. Venice and its lagoon fin de siecle. *Reg Env. Chang.* **2023**, *23*, 125. [CrossRef]
9. Mitsch, W.J.; Gosselink, J.G. *Wetlands*, 5th ed.; John Wiley & Sons: Hoboken, NJ, USA, 2015; pp. 527–562.
10. Barbier, E.B. The Value of Coastal Wetland Ecosystem Services. In *Coastal Wetlands*, 2nd ed.; Gerardo, M.E., Perillo, G.M.E., Wolanski, E., Cahoon, D.R., Hopkinson, C.S., Eds.; Elsevier: Amsterdam, The Netherlands, 2019; pp. 947–964. ISBN 9780444638939. [CrossRef]
11. Suman, D.; Guerzoni, S.; Molinaroli, E. Integrated coastal management in the Venice lagoon and its watershed. *Hydrobiologia* **2005**, *550*, 251–269. [CrossRef]
12. Sarretta, A.; Pillon, S.; Molinaroli, E.; Guerzoni, S.; Fontolan, G. Sediment budget in the lagoon of Venice, Italy. *Cont. Shelf Res.* **2010**, *30*, 934–949. [CrossRef]
13. Trincardi, F.; Barbanti, A.; Bastianini, M.; Cavaleri, L.; Chiggiato, J.; Papa, A.; Pomaro, A.; Sclavo, M.; Tosi, L.; Umgiesser, G. The 1966 flooding of Venice: What time taught us for the future. *Oceanography* **2016**, *29*, 178–186. [CrossRef]
14. Faranda, D.; Ginesta, M.; Alberti, T.; Coppola, E.; Anzidei, M. Attributing Venice Acqua Alta events to a changing climate and evaluating the efficacy of MoSE adaptation strategy. *NPJ Clim. Atmos. Sci.* **2023**, *6*, 181. [CrossRef]
15. Giupponi, C.; Bidoia, M.; Breil, M.; Di Corato, L.; Gain, A.K.; Leoni, V.; minooei Fard, B.; Pesenti, R.; Umgiesser, G. Boon and burden: Economic performance and future perspectives of the Venice flood protection system. *Reg. Environ. Chang.* **2024**, *24*, 44. [CrossRef]
16. Ferrarin, C.; Ghezzo, M.; Umgiesser, G.; Tagliapietra, D.; Camatti, E.; Zaggia, L.; Sarretta, A. Assessing hydrological effects of human interventions on coastal systems: Numerical applications to the Venice Lagoon. *Hydrol. Earth Syst. Sci.* **2013**, *17*, 1733–1748. [CrossRef]
17. Leoni, S.; Dominik, J.; Cassin, D.; Manfè, G.; Tagliapietra, D.; Acri, F.; Zonta, R. Sediment oxygen demand rate in a flow regulated lagoon (Venice, Italy). *Front. Environ. Sci.* **2022**, *10*, 1000665. [CrossRef]
18. Tognin, D.; D'Alpaos, A.; Marani, M.; Carniello, L. Marsh resilience to sea-level rise reduced by storm-surge barriers in the Venice Lagoon. *Nat. Geosci.* **2021**, *14*, 906–911. [CrossRef]
19. Tognin, D.; Finotello, A.; D'Alpaos, A.; Viero, D.P.; Pivato, M.; Mel, R.A.; Defina, A.; Bertuzzo, E.; Marani, M.; Carniello, L. Loss of geomorphic diversity in shallow tidal embayments promoted by storm-surge barriers. *Sci. Adv.* **2022**, *8*, eabm8446. [CrossRef]
20. Baldassarre, L.; Natali, V.; De Pascale, F.; Vezzi, A.; Banchi, E.; Bazzaro, M.; Relitti, F.; Tagliapietra, D.; Cibic, T. The Impact of MOSE (Experimental Electromechanical Module) Flood Barriers on Microphytobenthic Community of the Venice Lagoon. *Microorganisms* **2023**, *11*, 936. [CrossRef] [PubMed]
21. Zuliani, A.; Zaggia, L.; Collavini, F.; Zonta, R. Freshwater discharge from the drainage basin to the Venice Lagoon (Italy). *Environ. Int.* **2005**, *31*, 929–938. [CrossRef]
22. Zonta, R.; Botter, M.; Cassin, D.; Bellucci, L.G.; Pini, R.; Dominik, J. Sediment texture and metal contamination in the Venice Lagoon (Italy): A snapshot before the installation of the MOSE system. *Estuar. Coast. Shelf Sci.* **2018**, *205*, 131–151. [CrossRef]
23. Cappucci, S.; Amos, C.L.; Hosoe, T.; Umgiesser, G. SLIM: A numerical model to evaluate the factors controlling the evolution of intertidal mudflats in Venice Lagoon, Italy. *J. Marine Syst.* **2004**, *51*, 257–280. [CrossRef]
24. Silvestri, S.; D'Alpaos, A.; Nordio, G.; Carniello, L. Anthropogenic modifications can significantly influence the local mean sea level and affect the survival of salt marshes in shallow tidal systems. *J. Geophys. Res. Earth* **2018**, *123*, 996–1012. [CrossRef]
25. Carbognin, L.; Teatini, P.; Tosi, L. Eustacy and land subsidence in the Venice Lagoon at the beginning of the new millennium. *J. Mar. Syst.* **2004**, *51*, 345–353. [CrossRef]
26. Tosi, L.; Teatini, P.; Strozzi, T. Natural versus anthropogenic subsidence of Venice. *Sci. Rep.* **2013**, *3*, 2710. [CrossRef]
27. Rapaglia, J.; Zaggia, L.; Parnell, K.; Lorenzetti, G.; Vafeidis, A.T. Ship-wake induced sediment remobilization: Effects and proposed management strategies for the Venice Lagoon. *Ocean Coast Manag.* **2015**, *110*, 1–11. [CrossRef]
28. Sfriso, A.; Facca, C.; Marcomini, A. Sedimentation rates and erosion processes in the lagoon of Venice. *Environ. Int.* **2005**, *31*, 983–992. [CrossRef]

29. Defendi, V.; Kovačević, V.; Arena, F.; Zaggia, L. Estimating sediment transport from acoustic measurements in the Venice Lagoon inlets. *Cont. Shelf Res.* **2010**, *30*, 883–893. [CrossRef]
30. Collavini, F.; Bettiol, C.; Zaggia, L.; Zonta, R. Pollutant loads from the drainage basin to the Venice Lagoon (Italy). *Environ. Int.* **2005**, *31*, 939–947. [CrossRef]
31. MOSE. Available online: https://www.mosevenezia.eu/il-mose-in-funzione (accessed on 16 October 2024). (In Italian).
32. Ruol, P.; Favaretto, C.; Volpato, M.; Martinelli, L. Flooding of Piazza San Marco (Venice): Physical model tests to evaluate the overtopping discharge. *Water* **2020**, *12*, 427. [CrossRef]
33. Zonta, R.; Costa, F.; Collavini, F.; Zaggia, L. Objectives and structure of the DRAIN project: An extensive study of the delivery from the drainage basin of the Venice Lagoon (Italy). *Environ. Int.* **2005**, *31*, 923–928. [CrossRef]
34. Nordest Ingegneria S.R.L. Idrografia. In *Piano delle Acque del Comune di Castelfranco Veneto—Relazione Generale*; Comune di Castelfranco Veneto, Ed.; Franco Angeli: Castelfranco Veneto, Italy, 2018; pp. 62–71. (In Italian)
35. Landers, M.N.; Straub, T.D.; Wood, M.S.; Domanski, M.M. Sediment acoustic index method for computing continuous suspended-sediment concentrations. In *U.S. Geological Survey Techniques and Methods, Book 3 Applications of Hydraulics*; USGS: Reston, VA, USA, 2016; pp. 1–63. [CrossRef]
36. Moskalski, S.; Floc'h, F.; Verney, R. Suspended sediment fluxes in a shallow macrotidal estuary. *Mar. Geol.* **2020**, *419*, 106050. [CrossRef]
37. WMO. *Manual on Stream Gauging, Vol. II: Computational of Discharge—WMO-No. 1044*; World Meteorological Organization: Geneva, Switzerland, 2010; p. 198. ISBN 978-92-63-11044-2.
38. Simpson, M.R.; Bland, R. Methods for accurate estimation of net discharge in a tidal channel. *IEEE J. Oceanic Eng.* **2000**, *25*, 437–445. [CrossRef]
39. APAT; IRSA-CNR. Parametri fisici, chimici e chimico-fisici. In *Metodi Analitici per Le Acque*; APAT, Ed.; IGER SRL: Roma, Italy, 2003; Volume 1, pp. 164–166. (In Italian)
40. Domanski, M.M.; Straub, T.D.; Landers, M.N. *Surrogate Analysis and Index Developer (SAID) Tool No. 2015–1177*; U.S. Geological Survey: Reston, VA, USA, 2015; p. 3. [CrossRef]
41. Stanghellini, G.; Del Bianco, F.; Gasperini, L. OpenSWAP, an Open Architecture, Low Cost Class of Autonomous Surface Vehicles for Geophysical Surveys in the Shallow Water Environment. *Remote Sens.* **2020**, *12*, 2575. [CrossRef]
42. Zonta, R.; Zaggia, L.; Argese, L. Heavy metal and grain-size distributions in estuarine shallow water sediments of the Cona Marsh (Venice Lagoon, Italy). *Sci. Tot. Environ.* **1994**, *151*, 19–28. [CrossRef]
43. De Groot, A.J. Processes affecting heavy metals in estuarine sediments. In *Estuarine Chemistry*; Burton, J.D., Liss, P.S., Eds.; Academic Press: London, UK; Cambridge, MA, USA, 1976; pp. 131–157. ISBN 10:0121473503.
44. Raji, Z.; Karim, A.; Karam, A.; Khalloufi, S. Adsorption of Heavy Metals: Mechanisms, Kinetics, and Applications of Various Adsorbents in Wastewater Remediation—A Review. *Waste* **2023**, *1*, 775–805. [CrossRef]
45. Collavini, F.; Zonta, R.; Bettiol, C.; Fagarazzi, O.; Zaggia, L. Metal and nutrient loads from the drainage basin to the Venice Lagoon. In *DRAIN Project Workshop, Determination of the Pollutant Load Discharged into the Venice Lagoon by the Drainage Basin*; Thetis, Ed.; Thetis: Messa, Italy, 2001; pp. 53–60.
46. Walling, D.E.; Fang, D. Recent trends in the suspended sediment loads of the world's rivers. *Glob. Planet. Chang.* **2003**, *39*, 111–126. [CrossRef]
47. Vercruysse, K.; Grabowski, R.C.; Rickson, R.J. Suspended sediment transport dynamics in rivers: Multi-scale drivers of temporal variation. *Earth Sci. Rev.* **2017**, *166*, 38–52. [CrossRef]
48. Horowitz, A.J.; Clarke, R.T.; Merten, G.H. The effects of sample scheduling and sample numbers on estimates of the annual fluxes of suspended sediment in fluvial systems. *Hydrol. Process.* **2015**, *29*, 531–543. [CrossRef]
49. De Girolamo, A.M.; Ricci, G.F.; Abdelwahab, O.M.M.; Lo Porto, A.; Milillo, F.; Netti, A.M.; Gentile, F. Suspended Sediment Transport in Mediterranean Streams: Monitoring and Load Estimation. *Water* **2023**, *15*, 2715. [CrossRef]
50. Zabaleta, A.; Martínez, M.; Uriarte, J.A.; Antigüedad, I. Factors controlling suspended sediment yield during runoff events in small headwater catchments of the Basque Country. *Catena* **2007**, *71*, 179–190. [CrossRef]
51. Nadal-Romero, E.; Regüés, D.; Latron, J. Relationships among rainfall, runoff, and suspended sediment in a small catchment with badlands. *Catena* **2008**, *74*, 127–136. [CrossRef]
52. Gonzalez-Hidalgo, J.C.; Batalla, R.J.; Cerda, A. Catchment size and contribution of the largest daily events to suspended sediment load on a continental scale. *Catena* **2013**, *102*, 40–45. [CrossRef]
53. Lloyd, C.E.M.; Freer, J.E.; Johnes, P.J.; Collins, A.L. Using hysteresis analysis of high-resolution water quality monitoring data, including uncertainty, to infer controls on nutrient and sediment transfer in catchments. *STOTEN* **2016**, *543*, 388–404. [CrossRef]
54. Zonta, R.; Ghermandi, G.; Marotta, L.; Zaggia, L. Palude di Cona: Circolazione e trasporto dei metalli pesanti. In *Metodologie di Sperimentazione e di Rilevamento—Studio di Processi*; Istituto Veneto di Scienze, Lettere ed Arti: Venice, Italy, 2000; Volume II, pp. 289–303. ISBN 8886166826. (In Italian)
55. Burchard, H.; Baumert, H. The Formation of Estuarine Turbidity Maxima Due to Density Effects in the Salt Wedge. A Hydrodynamic Process Study. *J. Phys. Oceanogr.* **1998**, *28*, 309–321. [CrossRef]
56. Fan, H.; Yan, H.; Teng, L.; Liu, R.; Li, Z.; Cheng, H.; Zhang, E. The effects of extreme flood events on the turbidity maximum zone in the Yangtze (Changjiang) Estuary, China. *Mar. Geol.* **2023**, *456*, 106993. [CrossRef]

57. ARPAV—Dipartimento Regionale per la Sicurezza del Territorio. Servizio Meteorologico. Bollettini. Available online: https://wwwold.arpa.veneto.it/bollettini/storico/precmax/ (accessed on 11 July 2024). (In Italian)
58. Gumbel, E.J. Statistical theory of floods and droughts. *J. Inst. Water Eng.* **1958**, *12*, 157–184.
59. Uncles, R.J.; Stephens, J.A.; Harris, C. Freshwater, tidal and wave influences on a small estuary. *ECSS* **2014**, *150*, 252–261. [CrossRef]
60. Heidel, S.G. The progressive lag of sediment concentration with flood waves. *Eos Trans. Am. Geophys. Union* **1956**, *3*, 56–66.
61. Williams, G.P. Sediment concentration versus water discharge during single hydrologic events in rivers. *J. Hydrol.* **1989**, *111*, 89–106. [CrossRef]
62. Evans, C.; Davies, T.D. Causes of concentration/discharge hysteresis and its potential as a tool for analysis of episode hydrochemistry. *Water Resour. Res.* **1998**, *34*, 129–137. [CrossRef]
63. Seeger, M.; Errea, M.-P.; Begueríia, S.; Arnaíez, J.; Martíı, C.; Garcíia-Ruiz, J.M. Catchment soil moisture and rainfall characteristics as determinant factors for discharge/suspended sediment hysteretic loops in a small headwater catchment in the Spanish pyrenees. *J. Hydrol.* **2004**, *288*, 299–311. [CrossRef]
64. Malutta, S.; Kobiyama, M.; Chaffe, P.L.B.; Bonumá, N.B. Hysteresis analysis to quantify and qualify the sediment dynamics: State of the art. *Water Sci. Technol.* **2020**, *81*, 2471–2487. [CrossRef] [PubMed]
65. Eder, A.; Strauss, P.; Krueger, T.; Quinton, J.N. Comparative calculation of suspended sediment loads with respect to hysteresis effects (in the Petzenkirchen catchment, Austria). *J. Hydrol.* **2010**, *389*, 168–176. [CrossRef]
66. Onderka, M.; Krein, A.; Wrede, S.; Martínez-Carreras, N.; Hoffmann, L. Dynamics of storm-driven suspended sediments in a headwater catchment described by multivariable modeling. *J. Soils Sediments* **2012**, *12*, 620–635. [CrossRef]

environments MDPI

Article

Climatic Changes Shift Macroalgal Assemblages from Cold- to Warm-Adapted Species: The Venice Lagoon as a Study Case

Adriano Sfriso [1,*], Yari Tomio [1] and Andrea Augusto Sfriso [2]

[1] Department of Environmental Sciences, Informatics and Statistics, University Ca' Foscari of Venice, 30172 Venice, Italy; yari.tomio@unive.it

[2] Department of Life Science and Biotechnology, University of Ferrara, via Luigi Borsari 46, 44121 Ferrara, Italy; sfrndr@unife.it

* Correspondence: sfrisoad@unive.it

Abstract: Temperature increase is one of the main effects of climate change occurring worldwide, with drastic impacts on both terrestrial and aquatic biota. Changes in the dominant macroalgal taxa in the Venice Lagoon have been analyzed in relation to the rise in air temperature recorded since 1973, highlighting the significant decline in cold-adapted species, which have been replaced by taxa more tolerant of higher temperatures. Cold-adapted species such as the native *Fucus virsoides, Punctaria latifolia, Scytosiphon lomentaria,* and many other Phaeophyceae are in decline, whereas thermophilic species such as the non-indigenous species (NIS) *Gracilaria vermiculophylla, Agardhiella subulata, Solieria filiformis, Hypnea cervicornis, Caulacanthus okamurae,* and many others have replaced the species that once dominated the lagoon. These changes have been associated with an average air temperature increase of approximately 2.5 °C. The highest increase has mostly been recorded for average minimum temperatures (+2.8 °C), compared to average maximum temperatures (+2.0 °C). As a result, Phaeophyceae have declined, while Rhodophyceae, especially recent NIS introductions, have colonized the lagoon bottoms. Changes in Chlorophyceae, on the other hand, appear to be more linked to the reduction of the lagoon's trophic conditions, although the currently dominant species is *Ulva australis*, a NIS that has replaced the native *Ulva rigida* almost everywhere.

Keywords: macroalgae; temperature increase; *Fucus virsoides; Punctaria latifolia; Scytosiphon lomentaria; Gracilaria vermiculophylla; Agardhiella subulata; Solieria filiformis; Hypnea cervicornis; Caulacanthus okamurae*

1. Introduction

Climate change, driven by a global rise in temperature, is having severe consequences on life on Earth, affecting both terrestrial and marine environments. These impacts include heat waves, catastrophic rainfall events, and rising sea levels due to ice melting. Global temperatures in 2024 were approximately 1.47 °C higher than the late 19th century preindustrial average (1850–1900), and the past ten years have been the warmest on record [1]. It is highly likely that the global mean sea surface temperature increased by 0.88 °C between 1850 and 1900 and 2011 and 2020, due to global warming, with most of this increase (0.60 °C) occurring between 1980 and 2020. However, the rise in water temperature is not uniform across regions, with some areas experiencing more significant changes. This is particularly evident in the Mediterranean Sea, where in 2023, water temperatures reached the highest levels recorded in modern history, with an average surface temperature increase of over 1 °C in just 25 years [2].

In choked and shallow environments such as the Venice Lagoon (Figure 1), temperature increase has been even more pronounced, affecting both sea levels and local biota. In fact, global mean sea level rose by 17 cm over the 20th century and keeps rising at an accelerating rate ([3–5]. Moreover, tide-gauge data, after accounting for subsidence, show that the sea level rise in the Venice Lagoon accelerated from 1.23 $\pm$ 0.13 mm y^{-1} between 1872 and 2019 to 2.76 $\pm$ 1.75 mm y^{-1} between 1993 and 2019 [6]. Beyond the evident consequences of rising sea levels, which by the end of the century could range from 32 to 62 cm, in the most optimistic scenario, to 58 to 110 cm, in the most extreme scenario, the increase in water temperature could have catastrophic effects on the lagoon's biota. This different forecast depends on contributions from the melting of the Greenland and Antarctic ice sheets as well as water mass exchange through the Strait of Gibraltar.

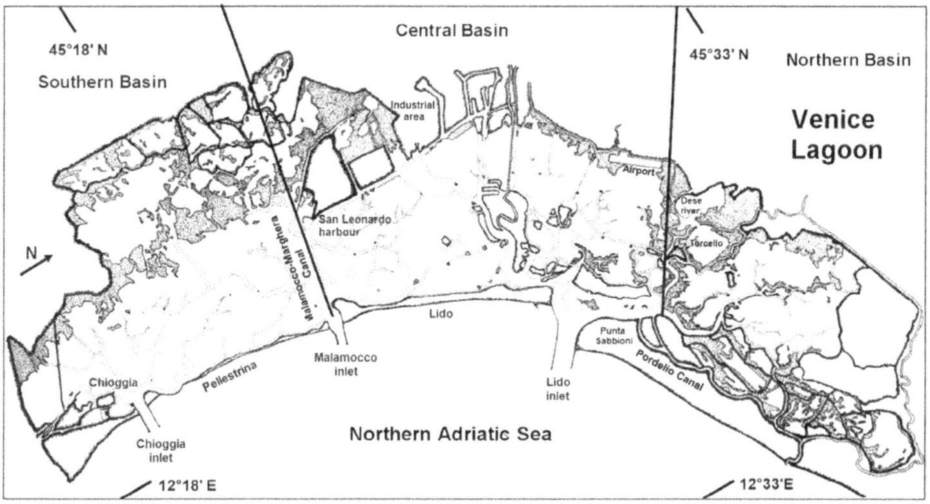

Figure 1. Map of the Venice Lagoon.

Both plant and animal communities are already undergoing significant changes, with thermophilic species becoming more prevalent while cold-adapted species are declining or disappearing. Species change in response to global warming is further enhanced by the introduction of exotic species [7], which compete with native species and threaten biodiversity conservation. One particularly evident consequence in the Venice Lagoon is the sharp decline of species sensitive to rising minimum temperatures, which no longer allow the waters to freeze as regularly as they did until the mid-2010s. This is the case for the endemic species *Fucus virsoides* J. Agardh, which had already experienced a significant decline by 2015 and has nearly disappeared from the entire Adriatic Sea since the late 2000s [8]. A similar trend has been observed in many other Phaeophyceae, which are becoming increasingly rare. In contrast, thermophilic species, especially certain non-indigenous species (NIS), primarily introduced through aquaculture and wholesale fish markets [7–9], have found favorable conditions for their establishment and expansion in the lagoon.

The aim of this paper is to analyze the major changes in dominant macroalgal taxa in the Venice Lagoon, with a particular focus on the past twenty years. The study examines taxa that have significantly declined as well as those that have increased since the early 2010s, replacing previously dominant species, in relation to temperature changes recorded in the lagoon over the past 51 years.

In addition, a checklist of taxa recorded during this period is provided, highlighting the sharp decline of many cold-adapted Phaeophyceae, particularly the contraction of *Fucus virsoides*, which is disappearing from the entire lagoon and the increase in thermophilic Rhodophyceae, mainly NIS that tolerate higher temperatures and have colonized the lagoon bottoms, reducing the abundance of native taxa.

2. Materials and Methods

2.1. Study Area

The Venice Lagoon (Figure 1) is a large coastal basin covering approximately 549 km^2, subdivided into three main sub-basins.

The Malamocco-Marghera Canal separates the South Basin from the Central Basin, while the North Basin is delineated by the Dese River and the Burano-Torcello tidal lands. These basins exhibit distinct morphological and ecological conditions, creating a diverse range of habitats that support high biodiversity [10–12]. Shallow, choked areas, characterized by limited water exchange, extreme salinity fluctuations (0–43), and temperature variations ranging from below 0 °C to 33 °C, with peaks up to 43 °C, alternate with deeper or shallow marine-influenced zones that are strongly affected by seawater inflows.

Water exchange occurs daily in areas near the three lagoon inlets (Lido, Malamocco, and Chioggia) and along the main canals, while in the most isolated areas, water renewal can take up to 40 days [13]. Currently, the lagoon has an average depth of approximately 1.2 m, whereas the main canals and lagoon inlets reach depths of 8–15 m, with some deeper pits (15–20 m) at canal intersections. The Malamocco inlet features a chasm exceeding 50 m in depth. The average tidal range in Venice is around 70–80 cm. However, the tidal range can vary significantly throughout the year, with high tide peaks exceeding 140 cm or even reaching much higher levels, while low tides can drop below 50 cm [14].

2.2. Macroalgal Changes

This study presents findings from 40 years of observations and research on macroalgal distribution, production and taxonomy ([15,16] and references therein) in the Venice Lagoon, with a focus on taxa that were once widespread and abundant but are now in decline, as well as newly introduced NIS that have thrived due to rising temperatures.

In this paper, we report the distribution of the cold-adapted species *Fucus virsoides*, an endemic relict species from the Tethys Ocean, during the early 2000s, when it was still abundant, and again in 2024, following its dramatic decline. Numerous studies conducted at lagoon soft and hard substrata in the whole lagoon on the framework of various projects [17–24] also reveal the near-total disappearance of other formerly common Phaeophyceae, such as *Scytosiphon lomentaria* and *Punctaria latifolia*.

At the same time, many Rhodophyceae, particularly NIS which tolerate or exhibit a preference for higher temperatures, allowing them to establish and spread throughout the lagoon [9], have successfully acclimated and expanded, replacing the previously dominant species.

2.3. Temperature Variation

In the Venice Lagoon, water temperatures are recorded through the SAMANET monitoring network (Venice Lagoon Environmental Monitoring Systems, Venice, Italy), which consists of 10 stations that continuously record water parameters (temperature, dissolved oxygen, salinity, pH, chlorophyll, turbidity) at half-hour intervals since 2002 [25]. This system was established to collect real-time environmental data, with the aim of monitoring and analyzing the ecological conditions of the lagoon. However, the datasets for many years are incomplete, and often there are gaps in days or months, so the annual values are not suitable for detecting long-term changes. Therefore, we used the daily air temperature dataset of Tessera, in Venice Airport at the edge of the central lagoon (Figure 1), recorded since 1973 [26]. For each year, we have calculated the average annual mean temperature (T_{mean}), the average annual minimum temperature (T_{min}), and the average annual maximum temperature (T_{max}). The results have been plotted in several graphs, highlighting the changes that have occurred over the last 51 years, during the cold (November–March) and warm (June–September) periods, as well as the number of days with temperatures exceeding 30 °C.

2.4. Statistical Analysis

Air temperatures have been analyzed determining the average annual mean temperature (T_{mean}), the average annual minimum temperature (T_{min}) and the average annual maximum temperature (T_{max}) recorded daily between 1973 and 2024. In addition, the average annual maximum temperatures recorded in March and in April during *Fucus* receptacle formation were calculated as well as the average maximum temperatures recorded in the warmest period June–September and the number of days with a temperature >30 °C. To verify whether air and water temperatures were similar or significantly different, temperature values recorded at 118 (2011) and 88 (2014, 2021, and 2023) stations equally distributed throughout the entire lagoon were analyzed. Sampling was carried out in late spring–early summer and repeated in autumn. The data were compared by calculating the mean, standard deviation (SD), minimum (min), and maximum (max) values, and by performing a one-way ANOVA ($p < 0.05$) using STATISTICA software, version 10 (StatSoft Inc., Tulsa, OK, USA).

3. Results

3.1. Species Changes

Since the middle of the second decade of the 2000s, a general decrease in many Phaeophyceae species has been observed in the Venice Lagoon and its marine littoral, accompanied by a progressive decline in both biomass and distribution. The most notable decline was that of the Phaeophycea *Fucus virsoides*, which, until 2015, colonized at least 70 linear kilometers of rocky substrates along the artificial Istrian stone panels, lagoon jetties, the docks of the historical center of Venice, and the shores of islands of all sizes (Figure 2).

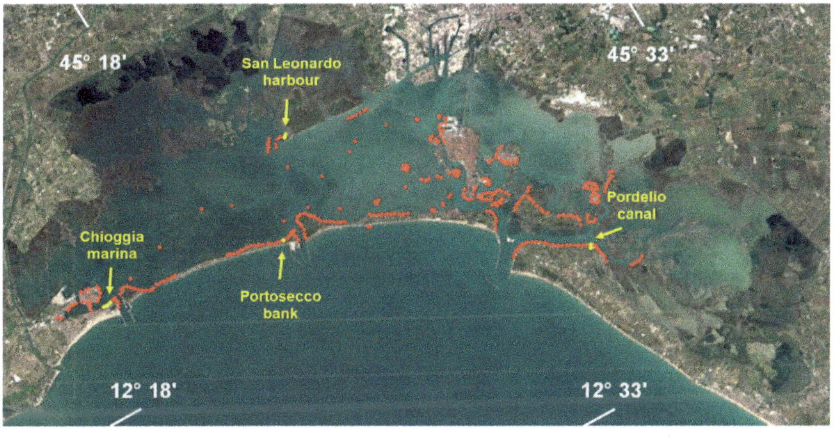

Figure 2. Map of the Venice Lagoon. Red points indicate locations with *Fucus virsoides* presence until 2015. Yellow points represent locations where *F. virsoides* is currently present.

In 2024, this species has almost disappeared everywhere, surviving only in a few small areas (approximately 2–3 linear km^2) where water exchange is higher, and temperature remains cooler. Similarly, *Punctaria latifolia* Greville and *Scytosiphon lomentaria* (Lyngbye) Link, two other taxa that previously formed dense belts in the mid-littoral zone and in shallow lagoon bottoms, are now very rare, with only occasional, isolated thalli observed. At the same time, some species that tolerate temperatures exceeding 30 °C for long periods without dying or decomposing, such as *Gracilaria vermiculophylla*, *Agardhiella subulata*, *Solieria filiformis*, and *Hypnea cervicornis* J. Agardh, have replaced the Ulvaceae, which were once dominant. A checklist of the taxa recorded since the mid-2000s with their abundance and temporal trends is reported in Table 1.

Table 1. Species recorded since the early 2000s in Venice Lagoon. Abundance (+), trends of decline (↓), increase (↑), or stability (≈) across three ranges. NIS species are marked with an asterisk (*).

Abundance	Trend	N°		Chlorophyceae
+++	≈	1	1	*Blidingia dowsonii* (Hollenberg and I.A. Abbott) S.C. Lindstrom et al.
++	≈	2	2	*Blidingia marginata* (J. Agardh) P.J.L. Dangeard *ex* Bliding
+++	≈	3	3	*Blidingia minima* (Nägeli *ex* Kützing) Kylin
++	≈	4	4	*Blidingia ramifera* (Bliding) Garbary *et* Barkhouse
++	≈	5	5	*Blidingia subsalsa* (Kjellman) Kornmann *et* Sahling *ex* Scagel et al.
+++	≈	6	6	*Bolbocoleon piliferum* Pringsheim
+	≈	7	7	*Bryopsis corymbosa* J. Agardh
+	≈	8	8	*Bryopsis cupressina* J.V. Lamouroux
++	≈	9	9	*Bryopsis cupressina* J.V. Lamouroux var. *adriatica* (J. Agardh) M.J. Wynne
+	↓	10	10	*Bryopsis duplex* De Notaris
r	↓	11	11	*Bryopsis feldmannii* Gallardo *et* G. Furnari
+	≈	12	12	*Bryopsis hypnoides* J.V. Lamouroux
+	≈	13	13	*Bryopsis muscosa* J. V. Lamouroux
+	↓	14	14	*Bryopsis plumosa* (Hudson) C. Agardh
r	↓	15	15	*Bryopsis secunda* J. Agardh
+++	≈	16	16	*Chaetomorpha aerea* (Dillwyn) Kützing
+	≈	17	17	*Chaetomorpha gracilis* Kützing
+++	≈	18	18	*Chaetomorpha ligustica* (Kützing) Kützing
++	≈	19	19	*Chaetomorpha linum* (O.F. Müller) Kützing
r	↓	20	20	*Chaetomorpha stricta* Schiffner
r	↓	21	21	*Cladophora aegagropila* (Linnaeus) Trevisan
++	≈	22	22	*Cladophora albida* (Nees) Kützing
r	≈	23	23	*Cladophora coelothrix* Kützing
+++	≈	24	24	*Cladophora dalmatica* Kützing
r	↓	25	25	*Cladophora echinus* (Biasoletto) Kützing
+	≈	26	26	*Cladophora fracta* (O.F. Müller *ex* Vahl) Kützing
+++	≈	27	27	*Cladophora glomerata* (Linnaeus) Kützing
+	≈	28	28	*Cladophora hutchinsiae* (Dillwyn) Kützing
++	≈	29	29	*Cladophora laetevirens* (Dillwyn) Kützing
+	≈	30	30	*Cladophora lehmanniana* (Lindenberg) Kützing
r	↓↓	31	31	*Cladophora liniformis* Kützing
r	↓↓	32	32	*Cladophora prolifera* (Roth) Kützing
r	≈	33	33	*Cladophora ruchingeri* (C. Agardh) Kützing
r	↓	34	34	*Cladophora rupestris* (Linnaeus) Kützing
+	≈	35	35	*Cladophora sericea* (Hudson) Kützing
+++	≈	36	36	*Cladophora vadorum* (Areschoug) Kützing
+++	≈	37	37	*Cladophora vagabunda* (Linnaeus) C. Hoek
+	≈	38	38	*Codium fragile* (Suringar) Hariot *
r	↓	39	39	*Derbesia tenuissima* (Moris *et* De Notaris) P. *et* H. Crouan

Table 1. *Cont.*

Abundance	Trend	N°		Chlorophyceae
+	≈	40	40	*Entocladia leptochaete* (Huber) Burrows
+++	≈	41	41	*Entocladia viridis* Reinke
++	≈	42	42	*Gayralia oxysperma* (Kützing) K.L. Vinogradova ex Scagel et al. *
r	↓	43	43	*Lamprothamnion papulosum* (Wallroth) J. Groves
+	≈	44	44	*Monostroma obscurum* (Kützing) J. Agardh
+	≈	45	45	*Neostromatella monostromatica* M.J. Wynne, G. Furnari, R. Nielsen
r	≈	46	46	*Pedobesia simplex* (Meneghini *ex* Kützing) M.J. Wynne *et* Leliaert
+++	≈	47	47	*Phaeophila dendroides* (Crouan frat.) Batters
r	≈	48	48	*Prasiola crispa* (Lightfoot) Kützing
+++	≈	49	49	*Pringsheimiella scutata* (Reinke) Höhnel *ex* Markewianka
+	≈	50	50	*Rhizoclonium tortuosum* (Dillwyn) Kützing
++	≈	51	51	*Ulothrix flacca* (Dilllwyn) Thuret
++	≈	52	52	*Ulothrix implexa* (Kützing) Kützing
+++	↓	53	53	*Ulva australis* Areschoug
+	≈	54	54	*Ulva clathrata* (Roth) C. Agardh
+++	≈	55	55	*Ulva compressa* Linnaeus
++	≈	56	56	*Ulva flexuosa* Wulfen
+	≈	57	57	*Ulva flexuosa* Wulfen subsp. *biflagellata* (Bliding) Sfriso *et* Curiel
+	↓	58	58	*Ulva kylinii* (Bliding) Hayden et al.
+++	↓	59	59	*Ulva intestinalis* Linnaeus
r	↓	60	60	*Ulva intestinalis* Linnaeus f. *cornucopiae* (Lyngbye) Sfriso *et* Curiel
+++	≈	61	61	*Ulva lactuca* Linnaeus
++	↓	62	62	*Ulva linza* Linnaeus
++	≈	63	63	*Ulva pilifera* (Kützing) Škaloud and Leliaert
+++	≈	64	64	*Ulva polyclada* Kraft
++	≈	65	65	*Ulva prolifera* O.F. Müller
+	≈	66	66	*Ulva prolifera* subsp. *blidingiana* Alongi, Cormaci and Furnari
+	≈	67	67	*Ulva ralfsii* (Harvey) Le Jolis
++	↓↓	68	68	*Ulva rigida* C. Agardh
+++	≈	69	69	*Ulvella lens* P. *et* H. Crouan
++	≈	70	70	*Ulvella setchellii* P. J. L. Dangeard
r	↓	71	71	*Valonia aegagropila* C. Agardh
Abundance	**Trend**	**N°**		**Rhodophyceae**
++	↑	72	1	*Acanthosiphonia echinata* (Harvey) Savoie and G.W. Saunders *
+++	↑	73	2	*Acrochaetium humile* (Rosenvinge) Garbary
++	≈	74	3	*Acrochaetium luxurians* (J. Agardh *ex* Kützing) Nägeli
+++	≈	75	4	*Acrochaetium microscopicum* (Nägeli *ex* Kützing) Nägeli
+	≈	76	5	*Acrochaetium secundatum* (Lyngbye) Nägeli
r	↓	77	6	*Acrosorium ciliolatum* (Harvey) Kylin
+++	↑	78	7	*Agardhiella subulata* (C. Agardh) Kraft *et* M.J. Wynne *
r	≈	79	8	*Aglaothamnion feldmanniae* Halos *
+++	↑↑	80	9	*Aglaothamnion halliae* (Collins) Aponte, D.L. Ballantine and J.N. Norris *
r	≈	81	10	*Aglaothamnion tenuissimum* (Bonnemaison) Feldmann-Mazoyer
++	≈	82	11	*Aglaothamnion tenuissimum* (Bonnemaison) Feldmann-Mazoyer var. *mazoyerae* G. Furnari et al.
+++	↑	83	12	*Alsidium corallinum* C. Agardh
r	↓	84	13	*Anotrichium furcellatum* (J. Agardh) Baldock
r	≈	85	14	*Anothrichium tenue* (C. Agardh) Nägeli
+++	≈	86	15	*Antithamnion cruciatum* (C. Agardh) Nägeli
++	≈	87	16	*Antithamnion nipponicum* Yamada et Inagaki
r	↓	88	17	*Antithamnionella elegans* (Berthold) J.H. Price *et* D.M. John
r	↓	89	18	*Antithamnionella spirographidis* (Schiffner) E.M. Wollaston

Table 1. *Cont.*

Abundance	Trend	N°		Rhodophyceae
+++	≈	90	19	*Bangia fuscopurpurea* (Dillwyn) Lyngbye
+++	≈	91	20	*Callithamnion corymbosum* (J.E. Smith) Lyngbye
++	≈	92	21	*Carradoriella elongata* (Hudson) Savoie and G.W. Saunders
++	≈	93	22	*Carradoriella elongella* (Harvey) Savoie and G.W. Saunders
r	≈	94	23	*Catenella caespitosa* (Withering) L.M. Irvine
+++	↑↑↑	95	24	*Caulacanthus okamurae* Yamada *
+++	↑↑	96	25	*Centroceras gasparrinii* subsp. *minus* M.A. Wolf, Buosi, Juhmani and Sfriso
++	≈	97	26	*Ceramium ciliatum* (J. Ellis) Ducluzeau
+++	↑	98	27	*Ceramium cimbricum* H.E. Petersen
r	≈	99	28	*Ceramium circinatum* (Kützing) J. Agardh
r	≈	100	29	*Ceramium codii* (H. Richards) Feldmann-Mazoyer
+	≈	101	30	*Ceramium connivens* Zanardini
+	≈	102	31	*Ceramium derbesii* Solier *ex* Kützing
r	≈	103	32	*Ceramium deslongchampsii* Chauvin *ex* Duby
r	≈	104	33	*Ceramium diaphanum* (Lightfoot) Roth
+	≈	105	34	*Ceramium inconspicuum* Zanardini
+++	↑↑	106	35	*Ceramium nodosum* (Kützing) A.W. Griffiths and Harvey
+++	≈	107	36	*Ceramium nudiusculum* (Kützing) Rabenhorst
+++	≈	108	37	*Ceramium polyceras* (Kützing) Zanardini
+	≈	109	38	*Ceramium rothianum* Wolf, Sciuto, Betto, Moro, Maggs and Sfriso
++	≈	110	39	*Ceramium siliquosum* (Kützing) Maggs and Hommersand
r	≈	111	40	*Ceramium tenerrimum* (G. Martens) Okamura
+	≈	112	41	*Chondrachantus acicularis* (Roth) Fredericq
+++	↑↑	113	42	*Chondria capillaris* (Hudson) M.J. Wynne
+	≈	114	43	*Chondria coerulescens* (J. Agardh) Falkenberg
+	≈	115	44	*Chondria dasyphylla* (Woodward) C. Agardh
++	↑	116	45	*Chylocladia verticillata* (Lightfoot) Bliding
r	≈	117	46	*Colaconema codicola* (Børgesen) Stegenga, J.J. Bolton and R.J. Anderson *
+++	≈	118	47	*Corallina berteroi* Montagne *ex* Kützing
++	≈	119	48	*Corallina officinalis* Linnaeus
+	≈	120	49	*Cruoria cruoriaeformis* (P. *et* H. Crouan) Denizot
+	≈	121	50	*Cryptonemia lomation* (A. Bertoloni) J. Agardh
r	≈	122	51	*Dasya corymbifera* J. Agardh
r	≈	123	52	*Dasya hutchinsiae* Harvey
++	≈	124	53	*Dasya pedicellata* (C. Agardh) C. Agardh
+	≈	125	54	*Dasya punicea* (Zanardini) Meneghini *ex* Zanardini
+	≈	126	55	*Dasysiphonia japonica* (Yendo) H.-S. Kim *
r	≈	127	56	*Dipterosiphonia rigens* (Schousboe *ex* C. Agardh) Falkenberg
+	≈	128	57	*Ellisolandia elongata* (J. Ellis and Solander) K.R. Hind and G.W. Saunders
++	≈	129	58	*Erythrocladia irregularis* Rosenvinge
+	≈	130	59	*Erythropeltis discigera* (Berthold) F. Schmitz
+++	≈	131	60	*Erythrotrichia carnea* (Dillwyn) J. Agardh
r	≈	132	61	*Erythrotrichia investiens* (Zanardini) Bornet
+	↑	133	62	*Gayliella flaccida* (Harvey *ex* Kützing) T.O. Cho *et* L. McIvor
r	≈	134	63	*Gayliella transversalis* (Collins *et* Harvey) T.O. Cho *et* Fredericq
++	↓	135	64	*Gelidium crinale* (Turner) Gaillon
++	↓	136	65	*Gelidium pusillum* (Stackhouse) Le Jolis
+	↓	137	66	*Gelidium spathulatum* (Kützing) Bornet
r	↓	138	67	*Gracilaria armata* (C. Agardh) Greville
+++	≈	139	68	*Gracilaria bursa-pastoris* (S.G. Gmelin) P.C. Silva
r	↓	140	69	*Gracilaria dura* (C. Agardh) J. Agardh
+++	≈	141	70	*Gracilaria gracilis* (Stackhouse) Steentoft et al.
+	≈	142	71	*Gracilaria longa* Gargiulo et al.

Table 1. *Cont.*

Abundance	Trend	N°		Rhodophyceae
++	↑↑	143	72	*Gracilaria vermiculophylla* (Ohmi) Papenfuss
+	≈	144	73	*Gracilaria viridis* A. Sfriso, M.A. Wolf, K. Sciuto, M. Morabito, C. Andreoli and I. Moro
+++	≈	145	74	*Gracilariopsis longissima* (S.G. Gmelin) Steentoft et al.
+	≈	146	75	*Grateloupia filicina* (J.V. Lamouroux) C. Agardh
r	≈	147	76	*Grateloupia minima* P. Crouan and H. Crouan *
+	≈	148	77	*Grateloupia turuturu* Yamada *
+	≈	149	78	*Grateloupia yinggehaiensis* H.W. Wang *et* R.X. Luan *
r	≈	150	79	*Griffithsia schousboei* Montagne
+	↓	151	80	*Gymnogongrus griffithsiae* (Turner) Martius
+	≈	152	81	*Halymenia floresii* (Clemente y Rubio) C. Agardh
r	≈	153	82	*Herposiphonia tenella* (C. Agardh) Ambronn
++	≈	154	83	*Hildenbrandia rubra* (Sommerfelt) Meneghini
+++	↑	155	84	*Hydrolithon boreale* (Foslie) Y.M. Chamberlain
++	≈	156	85	*Hydrolithon cruciatum* (Bressan) Y.M. Chamberlain
+	↓	157	86	*Hydrolithon farinosum* (J.V. Lamouroux) D. Penrose *et* Y.M. Chamberlain
+++	↑↑	158	87	*Hypnea cervicornis* J. Agardh *
++	≈	159	88	*Hypnea musciformis* (Wulfen) J.V. Lamouroux
+++	↑	160	89	*Kapraunia schneideri* (Stuercke and Freshwater) Savoie and G.W. Saunders *
++	↑	161	90	*Laurencia obtusa* (Hudson) J.V. Lamouroux
++	≈	162	91	*Lithophyllum pustulatum* (J.V. Lamouroux) Foslie
r	↓	163	92	*Lomentaria articulata* (Hudson) Lyngbye
r	↓	164	93	*Lomentaria clavellosa* (Turner) Gaillon
r	↓	165	94	*Lomentaria uncinata* Meneghini *ex* Zanardini
r	≈	166	95	*Lophosiphonia cristata* Falkenberg
r	≈	167	96	*Lophosiphonia obscura* (C. Agardh) Falkenberg
++	↑	168	97	*Melanothamnus japonicum* (Harvey) Díaz-Tapia and Maggs *
+	≈	169	98	*Melobesia membranacea* (Esper) J.V. Lamouroux
r	↓	170	99	*Nemalion helminthoides* (Velley) Batters
++	≈	171	100	*Nitophyllum punctatum* (Stackhouse) Greville
+++	↑↑	172	101	*Osmundea oederi* (Gunnerus) G. Furnari *
+	≈	173	102	*Osmundea truncata* (Kützing) K.W. Nam *et* Maggs
++	↑	174	103	*Palisada patentiramea* (Montagne) Cassano, Senties, Gil-Rodriguez *et* M.T. Fujii
+++	≈	175	104	*Pneophyllum fragile* Kützing
+	≈	176	105	*Polysiphonia breviarticulata* (C. Agardh) Zanardini
r	≈	177	106	*Polysiphonia fibrillosa* (Dillwyn) Sprengel
+++	↑↑	178	107	*Polysiphonia morrowii* Harvey *
r	↓	179	108	*Polysiphonia sanguinea* (C. Agardh) Zanardini
+	↓	180	109	*Polysiphonia scopulorum* Harvey
+	↓	181	110	*Polysiphonia sertularioides* (Grateloup) J. Agardh
r	≈	182	111	*Polysiphonia spinosa* (C. Agardh) J. Agardh
+	≈	183	112	*Porphyra linearis* Greville
r	↓	184	113	*Pterothamnion crispum* (Ducluzeau) Nägeli
r	↓	185	114	*Pterothamnion plumula* (J. Ellis) Nägeli
+++	≈	186	115	*Pyropia elongata* (Kylin) Neefus and J. Brodie
+++	≈	187	116	*Pyropia koreana* (M.S. Hwang and I.K. Lee) M.S. Hwang, H.G. Choi Y.S. Oh and I.K. Lee
r	≈	188	117	*Pyropia yezoensis* (Ueda) M.S. Hwang and H.G. Choi *
+++	↑↑	189	118	*Radicilingua mediterranea* Wolf, Sciuto and Sfriso
r	↓	190	119	*Radicilingua reptans* (Kylin) Papenfuss
+	↓	191	120	*Rhodophyllis divaricata* (Stackhouse) Papenfuss
++	≈	192	121	*Rhodymenia ardissonei* J. Feldmann

Table 1. *Cont.*

Abundance	Trend	N°		Rhodophyceae
r	↓	193	122	*Rhodymenia ligulata* Zanardini
+++	↑	194	123	*Rhodymenia pseudopalmata* (J.V. Lamouroux) P.C. Silva
+++	≈	195	124	*Sahlingia subintegra* (Rosenvinge) Kornmann
r	≈	196	125	*Scinaia furcellata* (Turner) C. Agardh
+++	↑↑	197	126	*Solieria filiformis* (Kützing) P.W. Gabrielson
r	≈	198	127	*Spermothamnion cymosum* (Harvey) De Toni *
r	≈	199	128	*Spermothamnion repens* (Dillwyn) Rosenvinge
r	≈	200	129	*Spermothamnion strictum* (C. Agardh) Ardissone
+++	↑↑	201	130	*Spyridia filamentosa* (Wulfen) Harvey
++	≈	202	131	*Stylonema alsidii* (Zanardini) K.M. Drew
r	≈	203	132	*Stylonema cornu-cervi* Reinsch
r	≈	204	133	*Vertebrata fucoides* (Hudson) Kuntze
r	↓	205	134	*Vertebrata furcellata* (C. Agardh) Kuntze
Abundance	**Trend**	**N°**		**Phaeophyceae**
r	≈	206	1	*Acinetospora crinita* (Carmichael) Sauvageau
r	↓↓	207	2	*Asperococcus bullosus* J.V. Lamouroux
r	↓	208	3	*Asperococcus ensiformis* (Delle Chiaje) M.J. Wynne
r	↓↓	209	4	*Asperococcus fistulosus* (Hudson) Hooker
r	↓	210	5	*Botrytella parva* (Takamatsu) H.S. Kim
r	≈	211	6	*Cladosiphon irregularis* (Sauvageau) Kylin
+	≈	212	7	*Cladosiphon zosterae* (J. Agardh) Kylin
+	≈	213	8	*Cystoseira aurantia* Kützing
r	↓↓	214	9	*Cystoseira compressa* (Esper) Gerloff *et* Nizamuddin
+	≈	215	10	*Desmarestia viridis* O.F. Müller
+++	↑	216	11	*Dictyopteris polypodioides* (A.P. De Candolle) J.V. Lamouroux
+	↓	217	12	*Dictyota dichotoma* (Hudson) J.V. Lamouroux
++	≈	218	13	*Dictyota dichotoma* (Hudson) J.V. Lamoroux var. *intricata* (C. Agardh) Greville
+++	↑	219	14	*Dictyota linearis* (C. Agardh) Greville
r	↓↓	220	15	*Ectocarpus fasciculatus* Harvey
+	↓↓	221	16	*Ectocarpus siliculosus* (Dillwyn) Lyngbye
+	↓↓	222	17	*Ectocarpus siliculosus* (Dillwyn) Lyngbye var. *arctus* (Kützing) Gallardo
r	↓↓	223	18	*Ectocarpus siliculosus* (Dillwyn) Lyngbye var. *hiemalis* (P. *et* H. Crouan *ex* Kjellman) Gallardo
+	↓↓	224	19	*Feldmannia irregularis* (Kützing) Hamel
r	↓↓↓	225	20	*Fucus virsoides* J. Agardh
+	↓	226	21	*Feldmannia mitchelliae* (Harvey) H.-S. Kim
+++	≈	227	22	*Gongolaria barbata* (Stackhouse) Kuntze
r	↓↓	228	23	*Hincksia granulosa* (J.E. Smith) P.C. Silva
r	↓↓	229	24	*Hincksia mitchelliae* (Harvey) P.C. Silva
r	↓↓	230	25	*Hincksia ovata* (Kjellman) P.C. Silva
r	↓↓	231	26	*Hincksia secunda* (Kützing) P.C. Silva
+++	↑	232	27	*Kuckuckia spinosa* (Kützing) Kornmann
+++	↑	233	28	*Myrionema magnusii* (Sauvageau) Loiseaux-de Goër, nom. inval.
++	↑	234	29	*Myrionema orbiculare* J. Agardh
+	↓↓	235	30	*Petalonia fascia* (O.F. Müller) Kuntze
r	↓	236	31	*Petalonia zosterifolia* (Reinke) Kuntze
+	↓	237	32	*Pilayella littoralis* (Linnaeus) Kjellman
r	↓↓↓	238	33	*Punctaria latifolia* Greville
r	↓	239	34	*Punctaria tenuissima* (C. Agardh) Greville
+++	↓	240	35	*Sargassum muticum* (Yendo) Fensholt *
+	↓↓	241	36	*Scytosiphon dotyi* M.J. Wynne *
+	↓↓↓	242	37	*Scytosiphon lomentaria* (Lyngbye) Link

Table 1. *Cont.*

Abundance	Trend	N°		Phaeophyceae
r	↓	243	38	*Stictyosiphon adriaticus* Kützing
r	↓	244	39	*Stictyosiphon soriferus* (Reinke) Rosenvinge
r	↓	245	40	*Stilophora tenella* (Esper) P.C. Silva
r	↓	246	41	*Striaria attenuata* (Greville) Greville
r	↓	247	42	*Taonia pseudociliata* (J.V. Lamouroux) Nizamuddin *et* Godeh
++	≈	248	43	*Undaria pinnatifida* (Harvey) Suringar *
Abundance	**Trend**	**N°**		**Xanthophyceae**
++	↓	249	1	*Vaucheria submarina* (Lyngbye) Berkeley

A list of 249 taxa was considered: 71 Chlorophyceae, 134 Rhodophyceae, 43 Phaeophyceae, and 1 Xanthophycea. Overall, 73 taxa showed a decrease (19 Chlorophyceae: −26.8%, 24 Rhodophyceae: −17.9%, 30 Phaeophyceae: −69.8%, and 1 Xanthophycea), while 30 taxa exhibited an increase in abundance (0 Chlorophyceae, 25 Rhodophyceae: +18.7%, and 5 Phaeophyceae: +11.6%). The other species remained almost unchanged with small variations depending on the year considered.

3.2. Temperature Changes

We recorded the mean, standard deviation, minimum and maximum values of air and water temperature in 118 (2011) and 88 (2014, 2021, and 2023) stations that are equally distributed in the lagoon (Table 2).

Table 2. Mean, standard deviation (std), minimum (min), maximum (max) values and one-way ANOVA of air and water temperature recorded in the entire lagoon.

	2011 (236 Samples)		2014 (176 Samples)	
mean	20.8	20.7	20.9	20.8
std	6.60	6.23	7.20	6.96
min	5.8	6.7	5.6	1.8
max	30.9	30.2	34.6	32.8
one-way ANOVA	$p < 0.767$ (NS)		$p < 0.961$ (NS)	
	2021 (176 Samples)		**2023 (176 Samples)**	
mean	21.6	20.2	20.4	19.4
std	7.36	7.29	9.02	8.22
min	4.6	3.9	4.8	4.1
max	33.3	31.6	34.3	32.0
one-way ANOVA	$p < 0.077$ (NS)		$p < 0.268$ (NS)	
	Total (764 Samples)			
mean	21.4	20.7		
std	7.70	7.32		
min	4.6	1.8		
max	34.6	32.8		
one-way ANOVA	$p < 0.205$ (NS)			

In 2011 and 2014 the difference between air and water temperature recorded in late spring-early summer and in autumn was 0.1 °C, whereas in 2021 and 2023 it ranged between 1.4 and 1.0 °C. The mean value of all data was 1.3 °C. The one-way ANOVA analysis shows that air and water temperature differences were not significant both in the single years and in all the data analyzed together.

The trends of air temperature over the last 51 years (1973–2024) at Tessera, in the Venice Lagoon, are shown in Figure 3.

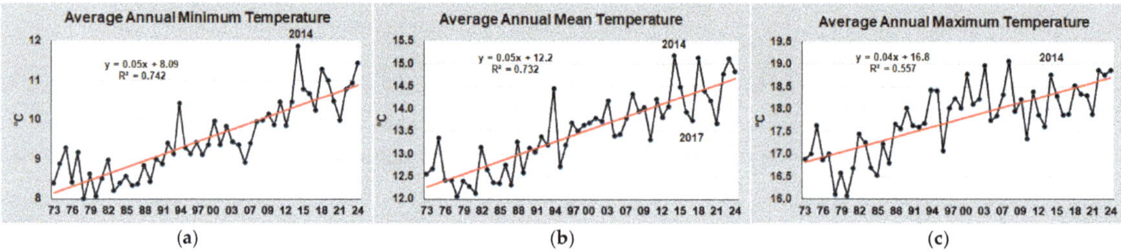

Figure 3. Trend of air temperature at Tessera between 1973 and 2024: (**a**) average annual mean temperatures (T_{mean}); (**b**) average annual minimum temperatures (T_{min}); (**c**) average annual maximum temperatures (T_{max}) recorded daily over this period. The red lines are the linear trend lines of data. The equation and R^2 of the linear lines are also reported in the graphs.

T_{mean} increased by approximately 2.5 °C (from 12.2 to 14.7 °C); T_{min} by approximately 2.8 °C (from 9.1 to 10.9 °C), and T_{max} by 2.0 °C (from 16.8 to 18.8 °C). The highest increase was recorded since 2014, when T_{min} was 11.9 °C compared to 9.5 °C for the entire period. In this last decade, T_{min} increased by 1.7 °C (from 9.2 to 10.9 °C). During the winter period, the lagoon waters have experienced only sporadic freezing events. Since 2020, freezing events have not occurred again, even in the more enclosed areas of the lagoon, where the water temperature rarely dropped below 3–4 °C. In previous years, significant freezing events were recorded in 1985 (air temperature −12 °C in January) and 2012 (−9 °C in February), when much of the lagoon was frozen. In the other years, freezing events occurred mainly in more enclosed areas.

The most significant temperature increases, recorded in March and April during the reproductive period of *Fucus virsoides*, are shown in Figure 4.

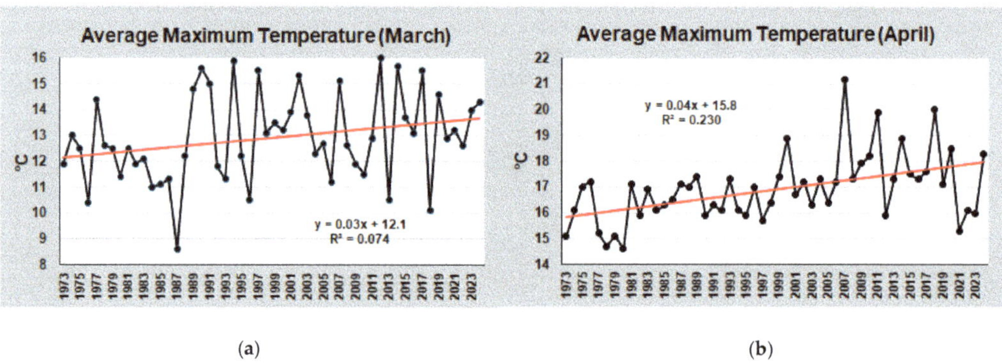

Figure 4. Trend of air temperature at Tessera between 1973 and 2024: average annual maximum temperatures in March (**a**) and in April (**b**), recorded daily over this period. The red lines are the linear trend lines of data. The equation and R^2 of the linear lines are also reported in the graphs.

Although annual values varied considerably, the average temperature in March increased by approximately 1.5 °C, and by more than 2.0 °C in April.

Similarly, the average maximum temperature during the warmest period (June–September) rose from approx. 28 °C to 30 °C, while the number of days with temperatures above 30 °C increased from 10 to 25 (Figure 5).

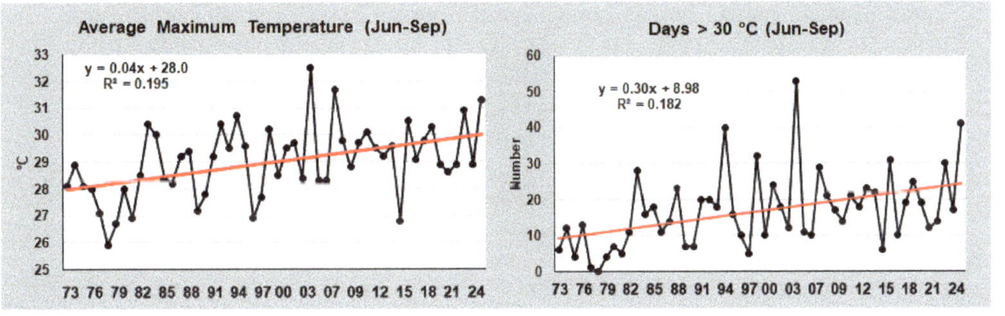

(a) (b)

Figure 5. (**a**) Trend of the average maximum air temperature in the period June–September at Tessera between 1973 and 2024; (**b**) Number of days with a temperature >30 °C in the period June–September. The red lines are the linear trend lines of data. The equation and R^2 of the linear lines are also reported in the graphs.

4. Discussion

In the last forty years, the vegetation of the Venice Lagoon has undergone significant changes, both in terms of biomass and species dominance [16]. The decrease in biomass observed since the 1990s was primarily due to trophic changes, with a significant reduction in phosphorus and nitrogen species in both the water column and surface sediments ([27] and references therein). In contrast, changes in species dominance were driven by both the combined effects of nutrient reduction and increased temperatures. The reduction in trophic status favored the progressive decline in the abundance of many Chlorophyceae (−26.8%), particularly Ulvaceae, and the replacement of *Ulva rigida* C. Agardh by *Ulva australis* Areschoug, a NIS that thrives in lower nutrient concentrations [16]. Indeed, *U. australis* colonizes marine littorals and lagoon areas characterized by high water renewal and moderate nutrient concentrations, whereas *U. rigida* dominates the eutrophic waters around the urban centers of the lagoon. The abundance of Chlorophyceae species has not increased, and the lagoon is now predominantly dominated by Rhodophyceae, particularly Gracilariaceae, Solieriaceae, and Cystocloniaceae.

This change is also clearly visible when observing the banks of the historical center of Venice and the islands of the lagoon. In the past, during low tide, the hard substrata and island shores were covered by dense belts of filamentous/laminar Chlorophyceae, mostly belonging to the genera *Blidingia*, *Ulva*, *Chaetomorpha*, and *Cladophora*. Currently, the green algae have almost disappeared everywhere. Therefore, the banks of the islands and the shores of the lagoon are now characterized by a dense stratification of Rhodophyceae, especially turf-forming algae such as *Caulacanthus okamurae* Yamada, another intertidal NIS of tropical origin (China, Korea, Taiwan) that has recently also colonized the Mar Piccolo of Taranto [28].

Climatic changes, with a general increase in temperature, have already favored the loss of many Fucales around the world [29–31]. The same negative trend was recorded in the Venice Lagoon with the decline of many Phaeophyceae, especially mid-littoral species such as *Fucus virsoides*, *Scytosiphon lomentaria*, and *Punctaria latifolia*, which, at low tides, are exposed to air, intense sunlight, and desiccation. Conversely, their decrease or disappearance favored the spread of taxa that are very resistant to high temperatures, many of which are NIS, such as *G. vermiculophylla*, *A. subulata*, *S. filiformis*, and *H. cervicornis* [9]. Generally, marine algae inhabiting the intertidal zone are tolerant to freezing and desiccation, showing higher photosynthetic rates during periods of air exposure. However, these rates drop dramatically when the algae are exposed to temperatures >30 °C and rapidly lose their water content over time, as reported for *Fucus virsoides* by [32] and *Fucus vesiculosus* Linnaeus [33], while changes in nutrient concentrations appear less important [9]. Similar results were obtained for other Phaeophyceae living in the intertidal zone, as found by [34] for *Hesperophycus harveyanus* (Decaisne) Setchell and N.L. Gardner (now recorded as *Fucus ceranoides* Linnaeus), and *Pelvetia fastigiata* f. *gracilis* (now recorded as *Silvetia compressa* (J. Agardh) E. Serrão, T.O. Cho, S.M. Boo and Brawley), or by [35] for *Mastocarpus papillatus* (C. Agardh) Kützing, and by [36] for *Fucus spiralis* Linnaeus.

Fucus virsoides is a glacial relict species endemic to the Adriatic Sea, growing on rocky substrata from Ancona to Venice and Trieste on the western coasts of Italy, and from the eastern coasts of Slovenia and Croatia up to Dürres in Albania [8,37–43]. The species was first reported in the Venice Lagoon by [44] and subsequently by all researchers who studied macroalgae in this environment [15].

In the Venice Lagoon, *F. virsoides* was present both inside the lagoon and along the marine shores of Lido and Pellestrina until the early 2000s. Subsequently, it disappeared from the seashores (panels and lagoon jetties) but remained very common inside the lagoon. Here, it colonized large areas with dense populations, particularly along the shores of Punta Sabbioni, Lido, Pellestrina, Chioggia, and the small islands scattered throughout the lagoon (Figure 2). It was even found in some embankments and inner canals of Venice's historical center (e.g., S. Barnaba Canal, Rio Novo), where high pollution levels made its presence unexpected.

The decline of *Fucus* populations in the lagoon was first observed in 2015–2016, when they rapidly disappeared from almost all areas. At present, only a few residual populations persist inside the harbor of San Leonardo in the central lagoon, in the Pordelio Canal at Punta Sabbioni in the northern lagoon (Figure 2), and along the shores of a marina in Chioggia, but every year they decrease more and more. These areas are protected from direct wave action and benefit from high seawater renewal, which mitigates extreme temperature fluctuations. Some thalli or very small populations can still be found in areas where *F. virsoides* was once abundant, but their presence is now only occasional.

Because systematic data on water temperatures in the lagoon or the adjacent sea littoral are not available, we considered the daily changes in air temperature recorded since 1973 at a weather station at Venice Airport (Tessera, [26]) located on the edge of the central lagoon basin. This choice was corroborated by an analysis of air and water temperatures recorded in 88–118 stations spread in the entire lagoon in 2011, and 2014 before the highest decline in Phaeophyceae, and subsequently in 2021 and 2023. Results show that air and water temperatures did not differ significantly (Table 1). Indeed, the lagoon's average depth is only 1.2 m, with extensive shallow areas around 0.5 m, and a mean tidal excursion of 70–80 cm. Consequently, the temperatures of both air and water quickly reach similar values. In addition, the macroalgal species showing the highest changes are mainly mid-littoral species that are exposed to air temperatures twice a day. Intense sunlight, desiccation, or freezing events can severely affect their presence or absence. Therefore, air temperature can play a major role in regulating the distribution of these species, providing information on their possible decline or spread.

The analysis of daily air temperatures over the past 51 years shows a significant increase, particularly in the average annual minimum values (approximately +2.8 °C). In the past, winter water temperatures were much colder, and numerous freezing events were recorded. This progressive increase in temperature suggests that, in the coming years, many other Phaeophyceae will also disappear, especially those in the mid-littoral zone, which are continuously exposed to the air during low tide. The species most at risk are those that grow in winter and early spring, such as the taxa belonging to the genera *Asperococcum*, *Petalonia*, *Stictyosiphon*, and *Striaria*, and disappear in May–June as temperatures rise.

Some authors [45] reconstructed freezing events in the Venice Lagoon over the past 1400 years. The first documented occurrence dates back to 604 AD, when the monk Paulus Diaconus (720–799), who lived in Aquileia (approximately 100 km northeast of Venice), reported that the winter was so harsh that it killed vineyards almost everywhere and even caused the lagoon to freeze over. He went on to document 57 additional freezing events. The most recent ones occurred in 1929, 1956, 1985, and 2012.

Between January and February 1929, air temperatures ranged from −8.5 °C to −13 °C, forming ice slabs 15–20 cm thick, allowing people to walk from one island to another. In January–February 1956 and 1985, arctic air masses and strong Bora winds caused temperatures to drop to −8 °C and −12 °C, respectively, freezing the entire lagoon once again.

In the following years, despite many canals regularly freezing (Figure 3), the last widespread freezing event affecting a large part of the lagoon occurred in 2012 when air temperatures dropped to −9 °C. Although this event was less severe than previous ones, the lagoon was partially frozen, and floating ice slabs invaded the canals of the city and surrounding islands. Between 2012 and 2015, occasional freezing events were recorded only in choked areas, salt marshes, and some canals. In the years since, until today, water temperatures in January–February have rarely dropped below 3–4 °C. At the same time, *Fucus virsoides* and other Phaeophyceae species, such as *Punctaria latifolia* and *Scytosiphon lomentaria*, which were abundant 10–20 years ago, have undergone a rapid decline and are currently disappearing.

Scytosiphon lomentaria is a species that presents tubular, filamentous gametophytic macrothalli in late winter to early spring and crustose, sporophytic microthalli in summer. In the past, in early spring, the macrothalli formed dense belts of tubular filaments, 40–60 cm long, in the mid-littoral zone of all hard substrata along marine and lagoon embankments, with a biomass of several kilograms of fresh weight per square meter. Currently, only isolated, occasional tubular filamentous macrothalli are present, accounting for only a few grams per square meter. In Denmark, the formation of plurilocular sporangia on macrothalli occurred at temperatures between 11 °C and 18 °C. Above 18 °C, macrothalli disappeared both in nature and in culture [46]. The sporophytic crusts produced unilocular sporangia at temperatures between 16 °C and 23.8 °C, thus functioning as the reproductive system during the summer. Higher temperatures inhibited the growth of both macrothalli and crusts, reducing the presence of this species, as also observed along Scotland's coast [47], except in the far northwest, where waters were colder.

Similarly, *P. latifolia*, was very abundant in the past, and in 1984 covered the lagoon bottoms around the historical center of Venice with luxuriant populations (2–3 kg FW m^{-2}). However, at present, this species has almost disappeared.

All these species have particularly suffered from global warming because they live in the mid-littoral zone and are exposed to the air during low tides. In particular, *Fucus virsoides* reproduces in late winter to early spring, and if temperatures are higher than 10–15 °C, reproduction is inhibited (Figure 4). A relevant study on the spawning conditions of six species of *Fucus*, carried out at the University of Groningen (The Netherlands) by [48], showed that reproduction depended on different conditions of photoperiod and temperature. The receptacles were not produced under short-day or long-day conditions. Generally, plants in spring initiate receptacle formation under a 12:12 h photoperiod [49]. Therefore, if temperatures are too high during this period, receptacle production is inhibited.

This seems to be the case for *F. virsoides* in the Venice Lagoon, which has almost disappeared in the last ten years. Figure 4 shows the average maximum temperatures recorded in March and April since 1973. In March, water temperatures fall within the range suitable for receptacle formation, but the photoperiod is too short. In contrast, in April, the photoperiod is optimal, but the temperature is too high, almost always exceeding 15 °C, with an increasing trend in recent years. These considerations can explain the decline of this species over the past decade, even though *F. virsoides* may have slightly different photoperiod and temperature thresholds.

If we consider the optimal growth temperature for *Fucus vesiculosus* in the North Sea, it ranges between 10 and 20 °C, with strongly reduced growth above 20 °C and an upper lethal limit of 28 °C after one week of exposure [50,51]. These results were confirmed by [33], who, in three-week exposure experiments, found that *F. vesiculosus* was able to grow and survive within a temperature range of 5 to 26 °C without any injury or visible damage to the apical growing meristem. However, at higher water temperatures ($\geq$27 °C), growth rapidly decreased from the third day onwards, and progressive necrosis was observed at 28 and 29 °C.

Since the average maximum temperature of the lagoon has also increased by approximately 2.0 °C, we cannot exclude the possibility that the decline of *F. virsoides* has been further exacerbated by prolonged periods with temperatures exceeding 28 °C (Figure 5) or, at the very least, that it has contributed synergistically to the species' decline.

Indeed, temperatures above 30 °C, heatwaves, and severe storm events are pushing environmental conditions beyond the species' tolerance limits, increasing thallus desiccation and damage while reducing photochemical activity [52,53].

In 1973, the average maximum temperature during the warm period (June–September) was approx. 28 °C. Over the following years, it progressively increased, reaching approx. 30 °C in 2024. Similarly, the average number of days with temperatures exceeding 30 °C was 9 in 1973, rising to 25 by 2024. However, studies on the optimal reproductive periods for this species are still necessary to confirm or refute these conclusions.

The most widespread species resistant to high water temperatures are NIS recorded in the lagoon since the early 2000s: *Agardhiella subulata* and *Solieria filiformis* in 2003; *Gracilaria vermiculophylla* in 2008; and *Hypnea cervicornis* in 2009. By 2019, these species accounted for 92.3% of the total NIS biomass in the lagoon (*G. vermiculophylla*: 45.3%, *A. subulata*: 25.1%, *H. cervicornis*: 19.3%, and *S. filiformis*: 2.57%).

Gracilaria vermiculophylla is a habitat-forming species that colonizes muddy substrata in the choked areas of the central and southern lagoon, particularly in more turbid regions where it tends to dominate. Owing to its resistance to high temperatures and high concentrations of phycoerythrin and phycocyanin, it has successfully colonized bare sediments where other species cannot survive. These habitats are particularly suitable for juvenile fish (nursery areas), shrimps, and numerous benthic macrofaunal species [54,55]. It is often found alongside other Gracilariaceae and Solieriaceae, especially the native species *Gracilariopsis longissima* (S.G. Gmelin) Steentoft, L.M. Irvine and Farnham, *Gracilaria gracilis* (Stackhouse) Steentoft, L.M. Irvine and Farnham, as well as the NIS *A. subulata* and *S. filiformis*. These species coexist across a range of habitats. *G. vermiculophylla* and *S. filiformis* occupy the most turbid and degraded areas, where even native species are unable to survive. In contrast, *H. cervicornis*, a NIS first recorded in 2009, prefers clearer waters. It forms aggregates with various other macroalgae, especially the *native Chondria capillaris* (Hudson) M.J. Wynne, *Spyridia filamentosa* (Wulfen) Harvey, and *Centroceras gasparrinii* subsp. *minus* M.A. Wolf, Buosi, Juhmani, and Sfriso, or inhabits aquatic angiosperm meadows in other shallow lagoon areas.

To these species, we must also add *Caulacanthus okamurae*, a NIS only recently distinguished through molecular analyses [28], previously confused with *C. ustulatus* (Turner) Kützing. This NIS has almost completely replaced *Gelidium* spp. and *Gymnogongrus griffithsiae* (Turner) Martius on intertidal hard substrata.

An exceptional case is *Grateloupia yinggehaiensis* H.W. Wang and R.X. Luan, first recorded in 2008. This species thrives in particularly warm waters and grows exclusively near the cooling water discharges (35 m^3 s^{-1}) of the Fusina thermoelectric power plant. These discharges are 10–14 °C warmer than natural waters, raising the temperature of the surrounding areas by 4–5 °C, with the greatest difference observed in winter [56]. This species is native to tropical regions such as Hainan Province (China), which experiences a tropical moist monsoonal climate with winter temperatures ranging from 16 to 21 °C [57]. In Venice, winter temperatures until the second half of the 2010s were generally close to 0 °C, with frequent freezing events. However, the thermal power plant has ensured a minimum temperature of 6–8 °C, allowing reproductive cells or basal crusts to survive even during the coldest periods.

5. Conclusions

Climate change, characterized by a progressive rise in air temperatures and significant impacts, particularly on minimum water temperatures, has had a major effect on lagoon vegetation over the past two decades. Freezing events are now rare, even in the most choked areas of the lagoon. These changes have severely affected native cold-water taxa, which are in sharp decline. At the same time, they have facilitated the establishment and spread of many NIS that are better adapted to elevated temperatures. As a result, almost all Phaeophyceae are declining, and some intertidal taxa are now very rare or even nearing extinction. In contrast, several thermophilic taxa, particularly exotic Rhodophyceae, have colonized both hard and muddy substrata, altering the dominant vegetation, which was previously composed largely of Ulvaceae.

However, the general decline of Chlorophyceae is also attributable to a significant reduction in nutrients in both the water column and surface sediments, as reported by ([15,16] and references therein). Thus, the current composition of lagoon vegetation is the result of a synergistic interaction between climate change and reduced anthropogenic pressures.

Similar results have been recorded in the lagoons of the Po Delta (Caleri, Goro, Barbamarco) and at Fattibello in the Valli di Comacchio, where Rhodophyceae have almost completely replaced Chlorophyceae. Indeed, recent sampling showed that the dominant taxa in these lagoons are currently the non-indigenous species *Gracilaria vermiculophylla*, *Solieria filiformis*, and *Ulva australis*. In addition, among the ten most frequently observed taxa, six are NIS species [58].

In the coming years, the ongoing decrease in nutrients [27] and rising temperatures [1,2,26] are expected to further transform lagoon vegetation. Species sensitive to elevated temperatures may decline or disappear, while the introduction of new NIS will continue to partially replace native species. This process has not led to a reduction in biodiversity, but rather a decline in the abundance of native species. Unlike other NIS, such as the ctenophore *Mnemiopsis leidyi* Agassiz [59–61] or the blue crab *Callinectes sapidus* Rathbun [62,63], which have had dramatic impacts on native species and the fisheries economy, allochthonous macroalgae have so far contributed only to an enrichment in the number of taxa present in the lagoon [9,54].

Author Contributions: Conceptualization, A.S.; methodology, A.S. and A.A.S.; validation, A.S., Y.T., and A.A.S.; formal analysis, A.S.; investigation, A.S., Y.T., and A.A.S.; resources, A.S.; data curation, A.S., Y.T., and A.A.S.; writing—original draft preparation, A.S.; writing—review and editing, A.A.S.

and Y.T.; supervision, A.S., Y.T., and A.A.S. All authors have read and agreed to the published version of the manuscript.

Funding: The preparation of this paper received no external funding. However, most of the results presented in this paper are based on observations collected over the last 15–20 years during the implementation of numerous projects. The main ones were those of the Veneto Region (references [17–21]), of CORILA (reference [22]), and the LIFE SERESTO and LIFE LAGOON REFRESH projects (references [23,24]).

Data Availability Statement: The original contributions presented in this study are included in the article. Further inquiries can be directed to the corresponding author.

Conflicts of Interest: The authors declare no conflicts of interest.

Abbreviations

The following abbreviations are used in this manuscript:

NIS	Non-indigenous species
T_{min}	Average minimum temperature
T_{mean}	Average mean temperature
T_{max}	Average maximum temperature.

References

1. NASA (National Aeronautics and Space Administration). Global Climate Change. Available online: https://climate.nasa.gov/vital-signs/global-temperature/?intent=121 (accessed on 25 February 2025).
2. ENEA (Agenzia Nazionale per le Nuove Tecnologie, l'Energia e lo Sviluppo Economico Sostenibile) and INGV (Istituto Nazionale di Geofisica e Vulcanologia). Environment: Mediterranean Temperature, 1 °C Warmer in Last 25 Years. 2025. Available online: https://www.ingv.it/en/press-and-urp/Press/Press-releases/5650-Mediterranean-environment-temperature-over-1-c-more-in-the-last-25-years (accessed on 20 February 2025).
3. Dangendorf, S.; Hay, C.; Calafat, F.M.; Marcos, M.; Piecuch, C.G.; Berk, K.; Jensen, J. Persistent acceleration in global sea-level rise since the 1960s. *Nat. Clim. Chang.* **2019**, *9*, 705–710. [CrossRef]
4. IPCC. Contribution of Working Group II to the Sixth Assessment Report of the Intergovernmental Panel on Climate Change. In *Climate Change 2022: Impacts, Adaptation, and Vulnerability*; Pörtner, H.-O., Roberts, D.C., Tignor, M., Poloczanska, E.S., Mintenbeck, K., Alegría, A., Craig, M., Langsdorf, S., Löschke, S., Möller, V., et al., Eds.; Cambridge University Press: Cambridge, UK; New York, NY, USA, 2022; p. 3056. [CrossRef]
5. Giupponi, C.; Bidoia, M.; Breil, M.; Di Corato, L.; Kumar Gain, A.; Leoni, V.; Fard, B.-M.; Pesenti, R.; Umgiesser, G. Boon and burden: Economic performance and future perspectives of the Venice flood protection system. *Reg. Environ. Change* **2024**, *24*, 44. [CrossRef]
6. Zanchettin, D.; Bruni, S.; Raicich, F.; Lionello, P.; Adloff, F.; Androsov, A.; Antonioli, F.; Artale, V.; Carminati, E.; Ferrarin, C.; et al. Sea-level rise in Venice: Historic and future trends. *Nat. Hazards Earth Syst. Sci.* **2021**, *21*, 2643–2678. [CrossRef]
7. Van der Loos, L.M.; Bafort, Q.; Samuel, S.; Ballesteros, E.; Bárbara, I.; Bercibar, E.; Blanfuné, A.; Bogaert, K.; Bouckenooghe, S.; Boudouresque, C.-F.; et al. Non-indigenous seaweeds in the Northeast Atlantic Ocean, the Mediterranean Sea, and Macaronesia: A critical synthesis of diversity, spatial and temporal patterns. *Eur. J. Phycol.* **2024**, *59*, 127–156. [CrossRef]
8. Descourvières, E.; Bandelj, V.; Sfriso, A.; Orlando-Bonaca, B.; Mačić, V.; Iveša, L.; Kipson, S.; Gljušćić, E.; Battelli, C.; Moro, I.; et al. Toward the first documented extinction of a marine macroalga in the Mediterranean Sea? *Reg. Environ. Change* **2024**, *24*, 132. [CrossRef]
9. Sfriso, A.; Buosi, A.; Wolf, M.A.; Sfriso, A.A. Invasion of alien macroalgae in the Venice Lagoon, a pest or a resource? *Aquat. Invasions* **2020**, *15*, 245–270. [CrossRef]
10. Gagič, M.; Solidoro, C. Lagoon of Venice. Circulation, water exchange and ecosystem functioning. *J. Mar. Syst.* **2004**, *51*, 1–4. [CrossRef]
11. Tagliapietra, D.; Zanon, V.; Frangipane, G. *Modello di Zonazione Gerarchica dei Bassofondali Della Laguna di Venezia, WP1, Tipologie Ambientali Lagunari (Habitat Acquatici Lagunari). Relazione Finale del Programma di Ricerca CORILA, 2004-2006, Linea 3.11: Indicatori ed Indici di Qualità Ambientale per la Laguna di Venezia*; CORILA: Venice, Italy, 2006; pp. 1–8.
12. Tagliapietra, D.; Volpi Ghirardini, A. Notes on coastal lagoon typology in the light of the EU Water Framework Directive: Italy as a case study. *Aquatic. Conserv. Mar. Freshw. Ecosyst.* **2006**, *16*, 457–467. [CrossRef]
13. Cucco, A.; Umgiesser, G. Modelling the Venice Lagoon residence time. *Ecol. Modell.* **2006**, *193*, 34–51. [CrossRef]

14. Variazioni del Livello Medio del Mare. Available online: https://www.comune.venezia.it/it/content/variazioni-livello-medio-mare (accessed on 20 February 2025).
15. Sfriso, A.; Curiel, D.; Rismondo, A. The Venice Lagoon. In *Flora and Vegetation of the Italian Transitional Water Systems*; Cecere, E., Petrocelli, A., Izzo, G., Sfriso, A., Eds.; CoRiLa: Venice, Italy, 2009; pp. 17–80.
16. Sfriso, A.; Buosi, A.; Tomio, Y.; Wolf, M.; Sciuto, K.; Sfriso, A.A. Macrophyte changes in transitional water systems: Role of water and sediment parameters, the Venice Lagoon as study case. *Ecol. Indic.* **2024**, *159*, 111623. [CrossRef]
17. Regione del Veneto. Monitoring Plan for the Venice Lagoon Pursuant to Directive 2000/60/EC Aimed at Defining the Ecological Status (Legislative Decree N. 152/2006 s.m.i). Line 2: EQB: Macrophytes, Mo.V.Eco. I, Final Report, 2012; pp. 23 + Tables and Figures. Available online: https://bur.regione.veneto.it/BurvServices/Pubblica/DettaglioDecreto.aspx?id=294424&highlight=true (accessed on 20 February 2025).
18. Regione del Veneto. Monitoring Plan for the Venice Lagoon Pursuant to Directive 2000/60/EC Aimed at Defining the Ecological Status (Legislative Decree N. 152/2006 s.m.i). EQB: Macrophytes, Mo.V.Eco. II, Final Report, 2015; pp. 27 + Tables and Figures. Available online: https://bur.regione.veneto.it/BurvServices/Pubblica/DettaglioDecreto.aspx?id=379241&highlight=true (accessed on 20 February 2025).
19. Regione del Veneto. Monitoring Plan for the Venice Lagoon Pursuant to Directive 2000/60/EC Aimed at Defining the Ecological Status (Legislative Decree N. 152/2006 s.m.i). EQB: Macrophytes, Mo.V.Eco. III, Final Report, 2019; pp. 26 + Tables and Figures. Available online: https://bur.regione.veneto.it/BurvServices/Pubblica/DettaglioDecreto.aspx?id=367174 (accessed on 20 February 2025).
20. Regione del Veneto. Monitoring Plan for the Venice Lagoon Pursuant to Directive 2000/60/EC Aimed at Defining the Ecological Status (Legislative Decree N. 152/2006 s.m.i). EQB: Macrophytes- Mo.V.Eco. IV, Final Report, 2022; pp. 19 + Tables and Figures. Available online: https://bur.regione.veneto.it/BurvServices/Pubblica/DettaglioDecreto.aspx?id=386914 (accessed on 20 February 2025).
21. Regione del Veneto. Monitoring plan for the Venice Lagoon Pursuant to Directive 2000/60/EC Aimed at Defining the Ecological Status (Legislative Decree N. 152/2006 s.m.i). EQB. Macrophytes- Mo.V.Eco. V, Final Report, 2024; pp. 26 + Tables and Figures. Available online: https://bur.regione.veneto.it/BurvServices/Pubblica/DettaglioDecreto.aspx?id=468004 (accessed on 20 February 2025).
22. CORILA (Consorzio per il coordinamento delle ricerche inerenti al sistema lagunare di Venezia). Venezia2021. Programma di Ricerca Scientifica per Una Laguna "Regolata. Linea 3.3". Produzione Primaria, Comunità Microbica, Bentonica, Planctonica e Nectonica Lagunare WP3.3.1. Aggiornamento Della Trofia, Della Speciazione e Produzione Primaria Delle Macrofite, Impatto dei Tassi di Sedimentazione e dei Processi di Erosione/Sedimentazione in Laguna di Venezia. 2022. Available online: https://www.corila.it/portfolio/venezia-2021-2/?doing_wp_cron=1743436511.5847239494323730468750 (accessed on 20 February 2025).
23. Life SERESTO. Habitat 1150* (Coastal Lagoon) Recovery by SEagrass RESTOration. A New Strategic Approach to Meet HD & WFD Objectives. LIFE12-NAT-IT-000331. Available online: https://una.city/nbs/venezia/life-seagrass-restoration (accessed on 20 February 2025).
24. Life LAGOON REFRESH. Coastal Lagoon Habitat (1150*) and Species Recovery by Restoring the Salt Gradient Increasing Fresh Water Input. LIFE16-NAT-IT-000663. Available online: www.lifelagoonrefresh.eu (accessed on 20 February 2025).
25. Ferrari, G.; Badetti, C.; Ciavatta, S. Real-Time Monitoring of the Venice Lagoon. The Magistrato alle Acque Develops a Monitoring Network to Study the Evolution of the Venice Lagoon Ecosystem. *Sea Technol.* **2004**, *45*, 22–24.
26. Archivio Meteo Storico. Meteo.it. Available online: https://www.ilmeteo.it/portale/archivio-meteo (accessed on 20 February 2025).
27. Sfriso, A.; Buosi, A.; Mistri, M.; Munari, C.; Franzoi, P.; Sfriso, A.A. Long-term changes of the trophic status in transitional ecosystems of the northern Adriatic Sea, key parameters and future expectations: The lagoon of Venice as a study case. *Nat. Conserv.* **2019**, *34*, 193–215. [CrossRef]
28. Petrocelli, A.; Wolf, A.M.; Cecere, E.; Sciuto, K.; Sfriso, A. Settlement and spreading of the introduced seaweed *Caulacanthus okamurae* (*Rhodophyta*) in the Mediterranean Sea. *Diversity* **2020**, *12*, 129. [CrossRef]
29. Thibaut, T.; Pinedo, S.; Torras, X.; Ballesteros, E. Long-term decline of the populations of Fucales (*Cystoseira* spp. and *Sargassum* spp.) in the Albères coast (France, North-western Mediterranean). *Mar. Pollut. Bull.* **2005**, *50*, 1472–1489. [CrossRef]
30. Orlando-Bonaca, M.; Mannoni, P.-A.; Poloniato, D.; Falace, A. Assessment of *Fucus virsoides* distribution in the Gulf of Trieste (Adriatic Sea) and its relation to environmental variables. *Bot. Mar.* **2013**, *56*, 451–459. [CrossRef]
31. Iveša, L.; Djakovac, T.; Devescovi, M. Long-term fluctuations in *Cystoseira* populations along the west Istrian Coast (Croatia) related to eutrophication patterns in the northern Adriatic Sea. *Mar. Pollut. Bull.* **2016**, *106*, 162–173. [CrossRef]
32. Gessner, F.; Hammer, L. Physiological Investigations on the Tolerance of *Fucus virsoides* (don) J. Ag. *Int. Rev. Gesamten Hydrobiol. Hydrogr.* **1971**, *56*, 581–597. [CrossRef]
33. Graiff, A.; Liesner, D.; Karsten, U.; Bartsch, I. Temperature tolerance of western Baltic Sea *Fucus vesiculosus*—Growth, photosynthesis and survival. *J. Exp. Mar. Bio. Ecol.* **2015**, *471*, 8–16. [CrossRef]

34. Oates, B.R.; Murray, S.N. Photosynthesis, dark respiration and desiccation resistance of the intertidal seaweeds *Hesperophycus harveyanus* and *Pelvetia fastigiata* f. *gracilis*. *J. Phycol.* **1983**, *19*, 371–380. [CrossRef]
35. Bell, E.C. Photosynthetic response to temperature and desiccation of the intertidal alga *Mastocarpus papillatus*. *Mar. Biol.* **1993**, *117*, 337–346. [CrossRef]
36. Madsen, T.V.; Maberly, S.C. A Comparison of air and water as environments for photosynthesis by the intertidal alga *Fucus spiralis* (*Phaeophyta*). *J. Phycol.* **1990**, *26*, 24–30. [CrossRef]
37. Linardić, J. Studije o jadranskom fukusu (*Fucus virsoides*). *Acta Bot. Croat.* **1949**, *12*, 7–131.
38. Munda, I.M. Seasonal and ecologically conditioned variations in the *Fucus virsoides* association from the Istrian coast (northern Adriatic). Dissertations. *SAZU* **1972**, *15*, 1–33.
39. Munda, I.M. Benthic marine algae as reflection of environment changes in the Northern Adriatic Sea. In Climate warming and related changes in Mediterranean marine biota. Helgoland, CIESM. In Proceedings of the CIESM Workshop Monographs, Menton, France, 1–4 October 2008; Volume 35, pp. 65–72.
40. Zavodnik, N.; Iveša, L.; Travizi, A. Note on recolonisation by fucoid algae Cystoseira spp. and Fucus virsoides in the North Adriatic Sea. *Acta Adriat.* **2002**, *43*, 25–32.
41. Rindi, F.; Battelli, C. Spatio-temporal variability of intertidal algal assemblages of the Slovenian coast (Gulf of Trieste, northern Adriatic Sea). *Bot. Mar.* **2005**, *48*, 96–105. [CrossRef]
42. Rindi, F.; Gavio, B.; Díaz-Tapia, P.; Di Camillo, C.G.; Romagnoli, T. Long-term changes in the benthic macroalgal flora of a coastal area affected by urban impacts (Conero Riviera, Mediterranean Sea). *Biodivers. Conserv.* **2020**, *29*, 2275–2295. [CrossRef]
43. Mačič, V. Distribution of seaweed *Fucus virsoides* J. Agardh in Boka Kotorska Bay (South Adriatic Sea). *Ann. Ser. Hist. Nat.* **2006**, *16*, 1–4.
44. Zanardini, G. Synopsis algarum in Mari Adriatico hucusque Collectarum cui accedunt monographia siphonearum nec non generales de algarum vita et structura disquisitiones cum tabulis auctoris manu ad vivum depictis. *Mem. Reale Accad. Sci. Torino. Ser. II Tomo IV* **1841**, *4*, 105–255.
45. Camuffo, D.; Bertolin, C.; Craievich, A.; Granziero, R.; Enzi, S. When the Lagoon was frozen over in Venice from A.D. 604 to 2012: Evidence from written documentary sources, visual arts and instrumental readings. *Méditerranée*. 2017. [Online], Varia, Online since 07 February 2017. Available online: http://mediterranee.revues.org/7983 (accessed on 1 January 2025).
46. Kristiansen, A.; Møller Pedersen, P.; Moseholm, L. Growth and reproduction of *Scytosiphon lomentaria* (*Fucophyceae*) in relation to temperature in two populations from Denmark. *Nord. J. Bot.* **1991**, *11*, 375–383. [CrossRef]
47. Scottish Association for Marine Science (SAMS). Significant Changes in Scotland's Rocky Intertidal Zones. 2025. Available online: https://www.cmscoms.com/?p=42868 (accessed on 20 February 2025).
48. Feis, M.E. Reproduction of the Genus Fucus. Bachelor's Thesis, Department of Marine Benthic Ecology and Evolution, Biological Centre, University of Groningen, Groningen, The Netherlands, 2010.
49. Berger, R.; Malm, T.; Kautsky, L. Two reproductive strategies in Baltic *Fucus vesiculosus* (*Phaeophyceae*). *Eur. J. Phycol.* **2001**, *36*, 265–273. [CrossRef]
50. Fortes, M.D.; Lüning, K. Growth Rates of North Sea Macroalgae in Relation to Temperature, Irradiance and Photoperiod. *Helgoländer Meeresunters.* **1980**, *34*, 15–29. [CrossRef]
51. Lüning, K. Temperature tolerance and biogeography of seaweeds: The marine algal flora of Helgoland (North Sea) as an example. *Helgoländer Meeresunters.* **1984**, *38*, 305–317. [CrossRef]
52. Frölicher, T.L.; Fischer, E.M.; Gruber, N. Marine heatwaves under global warming. *Nature* **2018**, *560*, 360–364. [CrossRef] [PubMed]
53. Rey, J.; Rohat, G.; Perroud, M.; Goyette, S.; Kasparian, J. Shifting velocity of temperature extremes under climate change. *Environ. Res. Lett.* **2020**, *15*, 34027. [CrossRef]
54. Sfriso, A.; Wolf, M.A.; Maistro, S.; Sciuto, K.; Moro, I. Spreading and autoecology of the invasive species *Gracilaria vermiculophylla* (*Gracilariales, Rhodophyta*) in the lagoons of the north-western Adriatic Sea (Mediterranean Sea, Italy). *Estuar. Coast. Shelf Sci.* **2012**, *114*, 192–198. [CrossRef]
55. Ramus, A.P.; Silliman, R.B.; Thomsen, M.S.; Long, Z.T. An invasive foundation species enhances multifunctionality in coastal systems. *Proc. Natl. Acad. Sci. USA* **2017**, *114*, 8580–8585. [CrossRef]
56. Wolf, M.A.; Sfriso, A.; Moro, I. Thermal pollution and settlement of new tropical alien species: The case of *Grateloupia yinggehaiensis* (*Rhodophyta*) in the Venice Lagoon. *Estuar. Coast. Shelf Sci.* **2014**, *147*, 11–16. [CrossRef]
57. Zhao, D.; Wang, H.; Sheng, Y.; Lü, J.; Luan, R. Morphological observation and rbcL gene sequences studies of two new species, *Grateloupia dalianensis* H.W. Wang et D. Zhao. Sp. Nov. and *G. yinggehaiensis* H.W. Wang et R.X. Luan, sp. nov. (*Halymeniaceae, Rhodophyta*) from China. *Acta Oceanol. Sin.* **2012**, *31*, 109–120. [CrossRef]
58. Life TRANSFER. Seagrass Transplantation for Transitional Ecosystem Recovery. LIFE19-NAT-IT-000264. Available online: www.lifetransfer.eu (accessed on 22 April 2025).

59. Haslob, H.; Clemmensen, C.; Schaber, M.; Hinrichsen, H.-H.; Schmidth, J.O.; Voss, R.; Kraus, G.; Köst, F.W. Invading *Mnemiopsis leidyi* as a potential threat to Baltic fish. *Mar. Ecol. Prog. Ser.* **2007**, *349*, 303. [CrossRef]
60. Roohi, A.; Fazli, H.; Rowshantabari, M.; Rahmati, R.; Khodaparast, N. Effect of an invasive species *Mnemiopsis leidyi* on the zooplankton community structure in the Caspian Sea. *J. Great Lakes Res.* **2024**, *50*, 102420. [CrossRef]
61. Piccardi, F.; Poli, F.; Sguotti, C.; Tirelli, V.; Borme, D.; Mazzoldi, C.; Barausse, A. Assessing the impact of the invasive ctenophore *Mnemiopsis leidyi* on artisanal fisheries in the Venice Lagoon: An interdisciplinary approach. *Hydrobiologia* **2025**, *852*, 2387–2405. [CrossRef]
62. Gavioli, A.; Mancinelli, M.; Turolla, E.; Lanzoni, M.; Paesanti, V.; Soana, E.; Eggleston, D.B.; Christian, R.R.; Castaldelli, G. Impacts of the invasive blue crab *Callinectes sapidus* on small-scale fisheries in a Mediterranean lagoon using fishery landing data. *Sci. Tot. Environ.* **2025**, *974*, 179236. [CrossRef] [PubMed]
63. Nardelli, F.; Fucilli, V.; Pinto, H.; Elston, J.N.; Carignani, A.; Petrontino, A.; Bozzo, F.; Frem, M. Socio-economic impacts of the recent bio-invasion of *Callinectus sapidus* on small-scale artisanal fishing in southern Italy and Portugal. *Front. Mar. Sci.* **2024**, *11*, 2024. [CrossRef]

environments

MDPI

Article

Phytoplankton Dynamics in a Large Lagoon: Nutrient Load Reductions, Climate Change, and Cold- and Heatwaves

Gerald Schernewski [1,2,*], Maria Schneider [1], Thomas Neumann [3] and Mario von Weber [4]

[1] Coastal Sea Geography Group, Leibniz-Institute for Baltic Sea Research, Seestrasse 15, D-18119 Rostock-Warnemünde, Germany
[2] Marine Research Institute, Klaipeda University, Universiteto Ave. 17, LT-92294 Klaipeda, Lithuania
[3] Biogeochemical Modelling Group, Leibniz-Institute for Baltic Sea Research, Seestrasse 15, D-18119 Rostock-Warnemünde, Germany
[4] Landesamt für Umwelt, Naturschutz und Geologie, Goldberger Straße 12b, D-18273 Güstrow, Germany
* Correspondence: gerald.schernewski@io-warnemuende.de; Tel.: +49-381-5197207

Abstract: The coastal Oder/Szczecin Lagoon is subject to multiple external changes, particularly the reduction in external nutrient loads and the impacts of climate change, including rising temperatures and more frequent heatwaves. By combining monitoring data covering the past 40 years with 3D ecosystem modelling, we assess changes in phytoplankton abundance and diversity across different temporal scales, ranging from long-term trends to the short-term effects. Despite strong reductions in external nutrient loads, neither the average annual phytoplankton biomass nor the long-term species composition changed significantly, although extreme summer blooms appear to have decreased. In summer, cyanobacteria, usually dominated by *Microcystis*, can reach a relative biovolume of up to 90%. Bacillariophyceae (diatoms) contribute up to 72% of the annual relative biovolume and dominate in spring. Both interannual and short-term variability in phytoplankton biomass and composition are pronounced. Heat- and coldwaves show no consistent immediate effects; however, results suggest that cyanobacteria, particularly *Microcystis*, benefit from hot summers. In contrast, diatoms appear less responsive to temperature, although they tend to contribute more in colder years, with distinct shifts in species composition observed between hot and cold springs. Model simulations indicate that a 1.5 °C increase in air temperature would, via elevated water temperatures, raise average monthly phytoplankton biomass by 4% in July and by 9% in August, further promoting cyanobacteria growth.

Keywords: cyanobacteria; diatoms; toxic blooms; nutrient loads; temperature; seasonality; model simulation

1. Introduction

Surface air temperatures in the Baltic Sea region have increased over the past century, with a clear trend visible across all seasons. At the same time, the number, duration, and intensity of extreme high- and low-temperature events have also increased [1,2]. Rising air temperature directly affects sea surface temperature [3,4], which in the Baltic Sea has increased by more than 0.5 °C per decade over the past 30 years [5,6]. Globally, marginal seas have warmed more rapidly than the open oceans, and the Baltic Sea is currently the fastest-warming marine region [7,8].

Model projections suggest that by the end of the 21st century, sea surface temperatures in the Baltic Sea will rise by 1.1–3.2 °C compared to the 1976–2005 baseline [9]. The

https://doi.org/10.3390/environments12100370

frequency of heatwaves is also expected to increase, particularly in the southern Baltic Sea [10]. Consequently, the Baltic Sea is considered a climate change hotspot, and ecological as well as management challenges related to rising temperatures are expected to intensify in the future.

Increased water temperature can have multiple ecological effects. For the Baltic Sea, Viitasalo & Bonsdorff [11] highlight several consequences, including enhanced bacterial growth, complex effects on microbial food webs and zooplankton, and the establishment of non-indigenous species. The most critical consequences concern phytoplankton, which form the base of marine food webs, drive nutrient cycling, and fuel the biological carbon pump [12,13]. Climate warming has multiple effects on phytoplankton, for example, it alters phytoplankton community composition by favoring harmful taxa, stimulates cyanobacterial blooms and toxin production, prolongs bloom duration, thereby increasing ecological and public health risks, and it causes temporal shifts in blooms (e.g., [14–16]). Coastal zones, and lagoons in particular, are considered especially sensitive to these climate change impacts [17].

Heatwaves, defined as temperature events exceeding either the difference between the 90th percentile and the climatological mean for several days [18] or a seasonally defined temperature threshold [19], are assumed to exacerbate climate change impacts [20]. They can have both short- and long-term ecological consequences in aquatic systems [21,22], such as altering physical structure and chemical properties, shifting species composition, and threatening biodiversity [11]. In Baltic coastal waters, observed impacts include increased oxygen deficiency [23] and higher emissions of carbon dioxide and methane [24]. Phytoplankton can respond rapidly to heatwaves through accelerated growth, shifts in community composition, and often an increased dominance of toxic, buoyant cyanobacteria, especially in lakes [25–27]. Recent in situ mesocosm experiments in a Mediterranean lagoon suggest similar effects [28].

In Baltic Sea research, climate change and heatwaves have become major focal points, largely due to advances in 3D hydrodynamic and ecosystem modelling, with much emphasis placed on physical processes [8,9]. However, as Chen et al. [29] note, such models are rarely applied to lakes and are usually unavailable for coastal lagoons, leaving comparative analyses scarce. Despite progress, Rutgersson et al. [30] conclude that significant knowledge gaps remain regarding the occurrence and effects of marine heatwaves. Moreover, existing biogeochemical monitoring programs are generally unsuitable for capturing short-term events like heatwaves [31,32]. For example, the German coastal monitoring program of the federal state of Mecklenburg-Vorpommern [33] collects data only every four weeks, at a depth of one meter and one meter above the sediment. While this provides a valuable 40-year dataset, its temporal and spatial resolution is inadequate for analyzing heatwaves and their ecological impacts. Additionally, temperature-driven changes in lakes and lagoons are often masked by other factors, such as external nutrient inputs [34].

To address these gaps, we implemented the 3D ecosystem model ERGOM in the Oder Lagoon and conducted simulations covering the past 40 years. These simulations were combined with long-term monitoring data to investigate the following questions: How have phytoplankton abundance and composition changed over the past four decades? What are the patterns of seasonal development and variability? What role do extreme weather events, particularly heatwaves, play? What are the implications for environmental monitoring and policy frameworks (e.g., the Water Framework Directive, Habitat Directive, and Baltic Sea Action Plan)? And finally, what lessons can be drawn for assessing and managing ecosystems under a warming climate?

2. Materials and Methods

2.1. The Oder River and the Oder/Szczecin Lagoon

The Oder Lagoon (German: Oderhaff; Polish: Zalew Szczeciński) is one of the largest coastal lagoons in Europe, located on the German–Polish border in the southern Baltic Sea (53°48′ N, 14°08′ E). It covers an area of 687 km^2, with an average depth of 3.8 m and a natural maximum depth of 8.5 m [35]. Around 40% of the lagoon, known as the Kleines Haff, lies within Germany, while the remaining 60%, called the Wielki Zalew, is located in Poland (Figure 1).

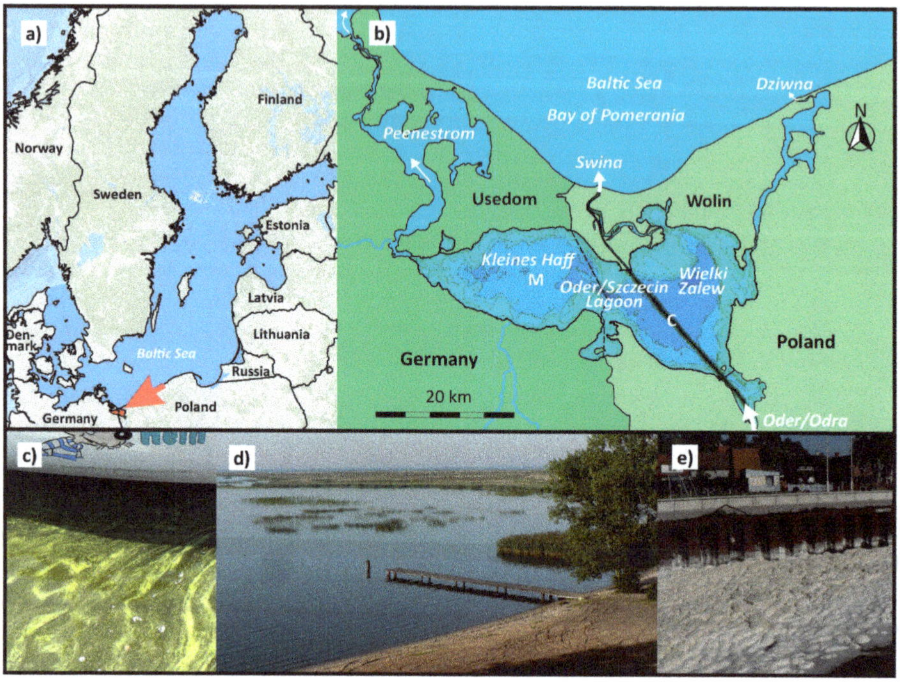

Figure 1. (**a**) The Baltic Sea region and location of the lagoon; (**b**) the Oder/Szczecin Lagoon at the German/Polish border with bathymetry and major rivers. M and C indicate major sampling stations; (**c**) heavy cyanobacteria bloom in July 2008, (**d**) view over the lagoon near Wolin and (**e**) accumulated foam after the cyanobacteria bloom in Sept 2009.

The lagoon's hydrology is strongly influenced by the Oder River (Odra in Polish), which has a catchment area of roughly 120,000 km^2 and an average discharge of 518 m^3/s. The lagoon connects to the Baltic Sea via three outlets: the Dziwna, the Peenestrom, and the Świna. The Świna is an artificially deepened and maintained shipping channel. Already in 1939, the channel had a depth of 9.6 m and was deepened to 10.5 m across the entire lagoon in 1984. From 2022, the maximum depth was increased to 12.5 m. The Świna accounts for around 70% of the lagoon's outflow. These connections allow occasional saltwater intrusions from the Baltic Sea, maintaining lagoon salinity between 1 and 3 PSU [35].

The Oder River is also the main source of nutrients, delivering more than 95% of the lagoon's nitrogen and phosphorus load—equivalent to 46,266 tons of total nitrogen (TN) and 1635 tons of total phosphorus (TP) annually [36]. As a result, the river is the primary driver of the lagoon's poor ecological condition. The OECD classifies the lagoon as hypertrophic, and the most recent HELCOM assessment [37] rated its eutrophication status as "bad." Under the European Water Framework Directive (WFD), which assesses ecological quality based on several biological elements, including benthic flora and fauna, the lagoon is currently classified as having "insufficient" ecological status [38].

2.2. The Long-Term Monitoring Data and Calculations

The official monitoring of water quality is a legal obligation under the Water Framework Directive (WFD) [39]. For the Oder/Szczecin Lagoon, the German–Polish Border Waters Commission coordinates a joint monitoring program. This program includes comparative measurements between German and Polish laboratories every 2–3 years to ensure the consistency and quality of results. At present, three stations are sampled monthly in both the German and Polish parts of the lagoon. Samples are taken 1 m below the water surface and 1 m above the bottom (in the German part, this applies only at station M). Nutrients such as nitrogen and phosphorus, as well as oxygen and other water quality parameters [33], are determined according to various DIN EN ISO standards. All laboratories analyzing border waters operate under an established quality system certified according to ISO 17025 and use reference or equivalent methods [40].

Phytoplankton and chlorophyll-a are determined according to HELCOM Guidelines 21/23 and DIN standard 16695 (since 2016). The number of algal cells counted is multiplied by a species-specific cell volume to calculate biovolume. Until the year 2009, the cell counting focused on dominant species and their biovolume. Other species were only grouped. From 2009, the counting became more detailed and included the abundance and biovolume of all species. Sampling takes place in winter, as well, but phytoplankton analyses are only carried out if chlorophyll concentrations exceed a threshold. In fact, phytoplankton data between December and February are lacking. For the years 2014, 2016 and 2018, data were missing. The full phytoplankton data is documented and attached as Supplementary Material (Table S1). Chlorophyll-a is measured using a photometric method. Because these methods differ fundamentally, the correlation between algal biovolume and chlorophyll-a is generally weak (Figure 2b).

For this study, only data from station M in the German part of the lagoon (Kleines Haff) were used. This station provides the most complete dataset, especially for phytoplankton. Additionally, station M is not directly influenced by the Oder River discharge. Our model simulations show that, on average, it takes river water about 100 days to reach this location (Figure 2c). Consequently, short-term changes in river discharge can be neglected when analyzing temperature-driven effects on timescales of days to months. By contrast, stations in the eastern lagoon (Wielki Zalew) are strongly influenced by the river.

Air temperature data were obtained from the Norwegian Meteorological Institute (MET Norway). The model-based data are provided on a spatial grid, with one grid point located close to station M. Near-surface water temperatures were derived from the ERGOM model. For analyses of hot and cold years, as well as heatwaves, we used air temperature data. Apart from temperatures below 4 °C, air and surface water temperatures are strongly correlated (Figure 2a), and air temperatures allow better comparability with climate model projections.

For each decade, monthly, seasonal, and annual mean air temperatures (measured at 2 m height above the water surface) were calculated, along with their standard deviations. Years (or months) with mean temperatures above the decadal mean + 1 SD were classified as warm, those below the decadal mean—1 SD as cold, and all others as normal. This classification was applied separately to monthly, seasonal, and annual values. Seasons were defined as follows: spring (March–May), summer (June–August), autumn (September–November), and winter (December–February). Extremes are compared to the mean value of all considered years, called "moderate years".

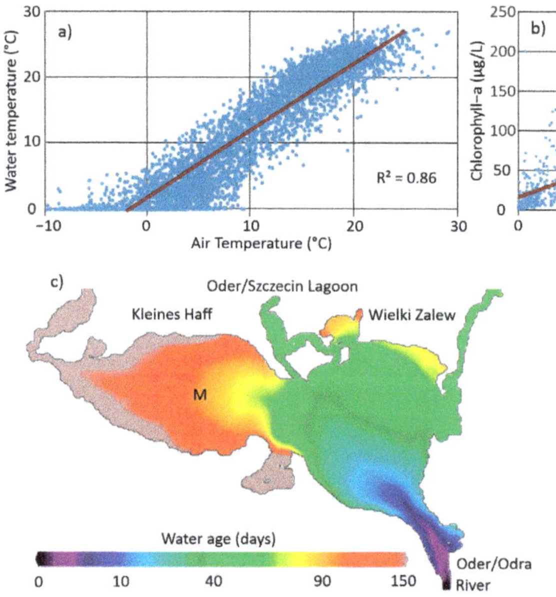

Figure 2. (**a**) Relationship between air temperature and surface water temperature at sampling station M (location indicated) in the Oder Lagoon based on a 30-year dataset. (**b**) Relationship between chlorophyll and biovolume data at station M based on long-term monitoring data. (**c**) ERGOM model calculations of the average annual time that entering river water needs to reach different parts in the lagoon, expressed as water age in days.

Heat- and coldwaves were identified using the 90th and 10th percentiles based on daily temperatures, respectively. Events lasting at least five consecutive days above (heatwave) or below (coldwave) these thresholds were considered. Only events between May and September were included, as these are most relevant for ecological analysis. The selected heatwave periods, focused mainly on August, were 8–28 August 1997, 28 July–1 August 2002, 2–11 July 2006, 14–18 August 2015, and 23 August–1 September 2019. Cold periods were identified in May: 7–22 May 1995, 5–9 May 2001, 2–11 May 2010, 14–18 May 2015, and 1–11 May 2017.

2.3. The Ecosystem Model Approach (ERGOM)

We applied the 3D biogeochemical model ERGOM (Ecological ReGional Ocean Model) developed by Neumann et al. [41]. The model was recently evaluated by the European Commission [42] and is recognized as one of the leading marine biogeochemical models in Europe. Its circulation component is based on the Modular Ocean Model (MOM5.1), while the biogeochemical module employs ERGOM version 1.2. The model was specifically adapted for the Oder Lagoon, using a high horizontal grid resolution of 150 m. Vertically, the water column was divided into 28 layers, with thicknesses ranging from 0.25 m near the surface to 0.5 cm near the bottom. Open boundary conditions from a 2 km resolution Baltic Sea model were applied at the lagoon's three outlets to the Baltic Sea.

The full Baltic Sea model, including its biogeochemical cycles, processes, and state variables, is described in Neumann [43]. The original model includes three functional phytoplankton groups: small algae (usually r-strategists, mainly diatoms), large algae (usually K-strategists, e.g., some flagellates), and N-fixing algae (mainly cyanobacteria). For the lagoon application, an additional freshwater phytoplankton group was introduced, which grows under very low salinity and has higher light sensitivity. Together, these four groups realistically reproduced the seasonal development and biomass of phytoplankton in the lagoon [36]. In the model, 2.5 g C/m^3 corresponds to approximately 100 µg/L chlorophyll-a.

The model accounts for temperature- and oxygen-dependent mineralization processes, which release dissolved inorganic nitrogen and phosphorus from organic matter in both the water column and sediments. In oxygen-rich conditions, phosphate binds with iron oxides to form particles that settle in the sediments. These can be resuspended by erosion and redistributed by currents. Under anoxic conditions, iron oxides are reduced, releasing dissolved phosphate back into the water. Oxygen is produced through primary production and consumed by respiration and mineralization. Furthermore, extracellular excretion by phytoplankton leads to non-Redfield carbon uptake. Model validation and application to the lagoon are described in Neumann et al. [36].

Meteorological forcing data were obtained from the coastDat-3 dataset [44] and the Norwegian Meteorological Institute. Wind data are provided as hourly averages at 10 m height with a 12.5 km spatial resolution. Discharge and nutrient load data for the Oder, Uecker, and Zarow Rivers were supplied by the Instytut Meteorologii i Gospodarki Wodnej, the Landesamt für Umwelt, Naturschutz und Geologie Mecklenburg-Vorpommern, and Friedland et al. [45]. The climate change model experiment covered six years, used the same input data, but assumed air temperatures to be 1.5 °C higher.

3. Results

3.1. Long-Term Developments and the Role of Changing External Loads

Annual, seasonal and short-term weather effects in a lagoon are usually superimposed by changes in external nutrient loads. Consequently, the analysis of weather-related effects requires knowledge about changes in nutrient loads on different temporal scales. This is especially the case in the Oder Lagoon, where the large Oder River contributes large amounts of water and nutrients to the lagoon.

Between 1985 and 2002, nutrient concentrations in the Oder River, and consequently the loads to the Oder Lagoon, declined (Figure 3a,b). Total phosphorus (TP) loads peaked in the late 1980s at about 9800 t TP/a. Due to improved urban and industrial wastewater treatment and economic restructuring, these loads decreased to about 1600 t TP/a by the late 2010s, with even lower values in dry years. Total nitrogen (TN) loads reached maxima above 100,000 t TN/a but have since fallen below 40,000 t TN/a, and in some dry years even below 30,000 t TN/a. The reductions in nitrogen and phosphorus loads differed in magnitude, resulting in a shift in the N/P ratio. While the N/P ratio was around 10 until the early 1990s, it has risen to above 25 in recent years. Consequently, the relative availability of nitrogen in external loads has increased over the past 40 years.

The long-term reduction in riverine nutrient loads has led to lower nutrient concentrations in the surface waters of the entire lagoon. Station M, located in the central western bay (Kleines Haff), more than 40 km from the Oder River, illustrates these changes. Over the past 40 years, concentrations of dissolved inorganic phosphorus (DIP) have decreased in all seasons, and phosphorus has consistently become a limiting nutrient at Station M during March and April (Figure 3c). Dissolved inorganic nitrogen (DIN) concentrations show a similar decline, and nitrogen shortages now occur in July and August. Despite these reductions, chlorophyll-a (chl-a), an indicator of total algal biomass, has still not shown a significant decline over the past 40 years (Figure 3e). Similarly, nutrient load reductions have not led to an increase in water transparency (Figure 3f). This indicates that productivity in the Kleines Haff remains unaffected by nutrient limitation, and Station M remains well-suited for analyzing the effects of weather-related phenomena.

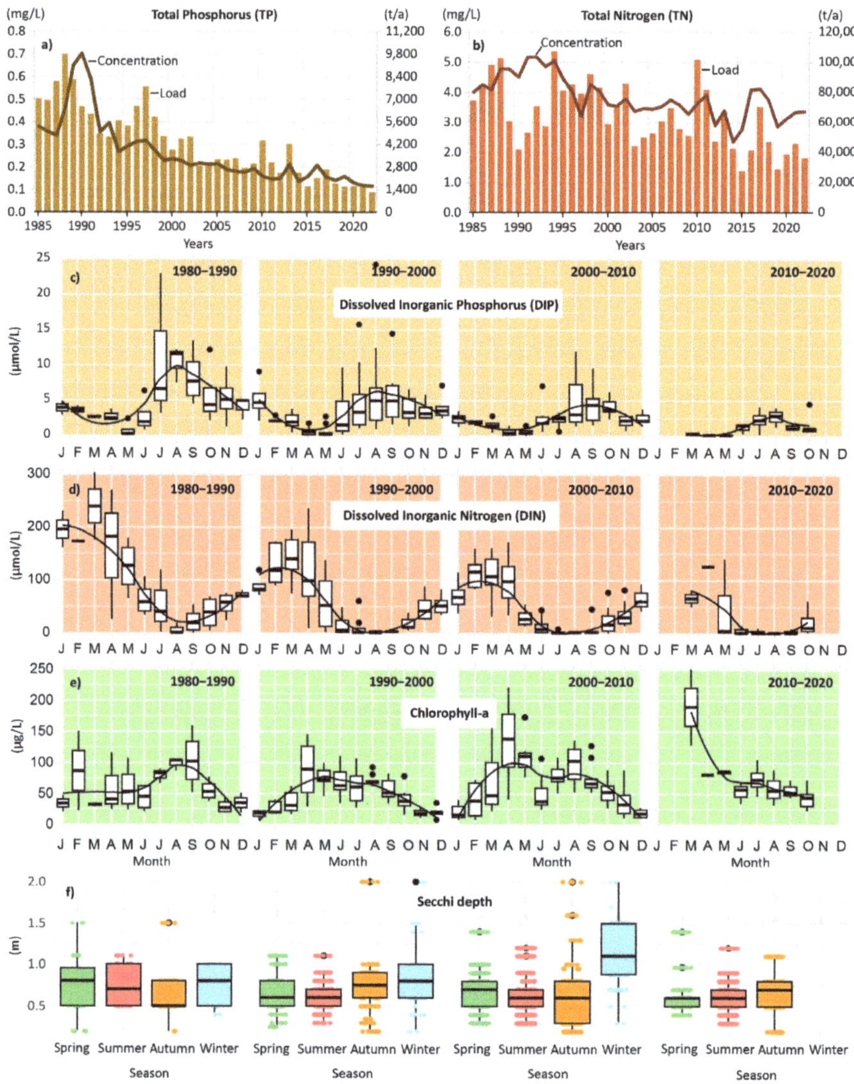

Figure 3. (**a**) Total phosphorus concentrations in the Oder/Odra River and loads to the Oder/Szczecin Lagoon and (**b**) total nitrogen concentrations in the Oder/Odra River and loads to the Oder/Szczecin Lagoon loads to the between 1885 and 2002 (the loads represent the situation north of the city of Szczecin); Decadal changes and annual concentration course of (**c**) dissolved inorganic phosphorus, (**d**) dissolved inorganic nitrogen and (**e**) chlorophyll-a as well as (**f**) decadal changes and seasonal course of water transparency (Secchi-depth) at station M in the central Kleines Haff.

3.2. Long-Term Phytoplankton Concentration

The monitoring data are limited by their temporal resolution, as sampling is conducted only once per month. To address this limitation, we applied a 3D ecosystem model to complement the monitoring data and capture short-term dynamics.

The model reproduces both the quantity and seasonal patterns of chlorophyll-a reasonably well compared to observations (Figure 4a). It indicates a decreasing trend in chlorophyll-a concentrations over the last 25 years at 1 m depth at Station M, consistent with reduced nutrient loads. However, this trend is overlain by strong interannual variability. In contrast, the monitoring data, due to spatial patchiness and limited temporal resolution, do not show a reliable long-term trend. Still, both the model and observations suggest that the magnitude of extreme blooms has decreased in the last decade, with strong blooms no longer occurring in recent years.

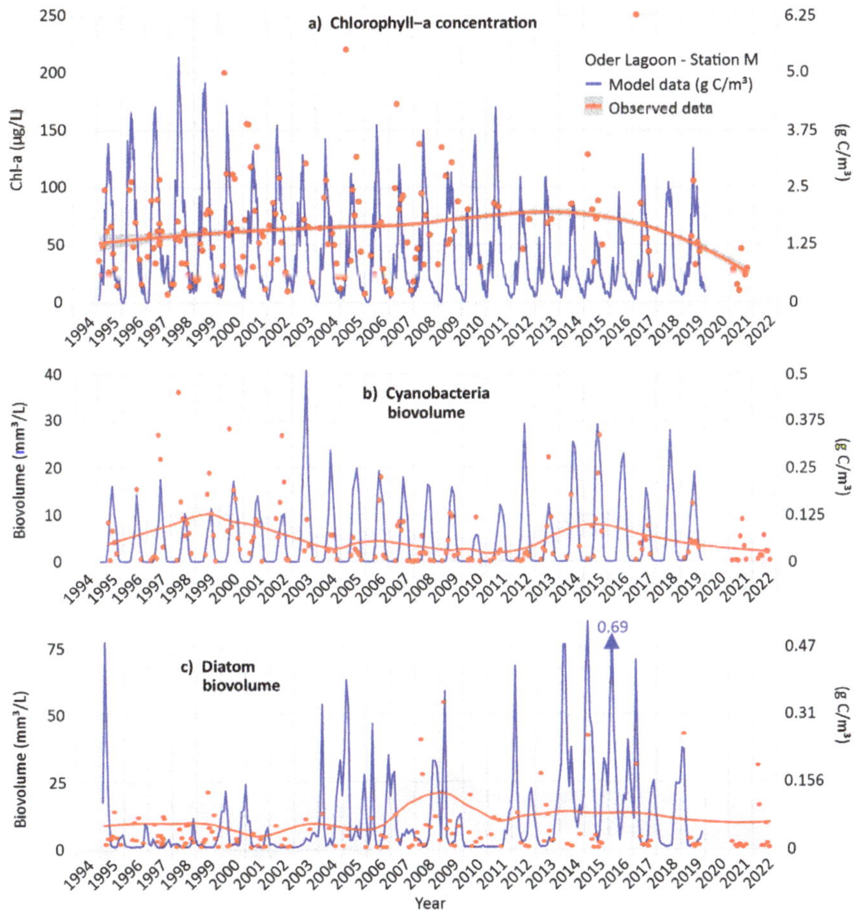

Figure 4. (**a**) chlorophyll-a concentrations and (**b**) cyanobacteria and (**c**) diatom biovolumes (left axes) in the Oder Lagoon (station M in 1 m water depth) between 1995 and 2022 based on observations and model simulations (monthly averaged). The model data is always expressed in mg C/m^3 (right axes). The red line indicates the floating average of the data.

Two phytoplankton groups are of particular interest in this context: cyanobacteria and diatoms. Both are represented as functional groups in the model. Cyanobacteria are quantitatively important (Figure 4b) and have the potential to form harmful algal blooms. Monitoring data suggest a decline in extreme summer cyanobacteria biovolume peaks over the last 25 years. In the 1990s the model underestimates these extreme peaks. During the last 15 years, the model data tends towards increased cyanobacteria peaks that are in agreement or even higher than the data. Higher recent cyanobacteria peaks in the model data may be explained by declining nitrogen concentrations in summer surface waters, which, in the model, favor a higher biovolume of N-fixing cyanobacteria. Nonetheless, both data and model results show high interannual variability in cyanobacterial peaks.

Comparing observations and model results for diatoms suggests that the model underestimates the data in the 1990s and overestimates biovolumes after 2003 (Figure 4c). The model indicates an increase over the decades, whereas the observed data show no clear trend. However, Figure 3 indicates an increase in chlorophyll-a concentrations in spring during the last decades and a shift in the spring maximum from April to March. The spring bloom is largely formed by diatoms.

Comparison between the data and the model is challenging because the model frequently predicts diatom blooms in winter, while data collection typically begins in March. The role of ice-cover has not been assessed. Therefore, the available data do not allow for reliable model validation for diatoms. Nonetheless, diatom biovolumes exhibit strong interannual variability.

3.3. Long-Term Phytoplankton Class Composition

Did the substantial reductions in nutrient loads affect the long-term species composition in the western lagoon (Kleines Haff, station M)?

Diatoms (Bacillariophyceae and Fragilariophyceae) consistently represent high relative biovolumes in the Oder Lagoon, reaching up to 72% in certain years (Figure 5a). Bacillariophyceae, together with Fragilariophyceae (pennate diatoms), which gained quantitative importance after 2005, form the characteristic spring bloom. Fragilariophyceae are today again considered as part of the Bacillariophyceae class.

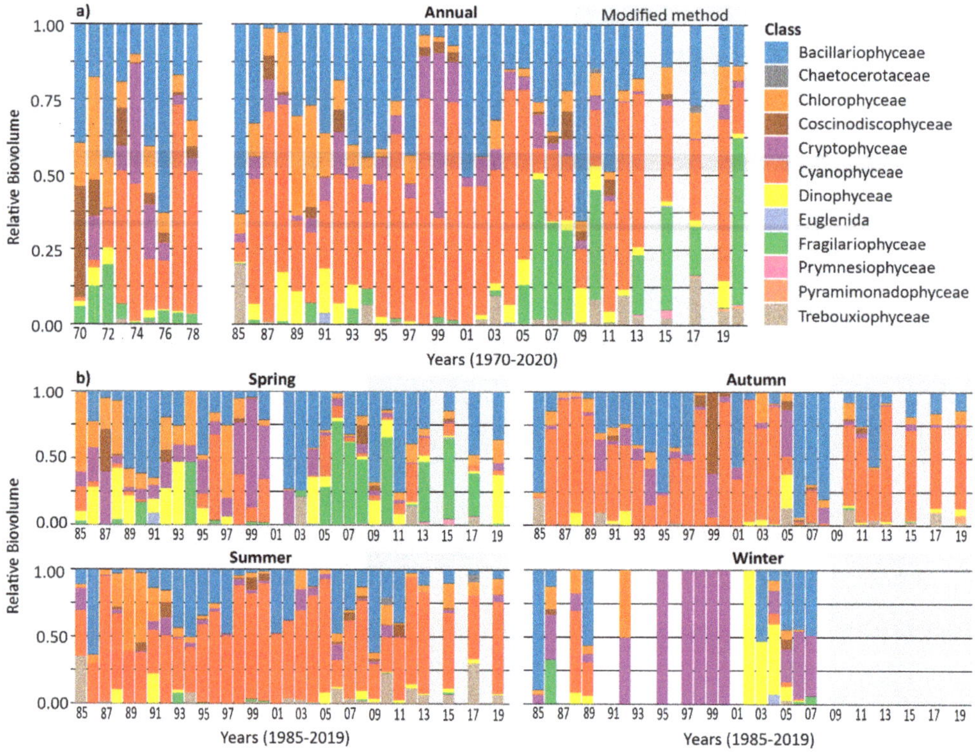

Figure 5. (**a**) Annual shares of relative biovolumes and (**b**) seasonal shares of relative biovolumes of 12 phytoplankton classes in the Oder Lagoon (station M, Kleines Haff, taken at 1 m water depth) over the years 1985 to 2019. Shaded grey areas indicate data based on the more detailed phytoplankton determination method.

The raw data (Table S1) shows that in seven years, maximum diatom biovolumes during spring blooms in March and April exceeded 20 mm^3/L, corresponding to up to 8 million individual cells per liter. This spring bloom is usually dominated by *Diatoma tenuis*, a Fragilariophyceae species. Biovolumes above 4 mm^3/L during spring blooms are common.

Outside of spring, high species-specific diatom biovolume concentrations have also been observed in other seasons. For example, in September 1987 (74 mm^3/L) and June 1999 (20 mm^3/L), the large Bacillariophyceae *Actinocyclus normanii*, which frequently occurs in summer and autumn, formed blooms. In October 1978, *Skeletonema subsalsum*, which occurs in all seasons, reached a biovolume concentration of 20 mm^3/L.

The classes Chlorophyceae and, less important, Trebouxiophyceae (green algae) had a high annual share until the mid-1990s, reaching up to 34% in some years (Figure 5a). They were particularly abundant in late spring and early summer, but their quantitative importance has declined in recent decades. The genus dominating in terms of biovolume is *Scenedesmus*, which reached biovolumes above 20 mm^3/L in certain years, such as June 1992 and September 1987. Another species observed in higher quantities in a few years is *Planctonema lauterbornii*, presently called *Binuclearia lauterbornii.*

Cryptophyceae, usually *Rhodomonas lacustris (Plagioselmis lacustris)*, exhibit high variability in occurrence and were observed mainly in spring and winter (Figure 5b). Since winter data are incomplete, their relevance may be underestimated. The same applies to Dinophyceae (dinoflagellates), a class that includes species capable of forming harmful algal blooms. The raw data (Table S1) shows that several species occur in the Oder Lagoon (e.g., Peridiniales spp., Amphidinium spp., and *Gonyaulax spinifera*), with biovolumes occasionally exceeding 2 mm^3/L, particularly in spring.

In some years after 2015, between May and July, Prymnesiophyceae (golden algae) were observed (Figure 5b). Typically, biovolumes were well below 1 mm^3/L but reached nearly 2 mm^3/L in May 2015. These algae gained public attention after causing a large-scale organism kill (mussels, fish) in the Oder River in summer 2022, suggesting that significant amounts may have subsequently entered the lagoon.

Cyanophyceae (cyanobacteria) are often the quantitatively dominant phytoplankton class in the Oder Lagoon, accounting for up to 69% of the annual relative biovolume. They dominate in summer and autumn and can comprise up to 90% of the phytoplankton biovolume during these seasons, as observed in summer 2014 (Figure 5). Cyanobacteria in the Oder Lagoon frequently reach biovolumes above 5 mm^3/L and cell numbers exceeding 200 million per liter during summer blooms (Figure 3b).

The most important cyanobacteria species in the lagoon include *Microcystis* spp. (e.g., *Microcystis flos-aquae, Microcystis aeruginosa*), but other species such as *Aphanothece* spp., *Aphanizomenon flos-aquae, Limnothrix redekei, Planktothrix agardhii, Woronichinia* spp., and *Synechococcus* also occasionally reach high concentrations. Of the 50 samples with cyanobacteria biovolumes above 10 mg/L, only two were dominated by species other than *Microcystis*: *Planktothrix agardhii* in late August 2000 and *Aphanizomenon flos-aquae* in late August 2002. In summer 1979, *Microcystis* biovolumes exceeded 50 mm^3/L on two occasions.

All three dominant species can potentially produce cyanotoxins. Many *Microcystis* strains, in particular, produce toxic microcystins and are frequently associated with harmful algal blooms. Biovolumes above 30 mm^3/L were observed exclusively before 1990.

Overall, the substantial reduction in nutrient loads over the past 25 years did not result in strong changes in phytoplankton composition at station M in the western lagoon, taking into account that the determination method was modified after 2008. However, within the diatoms, a shift took place with an increase in Fragilariophyceae after 2005 and a decrease in Bacillariophyceae.

3.4. Phytoplankton in Hot and Cold Years

In addition to eutrophication levels and external nutrient loads, parameters such as salinity, basin morphometry, and hydrodynamics influence phytoplankton abundance and composition. These factors are largely lagoon-specific and are particularly relevant when comparing different lagoons. An important parameter affecting phytoplankton within a lagoon across temporal scales is temperature.

During the 1980s, the mean air temperature above the water surface at station M was 8.2 °C. Between 2010 and 2019, the mean temperature rose to 9.44 °C, an increase of more than 1.2 °C over 30 years. Long-term changes in temperature and nutrient loads overlap, making it difficult to isolate temperature effects from the sparse monitoring data. Nevertheless, species composition exhibits strong interannual and seasonal variability (Figure 6), and differences in weather conditions between seasons and years likely play a major role.

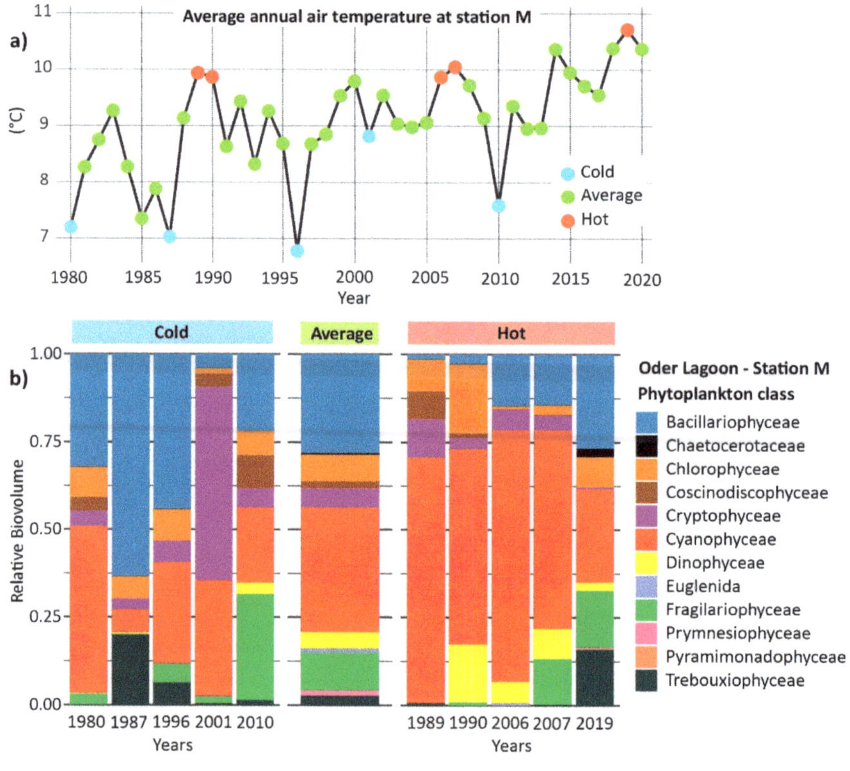

Figure 6. (**a**) Annual average air temperatures at station M and the five years that were selected to represent hot and cold years in every decade. All remaining years are used to calculate the 'normal' year. (**b**) Composition of phytoplankton classes in hot, cold and normal years (station M, 1 m water depth).

To analyze temperature effects on phytoplankton composition, hot and cold years were selected based on average annual temperatures. The five coldest years had an average annual temperature of 7.5 °C, while the five hottest years averaged 10.1 °C (Figure 5a). The years are distributed across the decades to minimize trend effects. One key question is what insights hot years can provide for the future under ongoing climate change and rising temperatures.

During average and cold years, diatom biovolumes (Bacillariophyceae together with Fragilariophyceae) account for approximately 40% of the total phytoplankton. In contrast, during hot years, diatoms represent only about 18% of the total.

Cyanobacteria are known to benefit from high summer temperatures, strong irradiance, and calm weather conditions that reduce vertical mixing in the surface water layer. This pattern is also evident in our data. Both Dinophyceae and Cyanophyceae appear to benefit from higher temperatures. Cyanophyceae contribute roughly 60% of the relative biovolume during hot years, 35% during moderate years, and about 30% during cold years.

Dinophyceae account for approximately 7% in hot years, 5% in moderate years, and 1% in cold years. However, these patterns are not consistent across all hot and cold years, and in some years, other phytoplankton classes dominate. For example, Cryptophyceae reached a relative share above 50% only in the slightly cooler year 2001.

In summary, the data suggest that cyanobacteria could become increasingly dominant under warmer climatic conditions, whereas diatoms may decline in relative abundance.

3.5. Phytoplankton in Hot and Cold Seasons

Comparisons between hot and cold years have limitations because average annual temperatures do not reflect seasonal temperature fluctuations. A single very hot or cold season can strongly influence the annual average, while hot and cold seasons may partially offset each other. Considering this and the pronounced annual succession of phytoplankton, a more detailed analysis of hot, cold, and average seasons is warranted.

Figure 5b indicates that phytoplankton composition in autumn is comparable to that in summer. For winter, the available data are insufficient for reliable analysis. Consequently, this study focuses on spring and summer. Figure 7 provides a comparison of phytoplankton composition between years with hot and cold springs and summers.

At the Oder Lagoon, the trend of rising air temperatures is evident in every month, but most pronounced during summer (July and August), when average temperatures increased from 15.6 °C to 17.5 °C—an increase of more than 1.9 °C over the 30-year period. In years classified as having a cold summer, average air temperatures were about 16.5 °C, whereas hot summers averaged 19.7 °C (Figure 7b). For spring, cold years showed average air temperatures of about 6 °C, while hot years reached nearly 10 °C (Figure 7a). The temperature differences between hot and cold seasons are substantial. Despite this, no systematic differences at the phytoplankton class level (Figure 7c,d) are apparent between hot and cold seasons. Strong interannual variability and limited data availability prevent the detection of reliable trends.

The annual phytoplankton succession in spring is typically dominated by diatoms, while cyanobacteria dominate in summer. Therefore, Figure 7e,f focuses on the species and genera of these two major groups. At the species level, differences between hot and cold springs are visible. In cold springs, species of the genus *Diatoma* (family Fragilariaceae) account for about 75% of all diatoms, whereas in hot springs they occur only sporadically and at much lower proportions. In contrast, hot springs are often characterized by the presence of *Skeletonema*, a cylindrical diatom with a silica frustule, as well as *Melosira*, which can become dominant.

For Cyanophyceae, *Microcystis* occasionally dominates during cold summers but becomes consistently dominant during hot summers, suggesting that high temperatures favor its growth. In cold summers, species of the diverse genus *Planktothrix* occur in several years, but they are absent in hot years. Conversely, species of the genus *Aphanizomenon* reach relative biovolume shares above 0.25 only in hot summers, in two observed years. Overall, species diversity appears higher in cold summers, whereas hot summers are strongly dominated by *Microcystis*.

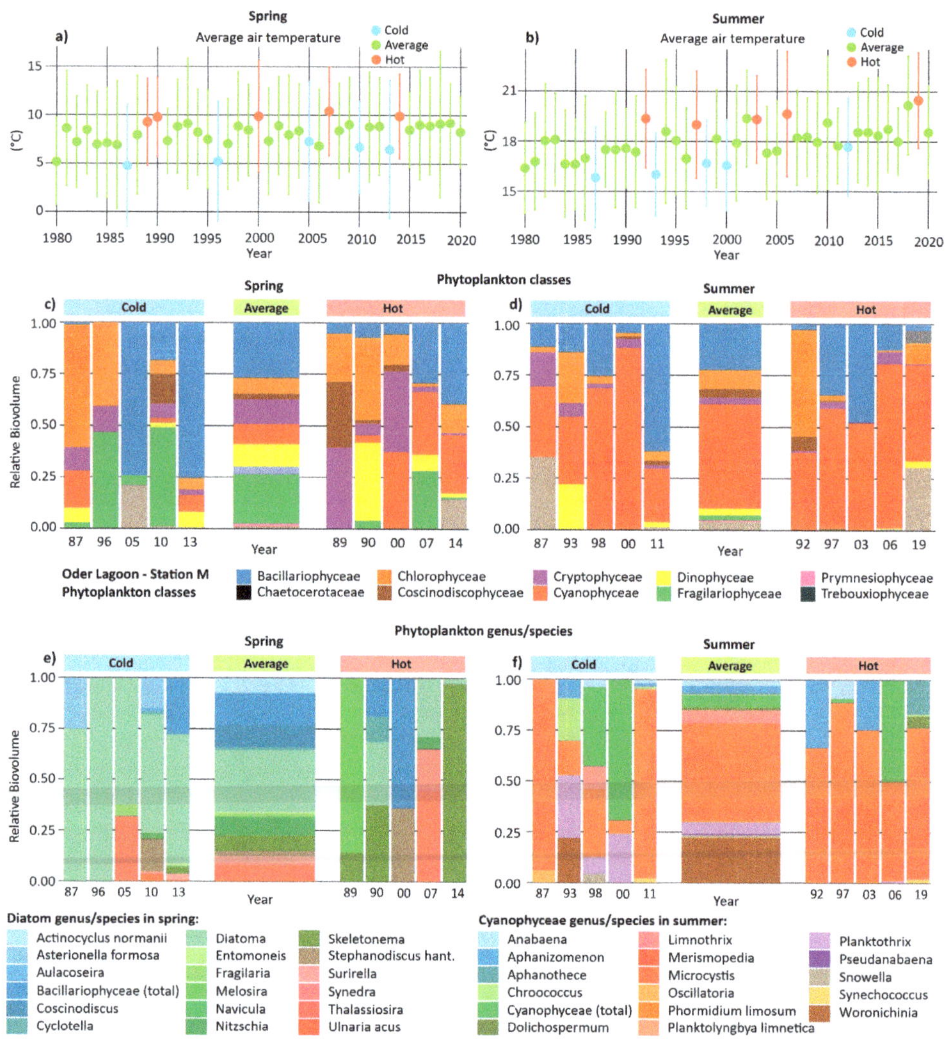

Figure 7. Average seasonal air temperatures in (**a**) spring and (**b**) summer at station M between 1980 and 2020. Hot and cold seasons are indicated. Relative biovolume of phytoplankton classes for selected years with a hot and a cold (**c**) spring and (**d**) summer. Relative biovolume of (**e**) diatom species and genus in spring and (**f**) cyanophyceae species in summer for the same selected years with hot and cold spring and summer seasons at Oder Lagoon station M.

3.6. Cyanobacteria and Diatoms During Hot and Cold Months

Because phytoplankton can rapidly adapt to changing environmental conditions, we compared hot and cold months between March and October (Figure 8a), focusing on cyanobacteria biomass (Figure 8b), diatom biomass (Figure 8d), and corresponding model results (Figure 8c,e). The model suggests that cyanobacteria benefit from hot months between June and October, developing higher biomasses than in normal or cold months, particularly in late summer and autumn. The average biomass difference between hot and cold months is around 10–20%. Observed biovolume data (Figure 8b) show a similar trend, but high variability and limited sample size reduce consistency. Nevertheless, the results indicate that between May and October, cyanobacteria benefit from higher temperatures, with maximum biomasses occurring in August and September.

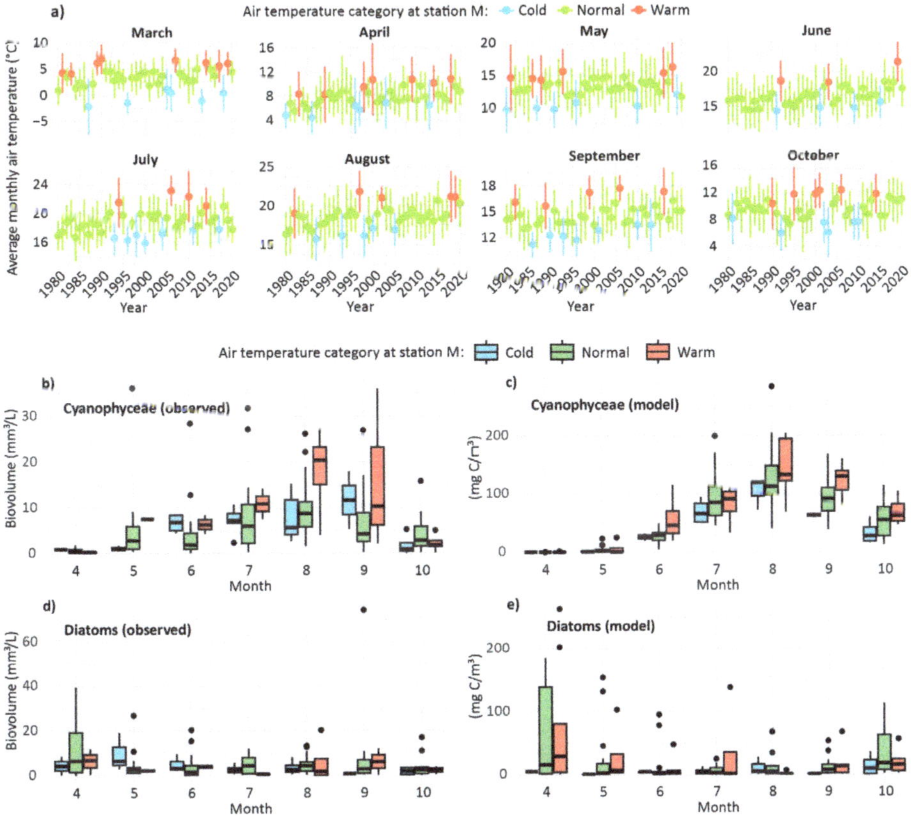

Figure 8. (**a**) Average monthly air temperature above the water surface at station M in the Oder Lagoon for the months March until October and the years between 2080 and 2020. The selected hot and cold months of the years are indicated by red resp. blue colors. Observed (**b**) and modeled (**c**) cyanobacteria biomass for the selected hot, normal/moderate and cold years. (**d**,**e**) show the same for diatoms. The box-and-whisker plots display the median, quartiles, and outliers (extremes) for every month.

For diatoms, the effects of hot months are less clear. While the model predicts higher diatom biomasses in hot spring months (April and May), this is not supported by the observational data. During summer, differences between hot and cold months are not well defined. It appears that diatoms may benefit from higher temperatures in spring, but temperature is not a major factor for their biomass in summer and autumn.

3.7. Phytoplankton During Heat and Cold Waves

We go further into detail and assess the effects of heatwaves and coldwaves, lasting several days to weeks, on cyanobacteria and diatoms. Five years with heatwaves in August (with exception of July 2006) and five years with coldwaves in May were selected (Figure 9a,b).

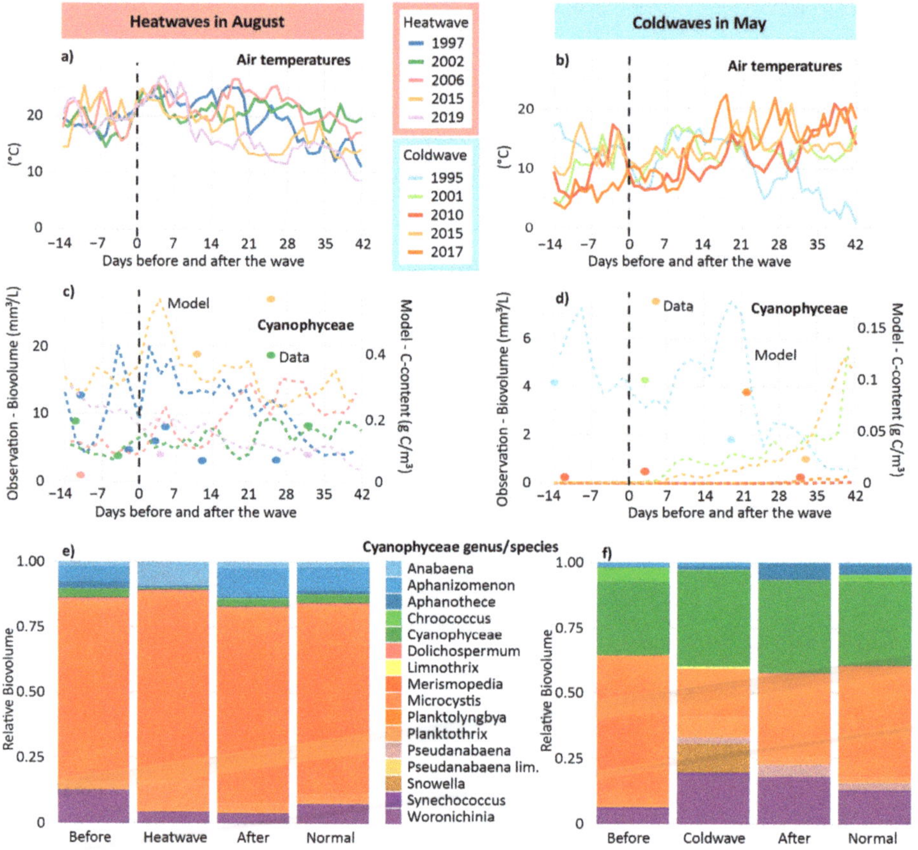

Figure 9. Air temperature above the water surface at station M in the Oder Lagoon, the weeks before, during and after the heatwaves in August (**a**) and coldwaves in May (**b**). Cyanophyceae biomass at station M in the Oder Lagoon the weeks before, during and after the heatwaves in August (**c**) and coldwaves in May (**d**). The data is indicated as dots and expressed as biovolume, and the model results are indicated as a dotted line and expressed in mg C/m^3. Cyanophyceae genus and species composition before, during and after the heatwave compared to the common "normal" situation during heatwaves in August (**e**) and coldwaves in May (**f**).

The model suggests that in most years after a heatwave in August, the cyanobacteria biomass increases, but usually decreases the week after again. Usually, coldwaves in May do not have an effect on cyanobacteria, because their biomass is very low at that time. Only in 1995, the model suggests a decline (Figure 9c,d). The available monitoring data is indicated, but not suitable for this analysis.

The cyanobacteria species composition before, during and after heatwaves in August does not show significant differences and is similar to the normal situation in August, with a strong dominance of *Microcystis*.

With respect to diatoms, the model suggests neither during heatwaves in August nor during coldwaves in May immediate and significant changes in biomass (Figure 10a,b). During the coldwave in May 2015 and 2017 declines in diatom biomass, but since it starts already before the coldwaves, it remains uncertain if this results from temperature changes.

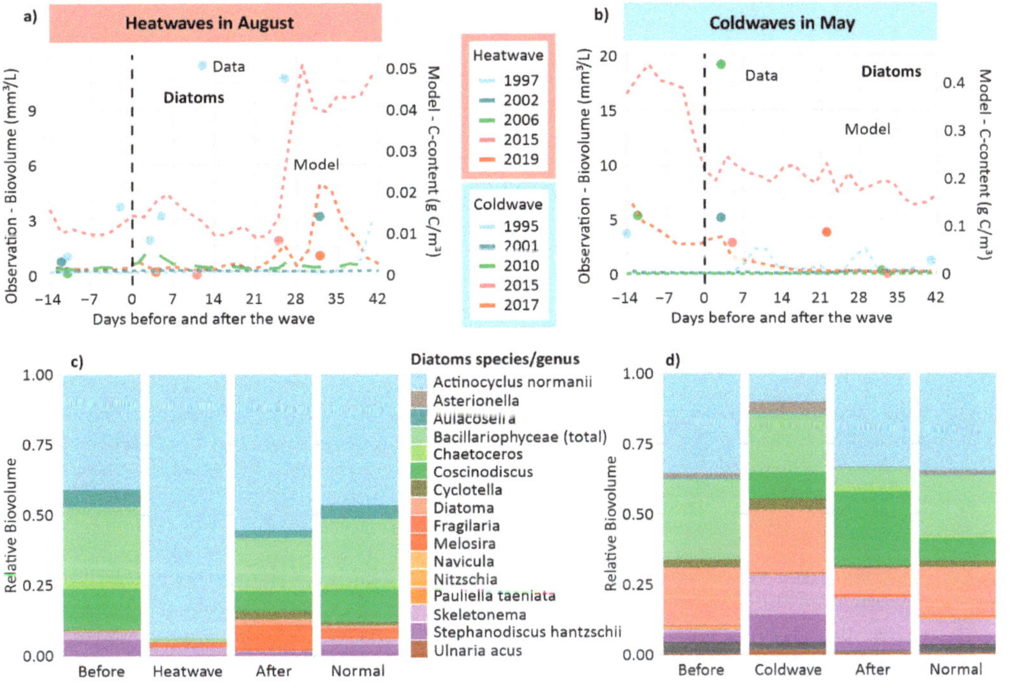

Figure 10. Diatom biomass at station M in the Oder Lagoon the weeks before, during and after the heatwaves in August (**a**) and coldwaves in May (**b**). The data is indicated as dots and expressed as biovolume, and the model results are indicated as a dotted line and expressed in mg C/m^3. Diatom genus and species composition before, during and after the heatwave compared to the common "normal" situation during heatwaves in August (**c**) and coldwaves in May (**d**).

The species composition before, during and after the heat and coldwaves are is not very different from the normal situation (Figure 10c,d). During heatwaves in August, the diatom species *Actinocyclus normanii* is dominant and contributes over 95% of the biovolume, but this might be an artefact resulting from the few available data.

3.8. Phytoplankton Model Simulations—Effects of Increased Temperature

To obtain a better insight into the role of temperature on phytoplankton biomass, especially in different seasons of the year, we carried out an additional model simulation study for the years 2010–2015, where the air temperatures were increased by 1.5 °C (Figure 11). The relative difference between the model run representing the present situation for the water surface layer compared to the run with increased temperatures shows that, on average, the monthly near-surface phytoplankton biomass over the 6 years is increased by 4% when the temperature is increased by 1.5 °C. However, the interannual variability is high, and the increase differs between the months of the year, from −3% in May to +32% in February. In June, August and Sept the increase is about 5–6% and in July 9%.

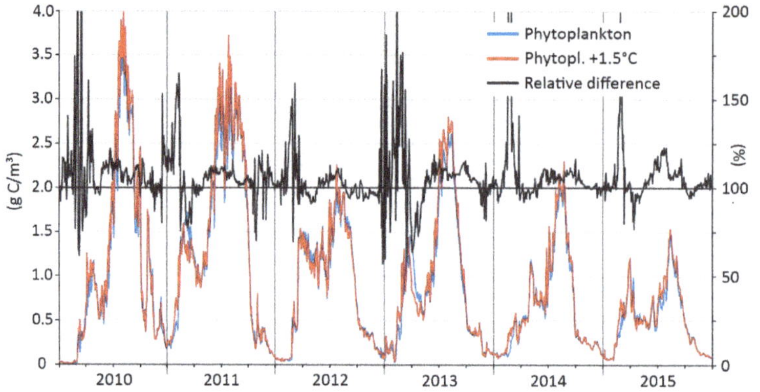

Figure 11. Daily model simulation results for the water surface layer at station M between the years 2010 and 2015. The blue line indicates the total phytoplankton concentration in g C/m^3. The red line shows similar results but assumes a 1.5 °C increase in air temperature. The black line shows the relative difference (%) between both simulations, where values above 100% indicate higher phytoplankton concentration at higher air temperatures.

Having a look at the absolute phytoplankton biomass increases, the average increase is highest in July and August (110 mg C/m^3), while the model suggests an absolute increase of 5 mg C/m^3 in February and 60 mg C/m^3 in March.

4. Discussion

4.1. Long-Term Development of the Western Oder Lagoon

Strong reductions in external nutrient loads over recent decades have led to a decline in nitrogen and phosphorus availability in the western lagoon. In recent years, dissolved inorganic phosphorus has shown very low concentrations in March and April, while dissolved inorganic nitrogen has been depleted in July and August. Despite these changes, the lagoon remains in a highly eutrophic state, with major indicators such as chlorophyll-a and water transparency (Secchi depth) showing no clear improvements. Nutrient limitation seems not pronounced, as rapid nutrient remobilization processes in this shallow, well-mixed system ensures sufficient supply. This is also supported by model simulations. Water transparency in spring and summer averages 0.5–0.7 m, indicating that light availability remains a key factor controlling primary production.

Results from other Baltic lagoons demonstrate that external nutrient inputs maintain eutrophic or hypertrophic conditions, leading to elevated average Chl-a concentrations and shifts in community composition toward summer cyanobacteria dominance [46–48]. Once a system becomes highly eutrophic, as in the Oder Lagoon, weak or delayed phytoplankton responses to nutrient load reductions are common [49], particularly in systems with long water residence times [50], where internal recycling and sediment–water exchange sustain nutrient availability [46].

4.2. Phytoplankton in the Oder Lagoon—A Comparison

Although detailed studies on phytoplankton succession in the Oder/Szczecin Lagoon are lacking, the observed annual succession resembles patterns in other shallow, eutrophic Baltic lagoons (Curonian, Vistula, and Darss-Zingst) [47,48,51]. Similarly to the Oder Lagoon, these systems show light-limited diatoms and cryptophytes with low biomass in winter, a diatom- and green algae-dominated spring bloom, cyanobacteria dominance in summer and autumn, and dinoflagellates and cryptophytes toward winter. Data from station M indicate a recent increase in the average annual contribution of diatoms, particularly in spring, along with a shift toward greater representation of Fragilariophyceae. The underlying causes remain unclear.

Phytoplankton composition in the Oder Lagoon shows strong interannual variability, a feature also reported for other coastal lagoons [51–53]. This variability is usually linked to environmental drivers such as hydrology, nutrient availability, temperature, and salinity [54,55]. In the Oder Lagoon, high interannual variability masks long-term trends resulting from nutrient reductions. The ~40 km distance between station M (central western bay) and the Oder River (the main nutrient source), together with the long average water residence time of ~100 days, reduces the impact of short-term changes in riverine nutrient loads. Moreover, the lagoon's shallowness and frequent mixing minimize the role of stratification, making temperature a likely key factor driving phytoplankton variability.

In the Oder Lagoon, warmer years are typically associated with higher cyanobacteria abundance and reduced diatom proportions, a trend also observed in other eutrophic Baltic lagoons, where hot summers extend the growing season and enhance cyanobacteria dominance [56,57].

However, nutrient load reductions in the Oder Lagoon have generally diminished the intensity of summer blooms and reduced the frequency of extreme events. In recent years, intense blooms have not been observed. Thus, although cyanobacteria may increase in relative importance under warmer conditions, their absolute biomass has declined due to reduced external nutrient inputs.

At station M, phosphorus availability in spring and early summer and nitrogen availability in summer have declined over recent decades, indicating an ongoing trend toward nutrient limitation. Many studies show that summer depletion of dissolved inorganic nitrogen promotes diazotrophic, N-fixing filamentous cyanobacteria (e.g., *Aphanizomenon, Dolichospermum, Nodularia*), leading to substantial N_2 fixation [57,58], which can partly compensate for reduced external nitrogen loads. This is well documented in the Baltic Sea [59] and has also been suggested for shallow coastal lagoons such as the Curonian Lagoon [60]. In the Oder Lagoon, *Microcystis*, a non-nitrogen-fixing cyanobacterium that relies on other available nitrogen sources, is still dominating. This is another hint that there is no lasting N-limitation. The model, too, suggests only a very minor N-fixation in the lagoon.

Uncertainty remains regarding how changes in cyanobacteria abundance, diversity, and regulation affect toxin production and potential health risks. Only strains carrying toxin biosynthesis genes (e.g., *mcy* for microcystins) produce toxins, and toxin presence is not directly linked to biomass. For example, blooms in the Curonian Lagoon often include *Microcystis, Planktothrix agardhii, Dolichospermum*, and *Anabaena* carrying *mcyE* genes, while abundant *Aphanizomenon flos-aquae* lacks them [56]. Nutrient stress may upregulate microcystin production [61], and both light intensity [62] and temperature [63] also contribute.

Temperature strongly influences lagoon phytoplankton dynamics. Larger diatom-dominated spring blooms often follow cold winters due to reduced grazing, while mild winters may weaken blooms and favor smaller taxa [51,56]. Rapid warming over days to weeks frequently initiates blooms in shallow lagoons because phytoplankton respond faster than grazers. Short warm spells can also shift community composition toward faster-growing taxa, including cyanobacteria [64].

Our results indicate that annual diatom biovolumes (Bacillariophyceae and Fragilariophyceae) account for about 40% of phytoplankton in cold years compared to only 18% in hot years. In cold springs, *Diatoma* constitutes about 75% of all diatoms, whereas hot springs show a shift toward other genera, such as *Skeletonema* and *Melosira*. Similar trends are documented in the Baltic Sea and lakes [65,66]. We are lacking sufficient phytoplankton data from winter, and shifts in species composition are not fully captured. The increased chlorophyll-concentrations in spring during the last decades and the shift in the spring maximum from April to March indicate a recently stronger dominance of diatoms, possibly resulting from increasing temperatures and the lack of ice cover. Model results, however, suggest that a 1.5 °C temperature increase alone does not substantially increase spring diatom biomass. Neither chlorophyll data nor the model indicate strong shifts in bloom timing toward January or February, supporting the view that in temperate lakes, bloom onset is determined by light availability, while temperature primarily controls growth rates [67,68].

4.3. Limitations of the Phytoplankton Data

Long-term phytoplankton datasets at the species level with monthly resolution over more than 40 years are rare, available only for a few lagoons such as the Darss-Zingst Bodden [51] and, to some extent, the Curonian Lagoon [54]. The dataset from station M (central Kleines Haff, 1 m depth) enables analyses of long-term and seasonal changes. However, the coarse sampling frequency and high variability limit reliable assessments on shorter timescales, such as individual months or extreme events (cold- and heatwaves). Our approach of aggregating data into hot and cold periods only partly compensates for this weakness.

The dataset captures interannual variability, but its spatial representativeness remains uncertain. Phytoplankton often exhibit vertical gradients, which can change daily. For example, cyanobacteria (*Microcystis, Planktothrix*) regulate buoyancy and migrate vertically on diel cycles [69,70], and similar behavior occurs in motile phytoflagellates and dinoflagellates [71]. Furthermore, single-site data may not capture spatial heterogeneity caused by wind, currents, mixing, and surface convergence [72], or biological interactions such as grazing [73]. Additional uncertainty arises from methodological aspects, including counting protocols, subsampling bias [74], reliance on taxonomic expertise, and phytoplankton taxonomy has undergone substantial revisions during this period, with many nomenclatural changes [75]. Methodological changes after 2008 further limit comparability between our earlier and later data.

4.4. The Model Approach—Opportunities and Weaknesses

Our model ERGOM distinguishes four phytoplankton functional groups, which broadly reproduce annual biomass dynamics. These groups capture key traits—morphological, physiological, or behavioral characteristics influencing fitness and ecosystem function [76,77]. However, the groups cannot be directly equated to taxonomic groups used in monitoring, as taxonomic categories often include diverse species with different seasonalities, behaviors, and ecological roles [76,78]. However, our results provide an opportunity for further development of the model.

Two model groups—diatoms and cyanobacteria—are reasonably comparable to taxonomic groups and can be used as proxies. A key advantage of complementing field data with model simulations is that the model provides a full three-dimensional, high-resolution spatio-temporal picture of the lagoon, enabling analyses where field data are lacking, revealing spatial gradients, and allowing critical evaluation of extreme field values.

A general challenge is the use of different biomass metrics, such as biovolume and chlorophyll-a. The relationship between these metrics is weak, as taxonomic groups (e.g., chlorophytes, diatoms, dinoflagellates, cyanobacteria) differ systematically in pigment content per unit volume or carbon, meaning identical biovolumes can yield very different Chl-a concentrations [79]. Surface scums of buoyant *Microcystis* colonies are also reflected differently in Chl-a versus biovolume [80].

ERGOM adds another measure, expressing biomass as carbon content. However, conversion between carbon, Chl-a, and biovolume depends on environmental conditions, season, and species [81]. Commonly used constant conversion factors oversimplify these relationships. To avoid misrepresentation, we refrained from recalculating and instead displayed two independent axes (Figure 4). However, the implied conversion factor from $g\,C/m^3$ to $\mu g/L$ Chlorophyll-a is 40 and well in agreement with literature [81].

4.5. Towards an Improved Monitoring of Phytoplankton

The European Water Framework Directive (2000/60/EC) requires all EU surface waters, including transitional and coastal waters, to achieve "good ecological status". Phytoplankton is a key biological quality element, requiring monitoring of biomass (Chl-a), abundance, and community composition (species or functional groups). The challenge lies in addressing high temporal and spatial variability without excessive costs. Beyond taxonomic identification, trait-based classification would add value.

A more flexible sampling design is also desirable. During extreme events (heatwaves, floods, droughts), temporarily increasing sampling frequency and coverage (both horizontal and vertical) could provide valuable insights.

Ideally, daily data with 1 m vertical resolution and <1 km horizontal resolution would capture spatio-temporal phytoplankton dynamics in shallow systems like the Oder Lagoon. Since horizontal distribution is strongly driven by physical conditions, 3D models can reveal spatial patterns and help assess how representative single-station data are.

Remote sensing can complement in situ data. Multispectral satellite imagery with spectral indices and pigment-specific algorithms, supplemented by airborne hyperspectral or UAV-based imaging, could map surface phytoplankton patterns.

Moored profiling systems at station M, equipped with optical sensors (fluorometers, backscatter, conductivity, temperature, depth (CTD), photosynthetically active radiation), or more affordable multi-depth sensor strings, could provide high-frequency temporal resolution. Additional instruments for imaging or cytometric analysis of individual phytoplankton cells, onboard classification algorithms, and molecular sensors for species and toxin detection would further enhance monitoring.

Current phytoplankton identification relies on traditional light microscopy. New molecular methods, such as DNA metabarcoding, would provide more robust and accurate taxonomic characterization of algal communities [82,83].

Since Baltic lagoons share similar phytoplankton dynamics, insights from one lagoon can largely be transferred to others.

Supplementary Materials: The following supporting information can be downloaded at: https://www.mdpi.com/article/10.3390/environments12100370/s1, Table S1 contains the full phytoplankton species data for station M/KHM in the central German part of the Oder/Szczecin Lagoon (53.825° N, 14.100° E) covering the years 1972 until 2022.

Author Contributions: Conceptualization, G.S. and M.S.; methodology and software, G.S., M.S. and T.N.; formal analysis and investigation, G.S., M.S. and T.N.; data curation, M.v.W.; writing—original draft preparation, G.S.; writing—review and editing, M.S., T.N. and M.v.W.; supervision and funding acquisition, G.S. All authors have read and agreed to the published version of the manuscript.

Funding: This work was financially supported by the German Federal Ministry of Research, Technology and Space (BMFBR), projects "DAM Schutz und Nutzen: Coastal Futures II" (grant number 03F0980B), "DAM Extrem: Prime Prevention" (grant number 03F0953D) and UBA-MoSea (FKZ 3723252040).

Data Availability Statement: The original phytoplankton dataset presented in this study is included in the Supplementary Material (Table S1). The most recent field data can be obtained from

Landesamt für Umwelt, Naturschutz und Geologie, Güstrow, Germany. The bio-chemical data of station M can additionally be downloaded from the UBA-MUDAB (https://geoportal.bafg.de/MUDABAnwendung/). The full model documentation is available under: https://doi.org/10.5281/zenodo.7252134. The model simulation output represents an intermediate state, and improvements are still taking place. Consequently, this data is not permanently available yet.

Acknowledgments: We thank Sarah Piehl and Rene Friedland for technical support.

Conflicts of Interest: The authors declare no conflicts of interest.

References

1. Meier, H.E.M.; Kniebusch, M.; Dieterich, C.; Gröger, M.; Zorita, E.; Elmgren, R.; Myrberg, K.; Ahola, M.P.; Bartosova, A.; Bonsdorff, E.; et al. Climate change in the Baltic Sea region: A summary. *Earth Syst. Dyn.* **2022**, *13*, 457–593. [CrossRef]
2. Rutgersson, A.; Jaagus, J.; Schenk, F.; Stendel, M. Observed changes and variability of atmospheric parameters in the Baltic Sea region during the last 200 years. *Clim. Res.* **2014**, *61*, 177–190. [CrossRef]
3. Kniebusch, M.; Meier, H.E.M.; Neumann, T.; Börgel, F. Temperature variability of the Baltic Sea since 1850 and attribution to atmospheric forcing variables. *J. Geophys. Res. Oceans* **2019**, *124*, 4168–4187. [CrossRef]
4. Dutheil, C.; Meier, H.E.M.; Gröger, M.; Börgel, F. Understanding past and future sea surface temperature trends in the Baltic Sea. *Clim. Dyn.* **2021**, *58*, 3021–3039. [CrossRef]
5. Siegel, H.; Gerth, M. Sea Surface Temperature in the Baltic Sea 2018. HELCOM Baltic Sea Environment Fact Sheets 2019; 7p. Available online: https://helcom.fi/wp-content/uploads/2020/07/BSEFS-Sea-Surface-Temperature-in-the-Baltic-Sea-2018.pdf (accessed on 17 February 2022).
6. Stramska, M.; Białogrodzka, J. Spatial and temporal variability of sea surface temperature in the Baltic Sea based on 32 years (1982–2013) of satellite data. *Oceanologia* **2015**, *57*, 223–235. [CrossRef]
7. Belkin, I.M. Rapid warming of large marine ecosystems. *Prog. Oceanogr.* **2009**, *81*, 207–213. [CrossRef]
8. Meier, H.E.M.; Dieterich, C.; Gröger, M.; Dutheil, C.; Börgel, F.; Safonova, K.; Christensen, O.B.; Kjellström, E. Oceanographic regional climate projections for the Baltic Sea until 2100. *Earth Syst. Dyn.* **2022**, *13*, 159–199. [CrossRef]
9. Gröger, M.; Arneborg, L.; Dieterich, C.; Höglund, A.; Meier, H.E.M. Summer hydrographic changes in the Baltic Sea, Kattegat and Skagerrak projected in an ensemble of climate scenarios downscaled with a coupled regional ocean–sea ice–atmosphere model. *Clim. Dyn.* **2019**, *53*, 5945–5966. [CrossRef]
10. Gröger, M.; Dieterich, C.; Meier, H.E.M. Is interactive air–sea coupling relevant for simulating the future climate of Europe? *Clim. Dyn.* **2021**, *56*, 491–514. [CrossRef]
11. Viitasalo, M.; Bonsdorff, E. Global climate change and the Baltic Sea ecosystem: Direct and indirect effects on species, communities and ecosystem function. *Earth Syst. Dyn.* **2022**, *13*, 711. [CrossRef]
12. Naselli-Flores, L.; Padisák, J. Ecosystem services provided by marine and freshwater phytoplankton. *Hydrobiologia* **2023**, *850*, 2691–2706. [CrossRef]
13. Hochfeld, I.; Hinners, J. Phytoplankton adaptation to steady or changing environments affects marine ecosystem functioning. *Biogeosciences* **2024**, *21*, 5591–5611. [CrossRef]
14. Hayes, N.M.; Haig, H.A.; Simpson, G.L.; Leavitt, P.R. Effects of lake warming on the seasonal risk of toxic cyanobacteria exposure. *Limnol. Oceanogr. Lett.* **2020**, *5*, 10164. [CrossRef]
15. Richardson, R.; Feuchtmayr, H.; Miller, C.; Hunter, P.D.; Maberly, S.C.; Carvalho, L. Response of cyanobacteria and phytoplankton abundance to warming, extreme rainfall events and nutrient enrichment. *Glob. Change Biol.* **2020**, *26*, e14701. [CrossRef] [PubMed]
16. Chatterjee, S.; More, M. Cyanobacterial harmful algal bloom toxin microcystin and increased *Vibrio* occurrence as climate-change-induced biological co-stressors: Exposure and disease outcomes via their interaction with gut–liver–brain axis. *Toxins* **2023**, *15*, 289. [CrossRef] [PubMed]
17. Brito, A.C.; Newton, A.; Tett, P.; Fernandes, T.F. How will shallow coastal lagoons respond to climate change? A modelling investigation. *Estuar. Coast. Shelf Sci.* **2011**, *95*, 98–104. [CrossRef]
18. Hobday, A.J.; Alexander, L.V.; Perkins, S.E.; Smale, D.A.; Straub, S.C.; Oliver, E.C.J.; Benthuysen, J.A.; Burrows, M.T.; Donat, M.G.; Feng, M.; et al. Categorizing and naming marine heatwaves. *Oceanography* **2018**, *31*, 162–173. [CrossRef]
19. Galli, G.; Solidoro, C.; Lovato, T. Marine heat waves hazard 3D maps and the risk for low motility organisms in a warming Mediterranean Sea. *Front. Mar. Sci.* **2017**, *4*, 136. [CrossRef]
20. Woolway, R.I.; Jennings, E.; Shatwell, T.; Golub, M.; Pierson, D.C.; Maberly, S.C. Lake heatwaves under climate change. *Nature* **2021**, *589*, 402–407. [CrossRef]
21. Benthuysen, J.A.; Oliver, E.C.J.; Chen, K.; Wernberg, T. Editorial: Advances in Understanding Marine Heatwaves. *Front. Mar. Sci.* **2020**, *7*, 147. [CrossRef]

22. Oliver, E.C.J.; Burrows, M.T.; Donat, M.G.; Sen Gupta, A.; Alexander, L.V.; Perkins-Kirkpatrick, S.E.; Benthuysen, J.A.; Hobday, A.J.; Holbrook, N.J.; Moore, P.J.; et al. Marine heatwaves. *Annu. Rev. Mar. Sci.* **2021**, *13*, 313–342. [CrossRef] [PubMed]
23. Safonova, K.; Meier, H.E.M.; Gröger, M. Summer heatwaves on the Baltic Sea seabed contribute to oxygen deficiency in shallow areas. *Commun. Earth Environ.* **2024**, *5*, 106. [CrossRef]
24. Humborg, C.; Geibel, M.C.; Sun, X.; McCrackin, M.; Mörth, C.-M.; Stranne, C.; Jakobsson, M.; Gustafsson, B.; Sokolov, A.; Norkko, A.; et al. High emissions of carbon dioxide and methane from the coastal Baltic Sea at the end of a summer heat wave. *Front. Mar. Sci.* **2019**, *6*, 493. [CrossRef]
25. Bergkemper, V.; Weisse, T. Phytoplankton response to the summer 2015 heat wave—A case study from pre-alpine Lake Mondsee, Austria. *Inland Waters* **2017**, *7*, 88–99. [CrossRef]
26. Li, N.; Zhang, Y.; Zhang, Y.; Shi, K.; Qian, H.; Yang, H.; Niu, Y.; Qin, B.; Zhu, G.; Woolway, R.I.; et al. The unprecedented 2022 extreme summer heatwaves increased harmful cyanobacteria blooms. *Sci. Total Environ.* **2023**, *896*, 165312. [CrossRef]
27. Jeppesen, E.; Pierson, D.; Jennings, E. Effect of extreme climate events on lake ecosystems. *Water* **2021**, *13*, 282. [CrossRef]
28. Soulié, T.; Vidussi, F.; Mas, S.; Mostajir, B. Functional and structural responses of plankton communities toward consecutive experimental heatwaves in Mediterranean coastal waters. *Sci. Rep.* **2023**, *13*, 8050. [CrossRef]
29. Chen, W.; Nielsen, A.; Andersen, T.K.; Hu, F.; Chou, Q.; Søndergaard, M.; Jeppesen, E.; Trolle, D. Modeling the ecological response of a temporarily summer-stratified lake to extreme heatwaves. *Water* **2020**, *12*, 94. [CrossRef]
30. Rutgersson, A.; Kjellström, E.; Haapala, J.; Stendel, M.; Danilovich, I.; Drews, M.; Jylhä, K.; Kujala, P.; Larsén, X.G.; Halsnæs, K.; et al. Natural hazards and extreme events in the Baltic Sea region. *Earth Syst. Dyn.* **2022**, *13*, 251. [CrossRef]
31. Aoki, L.R.; Brisbin, M.M.; Hounshell, A.G.; Kincaid, D.W.; Larson, E.I.; Sansom, B.J.; Shogren, A.J.; Smith, R.S.; Sullivan-Stack, J. Preparing aquatic research for an extreme future: Call for improved definitions and responsive, multidisciplinary approaches. *BioScience* **2022**, *72*, 508–520. [CrossRef]
32. Diamond, D.; Relyea, R.A.; McCaul, M. Understanding and mitigating global change with aquatic autonomous sensors: A fifty-year perspective. *Front. Sens.* **2023**, *4*, 1284043. [CrossRef]
33. LUNG. *Monitoringprogramm zur Überwachung der Oberflächengewässer und des Grundwassers in Mecklenburg-Vorpommern im Zeitraum 2022–2027*; Landesamt für Umwelt, Naturschutz und Geologie Mecklenburg-Vorpommern: Güstrow, Germany, 2024; Available online: https://www.lung.mv-regierung.de/static/LUNG/dateien/fachinformationen/wasser/Gewaesserueberwachung/monitoringprogramm_2022-2027.pdf (accessed on 15 April 2025).
34. Filiz, N.; Işkın, U.; Beklioğlu, M.; Öğlü, B.; Cao, Y.; Davidson, T.A.; Søndergaard, M.; Lauridsen, T.L.; Jeppesen, E. Phytoplankton community response to nutrients, temperatures, and a heat wave in shallow lakes: An experimental approach. *Water* **2020**, *12*, 3394. [CrossRef]
35. Radziejewska, T.; Schernewski, G. The Szczecin (Oder-) Lagoon. In *Ecology of Baltic Coastal Waters*; Schiewer, U., Ed.; Springer: Berlin/Heidelberg, Germany, 2008; pp. 115–129. [CrossRef]
36. Neumann, T.; Schernewski, G.; Friedland, R. Transformation Processes in the Oder Lagoon as Seen from a Model Perspective. *EGUsphere* **2025**. [CrossRef]
37. HELCOM. *State of the Baltic Sea. Third HELCOM Holistic Assessment 2016–2021*; Baltic Sea Environment Proceedings No. 194; HELCOM: Helsinki, Finland, 2023; Available online: https://helcom.fi/wp-content/uploads/2023/10/State-of-the-Baltic-Sea-2023.pdf (accessed on 15 April 2025).
38. IKSO. *Zweite Aktualisierung des Bewirtschaftungsplans für die Internationale Flussgebietseinheit*; IKSO: Wrocław, Poland, 2022; Available online: http://www.mkoo.pl/download.php?fid=6996&lang=DE (accessed on 15 April 2025).
39. Bundesregierung. *Verordnung zum Schutz der Oberflächengewässer (Oberflächengewässerverordnung—OGewV)*; Federal Ministry of Justice and Consumer Protection: Berlin, Germany, 2016. Available online: https://www.gesetze-im-internet.de/ogewv_2016/OGewV.pdf (accessed on 15 April 2025).
40. Grenzgewässerkommission. *Bericht über die Beschaffenheit der Deutsch–Polnischen Grenzgewässer 2022*; Wasser Blick: Loddin, Germany, 2024; Available online: https://www.wasserblick.net/servlet/is/1/ (accessed on 15 April 2025).
41. Neumann, T.; Radtke, H.; Cahill, B.; Schmidt, M.; Rehder, G. Non-Redfieldian Carbon Model for the Baltic Sea (ERGOM Version 1.2)—Implementation and Budget Estimates. *Geosci. Model Dev.* **2022**, *15*, 8473–8540. [CrossRef]
42. European Commission. *Marine Biodiversity Modelling Study*; Publications Office of the European Union: Luxembourg, 2022; Available online: https://data.europa.eu/doi/10.2777/213731 (accessed on 15 April 2025).
43. Neumann, T. *Model Code and Boundary Data for "Non-Redfield Carbon Model for the Baltic Sea (ERGOM Version 1.2)—Implementation and Budget Estimates" [Code]*; Zenodo: Geneva, Switzerland, 2022. [CrossRef]
44. Geyer, B.; Rockel, B. *coastDat-2 COSMO-CLM Atmospheric Reconstruction [Data Set]*; World Data Center for Climate (WDCC): Hamburg, Germany, 2013. [CrossRef]
45. Friedland, R.; Schernewski, G.; Gräwe, U.; Greipsland, I.; Palazzo, D.; Pastuszak, M. Managing Eutrophication in the Szczecin (Oder) Lagoon—Development, Present State and Future Perspectives. *Front. Mar. Sci.* **2019**, *5*, 521. [CrossRef]

46. Bartoli, M.; Žilius, M.; Ruginis, T.; Vybernaite-Lubiene, I.; Petkuvienė, J.; Daunys, D. Seasonal Nitrogen Cycling Fuels Cyanobacteria Blooms in a Hypertrophic Lagoon. *Front. Environ. Sci.* **2024**, *12*, 1384114. [CrossRef]
47. Pilkaitytė, R.; Razinkovas, A. Seasonal Changes in Phytoplankton Composition and Nutrient Limitation in a Shallow Baltic Lagoon. *Boreal Environ. Res.* **2007**, *12*, 551–559.
48. Kownacka, J.; Krzymińska, J.; Jasser, I.; Grabowska, M.; Hutorowicz, A. Phytoplankton Structure in the Vistula Lagoon (Southern Baltic Sea) and the Turning Point in Its Development. *Oceanologia* **2020**, *62*, 313–324. [CrossRef]
49. Saraiva, S.; Markus Meier, H.E.; Andersson, H.; Höglund, A.; Dieterich, C.; Gröger, M.; Hordoir, R.; Eilola, K. Baltic Sea ecosystem response to various nutrient load scenarios in present and future climates. *Clim. Dyn.* **2019**, *52*, 3369–3387. [CrossRef]
50. Cloern, J.E.; Foster, S.Q.; Kleckner, A.E. Phytoplankton Primary Production in the World's Estuarine–Coastal Ecosystems. *Biogeosciences* **2014**, *11*, 2477–2501. [CrossRef]
51. Schumann, R.; Schiewer, U.; Schubert, H. Phytoplankton Composition and Production during the Winter–Spring Transition in the Darss–Zingst Bodden Chain, Southern Baltic Sea. *Estuar. Coast. Shelf Sci.* **2005**, *62*, 169–181. [CrossRef]
52. Schiewer, U. Darß-Zingst Boddens, Northern Rügener Boddens and Schlei. In *Ecology of Baltic Coastal Waters*; Schiewer, U., Ed.; Springer: Berlin/Heidelberg, Germany, 2008; pp. 35–86. [CrossRef]
53. Poppeschi, C.; Charria, G.; Daniel, A.; Verney, R.; Rimmelin-Maury, P.; Retho, M.; Goberville, E.; Grossteffan, E.; Plus, M. Interannual variability of the initiation of the phytoplankton growing period in two French coastal ecosystems. *Biogeosciences* **2022**, *19*, 5667–5687. [CrossRef]
54. Gasiūnaitė, Z.R.; Daunys, D.; Olenin, S.; Razinkovas, A. The Curonian Lagoon. In *Ecology of Baltic Coastal Waters*; Schiewer, U., Ed.; Springer: Berlin/Heidelberg, Germany, 2008; pp. 197–216. [CrossRef]
55. Wasmund, N.; Tuimala, J.; Suikkanen, S.; Vandepitte, L.; Kraberg, A. Long-term trends in phytoplankton composition in the western and central Baltic Sea. *J. Mar. Syst.* **2011**, *87*, 145–159. [CrossRef]
56. Overlingė, D.; Toruńska-Sitarz, A.; Cegłowska, M.; Błaszczyk, A.; Szubert, K.; Pilkaitytė, R.; Mazur-Marzec, H. Phytoplankton of the Curonian Lagoon as a New Interesting Source for Bioactive Natural Products—Special Impact on Cyanobacterial Metabolites. *Biomolecules* **2021**, *11*, 1139. [CrossRef]
57. Vahtera, E.; Conley, D.J.; Gustafsson, B.G.; Kuosa, H.; Pitkänen, H.; Savchuk, O.P.; Tamminen, T.; Viitasalo, M.; Voss, M.; Wasmund, N.; et al. Internal Ecosystem Feedbacks Enhance Nitrogen-Fixing Cyanobacteria Blooms and Complicate Management in the Baltic Sea. *AMBIO* **2007**, *36*, 186–194. [CrossRef]
58. Karlson, A.M.L.; Duberg, J.; Motwani, N.H.; Hogfors, H.; Klawonn, I.; Ploug, H.; Svedén, J.B.; Garbaras, A.; Sundelin, B.; Hajdu, S.; et al. Nitrogen Fixation by Cyanobacteria Stimulates Production in Baltic Food Webs. *AMBIO* **2015**, *44*, 413–426. [CrossRef]
59. Olofsson, M.; Klawonn, I.; Karlson, B. Nitrogen Fixation Estimates for the Baltic Sea Indicate High Rates for the Previously Overlooked Bothnian Sea. *AMBIO* **2021**, *50*, 203–214. [CrossRef]
60. Žilius, M.; Vybernaite-Lubiene, I.; Vaiciute, D.; Overlingė, D.; Grinienė, E.; Zaiko, A.; Bonaglia, S.; Liskow, I.; Voss, M.; Andersson, A.; et al. Spatiotemporal Patterns of N_2 Fixation in Coastal Waters Derived from Rate Measurements and Remote Sensing. *Biogeosciences* **2021**, *18*, 1857–1871. [CrossRef]
61. Pimentel, J.S.M.; Giani, A. Microcystin Production and Regulation under Nutrient Stress in Cyanobacteria. *Appl. Environ. Microbiol.* **2014**, *80*, 5836–5843. [CrossRef]
62. Tonk, L.; Visser, P.M.; Christiansen, G.; Dittmann, E.; Snelder, E.O.; Wiedner, C.; Mur, L.R.; Huisman, J. The Microcystin Composition of *Planktothrix agardhii* Changes with Light Intensity. *Appl. Environ. Microbiol.* **2005**, *71*, 553–560. [CrossRef]
63. Silveira, S.B.; Odebrecht, C. Effects of Salinity and Temperature on Growth and Nodularin Production in *Nodularia spumigena*. *Front. Mar. Sci.* **2019**, *6*, 339. [CrossRef]
64. Trombetta, T.; Vidussi, F.; Mas, S.; Parin, D.; Simier, M.; Mostajir, B. Water Temperature Drives Phytoplankton Blooms in Coastal Waters. *PLoS ONE* **2019**, *14*, e0214933. [CrossRef]
65. Spilling, K.; Olli, K.; Lehtoranta, J.; Kremp, A.; Tamelander, T.; HELCOM Phytoplankton Group. Shifting Diatom–Dinoflagellate Dominance During Spring Bloom in the Baltic Sea. *Front. Mar. Sci.* **2018**, *5*, 327. [CrossRef]
66. Almén, A.K.; Tiselius, P.; Rydberg, L. Temperature-Related Timing of the Spring Bloom: Contrasts between Cold and Warm Years. *J. Sea Res.* **2020**, *160*, 101887. [CrossRef]
67. Reynolds, C.S. *The Ecology of Phytoplankton*; Ecological Studies 535; Springer: Dordrecht, The Netherlands, 2006. [CrossRef]
68. Sommer, U.; Gliwicz, Z.M.; Lampert, W.; Duncan, A. The PEG-Model of Seasonal Succession of Planktonic Events in Fresh Waters. *Arch. Hydrobiol.* **1986**, *106*, 433–471. [CrossRef]
69. Hašler, P.; Poulíčková, A. Diurnal Changes in Vertical Distribution and Morphology of a Natural Population of *Planktothrix agardhii* (Gom.) Anagnostidis et Komárek (Cyanobacteria). *Hydrobiologia* **2003**, *506*, 195–201. [CrossRef]
70. Ibelings, B.W.; Mur, L.R.; Walsby, A.E. Diurnal Changes in Buoyancy and Vertical Distribution in Populations of *Microcystis* in Two Shallow Lakes. *J. Plankton Res.* **1991**, *13*, 419–436. [CrossRef]

71. Hall, N.S.; Whipple, A.C.; Paerl, H.W. Vertical Spatio-Temporal Patterns of Phytoplankton Due to Migration Behaviors in Two Shallow, Microtidal Estuaries: Influence on Phytoplankton Function and Structure. *Estuar. Coast. Shelf Sci.* **2015**, *162*, 7–21. [CrossRef]
72. Cloern, J.E. Phytoplankton Bloom Dynamics in Coastal Ecosystems: A Review with Some General Lessons from Sustained Investigation of San Francisco Bay. *Rev. Geophys.* **1996**, *34*, 127–168. [CrossRef]
73. Borics, G.; Abonyi, A.; Krasznai, E.; Várbíró, G.; Grigorszky, I.; Szabó, S.; Deák, C.; Tóthmérész, B. Small-Scale Patchiness of the Phytoplankton in a Lentic Oxbow. *J. Plankton Res.* **2011**, *33*, 973–981. [CrossRef]
74. Jacobs, P.; Serre-Fredj, L.; Koeman, R.; van den Oever, A.; Peck, M.A.; Philippart, C.J.M. Impacts of Counting Protocols for Light Microscopy on Estimates of Biodiversity and Algal Density of Phytoplankton. *Limnol. Oceanogr. Methods* **2024**, *22*, 930–949. [CrossRef]
75. Cozzoli, F.; Stanca, E.; Selmeczy, G.B.; Francé, J.; Varkitzi, I.; Basset, A. Sensitivity of Phytoplankton Metrics to Sample-Size: A Case Study on a Large Transitional Water Dataset (WISER). *Ecol. Indic.* **2017**, *82*, 558–573. [CrossRef]
76. Litchman, E.; Klausmeier, C.A. Trait-Based Community Ecology of Phytoplankton. *Annu. Rev. Ecol. Evol. Syst.* **2008**, *39*, 615–639. [CrossRef]
77. Weithoff, G.; Beisner, B.E.; von Elert, E.; Reynolds, C.S.; Spaak, P. Measures and Approaches in Trait-Based Phytoplankton Community Ecology—From Freshwater to Marine Ecosystems. *Front. Mar. Sci.* **2019**, *6*, 40. [CrossRef]
78. Kruk, C.; Huszar, V.L.M.; Peeters, E.T.H.M.; Bonilla, S.; Costa, L.; Lürling, M.; Reynolds, C.S.; Scheffer, M. A Morphological Classification Capturing Functional Variation in Phytoplankton. *Freshw. Biol.* **2010**, *55*, 614–627. [CrossRef]
79. Felip, M.; Catalan, J. The Relationship Between Phytoplankton Biovolume and Chlorophyll in a Deep Oligotrophic Lake: Decoupling in Their Spatial and Temporal Maxima. *J. Plankton Res.* **2000**, *22*, 91–105. [CrossRef]
80. Gilabert, J.; Socorro, A.; Ferriol, C.; Sanpera, C.; Abad, M. Short-Term Variability of the Planktonic Size Structure in a Mediterranean Hypersaline Coastal Lagoon (Mar Menor). *J. Plankton Res.* **2001**, *23*, 219–226. [CrossRef]
81. Gui, J. Phytoplankton Carbon to Chlorophyll a Model Development: A Review and Perspective. *Front. Mar. Sci.* **2024**, *11*, 1466072. [CrossRef]
82. Baylón, M.; Ramirez, J.L. A Comparison of eDNA Metabarcoding and Microscopy Techniques to Analyze Algal Diversity in Lake Titicaca, Peru. *Diversity* **2025**, *17*, 560. [CrossRef]
83. Wang, R.; Dong, Z.; Lin, K.; Lv, S.; Liu, Y.; Lu, W.; Liu, L. Comparative Analysis of Light Microscopy and High-Throughput Sequencing for Phytoplankton Detection in Rivers Flowing into the Sea. *Water* **2025**, *17*, 1559. [CrossRef]

environments

MDPI

Article

Ecosystem-Model-Based Valuation of Ecosystem Services in a Baltic Lagoon: Long-Term Human Technical Interventions and Short-Term Variability

Gerald Schernewski [1,2,*], Thomas Neumann [3], Sarah Piehl [1] and Nicole M. Swer [1]

[1] Research Unit—Coastal Sea Geography, Leibniz-Institute for Baltic Sea Research, Seestraße 15, D-18119 Rostock-Warnemünde, Germany; sarah.piel@io-warnemuende.de (S.P.); nicoleswer01@gmail.com (N.M.S.)

[2] Marine Research Institute, Klaipeda University, Universiteto Ave. 17, LT-92294 Klaipeda, Lithuania

[3] Department of Physical Oceanography and Instrumentation, Leibniz-Institute for Baltic Sea Research, Seestraße 15, D-18119 Rostock-Warnemünde, Germany; thomas.neumann@io-warnemuende.de

* Correspondence: gerald.schernewski@io-warnemuende.de; Tel.: +49-381-5197207

Abstract: We conducted 3D ecosystem model simulations over a 10-year period, supplemented by socio-economic data, to evaluate the ecosystem services provided by the large, shallow Oder/Szczecin Lagoon. Our analysis focused on three scenarios reflecting the progressive deepening of the navigational waterway across the lagoon: from 6 m (1880) to 10.5 m (1984) and finally to 12.5 m (2023). For the 10.5 m scenario, the total value of all six ecosystem services was estimated at EUR 272 million/year, or approximately EUR 0.4 million/year/km^2. The individual contributions of each ecosystem service were as follows: nitrogen retention, EUR 166 million/a; phosphorus retention, EUR 5 million/a; carbon storage, EUR 0.4 million/a; active recreation, EUR 61 million/a; landscape aesthetics, EUR 36 million/a; wild fish catches, EUR 3.2 million/a; and transportation, EUR 32 million/a. Among these, denitrification emerged as the most economically important process, valued at EUR 178 million/year, or EUR 0.26 million/year/km^2. Regulating ecosystem services displayed substantial interannual variability and pronounced seasonality. Additionally, the two parts of the lagoon, Kleines Haff (Germany) and Wielki Zalew (Poland), exhibited distinct patterns. Our model indicates that channel deepening enhances sediment burial and significantly increases phosphorus and carbon retention. However, the associated increase in connectivity to the Baltic Sea appears to have a minor effect.

Keywords: Szczecin Lagoon; monetary valuation; coastal recreation; carbon storage; nutrient retention; cultural services; waterway deepening; socio-economic data; transportation; fisheries

1. Introduction

The Baltic Sea region is home to several important lagoons, including the Curonian, Vistula, and Szczecin/Oder Lagoons, which are some of the largest in Europe. For centuries, these lagoons have been focal points of human activity, particularly fishing, which has played a crucial economic and social role for the local residents [1,2]. In recent decades, however, there have been substantial transformations. Recreational activities and tourism have become the primary sources of income [3]. Additionally, there is growing recognition that these lagoons are an important part of Europe's natural heritage, providing essential ecological functions such as serving as resting places for migratory birds [4]. This awareness has led to extensive protection measures. Despite these efforts, the lagoons continue

https://doi.org/10.3390/environments12020035

to suffer from poor ecological conditions, including inadequate water quality and low transparency, which hinder nature protection as well as tourism development [5,6].

Another significant factor impacting the Baltic lagoons is the ongoing expansion of maritime transport. Over the past centuries, navigational waterways have been established and progressively deepened to accommodate larger ships [7]. This has had significant economic implications, increasing ship traffic, trade, and harbor development. However, artificial waterways also affect ecological processes in the lagoons. These effects include changes in water flow patterns and velocities, the distribution of riverine nutrients, and the alteration of water and matter exchange between the lagoons and the Baltic Sea [8,9].

Ongoing anthropogenic interventions and infrastructural measures, such as the deepening of navigational waterways, interact with other ongoing changes, particularly climate change [10]. Climate change impacts the Baltic Sea and its lagoons in various ways [11], notably by increasing the intensity and frequency of extreme events. These events include extreme storms and sea levels, river floods, and droughts [12], as well as heat waves, which are potentially associated with oxygen depletion and algal blooms [11]. An important question is how these technical human interventions influence the socio-economic and natural conditions of the lagoons and their capacity to provide ecosystem services, especially in the context of these extreme events.

The Millennium Ecosystem Assessment [13] defines ecosystem services as benefits that humans derive from ecosystems. Today, ecosystem services play a significant role in European coastal and marine policies. The expectation is that ecosystem service assessments provide an understanding of the interdependencies between humans and the environment [14] beyond existing disciplinary assessments.

Since coastal lagoons are world-wide considered as highly important ecosystems, many publications on ecosystem services exist and address a wide spectrum of questions [15–18]. This is true for Baltic lagoons as well [19–21]. Existing assessments of ecosystem services face several significant challenges. For instance, data are often insufficient, heterogeneous, and difficult to compare. Variations in data across individual ecosystem services, combined with inadequate spatio-temporal resolution, further exacerbate the problem. As a result, complementary estimations, statistical analyses, and expert knowledge are frequently required to fill these gaps. Additionally, the methods used to calculate individual ecosystem services vary in robustness and reliability, making cross-comparisons problematic. These shortcomings have been widely documented [22,23] and substantially limit the acceptance and practical applicability of ecosystem service assessment outcomes.

In the Baltic Sea, several three-dimensional ecosystem models exist. One example is the 3D biogeochemical model ERGOM (Ecological ReGional Ocean Model) [24], which was recently evaluated by the European Commission [25] and rated as one of the best marine biogeochemical and lower trophic level models in Europe. In general, the utilization of ecologic models is assumed to overcome many existing weaknesses of ecosystem service assessments [26]. A recent approach to assess ecosystem services based on the ERGOM model results covering the entire Baltic Sea over 150 years underlined the potential but also indicated limitations [27]. Among the challenges are an improved spatial model resolution, a better integration between socio-economic and model data, a higher ecosystem service assessment accuracy to enable shorter assessment time periods, and an increase in the number of addressed ecosystem services. To overcome the present limitations, smaller spatial units need to be addressed along with coastal lagoons. Oder Lagoon is not only one of the largest Baltic lagoons, but it is also of outstanding socio-economic importance and a prime example for the changes and transformations that have taken place and are taking place during the last centuries. Therefore, it can be considered as ideal test case.

Our objectives are to (a) provide a consistent database for the calculation of ecosystem services in Oder Lagoon by applying a new 3D ecosystem model in combination with socio-economic data; (b) analyze how ongoing human technical interventions during the last centuries, namely, the stepwise deepening of the navigational waterway in the lagoon, have affected different provisioning, regulating, and cultural ecosystem services; (c) assess the relevance of and reasons for interannual variabilities in ecosystem service values; (d) calculate the monetary values of the ecosystem services to enable direct comparisons; and (e) discuss the practical relevance of the results for managing the lagoon.

2. Study Site and Methods

2.1. *Oder/Szczecin Lagoon*

Oder/Szczecin Lagoon (53°48′ N, 14°08′ E) covers a surface area of 687 km^2. Approximately 40% (277 km^2) of the lagoon, referred as the "Kleines Haff" or "small bay", lies within Germany, while 60% (410 km^2), called the "Wielki Zalew" or "large bay", is situated in Poland. This shallow lagoon has an average depth of 3.8 m and reaches a natural maximum depth of 8.5 m [28]. Its total coastline measures 209 km, with 89 km in the Kleines Haff and 120 km in the Wielki Zalew (extending to Trzebież and excluding the riverine bay Roztoka Odrzańska) (Figure 1).

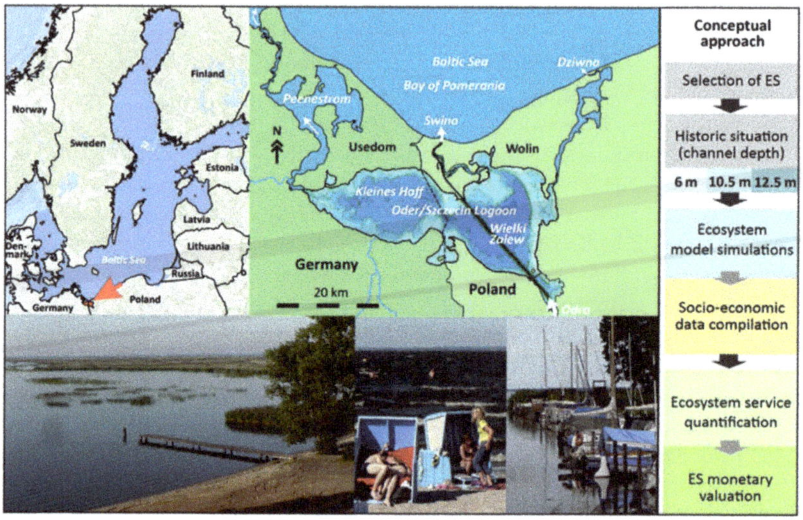

Figure 1. Oder/Szczecin Lagoon in the Baltic Sea region, map of the lagoon and impressions of the nature and human uses as well as the stepwise conceptual ecosystem service (ES) assessment approach of this study.

According to the OECD, the lagoon is classified as hypertrophic due to high nutrient inflows from rivers, primarily from the Oder (Polish: Odra). The Oder, with a catchment area of 120,000 km^2, accounts for approximately 98% of the water discharge as well as nitrogen (N) and phosphorus (P) inputs to the lagoon. In contrast, the Zarow and Uecker Rivers have only minor impacts. Around 1880, riverine nutrient loads were estimated at 14,000 t N/a and 1000 t P/a. These loads increased, peaking in the 1980s at 115,000 t N/a and 10,500 t P/a [29]. Between 1995 and 2019, the average water discharge into the lagoon was 518 m^3/s, with mean nutrient loads of 46,266 t N/a and 1635 t P/a [30].

The recent official HELCOM assessment rated both the integrated contamination status as well as the eutrophication status of Oder Lagoon as "bad" [31]. Similarly, under the European Water Framework Directive (WFD, which additionally considers benthic flora and fauna, the lagoon's ecological quality is classified as "insufficient" [32].

Despite these environmental challenges, the lagoon forms part of the Natura 2000 network, established to protect rare and threatened species. Large areas of the lagoon are designated as landscape protection zones, and the surrounding coastal zone features two national parks and several nature reserves. While fishing has historically been the primary economic activity, tourism is of increasing importance [33].

2.2. Human Interventions—The Navigation Waterway Across the Lagoon

Oder/Szczecin Lagoon connects to the Baltic Sea through three outlets: the Swina, Peenestrom, and Dziwna channels. The lagoon's theoretical average annual water residence time is 60 days but varies between 37 and 99 days, depending on river discharge quantities. Approximately 70% of the lagoon's water exits via the Swina channel, the major navigation waterway [34]. Periodic inflow of Baltic Sea water maintains the lagoon's oligohaline state, with salinity concentrations between 1 and 3 PSU.

The Swina channel represents the most significant technical human intervention in the lagoon. Spanning 67 km, it connects the Baltic Sea to the city of Szczecin through Oder Lagoon. Historically, the Swina River crossed the about 6 km wide Wolin/Usedom spit separating the lagoon from the Baltic Sea. Modifications to the Swina River date back to 1721, when the maximum depths ranged between 2 and 2.5 m [35]. In 1880, a shortened and deepened artificial channel was completed, featuring a fairway depth of 6 m and a river cross-section of approximately 1400 m^2 [36,37]. Subsequent deepening projects included an increase to 9.6 m in 1939 and to 10.5 m in 1984 across the entire lagoon. By 1984, the total cross-section of the Świna River/channel system (Piastowski Canal/Kaiserfahrt) had expanded to about 2500 m^2.

Between 2018 and 2023, the entire waterway across Oder Lagoon was further deepened to 12.5 m, increasing the total cross-section to about 2700 m^2. This deepening was part of the EU Strategy for the Baltic Sea Region, adopted by the European Council in 2009, aimed at enhancing the Baltic Sea region's potential, including improved internal and external transport [35]. The demand for enhanced shipping infrastructure reflects the ongoing growth in shipping activities. For example, cargo transshipments in the ports of Szczecin and Świnoujście increased sharply, from 20.8 million tons in 2010 to 36.8 million tons in 2022. Additionally, the number of Ro-Ro ship units (including vehicles, passengers, and cargo) and container reloading activities grew significantly, with container handling rising from 21,860 TEU in 2000 to 67,952 TEU by 2023 [38].

The channel's deepening to a fairway depth of 10.5 m initially enabled vessels with a maximum draft of 9.15 m and lengths of up to 215 m to access Szczecin. Following the deepening to 12.5 m, the channel can now accommodate container and bulk carrier ships up to 240 m in length, 32.4 m in width, with drafts of 11 m, as well as cruise ships up to 260 m long [35]. This upgrade supports larger ships, facilitates efficient two-way vessel traffic, improves navigation safety, and reduces transit times [39].

The increased cross-section of the channel enhances the interface between the mesohaline Baltic Sea and the oligohaline lagoon, enabling faster water exchange and further altering the lagoon's hydrodynamics.

2.3. The 3D Ecosystem Model

We utilized a modified version of the coupled circulation and biogeochemical model documented by Neumann et al. [24]. The circulation component is based on the Modular Ocean Model (MOM5.1), while the biogeochemical component employs ERGOM version 1.2 (Leibniz Institute for Baltic Sea Research, 2015). For this study, the model was specifically adapted for Oder Lagoon, featuring a high horizontal grid resolution of 150 m. The vertical structure consists of 28 layers, with thicknesses ranging from 25 cm at the surface to 50 cm near the bottom. Open boundary conditions from a 2 km resolution Baltic Sea model were applied at the lagoon's three outlets connecting to the Baltic Sea.

ERGOM simulates the biogeochemical cycles of nitrogen, phosphorus, carbon, oxygen, and partially sulfur. Primary production is driven by photosynthetically active radiation and is facilitated by four functional phytoplankton groups: large cells, small cells, limnic phytoplankton, and cyanobacteria. An optical sub-model calculates the underwater light climate based on chlorophyll and colored dissolved organic matter (CDOM) concentrations. Dead organic material accumulates as detritus, while zooplankton grazing on phytoplankton represents the highest trophic level in the model. Particulate organic carbon (POC) from sources such as phytoplankton and detritus can sink through the water column, accumulating in a sediment layer.

The model incorporates temperature- and oxygen-dependent mineralization processes, releasing dissolved inorganic nitrogen and phosphorus from organic matter in both the water column and sediment. In oxygen-rich conditions, phosphate binds with iron oxides to form particles that settle in sediments. These particles can be resuspended by erosion and redistributed by currents to deposition zones. Under anoxic conditions, iron oxides are reduced, releasing dissolved phosphate back into the water. Oxygen is produced through primary production and consumed during metabolism and mineralization. Additionally, extracellular excretion by phytoplankton leads to non-Redfield carbon uptake.

In contrast to Neumann et al. [24], we introduced a fourth phytoplankton group, limnic phytoplankton, designed for low-salinity, turbid coastal environments where growth is constrained by salinity and increased light sensitivity. All organic particles (e.g., phytoplankton and detritus) are modeled in nitrogen units. To compare modeled phytoplankton outputs with observed chlorophyll levels, we sum all phytoplankton groups and apply a constant chlorophyll-to-carbon ratio. The lagoon model validation and application are presented in [30].

Meteorological data were sourced from the coastDat-3 dataset [40] and the Norwegian Meteorological Institute. Data on discharge and riverine nutrient loads from the Oder, Uecker, and Zarow Rivers were supplied by the Instytut Meteorologii i Gospodarki Wodnej and Friedland et al. [29].

2.4. Model Simulations and Scenarios

We analyzed three scenarios: (1) the situation after the deepening in 1880, with a channel water depth of 6 m; (2) the situation between 1984 and 2022, with a channel depth of 10.5 m; and (3) the state after the recent deepening that was finished in 2023, representing a channel water depth of 12.5 m. It is called a future scenario. Historic changes in the waterways connecting the lagoon with the Baltic Sea were derived from historic maps. The cross-section and profile of the waterway newly deepened to 12.5 m were taken from Environmental Impact Assessment planning documents [41]. These morphometric changes were implemented in three distinct model bathymetries.

The ERGOM model simulations covered the years 2010–2019, representing a shipping depth/draft of 10.5 m in the Świna channel. Simulations for the same 10-year period were repeated under the assumed conditions of the future 12.5 m depth and the historic post-1880 depth of 6 m. While the historic and future scenarios employed different bathymetries, all simulations were based on the same input data (e.g., weather, river discharge) from the years 2010–2019. This approach overcomes the problem of the lack of reliable historical weather and discharge data and enables a direct analysis of the mere consequences of the channel deepening. From these simulations, we extracted two-day averages of relevant state variables, processes, and transports, which served as the basis for our assessments.

2.5. Socio-Economic Data Collection

The socio-economic data were gathered from various German and Polish sources. The population data were compiled for areas within two kilometers and 2–20 km of the lagoon coastline [42–44]. Existing data on tourism overnight stays were incorporated [45], with any gaps estimated using per capita metrics. The number of sport-boat berths was determined by analyzing the websites of individual sport-boat harbors around the lagoon, supplemented by Google Earth for missing information. Similarly, data on beach sizes were compiled, with details that were lacking being estimated through satellite imagery. Official statistics were utilized for passenger transport and commercial shipping volumes [46,47]. The numbers of fishing boats and fishermen were sourced from [48,49], while information on angling tourism was drawn from [50,51].

2.6. Ecosystem Service and Assessment Approach

Our assessments were based on the Common International Classification of Ecosystem Services (CICES), a comprehensive framework for ecosystem mapping, assessment, and natural capital accounting [52]. CICES is widely adopted internationally and aligns closely with European environmental policies. The system is hierarchical, with the top level distinguished by three sections: provisioning services, regulating and maintenance services, and cultural services. For this study, we focused on the most detailed classification level (classes), adapting definitions from CICES V5.1 [53].

Selected ecosystem services were quantified directly or indirectly using outputs from our 3D ecosystem model, supplemented with socio-economic data. To ensure granularity and country-specific accuracy, Oder Lagoon was subdivided into two basins: Kleines Haff (Germany) and Wielki Zalew (Poland). This subdivision also aligns with the water body delineations under the European Water Framework Directive (WFD).

2.7. Provisioning Ecosystem Services

Wild animal food and products: This service encompasses edible fish, seafood, fish meal, and fish oil. In Oder/Szczecin Lagoon, only edible wild fish is relevant. The data on fish catches were obtained from official statistics in Germany [54] and in Poland [55]. The average country-specific fish market prices were extracted from different official sources [56,57]. Aquaculture and benthic animal extraction are not significant in the lagoon. We assumed that the potential total commercially viable fish biomass depends on the lagoon's productivity, which is represented by phytoplankton concentrations.

Based on Scheffold and Hense [58], we used a phytoplankton-to-commercial wild fish biomass ratio of 2.2 (gC/m^2). The carbon-to-dry weight ratio (0.43) and fish dry weight-to-wet weight ratio (0.2) were applied to calculate fish wet weight [59,60]. Data from Thurow [61] validated the relationship between fish catches and wild fish biomass.

The ecosystems services "wild plants and products" were estimated using literature-based calculations. The detailed literature data are indicated in Section 3.3.

The ecosystem service "transportation" accounts for the lagoon's role in providing space and enabling shipping activities. The calculations are explained in Section 3.4 and were based on existing literature data and statistics [47,62].

2.8. Cultural Ecosystem Services

Active recreation: This service includes outdoor activities and tourism related to the local environment, such as sports, leisure, and outdoor pursuits. Coastal tourism and water-related activities at Oder Lagoon are concentrated during summer and are heavily influenced by water temperature and quality. We assumed that tourists perceive water quality as favorable when water transparency is high, bathing water meets official standards, and there are no visual pollutants or nuisances (e.g., toxic algae blooms, algae accumulations) [63–65]. Indices (I), with a minimum of 0 and a maximum of 100, were calculated for the modeled average summer water surface temperature (I = 77.7 ln (temperature) − 164.74), cyanobacteria concentrations in summer (I = 36.1 (ln (cyanobacteria) + 125), water transparency (Secchi-depth in summer) (I = 22.76 ln (Secchi-depth) + 52.4), and riverine phosphorus emissions (proxy for sewage) (I = −15.53 ln (P-emissions) + 110.77). These indices, equally weighted, were averaged into a total recreation index. All regression formulas scale the relevant upper and lower levels of the data to an index between 0 and 100. It is assumed that the benefit (increase in index value) is decreasing with higher values. For example, an increase in water transparency of 0.5 m has a higher index increase at low water transparencies than if the transparency is already high.

Monetarization exclusively considered annual guest overnight stays within 2 km of the lagoon coast in Germany [66] and Poland [67] multiplied by country-specific added value per overnight stay. Specifically, EUR 216 and EUR 148 per night were assumed for Germany and for Poland, respectively [68–70].

Landscape aesthetics (observational recreation): This service relates to the visual quality of landscapes and ecosystems, which enhance human well-being. The landscape aesthetics index included water transparency (Secchi-depth in summer) (I = 22.76 ln (Secchi-depth) + 52.4) and the eutrophication level (summer chl.a concentrations) (I = −36.07 ln (chl.a) + 183.05). Additionally, the coastal population density within 2 km from the coastline (habitants per km of coastline) (I = −13.96 ln (habitants) + 183.05) and tourism density, namely, overnight stay less than 2 km from the lagoon coastline (no. of tourists staying per km of coastline) (I = −9.241 ln (tourists) + 120.93), were included. Higher population and tourism densities were assumed to reduce landscape aesthetics. The indices, equally weighted, were averaged into a landscape aesthetics quality index.

For monetization, local studies and suitable transferable literature were unavailable. Thus, we based calculations on local inhabitants, assuming a per capita benefit equivalent to two times the country-specific added value per tourist overnight stay of EUR 432 per year in Germany and EUR 296 per year in Poland for residents living within 2 km of the coastline. For residents living 2–20 km away, we assumed 10% of these values.

2.9. Regulating Ecosystem Services

Water quality regulation and purification (nitrogen and phosphorus): For nitrogen, we considered processes in the water column and sediments, including nitrogen fixation, denitrification, and nitrogen burial. For phosphorus, we included only phosphorus burial in sediments. These data were directly extracted from model simulations for both basins. The costs of reducing nitrogen were based on average sewage treatment costs: EUR 15.8/kg N in Germany and EUR 12.7/kg N in Poland. For phosphorus, we used EUR 64.3/kg P in Germany and EUR 43.2/kg P in Poland. These figures, from Gren et al. [71], were adjusted to 2015 prices using an annual inflation rate of 1.5%.

Carbon sequestration/burial (climate regulation): Carbon burial data in sediments were also directly extracted from model simulations. Monetization was based on CO_2 emission prices in the European Emissions Trading System (EU-ETS). For 2019, a certificate cost of EUR 25 per ton of CO_2 (EUR 92/t C) was applied [72].

3. Results

3.1. Model Simulation Data—20-Year Averages

Table 1 provides a summary of the model simulation results that serve as the basis for the calculation of ecosystem services. The model data on nutrient and carbon burial in sediments are presented later. The channel deepening from 6 m to 12.5 m has no effect on water temperatures and causes only a minor increase in dissolved inorganic phosphorus (DIP) concentrations, remaining below 1 µmol/L, with negligible changes in dissolved inorganic nitrogen (DIN) concentrations. However, major nitrogen processes, such as denitrification and nitrogen fixation, are significantly impacted.

The model indicates that channel deepening increases nitrogen fixation by an average of 57 t/a (8%) in the Kleines Haff and 229 t/a (46%) in the Wielki Zalew, leading to a total increase of 286 t N/a (24%) across the entire lagoon. Nitrogen fixation also exhibits strong interannual variability. During hot summers, such as in 2015 and 2018, the model predicts high nitrogen fixation of 2849 t N/a and 2507 t N/a, respectively, under the 10.5 m scenario. Conversely, during the cooler year of 2010, nitrogen fixation is calculated at only 364 t N/a.

Denitrification, a key process for nitrogen removal from the lagoon, decreases in both basins due to channel deepening. The model estimates an average total decline of 12%, equivalent to 1595 t N/a. Unlike nitrogen fixation, the interannual variability in denitrification is relatively low, ranging from 9510 t N/a in 2015 to 15,573 t N/a in 2010.

The model also predicts reductions in summer chlorophyll-a concentrations and annual average phytoplankton concentrations by approximately 20% across all lagoon sections due to channel deepening. However, summer cyanobacteria concentrations increase by an average of 16%. This rise in cyanobacteria drives the observed increase in nitrogen fixation, as the model assumes that cyanobacteria are responsible for atmospheric nitrogen fixation.

The changes in phytoplankton and nutrient processes are strongly influenced by altered carbon and nutrient burial dynamics in the sediments of the deepened channel. The increased water depth and cross-section of the Świna channel potentially enhance water exchange between the lagoon and the Baltic Sea. This results in a relative salinity increase of 135% in the Wielki Zalew and 89% across the entire lagoon. Despite these percentages, the absolute salinity changes are minor—less than 1 PSU or 1 g/kg—and are ecologically negligible. In the Wielki Zalew, salinity ranges between 0.3 g/kg (2010) and 1.6 g/kg (2015), controlled primarily by Oder River discharge and sporadic saltwater intrusions from the Baltic Sea.

The Baltic Sea, being a micro-tidal sea with a tidal range below 0.2 m and salinity of approximately 8 g/kg in the Pomeranian Bay, does not contribute substantial saltwater quantities or significant salinity increases to the lagoon. The areas affected by Baltic water intrusions remain confined to the vicinity of the Świna channel, with no notable expansion resulting from the channel deepening.

Table 1. Over 20-year averaged 3D ecosystem model (ERGOM) simulation results for the three waterway deepening scenarios (6 m before 1880, 10.5 m representing 1984–2023 and 12.5 after 2024). The relative changes indicate the difference between the 6 m and the 12.5 m scenarios. The data are aggregated over the surface area of each part of Oder/Szczecin Lagoon, the German Kleines Haff and the Polish Wielki Zalew.

	Dissolved Inorganic N (Winter) [µmol/l]				**Dissolved Inorganic P (Winter) [µmol/l]**				**Chlorophyll-a (Summer) [µg/l]**				**Phytoplankton (Annual Mean) [mmol C/m³]**				**Cyanobacteria (Summer) [mmol C/m³]**			
Subbasin	6 m	10.5 m	12.5 m	Δ %	6 m	10.5 m	12.5 m	Δ %	6 m	10.5 m	12.5 m	Δ %	6 m	10.5 m	12.5 m	Δ %	6 m	10.5 m	12.5 m	Δ %
Kleines Haff	64.8	61.7	64.6	−0.4	1.4	2.0	2.0	40	51.2	42.9	40.5	−21	83.9	69.9	66.7	−20	15.4	18.2	16.7	9
Wielki Zalew	120.9	121.1	121.3	0.3	1.3	1.7	1.7	33	55.5	48.3	43.1	−22	102.7	89.1	82.1	−20	8.4	11.0	10.6	26
Total lagoon	97.3	96.3	96.9	−0.5	1.4	1.8	1.8	35	53.6	46.1	42.0	−22	94.6	81.0	75.6	−20	11.3	14.0	13.1	16
	N-Fixation [t/a]				**Denitrification [t/a]**				**Salinity [g/kg]**				**Water temperature (summer) [°C]**				**Secchi-depth (summer) [m]**			
Subbasin	6 m	10.5 m	12.5 m	Δ %	6 m	10.5 m	12.5 m	Δ %	6 m	10.5 m	12.5 m	Δ %	6 m	10.5 m	12.5 m	Δ %	6 m	10.5 m	12.5 m	Δ %
Kleines Haff	676	811	733	8	4886	4204	4121	−16	0.9	1.3	1.4	53	21.8	21.8	21.8	0.2	0.77	0.83	0.85	10
Wielki Zalew	498	721	728	46	8922	8759	8092	−9	0.4	0.9	1.0	135	20.6	20.6	20.5	−0.5	0.71	0.77	0.84	18
Total lagoon	1159	1510	1438	24	13716	12845	12098	−12	0.6	1.1	1.2	89	21.1	21.1	21.1	−0.2	0.74	0.82	0.84	14

3.2. Socio-Economic Data

The total number of inhabitants living within 2 km of the lagoon coastline is 76,244 (Table 2). With an average population density of approximately 70 inhabitants per km^2, the area surrounding the lagoon is relatively sparsely populated. Within a 2–20 km area of the lagoon coast, the population increases to about 273,000. This includes the seaside resorts along the Baltic Sea and approximately 72,000 residents in the northern outskirts of Szczecin. The city of Szczecin itself, with a population of 391,000, is not included in these figures.

Table 2. Socio-economic data of Oder/Szczecin Lagoon differentiated between the Kleines Haff (Germany) and the Wielki Zalew (Poland). The population number is compiled for a zone closer than 2 km and a zone closer than 20 km from the lagoon coastline.

Socio-economic data		Kleines Haff	Wielki Zalew	Total lagoon
Population	<2 km	15,616	60,628	76,244
Population	<20 km	73,223	199,803	273,026
Guest overnight stays	<2 km	239,316	61,702	301,017
Beach area	(m^2)	78,387	12,263	90,650
Sport boat berths	(no.)	804	520	1324
Ship passengers	(no.)	7472	14,917	22,389
Cargo ship arrivals	(no.)	50	3288	3338
Fishing boats	(no.)	80	126	206
Fishermen	(no.)	80	360	440

Tourism is a growing economic sector in the region, with approximately 300,000 overnight stays annually around the lagoon. This growth has been supported by regional infrastructure developments, including the creation and expansion of beaches, sport boat harbors, and cycling trails. The number of ship passengers primarily consists of tourists on boat trips, while most cargo ships transit through the lagoon on their route between the Baltic Sea and the Port of Szczecin.

In 2013, 80 fishing boats were registered in the Kleines Haff, supporting 34 full-time fishermen, 9 non-commercial fishermen, and 27 hobby fishermen. Data from 2007–2016 suggest that 126 fishing boats operate in the Wielki Zalew, providing livelihoods for 360 fishermen. Additionally, 110 individuals hold fishing permits. These relatively low numbers highlight the declining economic significance of fisheries in the region.

The calculated ecosystem services, based on the model simulation results and the complementing socio-economic data, are fully documented in Appendix A. The subsequent monetary ecosystem services are presented in Appendix B. Because of the methodological and conceptual heterogeneities, the following chapters address comparable groups of ecosystem services.

3.3. Provisioning Ecosystem Services—Wild Fish and Plants

Wild fish: Between 2010 and 2019, an average of approximately 2271 tons of wild fish was caught annually in the lagoon [54,55]. Of this, 79% was harvested from the Wielki Zalew in Poland, and 21% from the Kleines Haff in Germany (Table 3). The weighted average market price of fish differed between the two countries: EUR 1.90/kg in Germany and EUR 1.26/kg in Poland. Consequently, Germany accounted for 29% of the total annual economic value, while Poland contributed 71%. The overall economic value of the fish catch was approximately EUR 3.17 million, or EUR 4618 per square kilometer of lagoon surface area.

Table 3. Wild fish catches, market prices (in Euro) and economic value of fisheries in Oder/Szczecin Lagoon differentiated between the Kleines Haff (Germany) and the Wielki Zalew (Poland).

Species	Kleines Haff, Germany Catches [kg/a]	Price [EUR/kg]	Yield [EUR/a]	Wielki Zalew, Poland Catches [kg/a]	Price [EUR/kg]	Yield [EUR/a]
Herring	2115	0.66	1396	46,400	0.30	14,076
Salmon	896	8.22	7368	4283	6.90	29,539
Eel	3051	13.92	42,463	19,650	10.58	207,961
Pike perch	66,636	6.29	419,138	103,483	3.86	399,275
Pike	3354	2.80	9391	10,233	1.97	20,165
Perch	97,380	2.20	214,237	577,050	1.98	1,142,006
White fish	299,421	0.70	208,098	1,009,233	0.43	436,704
Others	5196	1.00	5196	22,400	0.70	15,680
Sum/mean:	478,049	1.90	907,286	1,792,733	1.26	2,265,407

Among the various species, eel, salmon (both *Salmo salar* and *Salmo trutta trutta*), and pike perch commanded the highest market prices. These species constituted only 9% of the total catch but generated nearly 50% of the economic value. In contrast, the more abundant bony white fish species, such as roach (*Rutilus rutilus*), bream (*Abramis brama*), whitefish (*Coregonus oxyrinchus*), silver bream (*Blicca bjoerkna*), and asp (*Aspius aspius*), are mainly used for soup or feed and have a lower market value [73].

Based on phytoplankton model simulations, the total commercially usable fish biomass in the lagoon during the same period was estimated at 13,497 tons of wet weight. Of this, 66% was located in the Wielki Zalew and 32% in the Kleines Haff. The larger share in the Wielki Zalew can be attributed to higher phytoplankton concentrations (Figure 2). The reported annual catch of 2271 tons indicates that approximately 16% of the total fish biomass is harvested each year. Fishing pressure is higher in the Wielki Zalew, where 20% of the fish stock is caught annually, compared to only 10% in the Kleines Haff.

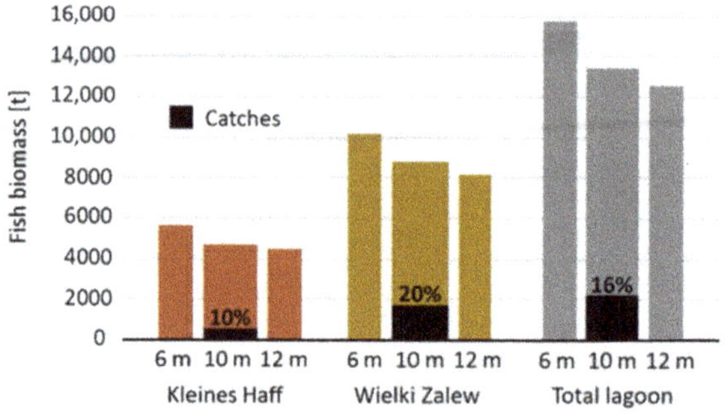

Figure 2. Calculated commercially usable fish biomass in different parts of Oder/Szczecin Lagoon and for the three channel deepening scenarios (6 m, 10.5 m, and 12.5 m). Additionally, the reported catches (t/a) during the period 2010–2019 are indicated in comparison.

Model simulations suggest that channel deepening from 6 m to 12.5 m would reduce phytoplankton biomass, leading to an approximate 20% decline in commercially usable fish stocks across the lagoon. Given the already high fishing pressure on valuable species, most of the remaining catchable biomass would, very likely, consist of white fish, with market values of EUR 0.43/kg in Poland and EUR 0.70/kg in Germany. Under these conditions, the total potential market value of the entire fish stock in the lagoon is estimated at EUR 15 million, equivalent to EUR 22,000 per square kilometer of lagoon surface area.

Wild plants and products: The only commercially utilized plant in Oder Lagoon is common reed (*Phragmites australis*), primarily used as a material for roof thatching [74]. Along the coastline of the lagoon, reed beds cover an area of approximately 6.0 km^2 in the

Kleines Haff and 8.3 km^2 in the Wielki Zalew [1,75]. The average width of the reed belt across the lagoon is around 69 m, potentially yielding a total dry weight of 14,320 tons: 5980 tons in the Kleines Haff and 8340 tons in the Wielki Zalew.

Market prices for reed are typically EUR 3.4 per 5 kg bundle or EUR 640/ton, giving the total reed stock an estimated value of EUR 9.2 million. However, local reed is no longer harvested or utilized around Oder Lagoon due to a lack of market demand. Reed harvesting typically occurs on land, involves smaller areas, and is not directly influenced by the lagoon itself. As a result, despite its potential economic value, reed provision was not further assessed. Other wild plants have no significant economic or ecological role in Oder Lagoon.

3.4. *Provisioning Ecosystem Services—Transportation*

The deepened waterway across the lagoon acts as trap [21] for approximately 425.000 t/a suspended particular material entering the lagoon via the Oder River as well as for resuspended and translocated lagoon sediments. Between 1949 and 2000, maintaining the shipping channel's depth required the annual dredging of an average of 1.5 million tons of sediment (wet weight) [76]. The recent channel deepening to a fairway depth of 12.5 m necessitated the removal of approximately 24 million cubic meters of sediment [77].

The dredging costs for the channel deepening were estimated at approximately EUR 335 million, with an additional EUR 470 million invested in complementary infrastructure [78]. This translates to a dredging cost of EUR 14 per cubic meter of wet sediment. It is assumed that the combined costs of channel deepening, amortized over 30 years, along with the annual maintenance dredging expenses, represent the monetary value of the ecosystem service "transportation."

Based on this assumption, the total annual monetary value of the waterway is estimated at EUR 32, comprising EUR 11 million per year for the deepening costs (spread over 30 years) and EUR 21 million per year for maintenance dredging. Based on ship arrival data and ship passenger numbers, about 95% of the value can be attributed to the Wielki Zalew.

3.5. *Cultural Ecosystem Services—Tourism, Recreation, and Aesthetics*

Cultural ecosystem services are grouped into two categories: those derived from tourism (recreational ecosystem services) and those benefiting the local population (observational ecosystem services).

For the recreation index, significant differences between the two parts of the lagoon are evident, driven by sub-indices such as algae blooms, water transparency, water temperature, and emissions. The low sub-index value for algae blooms is primarily due to frequent summer blooms of potentially toxic cyanobacteria, particularly in the Kleines Haff. Additionally, summer water transparency, typically below 1 m (Secchi depth), negatively impacts bathing tourism, leading to sub-index values below 50. Conversely, the shallow lagoon's summer water temperatures are higher than those of the neighboring Baltic Sea, averaging 20.6 °C between June and August from 2010 to 2019, which makes it favorable for swimming. However, the inflow of the Oder/Odra River into the Wielki Zalew negatively impacts microbial bathing water quality, as reflected in the very low emission sub-index values.

The overall recreation index, which averages the four sub-indices, shows no significant differences across the three channel deepening scenarios. The recreation index is slightly higher in the Kleines Haff compared to the Wielki Zalew (Figure 3a). Nonetheless, the eutrophic nature of the lagoon significantly limits its value for recreation, particularly for summer bathing tourism. The full data are shown in Appendix A.

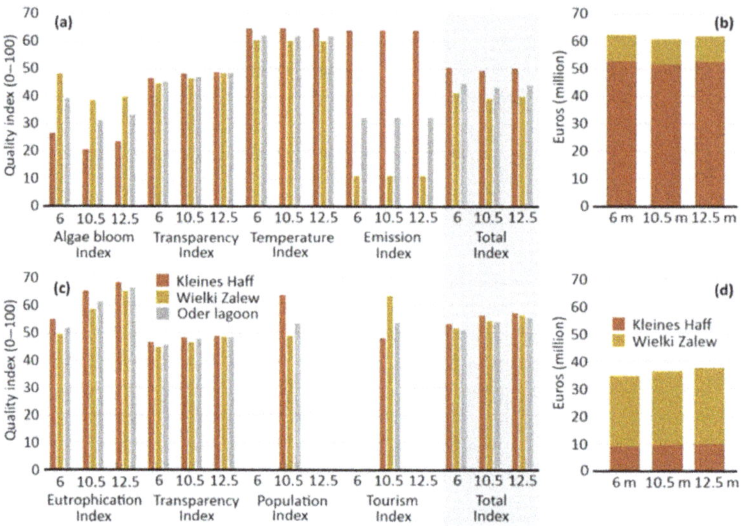

Figure 3. (**a**) Recreational ecosystem services: The values of the four sub-indices algae blooms, water transparency, water temperature, and emissions are averaged into the total recreation index for the three channel deepening scenarios and (**b**) the monetary recreation values. (**c**) Aesthetics ecosystem services: The values of the four sub-indices eutrophication, water transparency, population, and tourism density are averaged into the total aesthetics index for the three channel deepening scenarios and (**d**) the monetary aesthetics values. Tables 1 and 2 and Appendix A show the underlying data.

The monetary value of recreational ecosystem services is estimated at approximately EUR 60 million per year (Figure 3b). Tourism is more concentrated in the Kleines Haff, where overnight stays are four times higher than in the Wielki Zalew (Table 2). Furthermore, overnight stays in the Kleines Haff generate higher gross value added (EUR 216 per overnight stay compared to EUR 148 in the Wielki Zalew). Consequently, the monetary value of recreation in the Kleines Haff is about five times greater than in the Wielki Zalew.

The total aesthetics ecosystem index comprises sub-indices for eutrophication (summer chlorophyll-a concentration), water transparency, population, and tourism density. Population and tourism density reflect the real current state with a channel depth of 10.5 m and are assumed to be similar for the other two deepening scenarios. The differences in the sub-indices between the Wielki Zalew and Kleines Haff are minor. However, the eutrophication sub-index significantly increases with greater channel depth. As a result, the overall aesthetics ecosystem index also shows a slight increase with deeper channel scenarios due to the higher eutrophication sub-index, mainly resulting from lower annual average chlorophyll-a concentrations with increasing channel depth (Figure 3c).

The monetary value of the aesthetics ecosystem service is approximately EUR 35 million annually (Figure 3b). About 75% of this value is generated in the Wielki Zalew, reflecting the larger local population living along its coastline (Table 2).

3.6. Regulating Ecosystem Services—Nutrient and Carbon Retention

Nitrogen retention: On average over the lagoon, the model suggests a nitrogen retention of 17,862 kg N a^{-1} km^2 (100%) that results from 18,757 kg a^{-1} km^2 denitrification (105%) plus 1098 kg N a^{-1} km^2 burial in sediments (6%) minus 1993 kg N a^{-1} km^2 nitrogen fixation (−11%) (Figure 4a). Therefore, denitrification is the dominating process. The model suggests that presently, about 12,156 t N/a are kept back in the entire lagoon, with a monetary value of EUR 166 million/a (EUR 241,882 a^{-1} km^2). For the period 2010–2019, this means that on average, 39% of the annual total nitrogen loads to the lagoon were kept back.

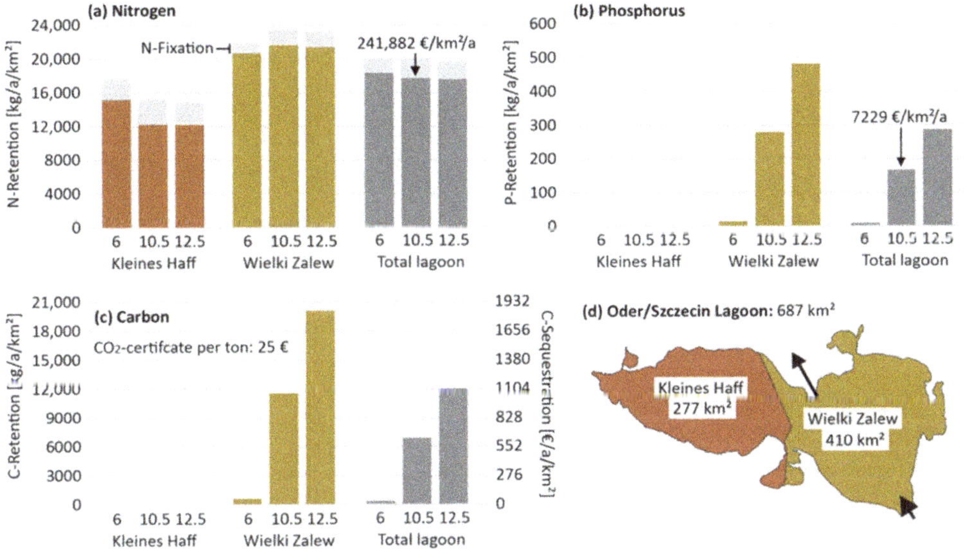

Figure 4. Model calculated annual nutrient and carbon retention in the two bays of Oder/Szczecin Lagoon. (**a**) Quantity and exemplary monetary value (Euro) of nitrogen retention, covering denitrification, burial in sediments, and N-fixation. The latter adds nitrogen to the lagoon. Quantity and exemplary monetary value of (**b**) phosphorus retention and (**c**) carbon retention or sequestration (burial in sediments). (**d**) Area of the lagoon and its basins.

According to the model, in the Kleines Haff, the channel deepening caused a decrease in N-retention of 20%, while in the Wielki Zalew, it suggests an increase of 3%. In total over the lagoon, this results in a decrease of more than 4%. The deepened channel serves as sediment trap and increases the nutrient burial, but, because of its location, this is only in the Wielki Zalew. The different behavior of both bays results from a complex interplay between water exchange, denitrification, and N-fixation.

Phosphorus retention: On average over the lagoon, the model suggests a minor P-retention of 9 kg P a^{-1} km^2 for the situation with a 6 m channel depth that increases with increasing channel depth to 287 kg P a^{-1} km^2 (Figure 4b). In the present situation (10.5 m channel depth), altogether 114 t P/a are kept back, and in the future (12.5 m channel depth), it is 197 t P/a. Retention exclusively takes place in the Wielki Zalew. The model suggests no permanent sediment burial in the Kleines Haff. The reason for the increase in the Wielki Zalew is the increasing sediment trap function of the deepened channel. The monetary value of the P-retention is EUR 5 million/a (10.5 m) resp. EUR 8.6 million/a (12.5 m). For the period 2010–2019, this means that on average, 12% of the annual total phosphorus loads to the lagoon were kept back in the lagoon.

Carbon retention/sequestration: Carbon retention shows a similar behavior compared to phosphorus since burial in sediments is the only sink-function (Figure 4c). On average over the lagoon, the model suggests a minor C-retention of 0.4 t C a^{-1} km^2 for the situation with 6 m channel depth that increases with increasing channel depth to 12 t C a^{-1} km^2. In the present situation (10.5 m channel depth), altogether 4740 t C/a are kept back, and in the future (12.5 m channel depth), the value is 8254 t P/a. The monetary value of the C-retention is about EUR 435,000/a (10.5 m) resp. EUR 757,000 million/a (12.5 m).

3.7. *Regulating Ecosystem Services—Inter-Annual Variability*

An advantage of a model-based ecosystem service assessment approach is flexibility with respect to the spatial and temporal resolution of the assessment. This enables addressing questions such as how strong is the spatio-temporal variability and what are reasonable spatial and temporal aggregation levels of assessments? Specifically, the regulating services nitrogen and phosphorus retention as well as carbon sequestration are suitable for this analysis. Concrete questions are as follows: Do these ecosystem services show an inter-annual variability, how strong is it, has it changed with changing channel depths, which parameters control the value of ecosystem services, and what are the consequences for temporally aggregated ecosystem service assessments?

The retention of nitrogen in the lagoon does not differ much between the scenarios over the period 2010 to 2019. This is very different for the nitrogen retention in the entire lagoon between single years. It ranges between 10,500 (2015) and 23,600 kg N a^{-1} km^2 (2010) for a channel depth of 10.5 m (Figure 5c). The year 2010 was a wet year, while 2015 was one of the years with the lowest ever reported water discharge (Figure 5b).

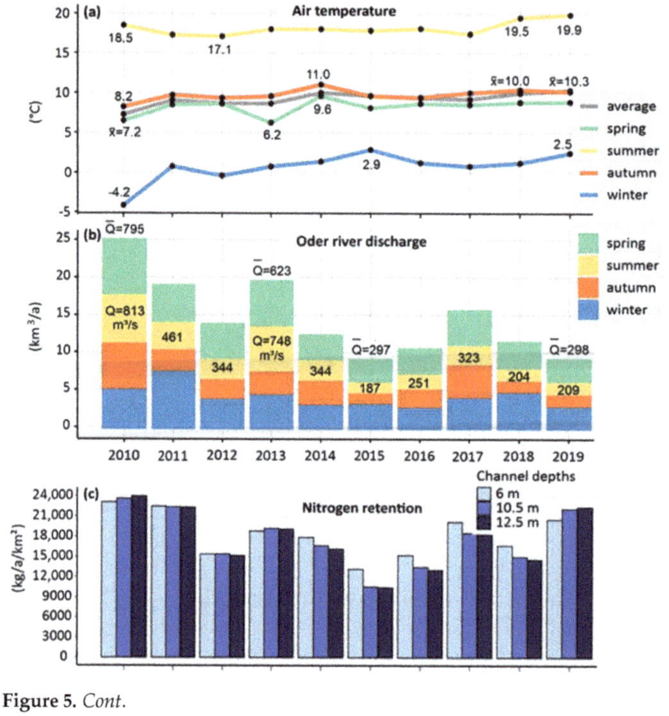

Figure 5. *Cont.*

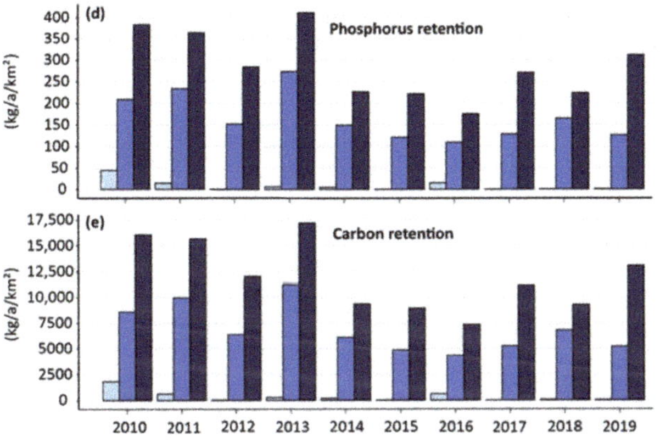

Figure 5. (**a**) Average annual air temperature above Oder Lagoon for different seasons (summer includes the months JJA) (data: Norwegian Meteorological Institute). (**b**) Annual Oder River water discharge (data: Instytut Meteorologii i Gospodarki Wodnej) separated into seasons. The Q-values indicate the average discharge for selected summer seasons (JJA) and years. Annual model-based nitrogen retention (including N-burial, N-fixation, and denitrification) (**c**), phosphorus retention (**d**), and carbon retention/sequestration (**e**) in the entire Oder Lagoon for the three channel deepening scenarios.

The relationship between annual water discharge and the annual nitrogen retention in the lagoon is close ($R^2 = 0.46$). This is not surprising since the annual riverine nitrogen load and the annual water discharge show a close positive relationship as well. This means the higher the riverine N-load is, the higher the retention in the lagoon. A clear dependency between average annual temperature and N-retention does not exist (Figure 5a,c).

For phosphorus and carbon, the difference between the three channel deepening scenarios is strong. The model generally suggests no burial in the sediment of the Kleines Haff. Storage only takes place in the Wielki Zalew, namely, in the shipping channel. The deeper the channel is, the higher the buried carbon and phosphorus (Figure 5d,e). The annual water discharge and the retention of phosphorus and carbon show a close positive correlation ($R^2 = 0.6$) as well. These elements also show a certain negative relationship ($R^2 = 0.27$) to temperature. This means that higher water discharge increases the quantitative retention, and higher temperature shows a tendency to reduce the quantitative retention.

The high interannual variabilities of these regulating ecosystem services indicate that assessment based on merely one year may be misleading and that an assessment over a decade is required to obtain reliable results that can be quantitatively compared to other ecosystem services.

3.8. Regulating Ecosystem Services—Seasonality and the Role of Extreme Events

Annual average temperatures or discharge cannot explain the strong differences in N-retention between the two dry years 2015 and 2019. An analysis of the annual course of the retention processes is required. It raises general questions about the seasonality of ecosystem services and how this impacts interannual variability.

Figure 6 shows the annual course of nitrogen, phosphorus, and carbon retention for several channel deepening scenarios for the entire lagoon. In general, the channel deepening does not affect the annual course of nutrient retention, but the patterns between the elements differ. While nitrogen retention is very low in winter, showing a maximum in May and June and later in the year a decline, the maximum burial of phosphorus and carbon takes place in spring, between March and May, and already by mid-summer, no burial in sediments takes place any more. In general, one can say that the retention that happens during three months of a year, between May and July with respect to nitrogen and between March and May with respect to carbon and phosphorus, largely determines the annual retention.

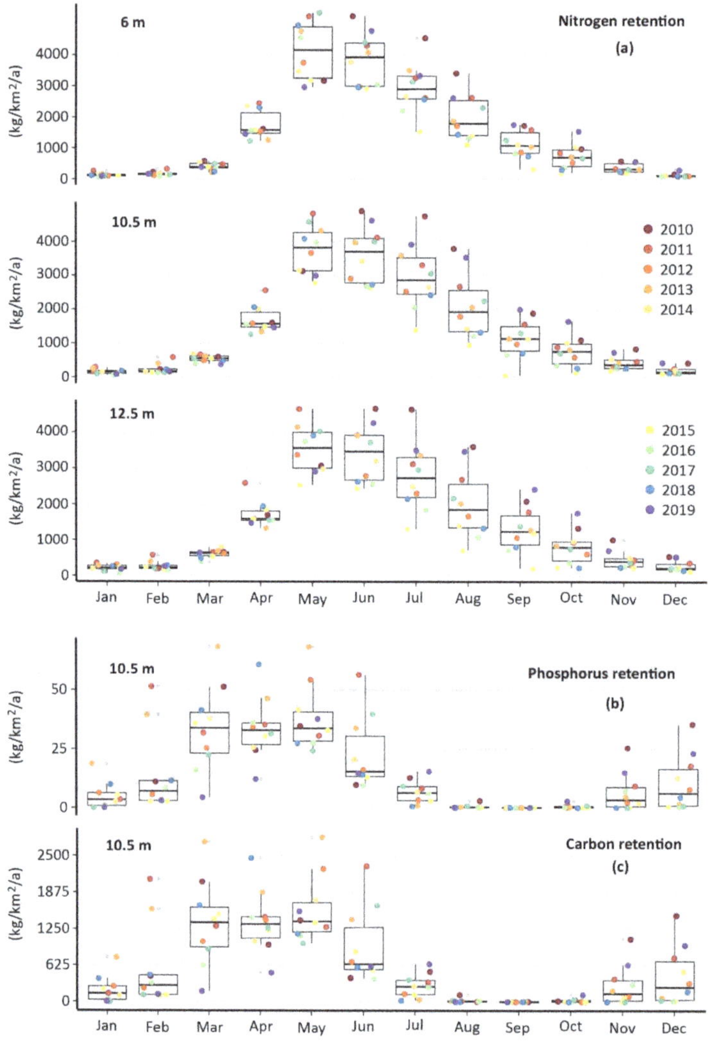

Figure 6. (**a**) Monthly model-based nitrogen retention (including N-burial, N-fixation, and denitrification) in the entire Oder Lagoon for the three channel deepening scenarios, as well as phosphorus retention (**b**) and carbon retention/sequestration (**c**) for a channel depth of 10.5 m.

The nutrient retention values for spring and summer do not show a strong relationship to the water discharge during this season. Assuming a delay in processes and analyzing the role of discharges in spring on the retention in summer gives a very good correlation (R = 0.91) for nitrogen retention, but only if the year 2019 is neglected.

Figure 6 clearly indicates a strong variability between the months of a year. This is especially true for carbon and phosphorus retention. For example, the carbon retention in March is only 177 kg C km^{-2} a^{-1} in 2019 and 2681 kg C km^{-2} a^{-1} in 2013 for a channel depth of 10.5 m, and this variability significantly affects annual values. For example, in the years 2018 and 2013, the retention during March accounts for 24% of the total annual retention, while in 2019, it is merely 3%. The retention in March 2013 alone is above 50% of the entire annual retention of the year 2019 (5245 kg C km^{-2} a^{-1}). This underlines the importance of seasonal and short-term events.

3.9. Synthesis

The monetary values of the assessed ecosystem services vary in a wide range. While the historically important provisioning ecosystem service "wild fish catches" generates a total annual value of EUR 3.2 million (10.5 m scenario), "transport" generates an annual value of EUR 32.1 million (Figure 7). In comparison, the cultural services "recreation" (EUR 60.8 million per year) and "landscape aesthetics/observation" (EUR 36.4 million per year) generate a high annual monetary benefit. However, the most important ecosystem service monetarily is nitrogen retention (EUR 166.2 million per year). The most important process for cleaning the lagoon from nitrogen is denitrification. "Phosphorus retention" (EUR 5.0 million per year) and "carbon storage" (EUR 0.4 million per year) are much less important. Both services merely depend on burial in sediments and on the assumed compensation costs per removed ton. Especially with respect to carbon, the CO_2-certificates are still comparatively cheap causing a low monetary carbon storge value. All monetary values are related to the 10.5 m scenario.

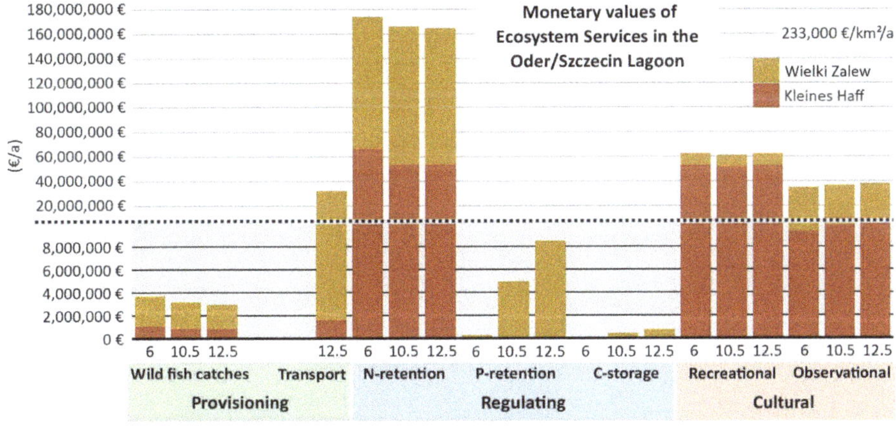

Figure 7. Monetary values (Euro) of ecosystem services in Oder/Szczecin Lagoon for three different Swina channel deepening scenarios: 6 m (historic situation after 1880), 10.5 m (after 1984), and 12.5 m (after 2024). The dashed line indicates a break in y-axis scale.

For most ecosystem services, the differences between the scenarios are minor. Strong differences only exist for phosphorus retention (EUR 0.3–EUR 8.6 million per year) and carbon storage (EUR 0.02–EUR 0.8 million per year). Some background for the differences is that the model calculates strongly increasing sediment burial in the channel with increasing channel depth. The full data are shown in Appendix B.

4. Discussion

4.1. Critical Evaluation of Approach and Quantifications

The reliability of our results largely depends on the quality and the predictive capacity of the 3D ecosystem model. This model has been applied to various scenarios in the Baltic Sea and undergone calibration, validation, and continuous development [24,30]. According to an evaluation by the European Commission [25], ERGOM is ranked among the top two European models in the category of "biogeochemical and lower trophic level models". For its application to Oder Lagoon, ERGOM has been specifically modified and applied in a very high spatial resolution. A recent study confirms the model's suitability, reliability, and long-term stability for Oder Lagoon [30]. One key lesson learned from Neumann et al. [30] is that the quality and temporal resolution of the input parameters—such as riverine discharge, nutrient loads, and wind conditions—are critical to model performance in lagoon environments. To address this, we utilized the most accurate and up-to-date official input data available.

The regulating services, nitrogen and phosphorus retention and carbon storage, were directly derived from model output. Their reliability is therefore closely tied to the model's robustness.

Our approach also integrates socio-economic and statistical data from a variety of sources, which collectively provide a strong foundation. However, challenges arose with certain ecosystem service calculation methods, which were diverse and partly subjective [27]. For instance, recalculating fish stock into biomass relied on literature data from the Baltic Sea [58,61]. The applicability and reliability of these data for the lagoon remain uncertain. Similarly, the development of the "recreation quality index" and "aesthetics quality index" involved subjective decisions in selecting, combining, and scaling parameters. These indices provide only a general indication of the lagoon's potential for recreation and tourism. Furthermore, some overlap exists between the two indices. However, the distinction was maintained to differentiate between benefits accruing to tourists and those relevant to the local population.

4.2. Monetarization of Ecosystem Services

Monetizing ecosystem services offers the advantage of making them directly comparable and easily understandable for a wide audience, including policymakers. However, this highly aggregated level of information has limitations and involves risks, as it relies on assumptions, simplifications, and various calculation methods [79–81]. The monetary values calculated are highly dependent on human demand for ecosystem services, existing policy frameworks, and cultural conditions, and they can change significantly over time.

Compared to broader studies encompassing large regions with multiple countries [27], our localized study benefits from a more consistent and comparable dataset. This allows us to account for country-specific market conditions and prices, reflecting the economic gradients between Germany and Poland.

The data on fish catches and their monetary value are considered reliable. Nonetheless, fisheries play a relatively minor role in the lagoon's economy today, in stark contrast to medieval times when they were a primary source of local income [33]. Even when secondary economic effects, such as fish processing, are included, this conclusion remains unchanged. Artisanal fishing in the lagoon has gained importance as a cultural heritage asset, attracting tourists. Recreational angling has also grown in significance and appears to generate economic values comparable to those of commercial fishing. However, due to insufficient spatially resolved statistics, these activities could not be assessed separately. The declining role of fishing is a trend observed across much of the Baltic Sea [82]. Despite growing protein demand, changing consumer preferences have led to only a few fish species being marketable, leaving the many bony whitefish species with little to no economic value [73].

Transportation, or the provision of space for commercial shipping, is assessed based on dredging and maintenance costs within the lagoon. It represents a major ecosystem service, particularly in the Polish part of the lagoon. However, this assessment does not account for significant secondary effects generated by nearby ports, such as Szczecin, since most economic activity occurs on land. Our analysis focuses on the current scenario and does not attempt to estimate historical transportation scenarios with shallower channel depths, as such estimates would be too speculative.

Cultural ecosystem services were separated into two categories: "recreation" and "observation/landscape aesthetics". Recreation was calculated based on the number and value of tourist overnight stays and is considered reliable. In contrast, the "observational" service, reflecting the lagoon's value to the local population, relies on simplified assumptions and provides little more than an approximate monetary value. No local surveys using common economic valuation methods, such as willingness-to-pay, travel cost, or hedonic pricing methods, were conducted. Additionally, transferring values from other case studies was not feasible due to various constraints [83], despite existing compilations by de Groot et al. [84]. However, cultural services at the lagoon generate comparatively high values, and their importance is increasing over time, consistent with trends across the Baltic Sea region [27].

Regulating services, including carbon storage/sequestration, nitrogen retention, and phosphorus retention, were also assessed. The monetary value of carbon storage in the lagoon is comparatively low (EUR 0.4 million per year for the 10.5 m scenario) and depends on the price of CO_2 emission certificates in the European Emissions Trading System (EU-ETS). We used the average value for 2019 of EUR 92/t CO_2 [73]. A rise to EUR 140/t [85] in the future would increase the value of this ecosystem service by approximately 50%. This illuminates the dependency of monetary ecosystem services on policy frameworks.

Phosphorus retention is a quantitatively important service in the lagoon (EUR 5 million per year for the 10.5 m scenario). It results from assuming avoidance costs of EUR 43/kg P for Poland and EUR 64/kg P for Germany [71]. Baltic Sea wide calculations used a lower averaged value of EUR 30/kg P, which met the order of magnitude of other studies by HELCOM [86].

Our model suggests that most organic material is trapped in the channel, leading to increased burial of phosphorus and carbon with greater channel depth. This finding aligns with field studies by Leipe et al. [87], who concluded that net sedimentation is negligible in other parts of Oder Lagoon due to frequent resuspension and transport processes. This view is further supported by the quantities of sediment removed from the channel during maintenance dredging in recent decades [76].

Nitrogen retention, valued at EUR 166 million per year for the 10.5 m scenario, is the lagoon's most economically significant ecosystem service. The lagoon acts as an effective sink for the large nutrient loads discharged by the Oder River [30]. This ecosystem service involves three mechanisms: sediment burial (EUR 10.5 million per year), denitrification (EUR 178 million per year), and nitrogen fixation, which adds nitrogen to the water body (−EUR 22 million per year). Denitrification is the most critical of these processes. These values are derived from assumed avoidance costs of EUR 12.7/kg N for Poland and EUR 15.8/kg N for Germany [71]. The quantity of these processes is realistic [30].

4.3. Consequences of Temporal Variabilities

The provisioning and cultural ecosystem services exhibit strong interannual variability, primarily due to inconsistencies in the data. Consequently, ecosystem service assessments require temporal integration of data to ensure reliable results. We consider our integration period of 10 years highly appropriate, as it provides stable results while also enabling the detection of long-term changes in ecosystem values over extended periods, such as several decades.

The quantification of regulating ecosystem services is entirely derived from model simulation outputs. These data feature high spatial and temporal consistency, allowing for the assessment of shorter periods, such as individual years, seasonal patterns, or specific extreme events (e.g., heatwaves, storms, droughts, or floods).

Interannual variability is particularly pronounced for all regulating services. For nitrogen retention, the variability ranges from 60% to 134%, while for phosphorus retention and carbon storage, it spans 63% to 163%, compared to the 10-year average. These variations highlight that assessments based on single years may lead to misleading conclusions. Therefore, assessments over a 10-year period are essential for generating reliable results that are meaningful compared to other monetary ecosystem services.

The model indicates that all regulating ecosystem services exhibit strong, albeit distinct, seasonality. In some years, over 50% of the total annual retention may occur within a single month. However, these seasonal patterns vary substantially between years, underscoring the influence of seasonal and short-term events. Extreme events lasting days to weeks can significantly impact annual ecosystem service values. Nonetheless, the role of such events is better understood when analyzed at the process level rather than the ecosystem service level.

Ecosystem service analyses on timescales shorter than a year are generally less meaningful, as human responses to changes in ecosystem services tend to occur over longer periods and through accumulated experience. However, recurring short-term phenomena, such as summer algal blooms, can have long-term implications for ecosystem service provision, particularly for services such as recreation.

4.4. Relevance, Transferability, and Research Needs

Ecosystem service assessments, particularly those involving comparable monetary evaluations, are increasingly required by numerous European policies [88]. The strengths and weaknesses of such assessments have been widely discussed [89]. For Oder Lagoon, several key policies are relevant and must be implemented, including the Water Framework Directive (WFD, 2000/60/EC), the Bathing Water Directive (2006/7/EC), and the Habitats Directive (FFH, 92/43/EEC). A management plan under the Habitats Directive exists for large parts of the lagoon [90]. The lagoon's currently insufficient environmental state, coupled with the requirements of these policies, underscores the need for implementing effective environmental management measures. However, the implementation of such measures is often hindered more by low public acceptance than by financial constraints. Our monetary ecosystem service evaluations quantify the benefits humans derive from the lagoon and help raise awareness of its significant economic value. de Groot et al. [85] provide summarized monetary estimates for ecosystem services in coastal systems (USD 28,917/ha/year) and rivers and lakes (USD 4267/ha/year) at 2007 price levels. Our calculated total value for all ecosystem services, approximately USD 4000/ha/year, falls within this range.

The full application of our approach requires a suitable and validated ecosystem model. However, some of our calculation methods can be adapted and transferred to other lagoon systems. For similar large lagoons in the Baltic region, monetary values for selected ecosystem services can be directly estimated using comparable methodologies.

The availability of suitable ecosystem models and their application provide a strongly improved and consistent database for the calculation of several provision and especially regulating services. Despite that, there is still space for further developments, since these models are usually restricted to lower trophic levels and focus on the water body and do not include benthic organisms and nearshore habitats. However, a major scientific challenge remains the inclusion of cultural ecosystem services. A recent review by Smardon [91] underscores their benefits and importance, with a focus on landscape aesthetics, but addresses shortcomings as well. One deficit exists with respect to quantifying, visualizing, and valuing cultural ecosystem services [92]. For coastal and marine cultural ecosystem services, a review by Garcia Rodrigues et al. [93] names several other deficits, such as a lack of integrated valuation assessments or insufficient links to benefits on human well-being. As a consequence, a generally applicable methodology for estimating different cultural ecosystem services does not exist. The few examples that address comparable lagoons address only some aspects and/or remain too conceptual [94] and suffer from the problem of a limited transferability [83].

5. Conclusions

Our model simulations suggest that channel deepening leads to a 20% decrease in annual average phytoplankton concentrations while simultaneously causing a 16% increase in summer cyanobacteria concentrations. This rise in cyanobacteria contributes to a 24% increase in nitrogen fixation within the lagoon. Additionally, a notable rise in nitrogen burial in sediments take place. These processes are compensated for by a 12% reduction in denitrification, which slightly reduces the lagoon's overall nitrogen retention. In the 10.5 m channel depth scenario, an average of 17,862 kg N a^{-1} km^2, equivalent to 39% of the annual total nitrogen load, is retained in the lagoon. This nitrogen retention is valued at EUR 166.2 million per year and is the lagoon's most important ecosystem service.

The interannual variability in nitrogen retention is substantial, ranging from 10,500 kg N a^{-1} km^2 (2015) to 23,600 kg N a^{-1} km^2 (2010) in the 10.5 m scenario. This highlights the importance of assessing ecosystem services over extended periods (e.g., a decade) rather than relying on single-year evaluations, as short-term assessments may provide misleading results.

The deepened channel also acts as a sediment trap, significantly increasing carbon and phosphorus burial rates. On average, 12% of the annual total phosphorus loads are retained in the lagoon. These regulatory ecosystem services exhibit pronounced seasonality and variability; for instance, carbon storage in March 2013 alone exceeded 50% of the total annual retention in 2019, emphasizing the critical role of seasonal and short-term events in ecosystem dynamics.

Phytoplankton-based simulations estimate the total commercially usable fish biomass in the lagoon at 13,497 tons of wet weight, with the reported annual catch of 2271 tons representing approximately 16% of this stock. However, channel deepening is projected to cause a 20% decline in commercially usable fish stocks. The total market value of the fish stock is estimated at EUR 15 million, while the annual monetary value of transportation associated with the deepened channel amounts to EUR 32 million. These findings highlight the diminishing economic significance of commercial fisheries, which once constituted the primary income source in the region.

The overall recreation index shows no significant differences across the three channel deepening scenarios. The lagoon's eutrophic condition limits its suitability for recreational activities such as summer bathing tourism. Despite these limitations, the monetary value of recreational ecosystem services is estimated at approximately EUR 60 million per year, underscoring their considerable and increasing importance.

For most ecosystem services, differences between the assessed channel deepening scenarios are minor, and the impact of ongoing human interventions appears limited. Changes in salinity remain below 1 PSU (1 g/kg) and are considered ecologically negligible. The model suggests that alterations in the lagoon–Baltic Sea connectivity resulting from channel deepening are of limited importance.

The high quantitative and monetary importance of regulating services, namely, nutrient retention, suggests that these ecosystem services and potential changes, resulting from the implementation of human interventions, need to be taken into account in the planning process of measures. They should be considered in Environmental Impact Assessments (EIAs).

Author Contributions: Conceptualization, G.S.; methodology, G.S. and T.N.; software, T.N. and S.P.; validation, G.S., T.N. and S.P.; formal analysis, G.S., T.N., S.P. and N.M.S.; investigation, G.S., T.N. and N.M.S.; writing—original draft preparation, G.S.; writing—review and editing, T.N., S.P. and N.M.S.; visualization, G.S., T.N., S.P. and N.M.S. All authors have read and agreed to the published version of the manuscript.

Funding: This work was financially supported by the German Federal Ministry of Education and Research, projects "Coastal Futures II" (grant number 03F0980B) and "Prime Prevention" (grant number 03F0911B).

Data Availability Statement: The original contributions presented in this study are included in the article. Further inquiries can be directed to the corresponding author.

Acknowledgments: We thank the following institutions for providing data: the Norwegian Meteorological Institute, the Instytut Meteorologii i Gospodarki Wodnej and the Landesamt für Umwelt, Naturschutz und Geologie—LUNG.

Conflicts of Interest: The authors declare no conflicts of interest.

Appendix A

Quantitative Ecosystem Services in Oder/Szczecin Lagoon.

Oder Lagoon	**Surface area** (km^2)	**Coast-line** (km)	**Provisioning services** **Wild fish catches** (kg wet weight/km^2)			**Wild fish biomass** (kg wet weight/km^2)			**Transportation**		
			6 m	**10.5 m**	**12.5 m**	**6 m**	**10.5 m**	**12.5 m**	**6 m**	**10.5 m**	**12.5 m**
Kleines Haff	277	89	2071	1726	1647	20,358	16,965	16,186			
Wielki Zalew	410	120	5037	4373	4031	24,910	21,624	19,933			
Total lagoon	687	209	3861	3305	3086	22,952	19,647	18,343			

Oder Lagoon	**Surface area** (km^2)	**Coast-line** (km)	**Regulating services** **Water quality regulation-nitrogen** (kg/a/km^2)			**Water quality regulation-phosphorus** (kg/a/km^2)			**Climate regulation-carbon sequestation** (kg/a/km^2)		
			6 m	**10.5 m**	**12.5 m**	**6 m**	**10.5 m**	**12.5 m**	**6 m**	**10.5 m**	**12.5 m**
Kleines Haff	277	89	15,199	12,249	12,229	0	0	0	9	1	0
Wielki Zalew	410	120	20,653	21,608	21,371	15	279	481	616	11,561	20,132
Total lagoon	687	209	18,341	17,695	17,551	9	166	287	371	6900	12,015

Oder Lagoon	**Surface area** (km^2)	**Coast-line** (km)	**Cultural services** **Active recreation-quality index** Index (0–100)			**Landscape aesthetics-quality index** Index (0–100)		
			6 m	**10.5 m**	**12.5 m**	**6 m**	**10.5 m**	**12.5 m**
Kleines Haff	277	89	50.3	49.3	50.2	53.2	56.2	57.1
Wielki Zalew	410	120	41.0	39.0	39.8	51.5	54.2	56.4
Total lagoon	687	209	44.8	43.2	44.0	51.1	54.0	55.5

Appendix B

Monetary (Euro) Ecosystem Services in Oder/Szczecin Lagoon.

Oder Lagoon	Surface area (km²)	Coast-line (km)	Provisioning services: Wild fish catches (EUR/a)			Provisioning services: Transportation (EUR/a)		
			6 m	10.5 m	12.5 m	6 m	10.5 m	12.5 m
Kleines Haff	277	89	1,088,718	907,286	865,627			1,605,208
Wielki Zalew	410	120	2,609,629	2,265,407	2,088,212			30,498,958
Total lagoon	687	209	3,698,347	3,172,693	2,953,839			32,104,167

Oder Lagoon	Surface area (km²)	Coast-line (km)	Provisioning services: Wild fish catches (EUR/km²/a)			Provisioning services: Transportation (EUR/a/km²)		
			6 m	10.5 m	12.5 m	6 m	10.5 m	12.5 m
Kleines Haff	277	89	3930	3275	3125			5795
Wielki Zalew	410	120	6365	5525	5093			74,388
Total lagoon	687	209	5383	4618	4300			46,731

Oder Lagoon	Surface area (km²)	Coast-line (km)	Regulating services: Water quality regulation–nitrogen (EUR/a)			Water quality regulation–phosphorus (EUR/a)			Carbon sequestration (EUR/a)		
			6 m	10.5 m	12.5 m	6 m	10.5 m	12.5 m	6 m	10.5 m	12.5 m
Kleines Haff	277	89	66,601,728	53,676,079	53,587,648	0	0	0	229	17	10
Wielki Zalew	410	120	107,352,392	112,320,539	111,088,233	266,946	4,939,794	8,521,102	23,164	434,500	756,641
Total lagoon	687	209	174,001,191	166,173,006	164,846,359	268,378	4,966,300	8,566,825	23,342	434,517	756,651

Oder Lagoon	Surface area (km²)	Coast-line (km)	Regulating services: Water quality regulation–nitrogen (EUR/a/km²)			Water quality regulation–phosphorus (EUR/a/km²)			Carbon sequestration (EUR/a/km²)		
			6 m	10.5 m	12.5 m	6 m	10.5 m	12.5 m	6 m	10.5 m	12.5 m
Kleines Haff	277	89	240,439	193,776	193,457	0	0	0	1	0	0
Wielki Zalew	410	120	261,835	273,953	270,947	651	12,048	20,783	56	1060	1845
Total lagoon	687	209	253,277	241,882	239,951	391	7229	12,470	34	632	1101

Oder Lagoon	Surface area (km²)	Coast-line (km)	Cultural services: Active recreation-quality index (EUR/a)			Landscape aesthetics-quality index (EUR/a)		
			6 m	10.5 m	12.5 m	6 m	10.5 m	12.5 m
Kleines Haff	277	89	52,825,238	51,724,004	52,678,264	9,386,207	9,915,370	10,075,479
Wielki Zalew	410	120	9,576,697	9,105,768	9,290,863	25,211,951	26,531,460	27,581,335
Total lagoon	687	209	62,401,934	60,829,772	61,969,126	34,598,159	36,446,830	37,656,814

Oder Lagoon	Surface area (km²)	Coast-line (km)	Cultural services: Active recreation-quality index (EUR/a/km²)			Landscape aesthetics-quality index (EUR/a/km²)		
			6 m	10.5 m	12.5 m	6 m	10.5 m	12.5 m
Kleines Haff	277	89	190,705	186,729	190,174	33,885	35,796	36,374
Wielki Zalew	410	120	23,358	22,209	22,661	61,493	64,711	67,272
Total lagoon	687	209	90,833	88,544	90,203	50,361	53,052	54,813

References

1. Wolnomiejski, N.; Witek, Z. *The Szczecin Lagoon Ecosystem*; De Gruyter Open Poland: Berlin, Germany, 2013. [CrossRef]
2. Bielecka, M.; Różyński, G. Management conflicts in the Vistula Lagoon area. *Ocean. Coast. Manag.* **2014**, *101*, 24–34. [CrossRef]

3. Schernewski, G.; Baltranaitė, E.; Kataržytė, M.; Balčiūnas, A.; Čerkasova, N.; Mėžinė, J. Establishing new bathing sites at the Curonian Lagoon coast: An ecological-social-economic assessment. *J. Coast. Conserv.* **2019**, *23*, 899–911. [CrossRef]
4. Marcinkevičiūtė, L.; Pranskūnienė, R.; Makutėnienė, D. Opportunities for Ecosystem Services in the Protected Areas in the Coastal–Rural Area of the Nemunas Delta and the Curonian Lagoon (Lithuania). *Sustainability* **2022**, *14*, 9647. [CrossRef]
5. Schernewski, G.; Friedland, R.; Buer, A.-L.; Dahlke, S.; Drews, B.; Höft, S.; Klumpe, T.; Schadach, M.; Schumacher, J.; Zaiko, A. Ecological-social-economic assessment of zebra-mussel cultivation scenarios for the Oder (Szczecin) Lagoon. *J. Coast. Conserv.* **2019**, *23*, 913–929. [CrossRef]
6. Różyński, G.; Bielecka, M.; Schönhofer, J. Application of Systems Approach Framework (SAF) to the Vistula Lagoon case in Poland. *Ocean. Coast. Manag.* **2019**, *168*, 90–102. [CrossRef]
7. Cieśliński, R.; Chlost, I.; Szydłowski, M. Impact of new, navigable canal through the Vistula spit on the hydrologic balance of the Vistula lagoon (Baltic Sea). *J. Mar. Syst.* **2024**, *241*, 103908. [CrossRef]
8. Stakėnienė, R.; Jokšas, K.; Kriaučiūnienė, J.; Jakimavičius, D.; Raudonytė-Svirbutavičienė, E. Nutrient Loadings and Exchange between the Curonian Lagoon and the Baltic Sea: Changes over the Past Two Decades (2001–2020). *Water* **2023**, *15*, 4096. [CrossRef]
9. Chubarenko, B.V.; Zakirov, R.B. Water Exchange of Nontidal Estuarine Coastal Vistula Lagoon with the Baltic Sea. *J. Waterw. Port Coast. Ocean Eng.* **2021**, *147*, 4. [CrossRef]
10. Reckermann, M.; Omstedt, A.; Soomere, T.; Aigars, J.; Akhtar, N.; Bełdowska, M.; Bełdowski, J.; Cronin, T.; Czub, M.; Eero, M.; et al. Human impacts and their interactions in the Baltic Sea region. *Earth Syst. Dynam.* **2022**, *13*, 1–80. [CrossRef]
11. Meier, H.E.M.; Kniebusch, M.; Dieterich, C.; Gröger, M.; Zorita, E.; Elmgren, R.; Myrberg, K.; Ahola, M.P.; Bartosova, A.; Bonsdorff, E.; et al. Climate change in the Baltic Sea region: A summary. *Earth Syst. Dynam.* **2022**, *13*, 457–593. [CrossRef]
12. Rutgersson, A.; Kjellström, E.; Haapala, J.; Stendel, M.; Danilovich, I.; Drews, M.; Jylhä, K.; Kujala, P.; Larsén, X.G.; Halsnæs, K.; et al. Natural hazards and extreme events in the Baltic Sea region. *Earth Syst. Dynam.* **2022**, *13*, 251–301. [CrossRef]
13. World Resources Institute. *Millennium Ecosystem Assessment. Ecosystems and Human Well-Being: Synthesis*; Island Press: Washington, DC, USA, 2005. Available online: https://www.millenniumassessment.org/documents/document.356.aspx.pdf (accessed on 18. November 2024).
14. European Commission. Commission Staff Working Document. EU Guidance on Integrating Ecosystems and Their Services into Decision-Making. *SWD (2019) 305 Final. Part 2/3.* Available online: https://data.consilium.europa.eu/doc/document/ST-11395-2019-ADD-1/en/pdf (accessed on 9 December 2024).
15. Newton, A.; Brito, A.C.; Icely, J.D.; Derolez, V.; Clara, I.; Angus, S.; Schernewski, G.; Inácio, M.; Lillebø, A.I.; Sousa, A.I.; et al. Assessing, quantifying and valuing the ecosystem services of coastal lagoons. *J. Nat. Conserv.* **2018**, *44*, 50–65. [CrossRef]
16. Maki Sy, M.; Rey-Valette, H.; Simier, M.; Pasqualini, V.; Figuières, C.; De Wit, R. Identifying Consensus on Coastal Lagoons Ecosystem Services and Conservation Priorities for an Effective Decision Making: A Q Approach. *Ecol. Econ.* **2018**, *154*, 1–13. [CrossRef]
17. Pérez-Ruzafa, A.; Pérez-Ruzafa, I.M.; Newton, A.; Marcos, C. Coastal Lagoons: Environmental Variability, Ecosystem Complexity, and Goods and Services Uniformity. *Coasts Estuaries* **2019**, 253–276. [CrossRef]
18. De Wit, R.; Leruste, A.; Le Fur, I.; Sy, M.M.; Bec, B.; Ouisse, V.; Derolez, V.; Rey-Valette, H. A Multidisciplinary Approach for Restoration Ecology of Shallow Coastal Lagoons, a Case Study in South France. *Front. Ecol. Evol.* **2020**, *8*, 108. Available online: https://www.frontiersin.org/journals/ecology-and-evolution/articles/10.3389/fevo.2020.00108 (accessed on 9 December 2024). [CrossRef]
19. Inácio, M.; Schernewski, G.; Nazemtseva, Y.; Baltranaitė, E.; Friedland, R.; Benz, J. Ecosystem services provision today and in the past: A comparative study in two Baltic lagoons. *Ecol. Res.* **2018**, *33*, 1255–1274. [CrossRef]
20. Schernewski, G.; Inacio, M.; Nazemtseva, Y. Expert Based Ecosystem Service Assessment in Coastal and Marine Planning and Management: A Baltic Lagoon Case Study. *Front. Environ. Sci.* **2018**, *6*, 19. Available online: https://www.frontiersin.org/journals/environmental-science/articles/10.3389/fenvs.2018.00019 (accessed on 9 December 2024). [CrossRef]
21. Schernewski, G.; Jekat, M.; Kösters, F.; Neumann, T.; Steffen, S.; von Thenen, M. Ecosystem Services Supporting Environmental Impact Assessments (EIAs): Assessments of Navigation Waterways Deepening Based on Data, Experts, and a 3D Ecosystem Model. *Land.* **2024**, *13*, 1653. [CrossRef]
22. Seppelt, R.; Fath, B.; Burkhard, B.; Fisher, J.L.; Grêt-Regamey, A.; Lautenbach, S.; Pert, P.; Hotes, S.; Spangenberg, J.; Verburg, P.H.; et al. Form follows function? Proposing a blueprint for ecosystem service assessments based on reviews and case studies. *Ecol. Indic.* **2012**, *21*, 145–154. [CrossRef]
23. Koundouri, P.; Alamanos, A.; Dellis, K.; Stratopoulou, A. Ecosystem Services into Water Resource Planning and Management 2022. Available online: https://EconPapers.repec.org/RePEc:aue:wpaper:2230 (accessed on 9 December 2024).
24. Neumann, T.; Radtke, H.; Cahill, B.; Schmidt, M.; Rehder, G. Non-Redfieldian carbon model for the Baltic Sea (ERGOM version 1.2)—Implementation and budget estimates. *Geosci. Model. Dev.* **2022**, *15*, 8473–8540. [CrossRef]

25. European Commission. Marine Biodiversity Modelling Study. In *Publications Office of the European Union*; European Comission: Brusells, Belgium, 2022. Available online: https://data.europa.eu/doi/10.2777/213731 (accessed on 9 December 2024).
26. Inacio, M.; Karnauskaitė, D.; Baltranaitė, E.; Kalinauskas, M.; Bogdzevič, K.; Gomes, E.; Pereira, P. Ecosystem services of the Baltic Sea: An assessment and mapping perspective. *Geogr. Sustain.* **2020**, *1*, 256–265. [CrossRef]
27. Schernewski, G.; Neumann, T.; Bučas, M.; von Thenen, M. Ecosystem Services of the Baltic Sea—State and Changes during the Last 150 Years. *Environments* **2024**, *11*, 200. [CrossRef]
28. Radziejewska, T.; Schernewski, G. The Szczecin (Oder-) Lagoon. In *Ecology of Baltic Coastal Waters*; Schiewer, U., Ed.; Springer: Berlin, Germany, 2008; pp. 115–129.
29. Friedland, R.; Schernewski, G.; Gräwe, U.; Greipsland, I.; Palazzo, D.; Pastuszak, M. Managing Eutrophication in the Szczecin (Oder) Lagoon-Development, Present State and Future Perspectives. *Front. Mar. Sci.* **2019**, *5*, 521. [CrossRef]
30. Neumann, T.; Schernewski, G.; Friedland, R. Transformation Processes in the Oder Lagoon as seen from a Model Perspective. *EGUsphere preprint*. **2025**. [CrossRef]
31. HELCOM. State of the Baltic Sea. *Third HELCOM Holistic Assessment 2016–2021. Baltic Sea Environment Proceedings 194*. Available online: https://helcom.fi/wp-content/uploads/2023/10/State-of-the-Baltic-Sea-2023.pdf (accessed on 9 December 2024).
32. IKSO. Zweite Aktualisierung des Bewirtschaftungsplans für die Internationale Flussgebietseinheit. *Int. Komm. Zum Schutz. Der Oder Gegen Verunreinig. Oder.* 2022. Available online: http://www.mkoo.pl/download.php?fid=6996&lang=DE (accessed on 9 December 2024).
33. Schernewski, G.; Friedland, R.; Paysen, S.; Bucas, M.; Dahlke, S.; von Weber, M. Macrophytes and water quality in a large Baltic lagoon: Relevance, development and restoration perspectives. *Front. Mar. Sci.* **2023**, *10*, 1049181. [CrossRef]
34. Mohrholz, V.; Lass, H.U. Transports between Oderhaff and Pomeranian Bight—A simple barotropic box model. *Ger. J. Hydrogr.* **1998**, *50*, 371–383. [CrossRef]
35. GDA SK S—SZCZECIN. Modernization Swinoujscie—Szczecin to a Depth of 12.5 m. *Volume II 2015*. Available online: https://atradiusdutchstatebusiness.nl/nl/documenten/poland_report_tom_ii.pdf (accessed on 19 August 2024).
36. Siedlik, K. Changes in the Szczecin Lagoon Shoreline as Determined from Selected Seventeenth and Nineteenth Century Maps. In *Coastline Changes of the Baltic Sea from South to East*; Harff, J., Furmańczyk, K., von Storch, H., Eds.; Springer: Cham, Switzerland, 2017; p. 19. [CrossRef]
37. Brandt, K. Ueber das Stettiner Haff. Kommission zur wissenschaftlichen Untersuchung der Deutschen Meere in Kiel und der Biologischen Anstalt auf Helgoland, Kiel und Leipzig. *Wiss. Meeresunters.* **1896**, *2*, 105–144. Available online: https://www.zobodat.at/pdf/MON-ALLGEMEIN_0294_0001-0640.pdf (accessed on 9 December 2024).
38. Port Szczecin. Summary Statistics. Available online: https://port.szczecin.pl/en/ports/szczecin/summary-statistics/ (accessed on 9 December 2024).
39. Filina-Dawidowicz, L.; Durczak, W. Organization of Vessel Traffic on Dredged Świnoujście Szczecin Fairway: VTS Operators' Viewpoint on the Use of Infrastructure for Ships Passing. *Eur. Res. Stud. J.* **2023**, *XXVI*, 152–169. [CrossRef]
40. Geyer, B.; Rockel, B. coastDat-2 COSMOCLM Atmospheric Reconstruction, 2013. Available online: https://www.wdc-climate.de/ui/entry?acronym=coastDat-2_COSMO-CLM (accessed on 9 May 2024).
41. MV. UVP-Dokumentation Für den Geplanten Bau Eines Containerterminals im Außenhafen in Swinemünde. Available online: https://www.regierung-mv.de/Landesregierung/wm (accessed on 19 August 2024).
42. Statistisches Bundesamt. Available online: https://www.destatis.de/DE/Themen/Laender-Regionen/Regionales/Gemeindeverzeichnis/Administrativ/Archiv/GVAuszugQ/AuszugGV1QAktuell.html (accessed on 9 December 2024).
43. Landesamt für innere Verwaltung. Available online: https://www.laiv-mv.de/Statistik/Zahlen-und-Fakten/Gesellschaft-&-Staat/Bevoelkerung/Statistische-Berichte (accessed on 9 December 2024).
44. Główny Urząd Statystyczny. Available online: https://bdl.stat.gov.pl/bdl/dane/podgrup/tablica (accessed on 19 August 2024).
45. Landesamt für innere Verwaltung. Available online: https://www.laiv-mv.de/Statistik/Ver%C3%B6ffentlichungen/Statistische-Berichte/G/ (accessed on 9 December 2024).
46. *Statistical Yearbook of Maritime Economy 2023*. Available online: https://stat.gov.pl/en/topics/statistical-yearbooks/statistical-yearbooks/statistical-yearbook-of-maritime-economy-2023,8,16.html (accessed on 9 December 2024).
47. Landesamt für innere Verwaltung. Available online: https://www.laiv-mv.de/Statistik/Zahlen-und-Fakten/Wirtschaftsbereiche/Verkehr/ (accessed on 9 December 2024).
48. Stybel, N.; Kleißler, K.; Schulz, N.; Gruzka, P. Fisheries Management in the Szczecin Lagoon. *Coastline Rep.* **2014**, *22*, 1–22. Available online: https://www.researchgate.net/publication/316991505_Fisheries_management_in_the_Szczecin_Lagoon (accessed on 9 December 2024).
49. GIOS. Update of the Initial Assessment of The Environmental Status of Marine Waters. Available online: https://rdsm.gios.gov.pl/images/Pliki/Update-of-initial-assessment-of-the-environmental-status-of-marine-waters.pdf (accessed on 9 December 2024).

50. Lewin, W.-C.; Weltersbach, M.S.; Haase, K.; Strehlow, H.V. Who travels how far: German Baltic sea anglers' travel distances as precondition for fisheries management and coastal spatial planning. *Ocean. Coast. Manag.* **2021**, *209*, 105640. [CrossRef]
51. Landesamt für Landwirtschaft, Lebensmittelsicherheit und Fischerei Mecklenburg-Vorpommern. Available online: https://www.lallf.de/fischerei/statistik/fischereidokumente/ (accessed on 9 December 2024).
52. Haines-Young, R.; Potschin-Young, M. Revision of the Common International Classification for Ecosystem Services (CICES V5.1): A Policy Brief. *One Ecosyst.* **2018**, *3*, e27108. [CrossRef]
53. Haines-Young, R.; Potschin, M. Common International Classification of Ecosystem Services (CICES) V5.1 and Guidance on the Application of the Revised Structure. 2018. Available online: www.cices.eu (accessed on 9 December 2024).
54. Landesamt für Landwirtschaft, Lebensmittelsicherheit und Fischerei Mecklenburg-Vorpommern. Available online: https://www.lallf.de/fischerei/statistik/fangstatistik-kuestengewaesser/ (accessed on 9 December 2024).
55. GIOS. Available online: https://rdsm.gios.gov.pl/en/6-year-assessments/second-update-of-the-initial-assessment-2016-2021 (accessed on 9 December 2024).
56. Landesamt für Landwirtschaft, Lebensmittelsicherheit und Fischerei Mecklenburg-Vorpommern. Available online: https://www.laiv-mv.de/Statistik/Ver%C3%B6ffentlichungen/Jahrbuecher/ (accessed on 9 December 2024).
57. European Commission. Economic and Social Analyses. Available online: https://stecf.ec.europa.eu/reports/economic-and-social-analyses_en (accessed on 9 December 2024).
58. Scheffold, M.; Hense, I. Quantifying Contemporary Organic Carbon Stocks of the Baltic Sea Ecosystem. *Front. Mar. Sci.* **2020**, *7*, 571956. [CrossRef]
59. CRESP. Wet to Dry Weight Conversations for Biota from Amchitka to Kiska. 2006. Available online: https://www.cresp.org/Amchitka/Final_WW_DW_3_13_06.pdf (accessed on 25 May 2024).
60. Czamanski, M.; Nugraha, A.; Pondaven, P.; Lasbleiz, M.; Masson, A.; Caroff, N.; Bellail, R.; Treguer, P. Carbon, nitrogen and phosphorus elemental stoichiometry in aquacultured and wild-caught fish and consequences for pelagic nutrient dynamics. *Mar. Biol.* **2011**, *158*, 2847–2862. [CrossRef]
61. Thurow, F. Estimation of the total fish biomass in the Baltic Sea during the 20th century. *ICES J. Mar. Sci.* **1997**, *54*, 444–461. [CrossRef]
62. Schiff und Hafen. Available online: https://www.schiffundhafen.de/nachrichten/schifffahrt/detail/industriehafen-bei-ueckermuende-wird-ausgebaut.html (accessed on 25 November 2024).
63. Dolch, T.; Schernewski, G. Hat Wasserqualität eine Bedeutung für Touristen? Eine Studie am Beispiel des Oderästuars. In *Aktuelle Ergebnisse der Küstenforschung. Berichte aus dem Forschung*; Daschkeit, A., Sterr, H., Eds.; Und Technologiezentrum Westküste der: Büsum, Germany, 2003; Volume 28, pp. 197–205.
64. Preißler, S. Wasserqualität an Europäischen Küsten und ihre Bewertung durch TOURISTEN—Eine Wahrnehmungsgeographische Untersuchung in Hamburg und auf Sylt. In *IKZM-Oder Berichte*; Universität Hamburg: Hamburg, Germany, 2008; Volume 54, p. 134; ISSN 1614-5968; Available online: https://databases.eucc-d.de/files/documents/00000738_IKZM-Oder_Berichte54.pdf (accessed on 13 September 2024)ISSN 1614-5968.
65. Bertram, C.; Ahtiainen, H.; Meyerhoff, J.; Pakalniete, K.; Pouta, E.; Rehdanz, K. Contingent Behavior and Asymmetric Pref-erences for Baltic Sea Coastal Recreation. *Environ. Resour. Econ.* **2020**, *75*, 49–78. [CrossRef]
66. Tourismus in Deutschland im Jahr 2023. Available online: https://www.destatis.de/DE/Presse/Pressemitteilungen/2024/02/PD24_053_45.html#:~:text=Die%20Zahl%20der%20%C3%9Cbernachtungen%20von,,9%20%25%20h%C3%B6her%20als%202022 (accessed on 9 December 2024).
67. Statistische Ämter. Available online: https://www.statistikportal.de/de/ugrdl/ergebnisse/wirtschaft-und-bevoelkerung/bipbws (accessed on 9 December 2024).
68. Bundesministerium für Wirtschaft und Energie (BMWi). Wirtschaftsfaktor Tourismus in Deutschland. Available online: https://www.bmwk.de/Redaktion/DE/Publikationen/Tourismus/wirtschaftsfaktor-tourismus-in-deutschland-lang.pdf?__blob=publicationFile&v=1 (accessed on 9 December 2024).
69. Europäische Union. Available online: https://european-union.europa.eu/principles-countries-history/eu-countries/germany_de (accessed on 9 December 2024).
70. Europäische Union. Available online: https://european-union.europa.eu/principles-countries-history/eu-countries/poland_de (accessed on 9 December 2024).
71. Gren, I.-M.; Jonzon, Y.; Lindqvist, M. *Costs of Nutrient Reductions to the Baltic Sea: Technical Report*; Department of Economics, Swedish University of Agricultural Sciences: Uppsala, Sweden, 2008; Working Paper.
72. Umweltbundesamt. Preisentwicklung für Emissionsberechtigungen (EUA) seit 2008. Available online: https://www.umweltbundesamt.de/daten/klima/der-europaeische-emissionshandel#teilnehmer-prinzip-und-umsetzung-des-europaischen-emissionshandels (accessed on 25 May 2024).
73. Institut für Binnenfischerei, e.V. Recherche und Analyse der Möglichkeiten der Weißfischverwertung in Nordbrandenburg und Mecklenburg-Vorpommern, Projektabschlussbericht, FFUS-004-2019. Available online: https://chara-seen.de/ (accessed on 9 December 2024).

74. Köbbing, J.F.; Thevs, N.; Zerbe, S. The utilisation of reed (*Phragmites australis*): A review. *Mires Peat* **2013**, *13*, 1–14. Available online: http://mires-and-peat.net/media/map13/map_13_01.pdf (accessed on 9 December 2024).
75. Copernicus. Available online: https://land.copernicus.eu/en/products/coastal-zones/coastal-zones-2012#download (accessed on 9 December 2024).
76. Minning, M. Der Schifffahrtskanal im Oderhaff—Eine Sediment-, Nähr- und Schadstofffalle? *Diplomarbeit zur Diplomprüfung im Fach Geographie an der Christian-Albrechts-Universität zu Kiel*. 2003. Available online: https://eucc-d-inline.databases.eucc-d.de/files/documents/00000695_Diplomarbeit_Minning2.pdf (accessed on 9 December 2024).
77. IADC. Modernisation of the Swinouscie—Szczecin Fairway. Available online: https://www.iadc-dredging.com/article/modernisation-of-the-swinoujscie-szczecin-fairway/#:~:text=With%20more%20approximately%2024%20million,the%20next%20generation%20of%20vessels (accessed on 9 December 2024).
78. Interreg Central Europe. The Action Plan for the Implementation of Environmentally Friendly Solutions in Freight Transport of Szczecin and Świnoujście Multimodal Node and Transport Connections. Available online: https://programme2014-20.interreg-central.eu/Content.Node/D.T-2.5.9-Action-Plan-Port-of-Szczecin----winouj-cie- -ENERG.pdf (accessed on 9 December 2024).
79. Mehvar, S.; Filatova, T.; Dastgheib, A.; De Ruyter van Steveninck, E.; Ranasinghe, R. Quantifying Economic Value of Coastal Ecosystem Services: A Review. *J. Mar. Sci. Eng.* **2018**, *6*, 5. [CrossRef]
80. Sagebiel, J.; Schwartz, C.; Rhozyel, M.; Rajmis, S.; Hirschfeld, J. Economic valuation of Baltic marine ecosystem services: Blind spots and limited consistency. *ICES J. Mar. Sci.* **2016**, *73*, 991–1003. [CrossRef]
81. Czajkowski, M.; Budziński, W.; Zandersen, M.; Zawadzki, W.; Aslam, U.; Angelidis, I.; Zagórska, K. The Recreational Value of the Baltic Sea Coast: A Spatially Explicit Site Choice Model Accounting for Environmental Conditions. *Environ. Resour. Econ.* **2024**, *87*, 135–166. [CrossRef]
82. Kulmala, S.; Haapasaari, P.; Karjalainen, T.P.; Kuikka, S.; Pakarinen, T.; Parkkila, K.; Romakkaniemi, A.; Vuorinen, P. Ecosystem services provided by Baltic salmon: A regional perspective to the socio-economic benefits associated with a keystone species. *Econ. Ecosyst. Biodiveristy Nord. Synth.* **2013**, 266–276. Available online: https://www.teebweb.org/wp-content/uploads/2013/01/TEEB-case_TEEBNordic_Ecosystem-services-provided-by-Baltic-salmon.pdf (accessed on 9 December 2024).
83. Liu, S.; Costanza, R.; Farber, S.; Troy, A. Valuing ecosystem services: Theory, practice, and the need for a transdisciplinary synthesis. *Ann. N. Y Acad. Sci.* **2010**, *1185*, 54–78. [CrossRef] [PubMed]
84. De Groot, R.; Brander, L.; Van Der Ploeg, S.; Costanza, R.; Bernard, F.; Braat, L.; Christie, M.; Crossman, N.; Ghermandi, A.; Hein, L.; et al. Global estimates of the value of ecosystems and their services in monetary units. *Ecosyst. Serv.* **2012**, *1*, 50–61. [CrossRef]
85. Homaio. 2030 EUA Price Predictions: Expert Analysis of 3 Scenarios. Available online: https://www.homaio.com/post/2030-eua-price-predictions-expert-analysis-of-3-scenarios (accessed on 20 August 2024).
86. HELCOM. Thematic Assessment of Economic and Social Analyses 2016–2021. *Baltic Sea Environment Proceedings 188*. Available online: https://helcom.fi/wp-content/uploads/2023/03/HELCOM-Thematic-assessment-of-economic-and-social-analyses-2016-2021.pdf (accessed on 25 May 2024).
87. Leipe, T.; Eidam, J.; Lampe, R. Das Oderhaff—Beiträge zur Rekonstuktion der holozänen geologischen Entwicklung und anthropogenen Beeinflussung des Oder-Ästuars. *Meereswissenschaftliche Berichte* **1998**, *28*, 1–61.
88. Bouwma, I.; Schleyer, C.; Primmer, E.; Winkler, K.J.; Berry, P.; Young, J.; Preda, E. Adoption of the ecosystem services concept in EU policies. *Ecosyst. Serv.* **2018**, *29*, 213–222. [CrossRef]
89. Bull, J.W.; Jobstvogt, N.; Böhnke-Henrichs, A.; Mascarenhas, A.; Sitas, N.; Baulcomb, C.; Lambini, C.; Rawlins, M.; Baral, H.; Zähringer, J.; et al. Strengths, Weaknesses, Opportunities and Threats: A SWOT analysis of the ecosystem services framework. *Ecosyst. Serv.* **2016**, *17*, 99–111. [CrossRef]
90. Staatliches Amt für Landwirtschaft und Umwelt Vorpommern. Managementplan. Available online: https://www.stalu-mv.de/ms/Themen/Naturschutz-und-Landschaftspflege/NATURA-2000/Managementplanung/?racr=a (accessed on 9 December 2024).
91. Smardon, R.C. Ecosystem Services for Scenic Quality Landscape Management: A Review. *Land* **2021**, *10*, 1123. [CrossRef]
92. Hermann, A.; Schleifer, S.; Wrbka, T. The concept of ecosystem services regarding landscape research; A review. *Living Rev. Landsc. Res.* **2011**, *5*, 1–37. Available online: http://www.livingreviews.org/lrlr-2011-1 (accessed on 9 December 2024). [CrossRef]
93. Garcia Rodrigues, J.; Conides, A.; Rivero Rodriguez, S.; Raicevich, S.; Pita, P.; Kleisner, K.; Pita, C.; Lopes, P.; Roldán, V.A.; Ramos, S.; et al. Marine and Coastal Cultural Ecosystem Services: Knowledge gaps and research priorities. *One Ecosyst.* **2017**, *2*, e12290. [CrossRef]
94. Marcinkeviciute, L.; Pranskuniene, R. Cultural Ecosystem "employing" available natural resources, i.e., the ecosystems. Services: The Case of Coastal-Rural Area (Nemunas Delta and Curonian Lagoon, Lithuania). *Sustainability* **2021**, *13*, 123. [CrossRef]

environments

MDPI

Article

Geomorphic Comparison of Three Globally Significant Wetland Landscapes

Jessica Sullivan [1,*], Michael Foster [1], James Chassereau [2] and Robert Sullivan, Jr. [3]

[1] Department of Biological, Environmental and Earth Sciences, University of South Carolina Aiken, Aiken, SC 29803, USA; mf30@usca.edu
[2] Savannah River Site Biomass Facility-Ameresco, Aiken, SC 29831, USA; jchassereau@ameresco.com
[3] Aiken County Sheriffs Office, Aiken, SC 29801, USA; rsullivan2@aikencountysc.gov
* Correspondence: jessicasu@usca.edu; Tel.: +1-803-641-3621

Abstract: Salt marshes are dynamic coastal environments that play a critical role in sediment transport, nutrient cycling, and carbon sequestration. However, the geomorphic factors that influence water flow and material exchange in and between marshes of different size, shape and type remain poorly understood. In this study, we compare the morphology of three distinct marsh landscapes within the ACE Basin, South Carolina: a natural salt marsh (Ethan's Island), a reclaimed agricultural marsh (Alligator Marsh), and a natural marsh at the upland forest interface (Hannah's Marsh). Using high-resolution digital elevation models and estimates for drainage density and drainage efficiency, we quantify the similarities and differences in morphology between the different marsh types, and discuss potential implications of our findings in regard to water flow and material exchange. We found that drainage density and drainage efficiency are not always positively correlated and, importantly, the agricultural marsh displays the highest drainage efficiency, despite a drainage density that is comparable to the natural marshes. This finding reveals that the unique linear, interconnected structure of creek networks in the agricultural marsh yields higher efficiency. Further, we found that the natural marsh with the most complex, meandering and fragmented creek networks displayed the lowest drainage efficiency, despite having the highest drainage density. Together, these findings suggest that both drainage density and drainage efficiency should be considered separately, and that drainage efficiency is largely influenced by the structure and spatial arrangement of creek networks with a marsh. Given the relatively higher drainage efficiency of the ag-marsh landscapes, we speculate that such landscapes may enhance the flow of water and sediment between inland areas and the coastal ocean, a process that can help marshes migrate landward as sea levels rise.

Keywords: salt marsh; geomorphology; drainage density; drainage efficiency; ag-marsh

1. Introduction

Salt marshes remain the focus of intensive study due to their ecosystem functions and, more recently, in response to the unknown effects of sea-level rise on their overall stability [1,2]. Salt marshes function as intertidal floodplains, storing water and sediment with each tidal cycle or storm event [1–3]. Typically, these low-relief landscapes serve as depositional environments for marine and terrestrial sediments, with complex overmarsh currents producing variable patterns of suspended and dissolved material exchange with the sedimentary surface [4–10]. Salt marshes are highly productive with a net primary production that ranges from 650 to 3700 g dry biomass $m^{-2}yr^{-1}$ [11,12] and are important nursery habitats for economically important fisheries [13–16]. They also help protect terrestrial lowlands from erosive tidal and wave energy [1,3,17] and have a high capacity for carbon storage and sequestration [18–26]. As such, the restoration, enhancement, and conservation of coastal marshes are becoming increasingly important [27].

The ecosystem services provided by marsh environments are sustained by process-form feedback that operates within the context of a complex geomorphic setting [28–34]. For instance, a study recently combined field observations and numerical modeling to show that current directions and speed vary substantially over relatively short time and space scales, and that this flow complexity arises from the submergence and emergence of subtle marsh topography [32,35]. Additionally, Lidar-based analyses have highlighted how microtopographic variations shape sediment deposition patterns, underscoring the need for precise geomorphic knowledge [36]. To accurately assess the function, structure and overall stability of marshes, we must deepen our understanding of the geomorphic controls on flow and sedimentation within and between marsh systems, and with adjacent upland regions. This is important because such interactions influence spatial and temporal aspects of marsh migration in response to sea level rise [37,38]. Recent studies have emphasized that the ability of marsh landscapes to migrate inland is a first-order determinant of their overall survival under accelerated sea-level rise, with models incorporating topographic and sediment dynamics showing variable migration success [37–39]. Nevertheless, there remains a paucity of robust data to characterize the geomorphic factors that influence inland marsh migration and, consequently, this process is not fully accounted for in predictive models which tend to overestimate marsh vulnerability to sea-level rise [37,40,41].

One reason for our poor constraint on overall marsh stability is that studies of inland marsh expansion are largely limited to forests, shrub land [42] and residential areas [43], although agricultural fields are abundant features of coastal plain landscapes worldwide and are more exposed to projected sea-level rise than other upland regions [44,45]. Furthermore, the low elevation and gentle slope of agricultural lands are likely to increase the potential for landward marsh expansion as waters rise [37,46–49]. As such, about half of the marsh restoration projects in the southeastern US [50,51], and in other coastal regions worldwide [50–55] occurred in areas with an agricultural past. Large-scale wetland restoration projects are also planned for coastal areas of Louisiana and Maryland, where broad expanses of agricultural lands are common [51]. In addition to these engineered "ag-marsh" systems, naturally restored marshes reclaimed by sea-level and land use changes are also common, particularly in the southeastern US.

Despite the rise in ag-marsh landscapes in the last decade, few studies have looked at how the unique morphology of ag-marshes affects water and sediment exchange [35,56]. Artificial drainage infrastructures in the form of ditches, channels and canals are ubiquitous features of ag-marshes, and these features may enhance hydraulic connectivity of water and sediment between terrestrial uplands and the sea [57–66], and increase inundation extent and frequency within adjacent uplands [63,64,67]. Hence, such marsh types may act as preferential corridors for inland marsh migration as waters rise [68]. Nevertheless, there is a paucity of work to quantify how ag-marsh landscapes compare to natural marsh systems regarding structure and function [69]. In this study, we provide a geomorphic comparison of three ACE Basin marsh sites along a salt marsh—forest continuum that encompasses a natural salt marsh, a reclaimed ag-marsh, and a forest-marsh transition zone. The proposed field sites are located within the Ashepoo, Combahee, and Edisto (ACE) River Basin, SC. We quantitatively compare the geomorphic structure of the marsh sites using drainage density and drainage efficiency as metrics and discuss our findings in the context of marsh structure and function. This work is important because it can help identify and prioritize coastal wetland landscapes that are most capable of facilitating the transfer of water, sediment, and nutrients in and between marsh systems, and between ocean and inland areas [70]. Such information is needed to inform coastal management and policy in the context of natural and anthropogenic change within our coastal margins [71].

2. Description of Study Sites

The study investigates the morphology of three distinct marsh sites located within the Ashepoo, Combahee, and Edisto (ACE) Basin of the southeastern North American coastal plain (Figure 1).

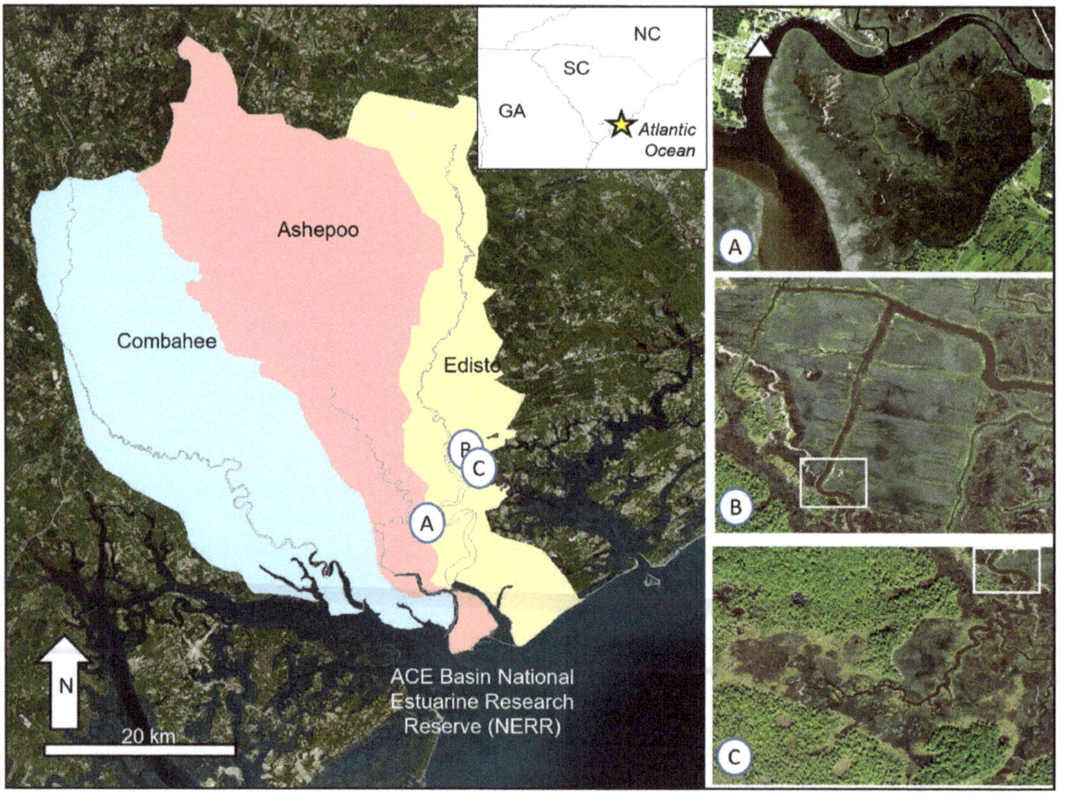

Figure 1. The ACE Basin, SC, consists of the Ashepoo, Combahee and Edisto River basins. The marsh study sites are (**A**) Ethan's Island, a natural salt marsh located closest to the coastal ocean, (**B**) Alligator Marsh, a reclaimed agricultural marsh, and (**C**) a natural marsh adjacent to the forested upland. Hannah's Marsh and Alligator marsh are connected by a main subtidal channel denoted with the white box.

The ACE Basin stands out as one of the largest undeveloped estuaries in the USA, with a drainage area of ~8000 km^2 [72]. The basin comprises diverse habitats, including pine and hardwood forests, freshwater, brackish, and saltwater tidal marshes, barrier islands, and beaches, establishing itself as the preeminent estuarine resource in South Carolina [73]. Despite its pristine condition, the ACE Basin has not escaped the impacts of both recent and historical land use practices such as agriculture, silviculture, aquaculture, impoundments, controlled burning, and forestry [74,75]. Agricultural activities in the basin ceased in the late 1800s, allowing natural processes to reclaim many of these landscapes [76,77]. In 1992, the ACE Basin was designated as a National Estuarine Research Reserve (NERR) site by the National Oceanic and Atmospheric Administration to promote long-term research, education, coastal training, and stewardship.

The first field site, Ethan's Island, is a ~0.15 km^2 natural salt marsh located ~10 km inland from the ocean (Figure 1A). The site receives semidiurnal tides along its northern border via a subtidal distributary channel of the South Edisto River. The salt marsh is unrestricted in flow along all other borders. Two primary intertidal creeks traverse the eastern and western perimeters, and each is flanked by natural levees. These natural levees gradually lose their distinctiveness as they extend into the marsh interior (Figure 1A). Discontinuities in the levees create hydraulic links, facilitating the flow of water between the primary intertidal creeks and the marsh platform through secondary intertidal creeks. The salinity fluctuates between 30 and 35 ppt.

The second site is Alligator Marsh, an abandoned rice field marsh that has been "reclaimed" by natural tidal processes for the past ~200 years (Figure 1B). The 0.30 km^2 "ag-marsh" is located 17 km from the coastal ocean. The marsh is open to flow along its entire perimeter, and receives diurnal tides via Fishing Creek, a subtidal channel which feeds the system from the east (Figure 1B). Within the study area, Fishing Creek splits into two distributed subtidal channels: one that flows along the eastern boundary of the marsh, and a second that forms the northern and western boundaries of the site. On the marsh platform, artificial drainages in the form of canals, ditches, and dikes incise the landscape. Most of these features are linear, while some form meanders and distributary creeks (Figure 1B). Salinity ranges from 22 to 32 ppt.

The third field site, Hannah's Marsh, is a 0.03 km^2 marsh-forest transition zone, located ~16 km from the ocean and located immediately southwest of Alligator Marsh (Figure 1C). Hannah's marsh is the smallest of the three sites and is unique in that it shares a direct connection to the ag-marsh landscape, Alligator Marsh, via the main subtidal channel that forms the southeastern edge of the site. The marsh shares its northwestern boundary with a forested upland area, forming a marsh-forest transition zone. Field observations of the site reveal freshwater vegetation die-off, indicating encroachment of the salt marsh into the forested area. The salinity ranges from 22 to 23 ppt.

3. Methods

3.1. Digital Elevation Models

A 2012 lidar point cloud dataset was obtained for Colleton and Charleston Counties, SC, sourced by the South Carolina Department of Natural Resources. The lidar dataset has a vertical accuracy of 0.08 m to 0.11 m root-mean-square error on open and vegetated terrain, with ground returns spaced nominally at 0.7 m. Lidar tiles were projected to NAD 1983 HARN State Plane South Carolina and North American Vertical Datum 1988 for horizontal and vertical projections, respectively. The raw point cloud data was processed to generate a bare-earth Digital Elevation Model (DEM) by first applying the LP360 Adaptive TIN Ground Filter algorithm to the lidar tiles. This tool classifies points as "ground returns" by creating and refining the TIN based on an initial set of ground seed points. It then iteratively adds new points to the TIN that are within a specified angular threshold and maximum distance from the current ground seed points [78,79]. Here, we prescribed an angular threshold of 6 degrees and a threshold distance of 0.1 m above the ground seed points. This meticulous filtering process is essential for removing dense vegetation canopy, which often obscures the true marsh surface [78–80]. Then, within the same software, triangulation was applied to generate bare-earth digital elevation models (DEMs) at a resolution of 1 m × 1 m for each of the marsh study sites.

3.2. Creek Network Extraction, Drainage Density and Drainage Efficiency

Elevation histograms were generated for each marsh site to determine the elevation range associated with geomorphic subclasses [32,80,81]. Using Spatial Analyst tools in ESRI ArcGIS version 10.8.2, the DEM surfaces were classified according to the histogram subclasses, and the percentage area of each landcover type was calculated. Creek networks were identified for each DEM using the elevation contour associated with the intertidal creek geomorphic subclass. The accuracy of the channel extraction was evaluated through visual inspection of the DEM and field observations. The length of the intertidal creek contour was then computed and divided by 2 to estimate the total creek length for each marsh site [32,33]. A sinuosity ratio was also computed for each intertidal creek by dividing the actual channel length (ACL) by the straight-line distance (SLD) between the creek mouth and the creek head (Equation (1)).

$$S = \frac{\text{ACL (m)}}{\text{SLD (m)}} \tag{1}$$

Excluding the area associated with subtidal channels, the drainage density of each marsh site was estimated by dividing the total creek length by the area of the marsh interior [26,56]:

$$Dd = \frac{\sum \text{intertidal creek length (m)}}{\text{Area of Marsh Interior } (\text{m}^2)} \tag{2}$$

Drainage efficiency was then estimated using the approach outlined by [82–84]. Specifically, the mean unchanneled path length, l, representing the average distance a water particle at a point on the marsh platform travels before reaching a conduit, was estimated [82,83,85]. This was achieved by using the Spatial Analyst Cost-Distance tool to estimate the shortest path from each grid cell of the DEM to the nearest intertidal creek edge, and assuming topographically driven flow. This tool generates a scalar field over the entire marsh area. The mean unchanneled path length is then divided by the inverse of drainage density, or the Hortonian length (Hl, after [82]), to yield a nondimensional metric for drainage efficiency [78]:

$$De = \frac{l\ (m)}{1/Dd} \tag{3}$$

4. Results

4.1. Marsh Morphology

The elevation histogram derived from Ethan's Island DEM shows values ranging from −4.7 m to 1.36 m (Figure 2).

The distribution is unimodal with a peak that is centered on a mode of 0.71 m. The mean elevation for the site is 0.45 ± 1.2 (standard deviation) which is less than the mode, indicating a negative skew. From left to right there are five breaks in the histogram slope which represent the elevation range of different geomorphic subclasses (Figure 2a). Specifically, the lowest elevations are associated with the subtidal channels that flank the study site to the west, north, and east (Figure 3), and extend to 0.09 m.

The subtidal area accounts for 14% of the total site area (Table 1).

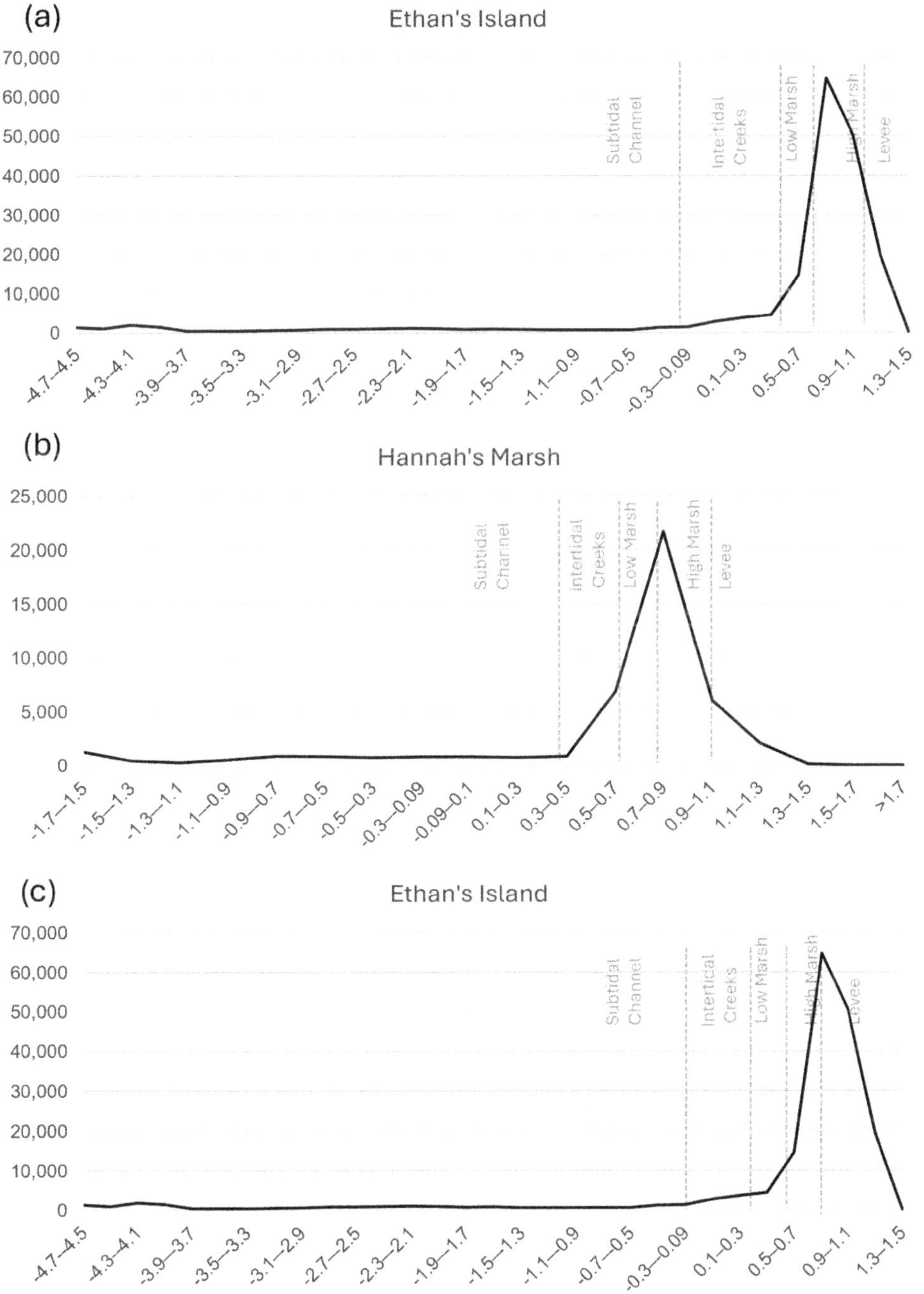

Figure 2. Elevation histograms for (**a**) Ethan's Island (**b**) Alligator marsh, and (**c**) Hannah's Marsh. The vertical and horizonal axis mark the frequency of elevation and elevation values, respectively. Vertical dashed lines mark the slope breaks associated with geomorphic classifications.

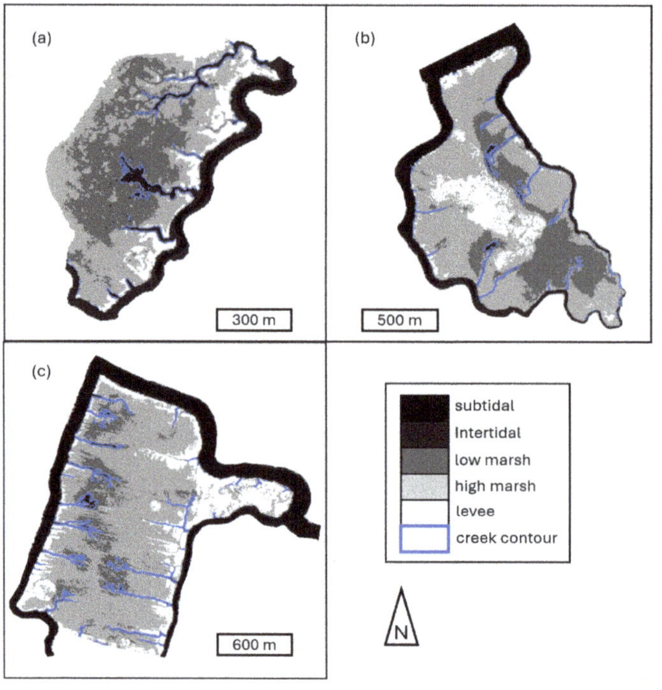

Figure 3. The DEMs for (**a**) Hannah's Marsh, (**b**) Ethan's Island, and (**c**) Alligator Marsh are shown. Shades of black to white represent the geomorphic subclasses for elevations associated with subtidal channels, intertidal creeks, low marsh, high marsh, and levees, as identified from histogram slope breaks. Note the linear nature of the ag-marsh landscape (**c**) relative to the natural marshes (**a**,**b**).

Table 1. Table showing metrics used for geomorphic comparison between the three study sites. The metrics include total creek length (m), average sinuosity, drainage density, unchanneled path distance, and drainage efficiency.

	Hannahs Marsh	Ethan's Island	Alligator Marsh
Min Elevation (m)	−1.5	−4.7	−3.87
Max Elevation (m)	1.7	1.36	1.52
Mean Elevation (m)	0.54 (±0.62)	0.45 (±1.2)	0.37 (±0.99)
Total Marsh Area (m^2)	45,014	181,526	311,460
Area of Marsh Interior (m^2)	42,825	155,876	257,079
Subtidal channel (m^2)	2189 (5%)	25,650 (14%)	54,381 (17%)
Intertidal creeks (m^2)	7314 (16%)	12,156 (7%)	24,037 (8%)
Low Marsh (m^2)	13,449 (30%)	36,422 (20%)	29,821 (10%)
High Marsh (m^2)	17,712 (39%)	87,580 (48%)	154,956 (50%)
Levee (m^2)	4350 (10%)	19,718 (11%)	48,265 (15%)
# Int Creeks	7	11	23
Total Creek Length (m)	717	1425	3430
Avg. S	1.20 (0.15)	1.10 (0.10)	1.07 (0.03)
Dd (m^{-1})	0.02	0.01	0.01
l (m)	15.9	23.58	22.23
De	3.14	4.24	4.50

The next break in slope marks the transition from subtidal channels to intertidal creeks, which extend to 0.65 m elevation (Figure 2a) and occupy 7% of the marsh area (Table 1). The intertidal creeks at Ethan's Island are relatively straight with an average sinuosity value of 1.1 ± 0.10 (Table 1) and tend to incise the marsh platform from the eastern and western flanks (Figure 3). However, these fragmented creek networks do not appear to form lateral connections across the marsh interior (Figure 3). Higher up in the tidal framework is the marsh platform, which can be subdivided into low marsh (0.65–0.79 m) and high marsh (0.79–1.1 m) (Figure 2a). The low marsh occupies 20% of the marsh area (Table 1) and forms an apparent trough connecting the heads of individual creek networks (map figure). Most of the site comprises high marsh, which takes up 48% of the total area. The highest elevations range from 1.1 to 1.37 (Figure 2a) and are associated with levees that occupy 11% of the area and tend to border subtidal channels and intertidal creeks (Figure 3).

The elevation distribution for Hannah's Marsh ranges from −1.5 m to 1.62 m with a mean value of 0.55 ± 0.62 (Figure 2b). The histogram shows a unimodal distribution with a peak centered at a mode of 0.74 m. The lowest elevations occupy 5% of the total area (Table 1) and form the main subtidal channel that borders the marsh to the southeast (Figure 3). Intertidal creeks form between 0.30 m and 0.60 m and occupy 16% of the site (Table 1). The creek features of Hannah's Marsh form along the subtidal channel to the east (Figure 3) and are more sinuous than those at Ethan's Island, with an average sinuosity ratio of 1.2 ± 0.15 (Table 1). The low marsh and high marsh platform range from 0.60 to 0.74 m, and 0.74 to 0.95 m, respectively. The low marsh occupies 30% of the site, while the high marsh takes up 39% (Table 1). The highest elevations are associated with levees, which range from 0.95 m to 1.62 m (Figure 2b) and take up 10% of the site area.

Alligator Marsh elevations range from −3.9 to 1.53 m (Figure 2c). The histogram shows a unimodal distribution with a mode of 0.76. The mean elevation is 0.37 m ± 0.99. Two main subtidal channels account for 17% of the study site (Table 1) and surround the marsh site along the northern, western, and eastern borders (Figure 3). Intertidal creeks form between −0.30 m and 0.60 m (Figure 2c) and occupy 8% of the marsh site. As stated earlier in the narrative, the spatial arrangement of intertidal creeks for Alligator Marsh is unique and reflects the agricultural history of the site. Specifically, intertidal creeks incise the marsh interior along predominantly linear paths (average sinuosity value of 1.07 ± 0.03), with an east–west orientation (Figure 3). In most cases, individual creek networks form apparent connections between the two separate subtidal systems flanking the marsh (Figure 3). The low marsh occupies 10% of the marsh site and ranges from 0.60–0.70 m (Figure 2), followed by the high marsh which extends from 0.70 to 0.96 m and occupies 50% of the site (Table 1). Levees form at elevations of 0.96–1.53 m and take up 15% of the marsh area.

4.2. Drainage Density and Drainage Efficiency

A total of 11 intertidal creeks were identified from the DEM for Ethan's Island (Table 1), with a total creek length of 1425 m (Table 1). Dividing the total creek length by the area of the marsh interior, excluding the area of the subtidal channel (155,876 m^2), yielded a drainage density of 0.01 m^{-1}. Given a Hortonian length, *Hl* of 35.05 and a mean unchanneled path length of 23.58 m (Table 1), the drainage efficiency was estimated to be 4.24.

The largest marsh site, Alligator marsh, also has the highest number of intertidal creeks. There is a total of 23 creeks with a total length of 3430 m. Dividing the total creek length by the area of the marsh interior (257,079 m^2) resulted in a drainage density of 0.01 m^{-1}. The *Hl* for Alligator marsh is 100 and the mean unchanneled path length is 22.23 m. This yielded a drainage efficiency of 4.50 (Table 1).

A total of 7 intertidal creeks were identified for Hannah's Marsh (Table 1). The creek networks form complex networks that have a southeast to northwest orientation (Figure 3) and have a total length of 717 m (Table 1). Dividing the total creek length by the area of the marsh interior (42,825 m^2) yielded a drainage density of 0.02 m^{-1}. Taking the inverse of the Dd yielded a Hortonian length, *Hl*, of 50. The mean unchanneled path length for the site was 15.9 m, revealing a drainage efficiency of 3.14 (Table 1).

A comparison of metrics between the three marsh sites reveals a drainage density of 0.01 m^{-1} for both Ethan's Island and the ag-marsh, Alligator marsh (Table 1). The smallest marsh site, Hannah's Marsh, exhibits the highest drainage density, with a value of 0.02 m^{-1}. On the other hand, Hannah's marsh has a drainage efficiency of 3.14, which is the lowest efficiency of the three sites (Table 1). Further, Alligator marsh displays the highest drainage efficiency (4.50), which is 6% higher than that of Ethan's Island, despite having the same drainage density (Table 1). This outcome is likely a result of the relatively shorter mean unchanneled path distance of Alligator Marsh with respect to Ethan's Island.

5. Discussion

Drainage density is commonly used to quantify a marsh's ability to move water in and out of a system [82,86–92], yet our results demonstrate that drainage density does not always correlate positively with drainage efficiency. While drainage density can provide a useful metric to describe how dissected a marsh landscape is, it does not account for the structure or spatial layout of the creek networks which play a key role in circulation pathways during flood and ebb [32,33,93,94]. On the other hand, drainage efficiency accounts for the spatial orientation of creek networks, particularly those that may enhance the potential for hydraulic connectivity across the system. Here, in the case Alligator Marsh, the unique morphology of the ag-marsh landscape, which is characterized by historically engineered, predominantly linear drainages, increased the potential for hydraulic connections from one side of the marsh to the other, resulting in a higher drainage efficiency than the other two marsh sites (Hannah's Marsh and Ethan's Island) even though the drainage density was lower than Hannah's Marsh and comparable to Ethan's Island (Table 1). This finding reinforces the assertion that drainage density and drainage efficiency should be considered separately in morphological studies of marsh landscapes, as the latter provides a more relevant measure of hydraulic function [90,92]. Indeed, recent studies have demonstrated that drainage efficiency, albeit a static metric, is relevant for processes (i.e., marsh plant invasion) that depend on how well water, sediments, seeds, etc., move across and throughout a marsh system [86], and that improving the drainage efficiency can yield better ecological/hydrological functioning over and above just having a certain density of channels [92].

This finding on the relatively higher drainage efficiency of the ag-marsh landscape with respect to its natural counterparts is important because it highlights a potentially important quality of ag-marsh landscapes in facilitating enhanced movement of water and sediment within and between marsh systems that is a direct result of their unique geomorphic structure. In particular, we speculate that the predominantly linear structure and spatial layout of the ag-marsh creeks, with creek heads that form apparent connections with the heads of creeks on the opposite sides of the system, may facilitate a unique type of hydraulic connectivity. This is important, given that such hydraulic connections have been shown to greatly impact overmarsh circulation patterns and water pathways between flood and ebb tide [33,90–95]. We further speculate that the increased hydraulic connectivity of the ag-marsh landscape may also reduce water residence times by creating "short-circuiting" pathways for water and sediment on the outgoing tide [33]. This potentially expedited transport enabled by the linear drainage features contrasts with the more sinuous, topographically controlled flow in natural marshes. Hence, we contend that the ag-marsh landscapes may play an important role in the efficient exchange of water and sediments within and between individual marsh systems. Future efforts are required, however, to further investigate our assertions through field assessment and numerical modeling.

This finding on the potential significance of ag-marsh landscapes in enhanced water and material exchange may also have important implications for temporal and spatial patterns of inland marsh migration in the context of sea level rise. Marsh survival is highly dependent on their ability to migrate to higher ground as waters rise [37,40,96–98]. Given their position between the coastal ocean and terrestrial uplands, ag-marsh landscapes are likely to share boundaries and/or hydraulic connections to marsh-forest transition zones. The engineered linear drainages and canal characteristics of former ag-marsh landscapes may act as preferential corridors for enhanced transmission of tidal water and sediment far inland towards terrestrial upland areas [61,62]. This enhanced inland water delivery, facilitated by a relatively higher drainage efficiency, could accelerate the salinization and inundation of adjacent uplands, thereby promoting marsh migration into abandoned agricultural fields and forest edges—a process increasingly recognized as critical for marsh persistence under the slow press of sea level rise [45,48,61,66,68,96].

Ag-marsh landscapes are abundant within the ACE Basin, SC [97], as well as in many other coastal areas worldwide [98,99], and often share a common boundary with terrestrial uplands [72,100–104]. If these landscapes are indeed more hydraulically efficient than their natural counterparts, they represent a critical, yet under-studied, resource for offsetting future marsh losses along the seaward margins. Their unique geomorphic structure could make them ideal targets for conservation and restoration efforts aimed at proactively facilitating inland transgression in the face of sea level rise.

Altogether, these findings suggest that ag-marsh landscapes may play an important, albeit unexplored, role in marsh function and stability. This study provides a necessary framework for future field and modeling studies that can further assess how the geomorphic legacy of ag-marsh landscapes influences the dynamics of circulation, material cycling, and water residence times in and between marsh systems. Such work will help support the findings presented herein and add accuracy to predictions of the fate of coastal wetlands worldwide in the context of natural and anthropogenic change.

6. Conclusions

This study provides a quantitative geomorphic comparison of three globally significant wetland landscapes in the ACE Basin, SC. Using high-resolution DEMs and drainage metrics for drainage density and drainage efficiency, we demonstrate that drainage density and drainage efficiency are not necessarily correlated, and that drainage efficiency is a more relevant metric for predicting marsh hydraulic function. In particular, we found the Alligator Marsh (ag-marsh) displayed the highest drainage efficiency (4.50) despite having a drainage density of 0.01 m^{-1}, comparable to the natural salt marsh, Ethan's Island. Conversely, Hannah's Marsh, with its more complex and fragmented creek networks, had the highest drainage density (0.02 m^{-1}), but the lowest efficiency (3.14). This demonstrates that the spatial structure and connectivity of the creek network, captured by the drainage efficiency metric, are the dominant controls on hydraulic performance. We contend that the unique, linear creek structure inherited from the historical artificial ditching in Alligator Marsh may promote more enhanced water and material exchange by minimizing the mean unchanneled path length, thereby increasing hydraulic efficiency compared to the more sinuous, fragmented creek networks of natural marshes. Future work should build upon the findings presented here by integrating field-based measurements and numerical modeling to further investigate the geomorphic controls of ag-marsh landscapes on the dynamics of water and sediment delivery in and between marsh systems, an important process influencing spatial and temporal patterns of inland marsh retreat as sea levels rise.

Author Contributions: Conceptualization, J.S.; methodology, J.S., M.F., J.C. and R.S.J.; formal analysis, J.S. and M.F.; investigation, J.S., M.F., J.C. and R.S.J.; writing—original draft preparation, J.S.; writing—review and editing, J.S. and M.F.; visualization, J.S.; supervision, J.S.; project administration, J.S.; funding acquisition, J.S. All authors have read and agreed to the published version of the manuscript.

Funding: This research was funded by the National Science Foundation CAREER Grant, award number: 2041366.

Data Availability Statement: The original data presented in the study are openly available in FigShare at https://doi.org/10.6084/m9.figshare.30148567.v1; https://doi.org/10.6084/m9.figshare.30148564.v1.

Acknowledgments: The authors would like to thank the SCDNR Staff for their project support and assistance with field work and site access.

Conflicts of Interest: Author Jessica Sullivan was employed by the University of South Carolina Aiken. The remaining authors declare that the research was conducted in the absence of any commercial or financial relationships that could be construed as a potential conflict of interest.

References

1. Allen, J.R.L. Morphodynamics of Holocene salt marshes: A review sketch from the Atlantic and southern North Sea coasts of Europe. *Quarternary Sci. Rev.* **2000**, *19*, 1155–1231. [CrossRef]
2. Hein, C.; Kirwan, M.L. Marine transgression in modern times. *Annu. Rev. Mar. Sci.* **2024**, *16*, 55–79. [CrossRef]
3. Wright, L.D.; Thom, B.G. Coastal morphodynamics and climate change: A review of recent advances. *J. Mar. Sci. Eng.* **2023**, *11*, 1997. [CrossRef]
4. Pestrong, R. The Development of Drainage Patterns on Tidal Marshes. Ph.D. Thesis, Stanford University, Stanford, CA, USA, 1965; p. 135.
5. Leonard, L.A. Controls on sediment transport and deposition in an incised mainland marsh basin, SE North Carolina. *Wetlands* **1997**, *17*, 263–274. [CrossRef]
6. Christiansen, T.; Wiberg, P.L.; Milligan, T.G. Flow and sediment transport on a tidal salt marsh. *Estuar. Coast. Shelf Sci.* **2000**, *50*, 315–331. [CrossRef]
7. Fagerhazzi, S.; Furbish, D.J. On the shape and widening of salt marsh creeks. *Front. Environ. Sci.* **2001**, *106*, 991–1003. [CrossRef]
8. Temmerman, S.; Bouma, T.J.; Govers, G.; Wang, Z.B.; de Vries, M.B.; Herman, P.J. Impact of vegetation on flow routing and sedimentation patterns: Three dimensional modelling for a tidal marsh. *J. Geophys. Res.* **2005**, *110*, F04019. [CrossRef]
9. Donatelli, C.; Passalacqua, P.; Jensen, D.; Oliver-Cabrera, T.; Jones, C.E.; Fagherazzi, S. Spatial variability in salt marsh drainage controlled by small scale topography. *J. Geophys. Res. Earth Surf.* **2023**, *128*, 11. [CrossRef]
10. Chen, S.; Gu, W.; Shi, B.; Chen, Y.; Chatzipavlis, A.; Ding, J.; Zhang, W.; Chen, Q.; Wang, Y.P. Sediment dynamic responses of coastal salt marsh to wind waves and swells in a semi-open tidal flat. *Sci. Total Environ.* **2024**, *954*, 176562. [CrossRef]
11. Williams, R.B.; Murdock, M.B. Annual production of Spartina alterniflora and Juncus roemerianus in salt marshes near Beaufort, North Carolina. *Assoc. Southeast. Biol. Bull.* **1966**, *49*, 13.
12. Odum, E.P.; De la Cruz, A.A. Particulate organic detritus in a Georgia salt marsh ecosystem. *Estuaries Publ. Am. Assoc. Adv. Sci.* **1967**, *83*, 333–388.
13. Weinstein, M.P. Shallow marsh habitats as primary nurseries for fishes and shellfish, Cape Fear River, North Carolina. *Fish Bull.* **1979**, *77*, 339–357.
14. Wenner, E.L.; Beatty, H.R. Utilization of shallow estuarine habitats in South Carolina, USA, by post larval and juvenile stages of Penaeus spp. (Decapoda: *Penaeidae*). *J. Crustac. Biol.* **1993**, *13*, 280–295. [CrossRef]
15. Cain, W.J.; Pomeroy, L.R.; Morgan, M.A.; Wang, Y.C. Oxygen and carbon dioxide mass balance for the estuarine-intertidal marsh complex of five rivers in the southeastern US. *Limnol. Oceanogr.* **1999**, *44*, 639–649.
16. Lipcius, R.N.; Seitz, R.D.; Seebo, M.S.; Colon-Carrion, D. Density, abundance and survival of the blue crab seagrass and unstructured salt marsh nurseries of Chesapeake Bay. *J. Exp. Mar. Biol. Ecol.* **2005**, *319*, 69–80. [CrossRef]
17. Moller, I.; Spencer, T.; French, J.R.; Leggett, D.J.; Dixon, M. Wave transformation over salt marshes: A field and numerical modeling study from North Norfolk, England. *Estuar. Coast. Shelf Sci.* **1999**, *49*, 411–426. [CrossRef]

18. Chmura, G.L.; Anisfel, S.C.; Cahoon, D.R.; Lynch, J.C. Global carbon sequestration in tidal, saline wetland soils. *Glob. Biogeochem. Cycles* **2003**, *17*, 1111. [CrossRef]
19. Bridgham, S.D.; Megonigal, J.P.; Keller, J.K.; Bliss, N.B.; Trettin, C. The carbon balance of North American Wetlands. *Wetlands* **2006**, *26*, 889–916. [CrossRef]
20. Pidgeon, E. Carbon sequestration by coastal marine habitats: Important missing sinks. In *The Management of Natural Coastal Carbon Sinks*; The International Union for Conservation of Nature (IUCN): Gland, Switzerland, 2009; p. 47.
21. Li, Y.L.; Wang, L.; Zhang, W.Q.; Zhang, S.P.; Wang, H.L.; Fu, X.H.; Le, Y.Q. Variability of soil carbon sequestration and microbial activity of different types of salt marsh soils at Chongming Dongtan. *Ecol. Eng.* **2010**, *36*, 1754–1760. [CrossRef]
22. McLeod, E.; Chmura, G.L.; Bouillon, S.; Salm, R.; Bjork, M.; Duarte, C.M.; Lovelock, C.E.; Schlesinger, W.H.; Silliman, B.R. A blueprint for blue carbon: Toward an improved understanding of the role of vegetated coastal habitats in sequestering CO_2. *Front. Ecol. Environ.* **2011**, *9*, 552–560. [CrossRef]
23. Pendleton, L.; Donato, D.C.; Murray, B.C.; Crooks, S.; Jenkins, W.A.; Sifleet, S.; Craft, C.; Fourqurean, J.W.; Kauffman, J.B.; Marba, N.; et al. Estimating global "blue carbon" emissions from conversation and degradation of vegetated coastal ecosystems. *PLoS ONE* **2012**, *7*, e43542. [CrossRef]
24. Ouyang, X.; Lee, S.Y. Updated estimates of carbon accumulation rates in coastal marsh sediments. *Biogeosciences* **2014**, *11*, 5057–5071. [CrossRef]
25. Emmer, I.; von Unger, M.; Needleman, B.; Crooks, S.; Emmett-Mattox, S. *Coastal Blue Carbon in Practice: A Manual for Using the VCS Methodology for Tidal Wetland and Seagrass Restoration VM0033*; Restore America's Estuaries and Silvestrum: Arlington, VA, USA, 2015.
26. Duarte, B.; Carreiras, J.; Cacador, I. Climate change impacts on salt marsh blue carbon, nitrogen, and phosphorous stocks and ecosystem services. *Appl. Sci.* **2021**, *11*, 1969. [CrossRef]
27. Rowland, P.I.; Wartman, M.; Bursic, J.; Carnell, P. Restored and created tidal marshes recover ecosystem services over time. *Environ. Sustain. Indic.* **2024**, *24*, 100539. [CrossRef]
28. French, J.R.; Stoddart, D.R. Hydrodynamics of salt marsh creek systems: Implications for marsh morphological development and material exchange. *Earth Surf. Process. Landf.* **1992**, *17*, 235–252. [CrossRef]
29. Lawrence, D.S.L.; Allen, J.R.L.; Havelock, G.M. Salt marsh morphodynamics: An investigation of tidal flows and marsh channel equilibrium. *J. Coast. Res.* **2004**, *20*, 301–316. [CrossRef]
30. Torres, R.; Styles, R. Effects of topographic structure on salt marsh currents. *J. Geophys. Res.* **2007**, *112*, F02023. [CrossRef]
31. Chen, S.; Torres, R. Effects of geomorphology on the distribution of metal abundance in salt marsh sediments. *Estuaries Coasts* **2012**, *35*, 1018–1027. [CrossRef]
32. Sullivan, J.C.; Torres, R.; Garrett, A.; Blanton, J.; Alexander, C.; Robinson, M.; Moore, T.; Amft, J.; Hayes, D. Complexity in salt marsh circulation for a semienclosed basin. *J. Geophys. Res. Earth Surf.* **2015**, *120*, 1973–1989. [CrossRef]
33. Sullivan, J.C.; Torres, R.; Garrett, A. Intertidal creeks and overmarsh circulation in a small salt marsh basin. *J. Geophys. Res. Earth Surf.* **2019**, *124*, 447–463. [CrossRef]
34. Adams, J.B.; Raw, J.L.; Riddin, T.; Wasserman, J.; Van Niekerk, L. Salt marsh restoration for the provision of multiple ecosystem services. *Diversity* **2021**, *13*, 680. [CrossRef]
35. Xin, P.; Wilson, A.; Shen, C.; Ge, Z.; Moffett, K.B.; Santos, I.R.; Chen, X.; Xu, X.; Yau, Y.Y.; Moore, W.; et al. Surface water and groundwater interactions in salt marshes and their impact on plant ecology and coastal biogeochemistry. *Rev. Geophys.* **2022**, *60*, 1. [CrossRef]
36. Shukla, T.; Tang, W.; Trettin, C.C.; Chen, S.E.; Allan, C. Determination of microtopography of low-relief tidal freshwater forested wetlands using LiDAR. *Remote Sens.* **2024**, *16*, 3463. [CrossRef]
37. Kirwan, M.L.; Temmerman, S.; Skeehan, E.E.; Guntenspergen, G.R.; Fagherazzi, S. Overestimation of marsh vulnerability to sea level rise. *Nat. Clim. Change* **2016**, *6*, 253–260. [CrossRef]
38. Nevermann, H.; AghaKouchak, A.; Shokri, N. Sea level rise implications on future inland migration of coastal wetlands. *Glob. Ecol. Conserv.* **2023**, *43*, e02421. [CrossRef]
39. Brinson, M.M.; Christian, R.R.; Blum, L.K. Multiple states in the sea level induced transition from terrestrial forest to estuary. *Estuaries* **1995**, *18*, 648–659. [CrossRef]
40. Fagherazzi, S.; Anisfeld, S.C.; Blum, L.K.; Long, E.V.; Feagin, R.A.; Fernandes, A.; Kearny, W.S.; Williams, K. Sea level rise and the dynamics of the marsh-upland boundary. *Front. Environ. Sci.* **2019**, *7*, 25. [CrossRef]
41. Coleman, D.J.; Schuerch, M.; Temmerman, S.; Guntenspergen, G.; Smith, C.G.; Kirwan, M.L. Reconciling models and measurements of marsh vulnerability to sea level rise. *Limnol. Oceanogr. Lett.* **2022**, *7*, 2. [CrossRef]
42. Wasson, K.; Woolfolk, A.; Fresquez, C. Ecotones as indicators of changing environmental conditions: Rapid migration of salt marsh—Upland boundaries. *Estuaries Coasts* **2013**, *36*, 654–664. [CrossRef]
43. Anisfeld, S.C.; Cooper, K.R.; Kemp, A.C. Upslope development of a tidal marsh as a function of upland land use. *Glob. Change Biol.* **2016**, *23*, 755–766. [CrossRef]

44. Feng, A.; Gao, J.; Wu, S.; Liu, L.; Li, Y.; Yue, X. Assessing the inundation risk resulting from extreme water levels under sea-level rise: A case study of Rongcheng, China. *Geomat. Nat. Hazards Risk* **2018**, *9*, 456–470. [CrossRef]
45. Guimond, J.A.; Michael, H.A. Effects of marsh migration on flooding, saltwater intrusion, and crop yield in coastal agricultural land subject to storm surge inundation. *Water Resour. Res.* **2020**, *57*, 2. [CrossRef]
46. Feagin, R.A.; Martinez, M.L.; Mendoza-Gonzalez, G.; Costanza, R. Salt marsh zonal migration and ecosystem service change in response to global sea level rise: A case study from an urban region. *Ecol. Soc.* **2010**, *15*, 14. [CrossRef]
47. Raabe, E.A.; Stumpf, R.P. Expansion of tidal marsh in response to sea-level rise: Gulf Coast of Florida, USA. *Estuaries Coasts* **2016**, *39*, 145–157. [CrossRef]
48. Sallenger, A.H.; Doran, K.S.; Howard, P.A. Hotspot of accelerated sea level rise on the Atlantic coast of North America. *Nat. Clim. Change* **2012**, *2*, 884–888. [CrossRef]
49. Kirwan, M.L.; Michael, H.A.; Gedan, K.B.; Tully, K.L.; Fagerhazzi, S.; McDowell, N.G.; Molino, G.D.; Pratt, D.; Reay, W.G.; Stotts, S. Feedbacks regulating the salinization of coastal landscapes. *Ann. Rev. Mar. Sci.* **2025**, *17*, 1. [CrossRef]
50. Steven, D.D.; Gramling, J.M. Diverse characteristics of wetlands restored under the wetlands reserve programs in the Southeastern United States. *Wetlands* **2012**, *32*, 593–604. [CrossRef]
51. Ardon, M.; Helton, A.M.; Scheulerell, M.D.; Berhardt, E.S. Fertilizer legacies meet saltwater incursion: Challenges and constraints for coastal plain wetland restoration. In *Elementa Science of the Anthropocene*; University of California Press: Oakland, CA, USA, 2017; p. 5.
52. Van Dijk, J.; Stroetenga, M.; Bos, L.; Van Bodegom, P.M.; Verhoef, H.A.; Aerts, R. Restoring natural seepage conditions on former agricultural grasslands does not lead to reduction of organic matter decomposition and soil nutrient dynamics. *Biogeochemistry* **2004**, *71*, 317–337. [CrossRef]
53. Calvo-Cubero, J.; Ibanez, C.; Rovira, A.; Sharpe, P.; Reyes, E. Changes in nutrient concentration and carbon accumulation in a Mediterranean restored marsh (Ebro Delta, Spain). *Ecol. Eng.* **2014**, *71*, 278–289. [CrossRef]
54. Bortolotti, L.E.; Vinebrooke, R.D.; St. Louis, V.L. Prairie wetland communities recover at different rates following hydrological restoration. *Freshw. Biol.* **2016**, *61*, 1874–1890. [CrossRef]
55. Velilla, E.; Snethlage, J.; Poelman, M.; van der Meer, I.M.; van der Werf, A.; Deolu-Ajayi, A.O.; van Belzen, J. Too salty to farm: Rethinking coastal land use in response to soil salinization. *Restor. Ecol.* **2025**, *33*, 4. [CrossRef]
56. Hughes, C.E.; Rodriguez, A.B. Salt marsh hydrogeology: A review. *Water* **2021**, *13*, 543. [CrossRef]
57. Hackney, C.T.; Yelverton, G.F. Effects of human activities and sea level rise on wetland ecosystems in the Cape Fear River Estuary, North Carolina, USA. *Wetl. Ecol. Manag. Case Stud.* **1990**, *25*, 55–61.
58. Turner, R.E. Wetland loss in the northern Gulf of Mexico: Multiple working hypotheses. *Estuaries Coasts* **1997**, *20*, 1–13. [CrossRef]
59. Doyle, T.W.; O'Neil, C.P.; Melder, M.P.; From, A.S.; Palta, M.M. Tidal freshwater swamps of the southeastern United States: Effects of land use, hurricanes, sea-level rise, and climate change. In *Ecology of Tidal Freshwater Forested Wetlands of the Southeastern United States*; Springer: Dordrecht, The Netherlands, 2007; pp. 1–28.
60. Manda, A.K.; Guiliano, A.S.; Allen, T.R. Influence of artificial channels on the source and extent of saline water intrusion in the wind tide dominated wetlands of the southern Albemarle estuarine system (USA). *Environ. Earth Sci.* **2014**, *71*, 4409–4419. [CrossRef]
61. White, E.; Kaplan, D. Restore or retreat? Saltwater intrusion and water management in coastal wetlands. *Ecosyst. Health Sustain.* **2017**, *3*, e01258. [CrossRef]
62. Bhattachan, A.; Emanual, R.E.; Marcelo, A.; Bernhardt, E.S.; Anderson, S.M.; Stillwagon, M.G.; Ury, E.A.; BenDor, T.K.; Wright, J.P. Evaluating the effects of land-use change and future climate change on vulnerability of coastal landscapes to saltwater intrusion. *Elem. Sci. Anthr.* **2018**, *6*, 62. [CrossRef]
63. Pasternack, G.B.; Brush, G.S.; Hilgartner, W.B. Impact of historic land-use change on sediment delivery to a Chesapeake Bay subestuarine delta. *Earth Surf. Process. Landf.* **2001**, *26*, 409–427. [CrossRef]
64. Alexander, C.R.; Hodgson, J.Y.S.; Brandes, J.A. Sedimentary processes and products in a mesotidal salt marsh environment: Insights from Groves Creek, GA. *Geo-Mar. Lett.* **2017**, *37*, 345–359. [CrossRef]
65. Zhang, Y.; Huang, C.; Zhang, W.; Chen, J.; Wang, L. The concept, approach, and future research of hydrological connectivity and its assessment at multiscales. *Environ. Sci. Pollut. Res.* **2021**, *28*, 52724–52743. [CrossRef]
66. Xin, P.; Zhan, L. Upland-marsh hydraulic connectivity affects marsh biogeochemistry. *Nat. Water* **2025**, *3*, 138–139. [CrossRef]
67. Montalvo, M.S.; Grande, E.; Braswell, A.E.; Visser, A.; Arora, B.; Seybold, E.C.; Tatariw, C.; Haskins, J.C.; Endris, C.A.; Gerbl, F.; et al. A fresh take: Seasonal changes in terrestrial freshwater inputs impact salt marsh hydrology and vegetation dynamics. *Estuaries Coasts* **2024**, *47*, 2389–2405. [CrossRef]
68. Fagherazzi, S.; Mariotti, G.; Leonardi, N.; Canestrelli, A.; Nardin, W.; Kearny, W.S. Salt marsh dynamics in a period of accelerated sea level rise. *J. Geophys. Res. Earth Surf.* **2020**, *125*, 8. [CrossRef]
69. Liu, Z.; Fagherazzi, S.; He, Q.; Gourgue, O.; Bai, J.; Liu, X.; Miao, C.; Hu, Z.; Cui, B. A global meta-analysis on the drivers of salt marsh planting success and implications for ecosystem services. *Nat. Commun.* **2024**, *15*, 3643. [CrossRef]

70. Snedden, G.A.; Smith, S.J. Seasonal variation in bay-marsh sediment exchange through a back-barrier salt marsh tidal creek. *Limnol. Oceanogr.* **2025**, *70*, 3143–3154. [CrossRef]
71. Armitage, A.R. Perspectives on maximizing coastal wetland restoration outcomes in anthropogenically altered ecosystems. *Estuaries Coasts* **2021**, *44*, 1699–1709. [CrossRef]
72. Noble, P.A.; Tymowski, R.G.; Fletcher, M.; Morris, J.T.; Lewitus, A.J. Contrasting patterns of phytoplankton community pigment composition in two salt marsh estuaries in Southeastern United States. *Appl. Environ. Microbiol.* **2003**, *69*, 4129–4143. [CrossRef]
73. Johnson, W.; Lewitus, A.; Fletcher, M.; Li, W. Linking bacterioplankton community structures to environmental state variables and phytoplankton assemblages in two South Carolina salt marsh estuaries. *Aquat. Microb. Ecol.* **2006**, *45*, 129–145. [CrossRef]
74. Buzzelli, C.; Holland, A.F.; Sanger, D.M.; Conrads, P.C. Hydrographic characterization of two tidal creeks with implications for watershed land use, flushing times, and benthic production. *Estuaries Coasts* **2007**, *30*, 321–330. [CrossRef]
75. Van Dolah, R.F.; Sanger, D.M.; Riekerk, G.H.M.; Crowe, S.E.; Levisen, M.V.; Bergquist, D.C.; Chestnut, D.E.; McDermott, W.; Fulton, M.H.; Wirth, E. *The Condition of South Carolina's Estuarine and Coastal Habitats During 2009–2010*; Technical Report 107; South Carolina Marine Resources Division: Charleston, SC, USA, 2013; p. 54.
76. Allen, J.; Lu, K. Modeling and prediction of future urban growth in the Charleston region of South Carolina: A GIS-based integrated approach. *Ecol. Soc.* **2003**, *8*, 2. Available online: http://www.consecol.org/vol8/iss2/art2/ (accessed on 17 November 2025). [CrossRef]
77. Purcell, A.D. Assessing the Value of South Carolina's Coastal Timber Resources. Master's Thesis, Clemson University, Clemson, SC, USA, 2020.
78. Ciou, T.S.; Lin, C.H.; Wang, C.K. Airborne Lidar point cloud classification using ensemble learning for DEM generation. *Sensors* **2024**, *24*, 6858. [CrossRef]
79. Chen, N.; Wang, N.; He, Y.; Ding, X.; Kong, J. An improved progressive triangular irregular network densification filtering algorithm for airborne Lidar data. *Front. Earth Sci.* **2023**, *6*, 1015153.
80. Chassereau, J.E.; Bell, J.; Torres, R. A comparison of GPS and Lidar Salt Marsh DEMs. *Earth Surf. Process. Landf.* **2011**, *36*, 1770–1775. [CrossRef]
81. Sullivan, J.C.; Wan, Y.; Willis, R. Modeling floodplain inundation, circulation and residence time under changing tide and sea-levels. *Estuaries Coasts* **2020**, *43*, 693–707. [CrossRef]
82. Marani, M.; Belluco, E.; D'Alpaos, A.; Defina, A.; Lanzoni, S.; Rinaldo, A. On the drainage and density of tidal networks. *Water Resour. Res.* **2003**, *39*, 1040. [CrossRef]
83. Kearny, W.; Fagherazzi, S. Salt marsh vegetation promotes efficient tidal channel networks. *Nat. Commun.* **2016**, *7*, 12287. [CrossRef]
84. Tucker, G.E.; Catani, F.; Rinaldo, A.; Bras, R.L. Statistical analysis of drainage density from digital terrain data. *Geomorphology* **2001**, *36*, 187–202. [CrossRef]
85. Horton, R.E. Erosional development of streams and their drainage basins: Hydrophysical approach to quantitative morphology. *Geol. Soc. Am. Bull.* **1945**, *56*, 275–370. [CrossRef]
86. Scheidegger, A.E. Horton's laws of stream lengths and drainage areas. *Water Resour. Res.* **1968**, *4*, 1015–1021. [CrossRef]
87. Gardner, L.R.; Gaines, E.F. A method for estimating pore water drainage from marsh soils using rainfall and well records. *Estuar. Coast. Shelf Sci.* **2008**, *79*, 51–58. [CrossRef]
88. Schwartz, J.M.; Cardenas, B.T.; Mohrig, D.; Passalacqua, P. Tributary channel networks formed by depositional processes. *Nat. Geosci.* **2022**, *15*, 216–221. [CrossRef]
89. Peck, E.K.; Waker, J.E.; Ackerman, K.V.; Carr, J.; Correll, M.D.; Defne, Z.; Deegan, L.A. Distribution and disturbances of ditches across salt marshes of the Northeast US with implications for management and restoration. *J. Environ. Manag.* **2025**, *376*, 124444. [CrossRef]
90. Zheng, S.; Shao, D.; Gao, W.; Nardin, W.; Ning, Z.; Liu, Z.; Cui, B.; Sun, T. Drainage efficiency and geometric nuances of tidal channel network mediate Spartina alterniflora landward invasion in marsh-channel system. *Front. Mar. Sci.* **2022**, *9*, 888597. [CrossRef]
91. Hautekiet, S.; Rossius, J.E.; Gourgue, O.; Kleinhans, M.; Temmerman, S. On the relative role of abiotic and biotic controls in channel network development: Insights from scaled tidal flume experiments. *Earth Surf. Dyn.* **2024**, *12*, 2. [CrossRef]
92. McKnown, J.G.; Burdick, D.M.; Moore, G.E.; Gibson, J.L.; Ferguson, W. Evaluation of drainage enhancement for vegetation recovery in New England salt marshes using public domain, high resultion aerial imagery. *J. Coast. Res.* **2024**, *40*, 6.
93. Chen, X.; Zhang, M.; Jiang, H. Morphological characteristics and hydrological connectivity evaluation of tidal creeks in coastal wetlands. *Land* **2022**, *11*, 1707. [CrossRef]
94. Regier Pl Ward, N.D.; Indivero, J.; Moore, C.W.; Norwood, M.; Myers-Pigg, A. Biogeochemical control points of connectivity between a tidal creek and its floodplain. *Limnol. Oceanogr. Lett.* **2021**, *6*, 3. [CrossRef]

95. Varugu, B.K.; Jones, C.E.; Oliver-Cabrera, T.; Simard, M.; Jensen, D.J. Study of hydrologic connectivity and tidal influence on water flow within Louisiana coastal wetlands using rapid-repeat-interferometric synthetic aperture radar. *Remote Sens.* **2025**, *17*, 459. [CrossRef]
96. Marcot, B.G.; Thorne, K.M.; Carr, J.A.; Guntenspergen, G.R. Foundations of modeling resilience of tidal saline wetlands to sea-level rise along the U.S. Pacific Coast. *Landsc. Ecol.* **2023**, *38*, 3061–3080. [CrossRef]
97. Hanks, R.D.; Baldwin, R.E.; Folk, T.H.; Wiggers, E.P.; Coen, R.H.; Gouin, M.L.; Agha, A.; Richter, D.D.; Fields-Black, E.L. Mapping Antebellum rice fields as a basis for understanding human and ecological consequences of the era of slavery. *Land* **2021**, *10*, 831. [CrossRef]
98. Liu, Y.; Ge, T.; Groenigen, K.J.; Yang, Y.; Wang, P.; Cheng, K.; Zhu, Z.; Wang, J.; Yong, L.; Guggenberger, G.; et al. Rice paddy soils are a quantitatively important carbon storage according to a global synthesis. *Commun. Earth Environ.* **2021**, *2*, 154. [CrossRef]
99. Verhoeven, J.T.A.; Setter, T.L. Agricultural use of wetlands: Opportunities and limitations. *Ann. Bot.* **2010**, *105*, 155–163. [CrossRef]
100. Alizad, K.; Morris, J.T.; Bilskie, M.V.; Passeri, D.L.; Hagen, S.C. Integrated modeling of dynamic marsh feedbacks and evolution under sea-level rise in a mesotidal estuary (Plum Island, MA, USA). *Water Resour. Res.* **2022**, *58*, 8. [CrossRef]
101. Smith, K.E.L.; Terrano, J.F.; Jenkins, R.L.; Pitchford, J.L.; Passeri, D.L.; Smith, C.G. Shifts in marsh erosion, migration and wave exposure over nearly two centuries of sea-level rise in the Gulf of Mexico. *Estuaries Coasts Shelf Sci.* **2025**, *313*, 109106. [CrossRef]
102. Powell, E.; Dubayah, R.; Stovall, A.E.L. Characterizing low-lying coastal upland forests to predict future landward marsh expansion. *Ecosphere* **2024**, *15*, 6. [CrossRef]
103. Jung, N.W.; Doe, T.A.; Jung, Y.; Dellapenna, T.M. Massive sea-level-driven marsh migration and implications for coastal resilience along the Texas coast. *Remote Sens.* **2024**, *16*, 2268. [CrossRef]
104. Anderson, C.P.; Carter, G.A.; Waldron, M.C.B. Precise elevation thresholds associated with salt-marsh upland ecotones along the Mississippi Gulf Coast. *Ann. Am. Assoc. Geogr.* **2022**, *112*, 1850–1865. [CrossRef]

environments

MDPI

Article

Groundwater/Surface Water Temperature Variations and Hydrogeological Implications in Doñana National Park

José Luis Yanes [1], Alejandro Jiménez-Bonilla [1], Marina Martínez-Caro [1], Ana Fernández-Ayuso [2] and Miguel Rodríguez-Rodríguez [1,*]

[1] Department of Physical, Chemical and Natural Systems, Pablo de Olavide University, Utrera Road, Km 1., 41013 Seville, Spain; jlyancon@upo.es (J.L.Y.); ajimbon@upo.es (A.J.-B.); mmarcar@upo.es (M.M.-C.)

[2] Faculty of Sciences, Campus of Cantoblanco, Autónoma University of Madrid, C/Francisco Tomás y Valiente 7, 28049 Madrid, Spain; ana.fernandeza@uam.es

* Correspondence: mrodrod@upo.es

Abstract: This study analyzes the evolution of surface water and groundwater temperatures at various depths in the sand-dune ponds of Doñana National Park (southern Spain) over eight hydrometeorological years (2016–2024). This research aims to characterize the water temperature regime, identify water temperature trends, and analyze patterns in groundwater flow dynamics. The results indicate that, in a recent dry–warm period (2020–2023), coinciding with a notable decrease in precipitation and an increase in the average air temperature in the area, there was an increase in the annual mean temperature of pond water and in shallow piezometers (~15 m depth). However, in deep piezometers, a decrease in water temperature was recorded during the dry–warm period, along with a reduction in temperature variability. A phase shift has also been observed between groundwater temperature extremes and air temperature variations, with the magnitude of this shift depending on sensor depth. These findings enable the analysis of the sensitivity of these wetlands to global environmental change and contribute to the characterization of recharge and discharge flows in the aquifer, both at local and regional scales, allowing for the evaluation of flow variability in hydrological years with below-average precipitation and above-average air temperature and shallow groundwater temperature.

Keywords: groundwater flow; global change; wetland; sand-dune pond; coastal aquifer

1. Introduction

Sand-dune ponds are among the most iconic ecosystems of Doñana National Park (DNP) in southwestern Spain. These wetlands play a critical role, not only supporting local biodiversity but also providing key ecosystem services, such as offering habitats for threatened species and contributing to global ecological balance [1–3]. The good ecological status of these ponds is closely tied to the Almonte-Marismas aquifer system, making them sensitive indicators of both global environmental change and anthropogenic pressures on the ecosystem [4].

The Almonte-Marismas detrital aquifer is a multilayered system administratively divided into six groundwater bodies, with complex hydrogeological dynamics that complicate sustainable management [2,5,6]. Since the 1970s, agricultural intensification and the growing demand for urban water have led to significant historical groundwater drawdowns, with water table declines of up to 20 m reported in some areas [2,7]. This overexploitation jeopardizes not only the aquifer's long-term sustainability but also the equilibrium of dependent ecosystems, such as ponds and marshes [2,8,9].

From a hydrological perspective, Doñana's sand-dune ponds function as local discharge or flow-through systems for one of the groundwater bodies comprising the Almonte-Marismas aquifer, known as the Coastal Aeolian Mantle of Doñana [10–13]. This close dependency on groundwater makes these wetlands particularly vulnerable to changes in the aquifer's water levels. Consequently, numerous scientific studies have focused on understanding how fluctuations in the aquifer's water regime influence pond dynamics and, by extension, the overall functionality of the ecosystem [4,9,12,14].

A particularly effective method for analyzing the interconnection between the surface environment and the behavior of an aquifer is the use of water temperature as a hydrological tracer ([15] and references therein). Ref. [16] emphasized the relevance of this method in the Almonte-Marismas aquifer, analyzing vertical temperature profiles and performing one-dimensional modeling of heat transport to investigate groundwater recharge and discharge processes. However, its initial applications were limited due to the lack of suitable technology for obtaining precise thermal records at the time. In recent decades, advances in instrumentation and numerical modeling have significantly enabled the use of temperature to study recharge, discharge, and groundwater transport processes in both karst [17,18] and detrital aquifers [19,20]. In fact, in [19], two-dimensional numerical modeling of heat transport in a detrital coastal aquifer was conducted, providing a detailed framework for understanding thermal transport in coastal aquifers, such as the Almonte-Marismas aquifer system. This model integrated field data and sensitivity analysis to evaluate how changes in input variables affect model outputs, to identify how parameters like hydraulic conductivity, thermal gradient, and aquifer geological properties influence temperature distribution, providing a solid foundation for future studies on the use of heat as a groundwater tracer in coastal aquifers. However, even in this study, the authors highlighted the need to calibrate such models with empirical data, particularly with thermal measurements from deep coastal wells (approx. 150 m).

Recent studies have provided methodological advancements in the modeling of heat transport in coastal aquifers, with particular relevance given to those that account for variability in the position of the saline interface [5,21–23]. These new models, beyond their methodological progress in describing the hydrogeological functioning of such aquifers, are particularly interesting from the perspective of using heat as a tracer of subsurface flow. However, no studies on thermal modeling or heat transport in groundwater have been conducted in the Almonte-Marismas aquifer since Custodio's work in 1996 [16], despite the significant uncertainty inherent in this single study due to the technological limitations at the time, which made it impossible to monitor surface water and groundwater temperature at key points within the aquifer. This expansion of the description of groundwater flows at both local and regional scales in the Almonte-Marismas aquifer becomes particularly relevant when considering the increasing impacts of climate change on Mediterranean wetlands, e.g., [24,25].

In 2015, the Guadalquivir River Basin Authority (Confederación Hidrográfica del Gualdalquivir, CHG), in collaboration with Pablo de Olavide University, launched a research project focused on the hydrological study of the sand-dune ponds of the DNP, which remains ongoing. As part of this initiative, a comprehensive monitoring network was established, incorporating pressure transducers and temperature sensors in the park's most significant ponds, along with piezometers installed near these water bodies. Data collection occurs at three-hour intervals, resulting in a rich dataset on time-dependent water level fluctuations and temperature dynamics both in the pond's surface water and in groundwater. The continuous thermal records since 2016 represent valuable information that may help calibrate the impacts of climate change on these ecosystems and contribute to the characterization of vertical and horizontal groundwater flow, providing important insights for understanding the complex hydrological dynamics of the Almonte-Marismas aquifer. Regardless of the results achieved by this research project in characterizing the hydrological functioning of the sand-dune ponds, a specific study was carried out [10] involving the use of a water temperature dataset obtained between 2016 and 2018. However, this thermal analysis was limited to the characterization of groundwater flow at a local scale (m to hm) and did not yet record the evolution of surface water and groundwater temperature during the extreme heat and drought periods that have occurred in recent years.

In this particular context, it is essential to assess the extent to which the increase in air temperature has affected the thermal regime of these aquatic ecosystems and what hydrogeological information can be extracted from high-resolution water temperature records during the period analyzed in this research.

Overall, the analysis of accurate water temperature records in these ponds is a crucial tool for monitoring the effects of global environmental change on these ecosystems and is also useful for the calibration and development of groundwater flow and heat transport models. Accordingly, the objectives of this study include, first, the creation of a database integrating the thermal records obtained during the research project. Second, this study aims to characterize the thermal regime of both the surface water and groundwater around the sand-dune ponds and identify recent trends related to rising temperatures and decreasing precipitation. Through this analysis, the water temperature records are intended to serve as indicators of global change impacts and help identify potential groundwater flow dynamics at both local (m to hm) and regional (>km) scales.

2. Study Zone and Hydrogeological Characteristics of the Almonte-Marismas Aquifer

Doñana National Park (DNP) is a large, protected area in southwestern Spain (Figure 1A). Located on the right bank of the Guadalquivir River, near its estuary where it meets the Atlantic Ocean, this area is one of the most important wetlands in Europe, recognized as a UNESCO World Heritage Site and a Ramsar Wetland of International Importance.

The DNP experiences a Mediterranean climate with hot, dry summers and mild, wet winters. During the summer months, temperatures can soar to over 40 °C, while winter temperatures typically range from 5 to 15 °C. The area receives most of its rainfall in the winter months. Despite significant spatial variability in rainfall, the average annual precipitation is 525 mm [26,27]. These weather conditions contribute to a highly diverse landscape in Doñana, which hosts a variety of habitats with high biodiversity, including ponds, seasonally flooded marshlands, dunes, and forests.

All these wetlands are dependent on the Almonte-Marismas aquifer, a hydrogeological system that extends beyond the boundaries of the DNP over an area of 2700 km^2 (Figure 1A). For a detailed description of the lithostratigraphy of the aquifer see [28,29]. It lies above a Miocene–Pliocene marl unit that forms its impervious basement, dipping southeastward from ~50 m above sea level in the north to ~200 m below sea level in the south. Overlying these marls, the aquifer consists of Plio-Quaternary siliciclastic sediments of variable thickness—gravels, sands, and silts—referred to as the "Almonte-Marismas basal sands" [30]. These sands overlie a coarser detrital unit that dips towards the Atlantic Ocean [16,28]. In the southeasternmost area, the basal sands are covered by silt and clay deposits from the marsh unit (Figure 1A). Consequently, this unit functions as an unconfined aquifer in the northern half, where it outcrops (Figure 1A), transitioning into a confined system beneath the low-permeability marsh deposits in the south. In the southernmost part of the area, a Quaternary aeolian sand unit further covers the basal sand and marsh deposits, forming an active dune belt (Figure 1A). The aquifer is primarily recharged through direct rainfall infiltration in the unconfined zone, with groundwater flow predominantly directed southward and southeastward [16].

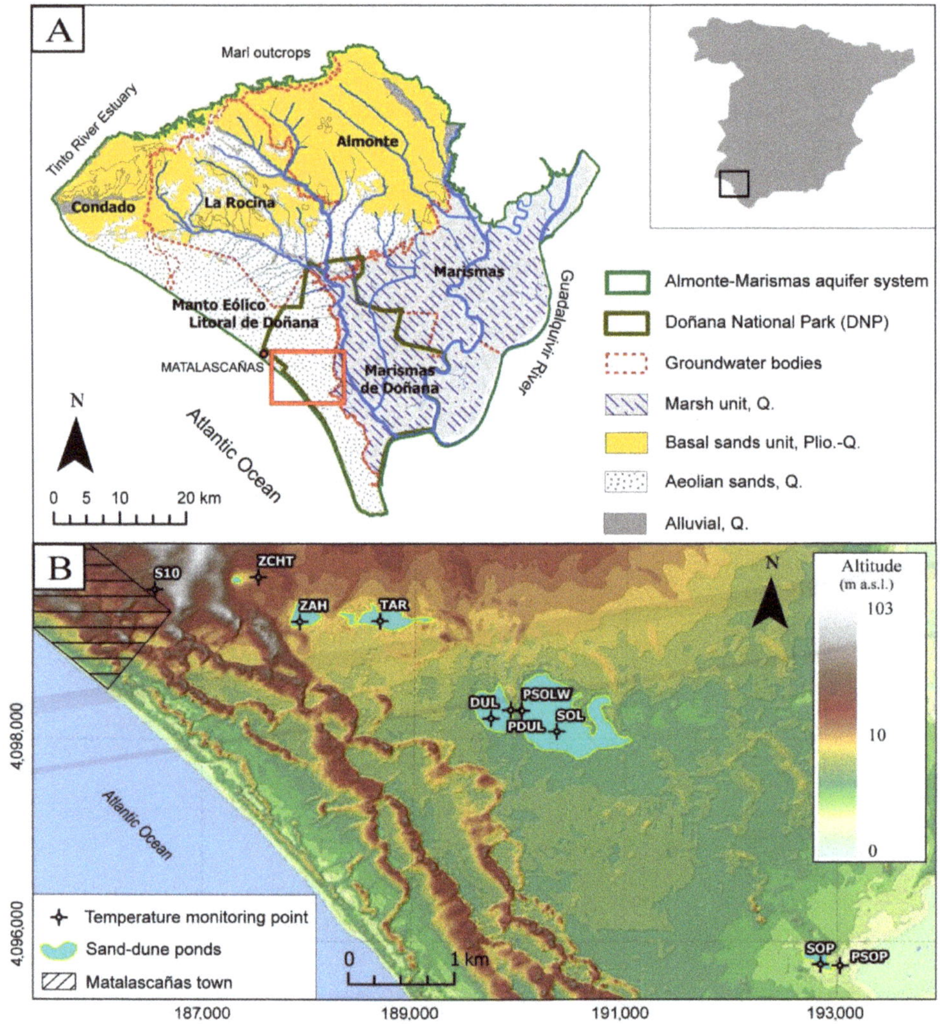

Figure 1. (**A**) Map of the Almonte-Marismas aquifer system, showing the boundaries of the groundwater bodies (Almonte, Condado, Rocina, Manto Eólico, Marismas, and Marismas de Doñana), the main lithostratigraphic units, and the location of Doñana National Park (DNP). (**B**) Location of monitoring points of the sand-dune pond system, Zacallón Charco del Toro (ZCHT), Zahíllo pond (ZAH), Taraje pond (TAR), Dulce pond (DUL), Santa Olalla pond (SOL), and Sopetón pond (SOP), and groundwater monitoring points, Sopetón pond piezometer (PSOP), Santa Olalla pond west piezometer (PSOLW), Dulce pond piezometer (PDUL), and borehole 10 (S10).

The Almonte-Marismas aquifer is a single hydrogeological system, but it is administratively divided into six groundwater bodies (Figure 1A), five managed by the Guadalquivir River Basin Authority (Marismas, Marismas de Doñana, Almonte, La Rocina, and Manto Eólico Litoral de Doñana) and one, called Condado, belonging to the Tinto–Odiel basin. The sand-dune ponds studied in this research are located on the edge of the mobile dune system of the aeolian sand unit, stretching from the town of Matalascañas to the boundary with the marshlands (Figure 1B). These ponds, mostly temporary, are primarily associated with the groundwater body of the aeolian sands and are directly supplied by rainfall and groundwater from this part of the aquifer. The specific characteristics of these ponds, as well as their location, are detailed in Table 1. In general, they are shallow lakes that usually do not exceed 2 m in depth. They are located at low altitudes, with the Charco del Toro pond being the highest (approx. 10.6 m a.s.l.). The flooded area occupied by these ponds varies greatly between wet and dry periods (Table 1). They are generally surrounded by sparse phreatophytic vegetation, except for the Taraje pond, which is densely surrounded by *Tamarix* sp. The Zacallón of Charco del Toro is also shaded by conifers most of the time.

Table 1. Characteristics of the monitoring points, including ponds and piezometers. Altitude refers to the ground elevation (m a.s.l.), either at the lowest point of the pond basin or at the piezometer location. * Depth indicates the maximum depth reached by the pond during the monitoring period or the total depth of the piezometer. * For ZCHT, dimensions of the trench are provided, with the depth representing the total trench depth and the flooded area corresponding to the Zacallón area.

	Monitoring Point	UTM X	UTM Y	Altitude (m a.s.l.)	Depth * (m)	Average Flooded Area (m^2)	
						Wet Periods	Dry Periods
PONDS	SOP	192,853	4,095,825	2.10	0.98	28,750	12,448
	SOL	190,342	4,098,107	4.60	1.20	269,410	97,852
	DUL	189,871	4,098,395	5.10	1.32	74,941	14,710
	TAR	188,694	4,099,221	7.50	0.80	106,998	2908
	ZAH	187,886	4,099,189	6.50	0.53	46,015	1706
PIEZOMETERS	ZCHT *	187,525	4,099,661	10.6	1.85	84	84
	PSOP	193,141	4,095,961	2.45	2		
	PSOLW	190,142	4,098,479	5.80	20		
	PDUL	190,037	4,098,486	5.75	20		
	S10	186,478	4,097,170	17.45	170		

The Charco del Toro pond (ZCHT), the closest to Matalascañas, is located less than one kilometer from the town (Figure 1B). It is a temporary pond that only fills sporadically after heavy rainfall and remained dry during the study period. Southeast of it are the Zahíllo (ZAH) and Taraje (TAR) ponds, which have historically maintained a seasonal hydroperiod but have only flooded in recent years following intense rain episodes [31]. In contrast, the Dulce (DUL) and Santa Olalla (SOL) ponds exhibit longer hydroperiods. Santa Olalla is the only sand-dune pond with a permanent hydroperiod, although in recent years, it has dried up during the summer droughts of 2022 and 2023, an unprecedented event since 1995. Finally, the Sopetón pond (SOP), located at the boundary between the aeolian sands and the marshlands, is the farthest from Matalascañas.

All the ponds mentioned are connected to the aquifer; therefore, they are not perched systems. The water temperature monitoring points in this study were located in these six ponds and in piezometers near the Dulce (PDUL), Santa Olalla (PSOLW), and Sopetón (PSOP) ponds (Figure 1B). Temperature was also monitored in borehole 10 (S10), a supply well for Matalascañas, located only 700 m from the Charco del Toro pond.

3. Material and Methods

Surface water and groundwater temperatures were meticulously recorded through high-frequency monitoring (every three hours) in the sand-dune ponds of the DNP and in piezometers situated nearby. This monitoring was carried out by installing Levelogger and TD-Divers® 10 m pressure transducers, which accurately measure both water level variations, based on hydrostatic pressure, and water temperature. The TD-Diver and

CTD-Diver sensors used in this study have a temperature accuracy of ±0.1 °C and a resolution of 0.01 °C. These sensors undergo a 30-point factory calibration, ensuring high precision, and do not require additional user calibration. However, monthly field visits were conducted throughout the monitoring period to collect temperature data. These visits allowed for the in situ calibration of the sensors using a portable multimeter (HACH-HQ40D, Hach Company, Loveland, Colorado, EE.UU.), ensuring consistency and accuracy in the recorded measurements.

Specifically, the monitoring encompassed six points for surface and subsurface waters and four points for groundwater (Figure 1B and Table 1).

Monitoring was conducted in each of the previously described ponds: Charco del Toro (ZCHT), Zahíllo (ZAH), Taraje (TAR), Dulce (DUL), Santa Olalla (SOL), and Sopetón (SOP). After bathymetric analyses of these ponds, the sensors were installed at their deepest points, where the staff gauge is generally located. However, in ponds like Zahíllo and Sopetón, significant water level reductions in recent years necessitated lowering the sensors to just a few centimeters below the ground surface. In these cases, the recorded temperature data represent shallow subsurface water temperatures, referred to here as "subsurface water". Furthermore, in the case of the Charco del Toro pond, which remained dry throughout the monitoring period, the sensor was installed in a nearby artificial trench locally known as a "Zacallón". These trenches are created to access groundwater for cattle when ponds dry out, ensuring the presence of water during drought periods. In this case, the sensor was also placed a few centimeters below the ground level, although it has not dried up as frequently as the ponds. Finally, the Dulce pond has no records between 2016 and 2020, as the sensor was installed in this pond in September 2020.

For groundwater monitoring, data were collected from four piezometers at different depths. The deepest of these, borehole 10 (S10), is a reserve water supply well located in Matalascañas, with a total depth of 170 m. However, the sensor was installed at a depth of 15 m below ground level. This well is located approximately 750 m from the Charco del Toro pond and about 1 km from the coastline (Figure 1). It is the only well in this study intermittently exploited, primarily during summer. Other piezometers, each around 20 m deep, are situated near the Dulce (PDUL) and Santa Olalla (PSOLW) ponds, with sensors installed at 9 m and 4 m depths, respectively. The shallowest piezometer, near the Sopetón pond, is a perforated PVC tube approximately 2 m deep, with the sensor positioned 1.5 m below ground level.

In addition to the temperature monitoring in piezometers, vertical temperature profiles were obtained in two piezometers using a Solinst Reelogger temperature-conductivity probe, with 0.1 °C temperature resolution. Temperature profiles were conducted with the aim of confirming the temperature profiles already presented in [16]. These two profiles were made in October 2024. Temperature measurements were taken at 1 m depth intervals, allowing the temperature to stabilize for a minimum of two minutes at each measurement point.

For data analysis, a database was created containing temperature records from each monitoring point. To identify and remove anomalous values, first, we graphically represented the temperature time series. Anomalous values were detected as clear outliers that deviated significantly from the expected trends in the dataset. These values were typically associated with sensor malfunctions and are very rare. Additionally, to ensure consistency, we applied a simple statistical outlier identification based on values that exceeded standard deviations considerably. Outliers were verified manually. Gaps in the dataset correspond to periods when the sensor was not functioning, usually spanning several days. These gaps were filled using linear interpolation when appropriate, ensuring minimal impact on the overall data structure.

The water temperature records covered eight full hydrometeorological years, from 1 September 2016 to 31 August 2024, in most locations, except for the Dulce pond, where monitoring began in 2020, and in the ZCHT pond, where water temperature was not recorded between 2018 and 2020. Similarly, the reliability of the data from well S10, which is intermittently exploited for urban supply to Matalascañas, was conditioned by the extraction frequency, leading to the exclusion of some years from the record due to insufficient reliability.

The temperature regime of the surface water, subsurface water, and groundwater was determined through a statistical descriptive analysis of the mean daily temperature from each monitoring point. After filtering and completing the database, preliminary analyses of daily average temperatures were conducted, followed by statistical descriptive evaluations to characterize the thermal regime at each monitored location. These statistical analyses were conducted using R Studio 2023.03.0+386 [32]. To assess temperature trends and variance comparison, we performed a t-test for independent samples to evaluate differences between different periods and ANOVA when comparing more than two groups. The significance level (α) was set at 0.05, and the obtained p-values were greater than 0.05, indicating no statistically significant differences.

Air temperature and precipitation data were also obtained from two meteorological stations in the study area: the Almonte station (37°08′53″ N, 06°28′35″ W) and the Palacio de Doñana station (36°59′19″ N, 06°26′35″ W). These meteorological stations are the closest to the ponds studied. Although there are local differences in precipitation distribution, this is not the case for air temperature. Therefore, the combined use of both stations in the climatic description of the study area is primarily determined by data availability. The Almonte station provides daily data starting from 2020, while the Palacio de Doñana station provides a longer dataset, dating back to 1978. These data were analyzed to contextualize the climatic conditions in the study area over the past two decades.

Considering the climatic context during the monitoring period, trend analyses were performed for 2020–2023, a period which represents the driest and warmest years in the dataset. Variance analysis (t-test) was also performed to compare the temperatures from the 2022/23 hydrometeorological year (end of the dry–warm period) with those of 2017/18, the coldest and wettest year in the series. We considered the month of September as the starting point of each hydrometeorological year.

During the development of the research project and up to the present, actual evapotranspiration has been measured using a Class A evaporimeter near the ponds. This has allowed for the precise determination of the hydrological balance variables in the studied ponds. However, in the Mediterranean climate, potential evapotranspiration always greatly exceeds annual precipitation. Therefore, we consider it relevant to highlight the evolution of annual precipitation when evaluating changes in water availability in these ecosystems.

4. Results

4.1. Climate Setting

Figure 2 shows the evolution of the annual average, maximum, and minimum air temperatures recorded at the Almonte weather station over the past 20 years. It also illustrates the annual thermal anomaly relative to the historical average air temperature in Doñana (i.e., 17.3 °C, based on the average temperature recorded at the Palacio de Doñana weather station between 1978 and 2024).

The average annual air temperature for the 2003–2024 period was 17.6 °C, slightly higher than the average for the 1978–2024 period (17.3 °C). While the annual average air temperature remained close to the historical average during the first decade of the 21st century, a sustained increase has been observed since 2014, with thermal anomalies reaching nearly 1.5 °C during the 2022/23 hydrometeorological year. In fact, between 2015 and 2024, the air temperature rose by up to 0.47 °C compared to the 2003–2014 period. This trend persisted throughout the monitoring period, except during the 2017/18 hydrometeorological year, which was relatively cold in this context (Figure 2).

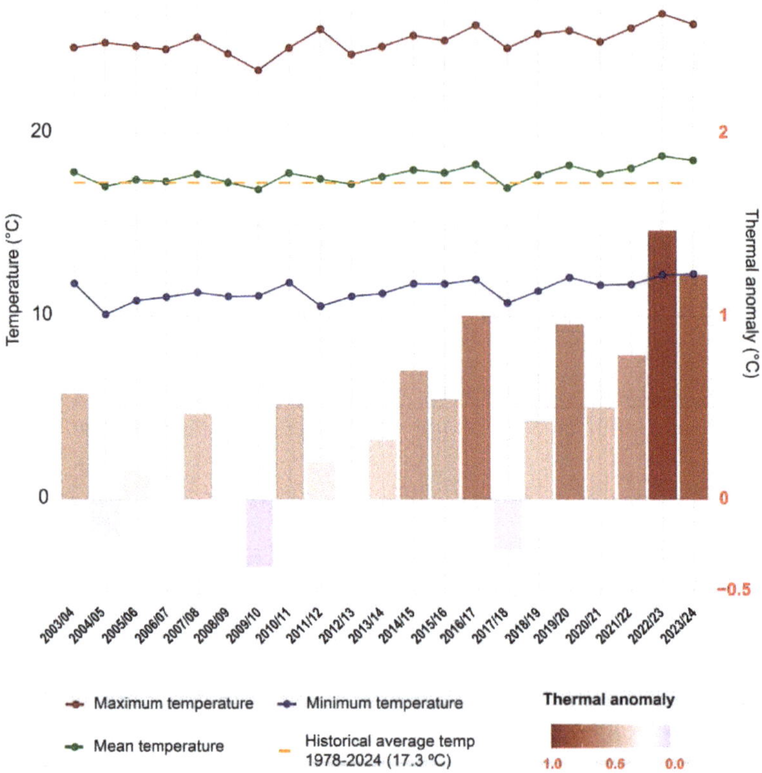

Figure 2. Annual maximum, minimum, and mean air temperatures (°C) recorded at the Almonte meteorological station (2003–2024), along with the thermal anomaly. The historical average air temperature (1978–2024) of 17.3 °C was calculated from the daily mean temperature data recorded at the Palacio de Doñana meteorological station.

The total annual precipitation for the last 20 hydrometeorological years recorded at the Almonte weather station is displayed (Figure 3). Like the air temperature, a breaking point is observed in 2014. After this year, precipitation fell below the historical average in Doñana (i.e., 525 mm/year, based on the average precipitation recorded at the Palacio de Doñana weather station between 1978 and 2024), once again with the exception of the 2017/18 hydrometeorological year, when precipitation slightly exceeded the average.

Based on these data, the monitoring period has been characterized as a climatically anomalous dry–warm period, especially between 2020 and 2023, during which the most significant thermal anomalies and markedly below-average precipitation (approx. 360 mm/year) were recorded. In this context, the 2017/18 hydrometeorological year stands out as an exception, being relatively wet and cold.

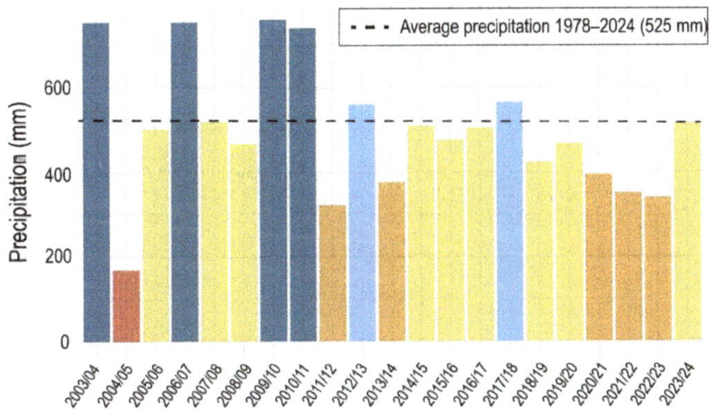

Figure 3. Annual precipitation (mm) recorded at the Almonte meteorological station (2003–2024). The historical average precipitation (1978–2024) of 525 mm was calculated from the daily precipitation data recorded at the Palacio de Doñana meteorological station. Blue bars indicate years with above-average precipitation, while red, orange, and yellow bars indicate below-average precipitation.

4.2. Thermal Regime of Surface and Subsurface Waters

The daily mean temperature in the surface and subsurface waters (at depths of less than 2 m) over the study period is presented in Figure 4. This figure also shows the mean air temperature recorded at the Almonte meteorological station during the same period (pale grey).

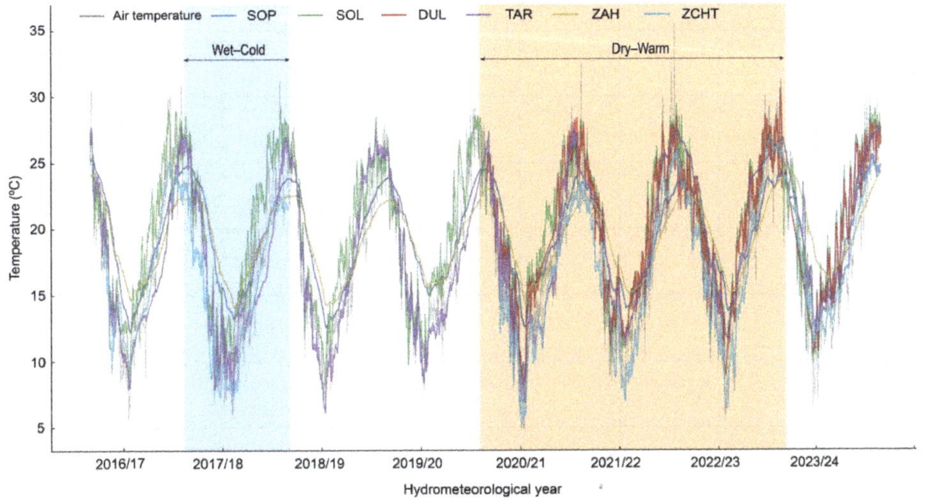

Figure 4. Daily mean temperature of surface and subsurface water for the hydrometeorological years from 2016/17 to 2023/24. SOP: Sopetón pond; SOL: Santa Olalla pond; DUL: Dulce pond; TAR: Taraje pond; ZAH: Zahíllo pond; ZCHT: Zacallón Charco del Toro. Wet–cold and dry–warm years are indicated with shaded color band.

Graphically, the variation in the water temperature in the ponds closely mirrors that of the air temperature, following a seasonal intra-annual evolution with maximum average temperatures in summer (July–August) and minimums in winter (December–January). However, the water temperature shows a degree of attenuation in extreme temperatures, both daily and seasonal. This attenuation is particularly pronounced in the Sopetón and Zahíllo ponds, where subsurface water temperatures were recorded.

Table 2 provides a statistical summary of the water temperature data recorded at each monitoring point during the study period, enabling the characterization of the thermal regime of each system. In addition, air temperature statistics from the Almonte weather station are shown.

Table 2. Statistical summary of temperature data for surface and subsurface water (less than 2 m), groundwater, and air. n: number of daily mean temperature measurements; mean: average temperature (°C); SD: standard deviation; min: minimum temperature; 25%, 50%, and 75%: first, second (median), and third quartiles; max: maximum temperature. SOP: Sopetón pond; SOL: Santa Olalla pond; DUL: Dulce pond; TAR: Taraje pond; ZAH: Zahíllo pond; ZCHT: Zacallón del Charco del Toro. PSOP: Sopetón pond piezometer; PSOLW: western piezometer at Santa Olalla pond; PDUL: Dulce pond piezometer; S10: borehole 10.

	n	Mean	SD	Min	25%	50%	75%	Max
Surface and subsurface water								
SOP	2922	19.29	3.59	12.25	16.02	19.28	22.80	25.56
SOL	2922	19.85	5.29	7.91	15.36	20.32	24.51	30.57
DUL	1461	19.80	5.05	8.23	15.41	19.86	24.16	30.76
TAR	2922	18.29	5.53	6.09	13.65	17.58	23.67	27.93
ZAH	2922	19.22	2.74	14.10	16.75	19.28	21.80	24.69
ZCHT	2098	16.75	5.27	5.20	12.48	16.48	21.63	27.03
Groundwater								
PSOP	2922	19.79	3.71	12.25	16.42	19.83	23.30	26.43
PSOLW	2922	18.44	1.81	15.20	16.92	18.27	20.12	21.64
PDUL	2922	18.04	0.39	17.39	17.73	18.00	18.30	18.97
S10	1453	20.26	0.16	19.56	20.17	20.27	20.37	20.57
Air	2920	18.05	5.83	4.80	13.20	17.80	22.70	35.50

The average temperature of surface and subsurface waters is approximately 19 °C. The Santa Olalla and Dulce ponds show the highest mean temperatures of 19.85 °C and 19.8 °C, respectively (Table 2). In contrast, the Taraje pond and the Zacallón del Charco del Toro exhibit the lowest mean temperatures, with the Zacallón water falling below 17 °C. The standard deviation reveals that the variability in the surface water temperature is similar to that in the air temperature (approx. 5 °C). The Santa Olalla and Dulce ponds closely align with the maximum air temperature values, with maximum water temperatures exceeding 30 °C. In contrast, in the Taraje pond and ZCHT, the maximum water temperature values are attenuated, resulting in lower average temperatures.

In the case of subsurface water, temperatures recorded in the Sopetón and Zahíllo ponds, while the mean temperature is similar to that of surface water, extreme values are significantly attenuated, as indicated by the lower standard deviation (Table 2).

4.3. *Thermal Regime of Groundwater*

Figure 5 depicts the daily mean groundwater temperature evolution measured at the four piezometers over the study period, along with the daily mean air temperature recorded at the Almonte weather station. For borehole 10 (S10), the data are limited due to the exclusion of records from hydrometeorological years when this borehole was more extensively used.

Groundwater temperature follows a seasonal intra-annual cycle influenced by air temperature variability, but these cycles are significantly attenuated (Figure 5). In shallower piezometers, particularly the PSOP (~1.5 m), the thermal signal follows a similar trend to air temperature, although with slightly reduced amplitude. Figure 5 graphically shows how the amplitude of water thermal variability decreases progressively in depth. The greatest stability in temperatures is observed at the deepest levels.

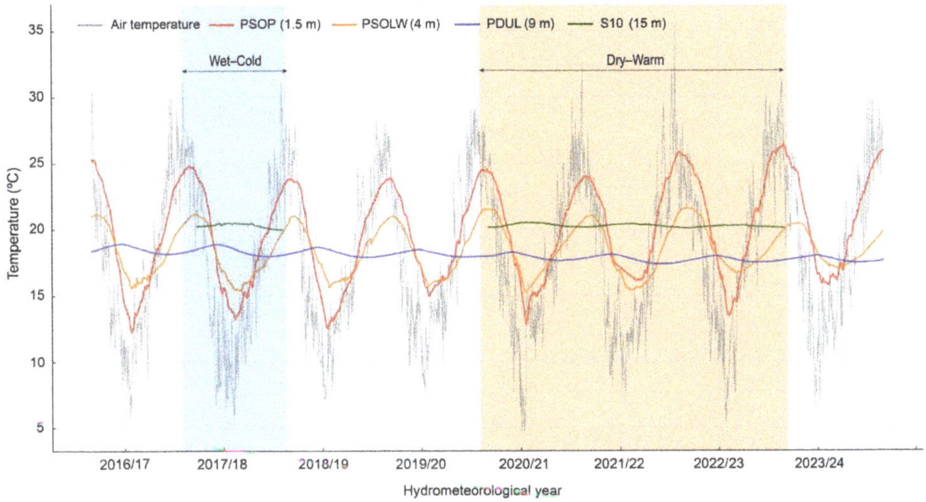

Figure 5. Daily mean temperature of groundwater for the hydrometeorological years from 2016/17 to 2023/24. PSOP: Sopetón piezometer; PSOLW: Santa Olalla piezometer; PDUL: Dulce piezometer; S10, borehole 10. The depth at which the temperature has been monitored inside the piezometer is indicated in parentheses. Wet–cold and dry–warm years are indicated with shaded color band.

According to Table 2, the average groundwater temperature at the PSOP is 19.79 °C, similar to surface and subsurface water temperatures. This temperature decreases slightly with depth, reaching 18.44 °C at the PSOLW (~4 m) and 18.04 °C at the PDUL (~9 m). However, at S10 (~15 m), the mean temperature increases to 20.26 °C, even surpassing the average surface water temperatures. In terms of thermal variability, as can also be observed in Figure 5, the standard deviation decreases with depth: the PSOP has a standard deviation of 3.71 °C, like subsurface systems, while the PSOLW and PDUL exhibit lower standard deviations of 1.81 °C and 0.39 °C, respectively, reflecting greater thermal stability. S10, with a standard deviation of 0.16 °C, exhibits the highest thermal stability among all analyzed piezometers.

The vertical temperature profiles shown illustrate how the attenuation of air temperature influence on groundwater temperature diminishes with depth (Figure 6). The profiles reveal that the shallow groundwater temperatures closely align with the air temperature recorded on the day the temperature profiles were taken (i.e., approximately 25 °C in October 2024). The groundwater temperature progressively decreases with depth, until it stabilizes at a depth of 12 m. Below this depth, the temperature remains constant at around 17.7 °C. This stratification in groundwater temperature defines two distinct zones: a shallow heterothermal zone extending to about 12 m and a deeper homothermal zone below this depth (Figure 6).

In addition to attenuation of thermal signal amplitude with depth, a temporal lag is observed in the annual temperature cycles of groundwater relative to air temperature. Figure 7 illustrates the sinusoidal function representing the variability in air and groundwater temperatures for each piezometer. In reference to the annual cycle, extreme groundwater temperatures occur with a delay relative to air temperature extremes, and this lag increases with depth. At the PSOP, the lag is 23 days, while at the PSOLW, it is 40 days. For the PDUL, the lag extends to 129 days, and at S10 (15 m depth), it reaches 160 days, indicating a complete inversion of the annual temperature cycle. Thus, in deeper piezometers, maximum groundwater temperatures are typically recorded in the winter, while minimum temperatures are observed in the summer.

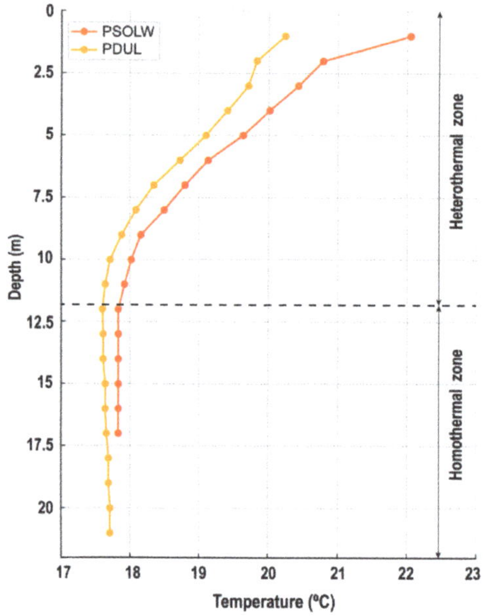

Figure 6. Vertical temperature profiles measured in the PSOLW and PDUL piezometers, highlighting the heterothermal and homothermal zones along the depth gradient.

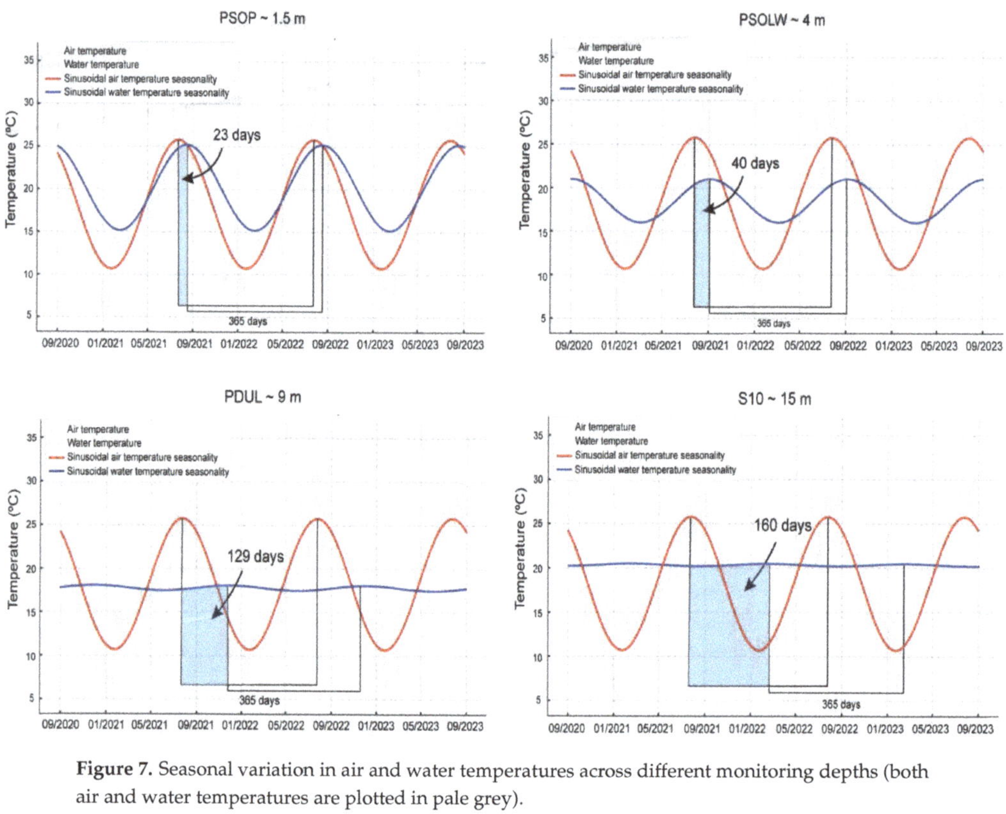

Figure 7. Seasonal variation in air and water temperatures across different monitoring depths (both air and water temperatures are plotted in pale grey).

4.4. Water Temperature Trend

The trend analysis references the climatic context during the monitoring period outlined in the Climate Setting Section (see Section 4.1). Figure 8 displays the annual water temperature in three dry–warm hydrometeorological years (2020/21 to 2022/23) compared to one wet–cold year (2017/18).

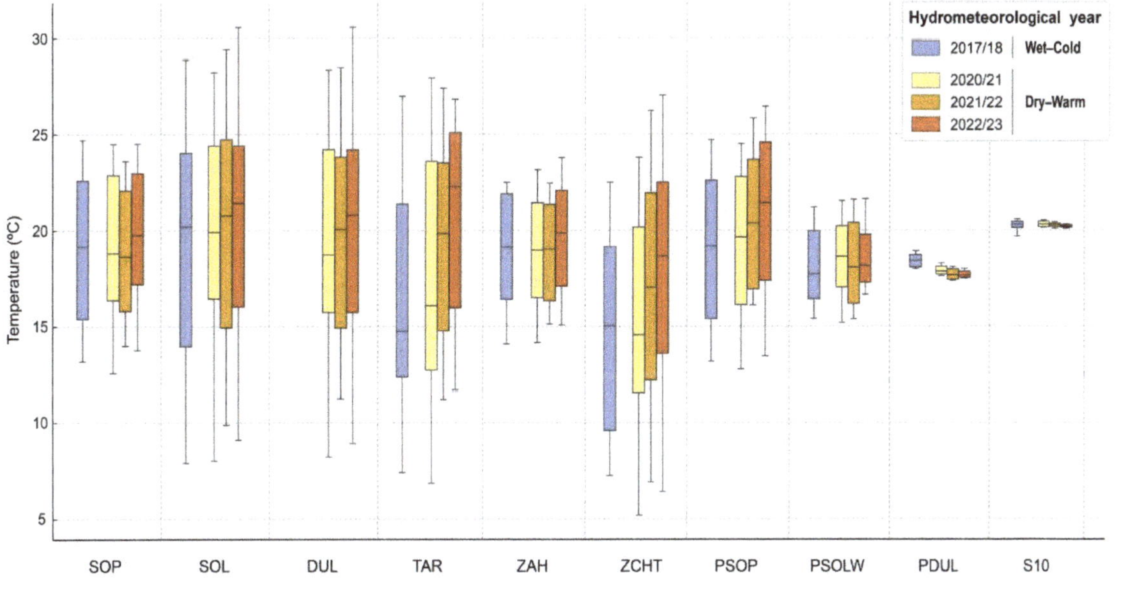

Figure 8. Boxplot of water temperature distributions across monitoring points during the wet–cold (2017/18) and dry–warm (2020–2023) periods. SOP (Sopetón pond), SOL (Santa Olalla pond), DUL (Dulce pond), TAR (Taraje pond), ZAH (Zahíllo pond), ZCHT (Zacallón del Charco del Toro), PSOP (Sopetón pond piezometer), PSOLW (western piezometer at Santa Olalla pond), PDUL (Dulce pond piezometer), and S10 (borehole 10).

Surface and subsurface waters experienced an increase in the annual mean water temperature during the dry–warm period of 2020–2023 (Figure 8). This rise was particularly pronounced in the Taraje pond and Zacallón del Charco del Toro, while it was lower in systems where subsurface water temperatures were recorded, such as the Zahíllo and Sopetón ponds. Most systems also experienced greater intra-annual water temperature variability because the minimum maintained and maximum temperatures increased. An analysis of variance comparing the mean water temperatures at the end of the dry–warm period (2022/23) with those from the wet–cold year (2017/18) revealed a significant increase ($p > 0.05$) in the mean water temperature during the dry–warm period. Specifically, for Laguna Taraje and Zacallón, the increase was 3.88 °C and 3.14 °C, respectively. For the Dulce and Santa Olalla ponds, the increase was more moderate, around 1.2 °C, while subsurface water systems showed a smaller increase of 0.5 °C.

For groundwater, the trend in the annual mean water temperature during the dry–warm period varies with the depth of temperature recording. In the shallowest piezometer, the PSOP, the trend is similar to that of surface and subsurface waters, with a significant increase in the water temperature (Figure 8). However, in deeper piezometers, the PDUL and S10, a significant downward trend ($p > 0.05$) in the water temperature has been observed, accompanied by reduced intra-annual variability. This reduction is particularly evident in the PDUL, where a notable decrease of more than 1 °C in the water temperature has been recorded between 2017/18 and 2022/23. In S10, the decrease was only 0.1 °C, but both cases show remarkable cooling considering the expected thermal stability at these depths. For the PSOLW, no statistically significant trend has been identified for the dry–warm period, although a slight decrease in the average temperature was observed (Figure 8).

5. Discussion

The statistical analysis of the extensive thermal record obtained from hydrological monitoring in the most important sand-dune ponds of the DNP has enabled the characterization of the thermal regime of the surface water, subsurface water, and groundwater. The results indicate that this regime is primarily influenced by the extent to which air temperature variability influences water temperature. Nevertheless, there are significant differences between surface water (ponds) and shallow groundwater (PSOP, PDUL, PSOLW), on the one hand, and deeper groundwater (S10), on the other.

5.1. Surface Water and Shallow Groundwater Temperature

Air temperature determines the intra-annual seasonal pattern in the surface water, characterized by maximum temperatures in the summer and minimums in the winter (Figure 4). The average surface water temperature is approximately 19 °C, slightly higher than the average air temperature during the monitoring period (~18 °C). The lower standard deviation values in the surface water indicate some attenuation of extreme environmental values (Table 2). However, given that the sensors are located in the deepest areas of ponds, the seasonal temperature pattern suggests that, due to the shallowness of these ponds, the temperature oscillations are not buffered, as in the case of deeper lakes [33,34], and it does not suggest that ponds sustain prolonged water column stratification. This implies that the water temperature in the studied ponds is heavily influenced by air thermal oscillation, with rapid heating and cooling of the entire mixing water column. This is a typical phenomenon in shallow lakes [33,35,36]. Thus, the characterization of these thermal dynamics is crucial because they influence the rates of chemical and biological processes [33,36,37] and determine their vulnerability to climate changes [26,34].

Although the pronounced influence of air temperature is consistent across all the monitored surface waters, some differences emerge, likely due to local heterogeneity in factors affecting the water temperature in shallow lakes, such as solar exposure and the presence of vegetation [38]. For instance, the Taraje pond and the Zacallón del Charco del Toro exhibit lower temperatures compared to other surface waters, likely due to reduced solar exposure caused by the surrounding vegetation. This could explain why the minimum temperatures in these systems are similar to the minimum air temperature (~5 °C, Table 2), yet their maximum temperatures are significantly lower than the environmental maximum (~27 °C versus 35 °C). Conversely, ponds without dense vegetation cover, such as Santa Olalla and Dulce, show less buffered maximum temperatures.

At the Santa Olalla pond, the only permanent pond, some progress had already been made in describing local-scale groundwater flow using temperature as a tracer, as indicated in the Introduction. A specific study was conducted between 2016 and 2018, which utilized thermal data obtained during the project [10]. This study contributed to understanding surface water–groundwater interactions in this pond system. The results allowed for the estimation of groundwater discharge rates of 0.36 hm^3/year during high water levels (February–May 2017) and significantly lower rates of 0.05 hm^3/year during low water levels (October–March 2018), which included periods of recharge processes from the pond to the aquifer.

In the Zahíllo and Sopetón ponds, the sensor was located in the hyporheic zone, recording subsurface water temperatures in the pond beds. In these cases, as reflected by the lower standard deviation of the water temperature (Table 2), the seasonal cycle is more attenuated (Figure 4), consistent with subsurface water dynamics [10,39]. This highlights the effect of soil on heat conduction within the unsaturated soil zone, even just a few centimeters below ground.

In fact, the soil's buffering effect is even more evident in the groundwater temperature records, particularly at greater depths (Table 2 and Figure 5). In the shallowest piezometer, the PSOP (~1.5 m), the thermal signal follows a seasonal trend similar to that of subsurface water. As the depth increases, the groundwater thermal regime shows a progressive attenuation of seasonal variability (Figure 5), as seen in deeper piezometers like the PDUL (~9 m) and S10 (~15 m), where temperatures are more stable and standard deviations significantly lower (Table 2). This results in a groundwater temperature stratification, with a heterothermal zone in shallow layers, where groundwater is more exposed to seasonal variations, and a homothermal zone below 12 m depth (Figure 6), where temperatures are nearly constant (~17.7 °C), irrespective of external climatic conditions.

This groundwater temperature stratification pattern, although common in detrital aquifers, varies significantly depending on the local factors affecting heat transport within the subsurface, e.g., [19] and references therein. While it is true that in our study we only present two vertical temperature profiles conducted in the autumn (Figure 6), our findings are consistent with the vertical temperature profiles described for the Almonte-Marismas aquifer in [16]. These profiles are valuable for characterizing the depth of the homothermal zone within this context.

In addition to the climatic impact on the temperature of shallow groundwater, ref. [16] modeled the long-term impact of anthropogenic influences, such as groundwater extraction, land-use changes, and agricultural practices, on the thermal regime of the entire Almonte-Marismas aquifer. Our results provide valuable insights into their localized effects on sand-dune ponds. The thermal record provided in our study for the sand-dune pond area is relevant for assessing the impacts described in [16] on the thermal regime of this part of the aquifer. However, although in recent years the climatic impact appears to have prevailed over these anthropogenic influences, it is crucial to continue detailed temperature monitoring to detect thermal changes associated with anthropogenic impacts.

Regardless of this, these conditions largely determine the sensitivity of surface and groundwater to environmental changes. Shallow groundwater and surface water exhibit rapid responses to prevailing climatic conditions, as observed during the dry–warm period (2020–2023). During this period, the average annual surface and subsurface water temperatures increased significantly, with more pronounced rises in systems like Zacallón del Charco del Toro and the Taraje pond, where temperatures rose by over 3 °C compared to the cool–wet year of 2017/18 (Figure 8). This increase is substantially greater than the air temperature anomalies (~1 °C, Figure 2) recorded during the same period. Given their proximity to Matalascañas town, where hydrological alterations due to water extractions for human use have been documented [9,14,40], the recorded temperature increases are likely a result of a combination between environmental and anthropogenic factors. The decreasing water levels observed in recent years [31] may have contributed to faster warming in systems with greater water volume reductions, such as in the Taraje pond.

5.2. Deep Groundwater Temperature

Unlike the surface and subsurface temperature trends, groundwater responses varied with depth (Figure 8). In the PSOP (~1.5 m), the thermal trend mirrored subsurface water, with a significant rise in the average annual temperature during the dry–warm period.

However, in deeper piezometers like the PDUL (~9 m) and S10 (~15 m), an opposite trend was observed, with a significant reduction in the average annual temperature.

For deeper piezometers, this variability might be linked to aquifer recharge and discharge processes rather than air temperature fluctuations. Recharge processes can influence the temperature in the superficial zone of the aquifer and contribute to more efficient heat transport to deeper zones via convective heat transfer through vertical recharge flow [16,19]. In the Doñana aquifer, this phenomenon is particularly notable due to the aquifer's hydrogeological characteristics (see Section 2). As previously mentioned, although the Doñana aquifer is a multilayer system, it can be considered to consist of a shallow layer of fine–medium sands, where water flow is nearly vertical, and a deeper layer of gravel and coarse sands, where flow is predominantly horizontal. It can be assumed that temperature is controlled from above (ground surface) and below (coarse layer), as modeled by [16]. In this way, a thermal gradient is present but attenuates when convective transport carries heat to deeper zones during recharge or elevates it from deeper zones in discharge areas. Thus, the thermal gradient reflects the dominance of convective heat transport in shallow layers, where groundwater flow is vertical, while in deeper zones, conduction prevails, leading to a more stable thermal regime. This distinction explains the progressive attenuation of temperature variability with depth (Figure 6).

These flows could explain why the maximum groundwater temperatures in the PDUL and S10 occur in the winter months (Figure 6), when aquifer recharge and discharge processes are more active. They also account for the highest annual average temperature recorded at the PDUL during the wettest, coolest hydrometeorological year (2017/18, Figure 8). Similarly, a cessation of heat transport due to reduced recharge flow would result in an increased thermal gradient by cooling in deeper levels. Thus, the significant temperature decline detected in the PDUL could be attributed to the reduction in recharge that has occurred in recent years. This cooling at deeper levels during drought periods was previously described theoretically by [16]. Our findings could empirically confirm this model.

However, the temperature variability observed in S10 does not seem to be associated with recharge flow, as the average water temperature at this point (20.26 °C, Table 2) does not correspond to what would be expected at the depth where the sensor is located (15 m). This anomaly may stem from different causes: (1) borehole pumping, as it is the only one in the study area; (2) borehole depth, as it is the deepest piezometer (~170 m); and (3) proximity to the coast.

Indeed, as stated by [16], the coastline acts as an abrupt thermal transition zone between the continent and the sea. This phenomenon could cause lower temperatures in the underlying area, generating a thermal gradient that extends towards zones farther from the coast. Consequently, the average temperature at S10 might even be lower than that of the PDUL. Figure 9 presents a hypothetical heat transport schematic in the Almonte-Marismas aquifer, based on previous models of heat transport in coastal detrital aquifers [16,19,20], which could explain the thermal regime and trends observed in groundwater temperature, as described in this study.

According to this schematic, both the variability in the water temperature at S10 and its average temperature could be linked to regional flow originating from zones farther inland, facilitated by horizontal transport in the deeper, coarser unit of the aquifer, where the bottom of S10 is located and where the majority of boreholes in the Almonte-Marismas aquifer takes groundwater from [41]. In this scenario, it would need to be assumed that no temperature stratification occurs within the borehole itself, possibly because occasional pumping homogenizes the entire water column. The inability to conduct vertical temperature profiles at this point has prevented confirmation of this assumption.

Nonetheless, a similar scenario to that modeled for the Motril-Salobreña aquifer by [21], also schematically represented in Figure 9, cannot be ruled out. In this case, deep regional flow could be influenced by the geothermal gradient, forcing heat to ascend near the saline interface, which could raise temperatures to shallower levels and reduce the expected thermal gradient in these zones (Figure 9). However, the exact position of the saline interface in the Doñana aquifer has yet to be identified, and a combined effect of these scenarios is also a possibility.

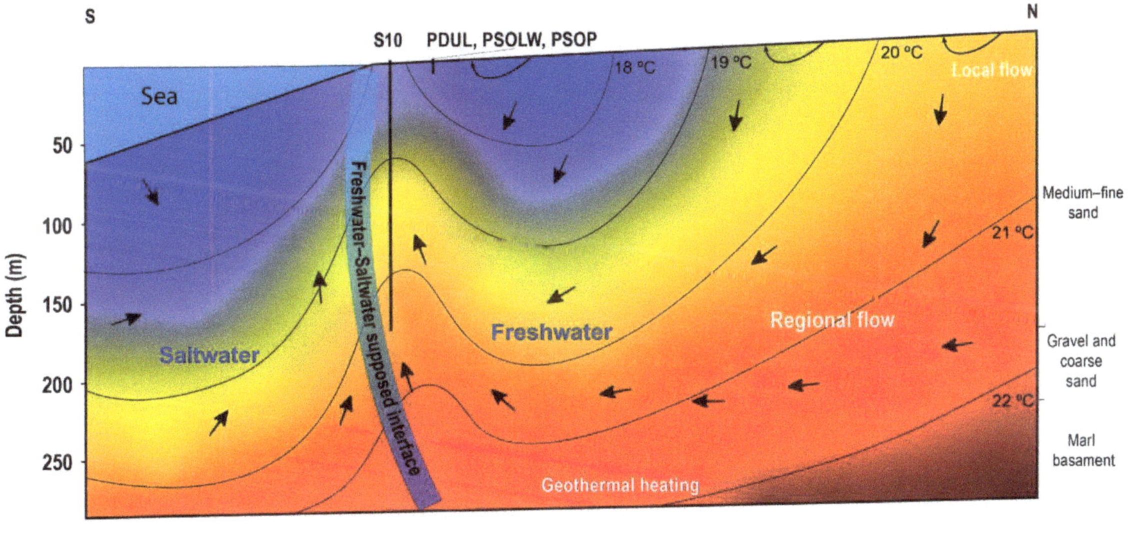

Figure 9. Simplified schematic of temperature distribution in Almonte-Marismas aquifer, based on water thermal records obtained in this study, along with models developed by [16,19]. Colors represent the temperature gradient defined by isotherms, while arrows indicate the direction of groundwater flow.

In any of these cases, this flow schematic (Figure 9) would explain the observed temperature decline at S10 and the reduced variability in the water temperature which can be attributed to a lower hydraulic gradient and changes in regional flow during the driest period. To confirm or rule out these scenarios, vertical temperature profiles in boreholes with similar characteristics to S10, but without any exploitation throughout the year, would be necessary.

Although the temperature data monitored at S10 and the PDUL carry a high degree of uncertainty due to the thermal disturbance caused by piezometer-based measurements, the observed temperature trends during drought periods provide insight into the variability in the dynamics of local and regional groundwater flows under these conditions. These trends also suggest a strong interconnection among the different aquifer levels, highlighting the complex interactions within the multilayered Almonte-Marismas aquifer system.

6. Conclusions

In this study, water temperature data recorded in several sand-dune ponds in Doñana National Park reveal an increase (e.g., >3 °C in the Taraje pond between 2017/18 and 2022/23) in the surface water temperature which is significantly higher than the increase in the air temperature (~1 °C) recorded during the same period. Water temperature dynamics in the different ponds are influenced by local factors such as depth, hydraulic connectivity with the aquifer, and exposure to climate influences. In the case of groundwater, temperature monitoring has allowed the characterization of water temperature trends at different depths. These trends are likely associated with local and regional groundwater flows, and a decrease in recharge rates during the dry periods could be behind the temperature drop (<1 °C) recorded in piezometers near the ponds. Despite the high degree of uncertainty involved, analyzing these trends allows for tracing these flows and provides a foundation for developing a heat flow and transport model for the Almonte-Marismas aquifer.

Moreover, the combined characterization of the water temperature regime of surface water and groundwater can aid at identifying and quantifying the influence of regional flow on the sand-dune ponds located farther from Matalascañas town, such as Dulce, Santa Olalla, and Sopetón. Considering that water extraction for Matalascañas town supply occurs at deep levels and agricultural extractions take place in areas distant from the ponds, outside the National Park boundaries, this characterization is crucial for assessing the degree of impact of these extractions on the ponds. It will help enable informed decision-making regarding groundwater resource management.

Author Contributions: Conceptualization, J.L.Y., M.R.-R. and A.J.-B.; methodology, J.L.Y., M.R.-R. and A.J.-B.; investigation, J.L.Y., M.R.-R., A.F.-A., A.J.-B. and M.M.-C.; resources, M.R.-R. and A.F.-A.; data curation, J.L.Y., M.M.-C. and A.J.-B.; writing—original draft preparation, J.L.Y.; writing—review and editing, J.L.Y., A.J.-B., A.F.-A.and M.R.-R.; supervision, J.L.Y., M.R.-R. and A.J.-B.; project administration, M.R.-R. All authors have read and agreed to the published version of the manuscript.

Funding: This work was possible thanks to the collaboration agreement between the Guadalquivir Hydrographic Confederation and the Pablo de Olavide University: "Study of hydrological monitoring and modeling of the pond-aquifer relationship in the Doñana aquifer" (BOE-A-2023-13925).

Data Availability Statement: All data supporting the results of this study are available upon request from any of the authors.

Acknowledgments: The logistical and technical support of the Singular Scientific—Technical Infrastructure of the Doñana Biological Reserve (DBR-ICTS) is gratefully acknowledged.

Conflicts of Interest: The authors declare no conflicts of interest.

References

1. Zorrilla-Miras, P.; Palomo, I.; Gomez-Baggethun, E.; Martín-López, B.; Lomas, P.L.; Montes, C. Effects of land-use change on wetland ecosystem services: A case study in the Doñana marshes (SW Spain). *Landsc. Urban Plan.* **2014**, *122*, 160–174. [CrossRef]
2. Green, A.J.; Guardiola-Albert, C.; Bravo-Utrera, M.A.; Bustamante, J.; Camacho, A.; Caro, J.; Casado, A.; González-Arbáez, E.; Espinar, J.L.; Teles da Gama, F.; et al. Groundwater abstraction has caused extensive ecological damage to the Doñana World Heritage Site, Spain. *Wetlands* **2024**, *44*, 20. [CrossRef]
3. De Felipe, M.; Amat, J.A.; Arroyo, J.L.; Rodríguez, R.; Díaz-Paniagua, C. Habitat changes at the local scale have major impacts on waterfowl populations across a migratory flyway. *Glob. Change Biol.* **2024**, *30*, e17600. [CrossRef]
4. Rodríguez-Rodríguez, M.; Aguilera, H.; Guardiola-Albert, C.; Fernández-Ayuso, A. Climate influence vs. anthropogenic drivers in surface water-groundwater interactions in eight ponds of Doñana Natural Area (southern Spain). *Wetlands* **2021**, *41*, 25. [CrossRef]
5. García-Bravo, N.; Guardiola-Albert, C. Upgrading aquifer test interpretations with numerical axisymmetric flow models using MODFLOW in the Doñana area (Spain). *Boletín Geológico Min.* **2012**, *123*, 41–54.
6. Green, A.J.; Bustamante, J.; Janss, G.F.E.; Fernández-Zamudio, R.; Díaz-Paniagua, C. Doñana Wetlands (Spain). In *The Wetland Book*; Finlayson, C., Milton, G., Prentice, R., Davidson, N., Eds.; Springer: Dordrecht, The Netherlands, 2018; pp. 1–10. [CrossRef]
7. Manzano, M.; Custodio, E.; Iglesias, M.; Soler, A. Groundwater resources and ecological impacts in Doñana National Park. *Hydrol. Process.* **2001**, *15*, 1431–1450. [CrossRef]
8. Díaz-Paniagua, C.; Fernández-Zamudio, R.; Green, A.J.; Bustamante, J. The impacts of climate variability on temporary ponds in Doñana National Park. *Wetl. Res. Lett.* **2024**, *12*, 56–70.
9. De Felipe, M.; Aragonés, D.; Díaz-Paniagua, C. Thirty-four years of Landsat monitoring reveal long-term effects of groundwater abstractions on a World Heritage Site wetland. *Sci. Total Environ.* **2023**, *880*, 163329. [CrossRef]

10. Rodriguez-Rodriguez, M.; Fernández-Ayuso, A.; Hayashi, M.; Moral-Martos, F. Using Water Temperature, Electrical Conductivity, and pH to Characterize Surface–Groundwater Relations in a Shallow Ponds System (Doñana National Park, SW Spain). *Water* **2018**, *10*, 1415. [CrossRef]
11. Fernández-Ayuso, A.; Rodríguez-Rodríguez, M.; Benavente, J. Assessment of the hydrological status of Doñana dune ponds: A natural World Heritage Site under threat. *Hydrol. Sci. J.* **2018**, *63*, 2048–2059. [CrossRef]
12. Fernández-Ayuso, A.; Aguilera, H.; Guardiola-Albert, C.; Rodríguez-Rodríguez, M.; Heredia, J.; Naranjo-Fernández, N. Unraveling the hydrological behavior of a coastal pond in Doñana National Park (Southwest Spain). *Ground Water* **2019**, *57*, 895–906. [CrossRef]
13. Treviño, J.; Rodríguez-Rodríguez, M.; Montes-Vega, M.J.; Aguilera, H.; Fernández-Ayuso, A.; Fernández-Naranjo, N. Wavelet Analysis on Groundwater, Surface-Water Levels andWater Temperature in Doñana National Park (Coastal Aquifer in Southwestern Spain). *Water* **2023**, *15*, 796. [CrossRef]
14. Dimitriou, E.; Mousoulis, E.; Díaz-Paniagua, C.; Serrano, L. Hydrodynamic numerical modelling of the water level decline in four temporary ponds of the Doñana National Park (SW Spain). *J. Arid. Environ.* **2017**, *147*, 92–100. [CrossRef]
15. Kurylyk, B.L.; Irvine, D.J.; Bense, V.F. Theory, tools, and multidisciplinary applications for tracing groundwater fluxes from temperature profiles. *Water* **2018**, *5*, e1299. [CrossRef]
16. Custodio, E.; Manzano, M.; Iglesias, M. Análisis térmico preliminar de los acuíferos de Doñana. *IV SIAGA* **1996**, *II*, 57–87.
17. Tamburini, A.; Menichetti, M. Groundwater temperature as a natural tracer to characterize hydraulic behaviour and geometry of carbonate aquifers: Mt. Nerone karst system, central Italy. *Rend. Online Soc. Geol. Ital.* **2019**, *47*, 121–125. [CrossRef]
18. Gruszczyński, T.; Szostakiewicz-Hołownia, M.; Zaszewski, D. Interpretation of spring-water temperature variability based on continuous monitoring and numerical modelling of heat transport and water mixing: Case study of Czarny Potok spring, Pieniny Mountains, southern Poland. *Hydrogeol. J.* **2023**, *31*, 1–12. [CrossRef]
19. Blanco-Coronas, A.M.; Duque, C.; Calvache, M.L.; López-Chicano, M. Temperature distribution in coastal aquifers: Insights from groundwater modeling and field data. *J. Hydrol.* **2021**, *603*, 126912. [CrossRef]
20. Blanco-Coronas, A.M.; Calvache, M.L.; López-Chicano, M.; Martín-Montañés, C.; Jiménez-Sánchez, J.; Duque, C. Salinity and temperature variations near the freshwater-saltwater interface in coastal aquifers induced by ocean tides and changes in recharge. *Water* **2022**, *14*, 2807. [CrossRef]
21. Guardiola-Albert, C.; Jackson, C.R. Potential impacts of climate change on groundwater supplies to the Doñana wetland, Spain. *Wetlands* **2011**, *31*, 907–920. [CrossRef]
22. Naranjo-Fernández, N.; Guardiola-Albert, C.; Montero, E. Applying 3D geostatistical simulation to improve the groundwater management modelling of sedimentary aquifers: The case of Doñana (Southwest Spain). *Water* **2018**, *11*, 39. [CrossRef]
23. Blanco Coronas, Á.M. Coastal Aquifers and Processes Related to Global Change: Wetlands, Heat Transport and Floodings. Ph.D. Thesis, Universidad de Granada, Granada, Spain, 2023. Available online: https://hdl.handle.net/10481/82172 (accessed on 17 February 2025).
24. Jiménez-Bonilla, A.; Rodríguez-Rodríguez, M.; Yanes, J.L.; Gázquez, F. Impact of climate change on permanent lakes in a semiarid region: Southwestern Mediterranean basin (S Spain). *Sci. Total Environ.* **2025**, *961*, 178305. [CrossRef] [PubMed]
25. Erostate, M.; Huneau, F.; Garel, E.; Ghiotti, S.; Vystavna, Y.; Garrido, M.; Pasqualini, V. Groundwater dependent ecosystems in coastal Mediterranean regions: Characterization, challenges and management for their protection. *Water Res.* **2020**, *172*, 115461. [CrossRef]
26. Montes-Vega, M.J.; Guardiola-Albert, C.; Rodríguez-Rodríguez, M. Calculation of the SPI, SPEI, and GRDI Indices for Historical Climatic Data from Doñana National Park: Forecasting Climatic Series (2030–2059) Using Two Climatic Scenarios RCP 4.5 and RCP 8.5 by IPCC. *Water* **2023**, *15*, 2369. [CrossRef]
27. Naranjo-Fernández, N.; Guardiola-Albert, C.; Aguilera, H.; Serrano-Hidalgo, C.; Rodríguez-Rodríguez, M.; Fernández-Ayuso, A.; Ruiz-Bermudo, F.; Montero-González, E. Relevance of spatio-temporal rainfall variability regarding groundwater management challenges under global change: Case study in Doñana (SW Spain). *Stoch. Environ. Res. Risk Assess.* **2020**, *34*, 1289–1311. [CrossRef]
28. Salvany, J.M.; Custodio, E. Características litoestratigráficas de los depósitos plio-cuaternarios del entorno de Doñana: Implicaciones hidrogeológicas. *Rev. Soc. Geol. Esp.* **1995**, *8*, 2–31.
29. Custodio, E.; Manzano, M.; Montes, C. Las aguas subterráneas en Doñana. In *Aspectos Ecológicos y Sociales*; Junta de Andalucía: Sevilla, Spain, 2009; p. 243.
30. Instituto Geológico y Minero de España. *Encomienda de Gestión para la Realización de Trabajos Científico-Técnicos de Apoyo a la Sostenibilidad y Protección de las Aguas Subterráneas: Demarcación Hidrográfica 051 Guadalquivir, Masa de Agua Subterránea 051.051 Almonte-Marismas del Guadalquivir*; Ministerio de Ciencia e Innovación, Ministerio de Medio Ambiente y Medio Rural y Marino, Dirección General del Agua: Madrid, Spain, 2024.
31. Yanes, J.L.; Rodríguez-Rodríguez, M.; Montes-Vega, M.J.; Jiménez-Bonilla, A.; Moral, F. Monitorización hidrológica de las lagunas peridunares de los mantos eólicos de Doñana en el periodo 2020–2023. *Geogaceta* **2024**, *76*, 55–58. [CrossRef]

32. R Core Team. *R: A Language and Environment for Statistical Computing;* R Foundation for Statistical Computing: Vienna, Austria, 2024; Available online: https://www.R-project.org/ (accessed on 17 September 2024).
33. Martinsen, K.T.; Andersen, M.R.; Sand-Jensen, K. Water temperature dynamics and the prevalence of daytime stratification in small temperate shallow lakes. *Hydrobiologia* **2018**, *830*, 87–103. [CrossRef]
34. Woolway, R.I.; Jennings, E.; Shatwell, T.; Golub, M.; Pierson, D.C.; Maberly, S.C. Lake heatwaves under climate change. *Nature* **2021**, *589*, 402–407. [CrossRef]
35. Christensen, J.P.A.; Sand-Jensen, K.; Staehr, P.A. Fluctuating water levels control water chemistry and metabolism of a charophyte-dominated pond. *Freshw. Biol.* **2013**, *58*, 1353–1365. [CrossRef]
36. Holgerson, M.A.; Richardson, D.C.; Roith, J.; Bortolotti, L.E.; Finlay, K.; Hornbach, D.J.; Gurung, K.; Ness, A.; Andersen, M.R.; Bansal, S.; et al. Classifying mixing regimes in ponds and shallow lakes. *Water Resour. Res.* **2022**, *58*, e2022WR032522. [CrossRef]
37. Yvon-Durocher, G.; Caffrey, J.M.; Cescatti, A.; Dossena, M.; Giorgio, P.D.; Gasol, J.M.; Montoya, J.M.; Pumpanen, J.; Staehr, P.A.; Trimmer, M.; et al. Reconciling the temperature dependence of respiration across timescales and ecosystem types. *Nature* **2012**, *487*, 472–476. [CrossRef] [PubMed]
38. Andersen, M.R.; Sand-Jensen, K.; Woolway, R.I.; Jones, I.D. Profound daily vertical temperature variation and mixing in a small, shallow, wind-exposed lake with submerged macrophytes. *Aquat. Sci.* **2017**, *79*, 395–405. [CrossRef]
39. Keery, J.; Binley, A.; Crook, N.; Smith, J.W.N. Temporal and spatial variability of groundwater–surface water fluxes: Development and application of an analytical method using temperature time series. *J. Hydrol.* **2007**, *336*, 1–16. [CrossRef]
40. Montes-Vega, M.J.; Rodríguez-Rodríguez, M. Análisis del hidroperiodo de tres lagunas de la Reserva Biológica de Doñana (2018–2020). *Geogaceta* **2021**, *70*, 43–46. [CrossRef]
41. Acreman, M.; Casier, R.; Salathé, T. Evidence-based risk assessment of ecological damage due to groundwater abstraction: The case of Doñana Natural Space, Spain. *Wetlands* **2022**, *42*, 63. [CrossRef]

environments

MDPI

Article

Toxic Phytoplankton in Mussel Farms in the Gulf of Trieste, Adriatic Sea (Italy): A Preliminary Analysis of Long-Term Data (2001–2022) in Relation to Environmental Conditions

Lisa Tondelli, Nicola Bettoso, Oriana Blasutto, Massimo Celio and Alessandro Acquavita *

Agenzia Regionale per la Protezione dell'Ambiente (ARPA FVG), Via Cairoli 14, 33057 Palmanova, Italy; lisa.tondelli@arpa.fvg.it (L.T.); nicola.bettoso@arpa.fvg.it (N.B.); oriana.blasutto@arpa.fvg.it (O.B.); massimo.celio@arpa.fvg.it (M.C.)

* Correspondence: alessandro.acquavita@arpa.fvg.it

Abstract: The present study utilised a comprehensive, long-term dataset of toxic phytoplankton (2001–2022) to analyse the relationships between the abundance, distribution, and seasonal trends of toxic phytoplankton, as well as the influence of various sampling methods, physico-chemical parameters, nutrients, and meteo-climatic parameters. The data were obtained through institutional monitoring at four selected sites dedicated to shellfish farming in the Gulf of Trieste (Adriatic Sea, Italy). The results show significant gradients in the spatial distribution of toxic phytoplankton and clear seasonal patterns in regard to the physico-chemical parameters and nutrients. Toxic phytoplankton abundance peaked in late winter/early spring and early autumn, depending on the genus considered. Significant correlations suggest that rainfall influences the nutrient levels and the proliferation of toxic phytoplankton. The time series analysis highlighted significant increases in temperature, salinity, and nitrogen species, during the study period, and decreases in silicon and phosphorous, while chlorophyll *a* and the overall phytoplankton abundance remained relatively stable, except for a significant decrease in *Lingulodinium* from 2015 to 2022. This preliminary assessment provides a valuable basis for further approaches (e.g., continuous in situ measurements, modelling, machine learning) to investigate the potential impact of climate variability on toxic phytoplankton dynamics in the Gulf of Trieste and to support mussel-farming management from both health and environmental perspectives.

Keywords: toxic phytoplankton; nutrients; physico-chemical parameters; temporal series; Gulf of Trieste; Adriatic Sea

1. Introduction

Toxic phytoplankton are microalgae that are of particular concern because they can negatively impact human health. They include numerous species associated with food poisoning and that are responsible for various syndromes (Paralytic Shellfish Poisoning—PSP, diarrhetic shellfish poisoning—DSP, Azaspiracid Shellfish Poisoning—AZP, Amnesic Shellfish Poisoning—ASP, Neurotoxic Shellfish Poisoning—NSP, Ciguatera Fish Poisoning—CFP, and cyanotoxin poisoning) [1]. In this study we focused on the dinoflagellate genera, *Alexandrium*, *Dinophysis*, and *Lingulodinium*, and the diatom genus, *Pseudo-nitzschia*, which contain several potentially toxic species [1,2]. These genera produce saxitoxin (PSP), okadaic acid and yessotoxin (DSP), and domoic acid (ASP), respectively [3–6]. The presence of harmful algal blooms (HABs) caused by these genera, with the exception of *Dinophysis*, has been shown to harm marine ecosystems, natural marine resources, and aquaculture [7], by inducing hypoxia and anoxia, toxicity, and causing mechanical damage to organisms. Toxic phytoplankton also poses a significant risk to human health via ingestion and direct contact [8], and represents a substantial economic threat to the shellfish industry, particularly mussel farming.

The Gulf of Trieste (GoT) has a long tradition of shellfish farming, with numerous facilities located along its coastline. The most exploited species is the Mediterranean mussel, *Mytilus galloprovincialis* (Lamarck, 1819). *M. galloprovincialis* is a lamellibranch bivalve mollusc and filter feeder, capable of retaining part of the suspended particles (in the order of μm) floating in the water column. It feeds primarily on plankton and organic particles, with the filtration rate and assimilation efficiency influenced by environmental factors, such as temperature and the concentration of dissolved substances [9,10].

In 1973, the local production of *M. galloprovincialis* in the GoT peaked at 6000 t year^{-1} (200 employment units) [11]. Following an outbreak and subsequent recovery, production was restored in 1990 (7500 t year^{-1}, 198 employment units) [11]. However, biotoxin events and competition from producers in Greece and Spain led to another decline in the late 1990s. Today, mussel farms cover approximately 300 ha along the coast (about 15 km) and are organised into 14 small cooperatives. Total production is in a serious constant decline (3249 to 1090 t year^{-1} in 2010 and 2023, respectively). According to Commission Implementing Regulation (EU) 2019/627 of 15 March 2019 regarding official controls on products of animal origin intended for human consumption, nearly all GoT farms are classified as zone A [12], meaning live bivalves may be harvested for direct human consumption provided they meet the *E. coli* limits specified in Regulation (EC) No 853/2004 (80% of mollusc samples collected during the review period shall not exceed 230 *E. coli* per 100 g of flesh and intravalvular liquid and the remaining 20% of samples shall not exceed 700 *E. coli* per 100 g of flesh and intravalvular liquid) (EC, 2019/627) [13].

Solidoro et al. investigated the dietary preferences of *M. galloprovincialis* in the GoT and found selectivity of most genera of Bacillariophyceae (including *Pseudo-nitzschia*) and Dinophyceae (including *Alexandrium*, *Dinophysis*, and *Lingulodinium*), as well as protozoa, foraminifera, and the larvae of various organisms, while rejecting colonial or setaceous diatoms and dinoflagellates with elongated shapes and remarkable dimensions [12]. This poses a potential risk through the ingestion pathway, especially from September to November, coinciding with the presence of DSP-producing dinoflagellates [14].

As required by EU Regulation, water samples are collected periodically from sources close to the mussel farms to monitor changes in the composition of toxic phytoplankton and its geographical distribution, and live bivalve molluscs are tested weekly during harvesting (or more frequently if needed) for the presence of biotoxins. No triggers in regard to class, genus, or toxic species have been reported [13]. In the Friuli Venezia Giulia Region (the regional authority on the Italian side of the GoT), an early warning system on the possible toxicity of mussels is in operation, thus optimising prevention measures.

Phytoplankton dynamics are strongly influenced by the surrounding environment [15], yet the drivers of a toxic phytoplankton outburst remain poorly understood, although some hypotheses have been proposed (i.e., aquaculture expansion, climate change, ballast water discharge, the translocation of shellfish stocks to new sites, or excess nutrients) [16–18]. In addition, the enrichment of water with nutrients (primarily nitrogen, silicon, and phosphorus) can fuel algal biomass growth [19], a phenomenon recently investigated using a machine learning approach [20], and documented in different areas of the Mediterranean Sea [21,22].

Covering 2001–2022, this study examines toxic phytoplankton in terms of its spatial distribution, seasonal patterns, and temporal trends, within the framework of regulated mussel farm monitoring. We also analyse the physico-chemical parameters, nutrient concentrations, and precipitation to assess their influence on phytoplankton dynamics. These preliminary findings will be useful for subsequent modelling research and the potential application of machine learning in regard to the dynamics of toxic phytoplankton within mussel-farming areas under climate change scenarios.

2. Materials and Methods

2.1. Study Area

The GoT is a semi-enclosed basin, located in the northernmost part of the Adriatic Sea (Mediterranean Sea). It lies east of the coastline joining Grado (Italy) to Cape Savudrija (Republic of Croatia), covering approximately 600 km^2 [23,24], with a maximum depth of 25 m; about 10% of the area is shallower than 10 m deep [25].

The Isonzo river (Soča in Slovenian), flowing from the Julian Alps into the northern Adriatic Sea (length = 135 km, basin area = 3430 km^2), is the GoT's primary freshwater source. The hydrological regime of the Isonzo River is characterised by a narrow flow range, with a mean discharge of 82 $m^3\ s^{-1}$ (1998–2008 data) [26,27]. Secondary inputs are from smaller rivers in the Italian (Timavo and Rosandra) and Slovenian (Rižana, Badaševica, Drnica, and Dragonja) sectors [28,29]. The GoT's hydrology is also influenced by temperature, wind, and precipitation [30,31]. In summer, the water column is predominantly stratified, while in winter, mixing prevails, aided by local winds. The Bora is notable, a cold northeasterly wind that blows especially in winter, which stirs the water column, cools surface waters, enhances oxygenation, and causes the sinking of water masses [32].

The anthropogenic pressures in the region include urban settlements (Trieste, Monfalcone, Koper) and the respective ports and industrial areas [24], as well as maritime traffic, fishing, aquaculture, and tourism [33]. Agricultural runoff delivers substantial nitrogen (N) and phosphorus (P) loads via the Isonzo river, supplemented by minor contributions from atmospheric deposition and urban/industrial wastewater discharge [34].

The GoT hosts several naturalistic areas of high ecological value and that are subject to special protection. At the mouth of the Isonzo river lies the Isonzo Nature Reserve, largely overlapping with the ZSC/ZPS Foce dell'Isonzo–Isola della Cona (IT3330005), which is categorised as a Natura 2000 site by the 92/43/EEC and 79/409/EEC European Directives (http://riservafoceisonzo.it/area-protetta/sito-natura-2000-zsc-zps/) (accessed on 10 February 2025). Moving eastwards, the Duino Cliffs Natural Reserve (https://www.falesiediduino.it/) (accessed on 10 February 2025) and the Miramare Marine Protected Area (SIC IT3340007; https://www.ampmiramare.it/) (accessed on 10 February 2025) are located on the Italian side, and are categorised according to the abovementioned Directives, while on the Slovenian side lies Punta Grossa Nature Park/Krajinski park Debeli Rtič (https://park-debelirtic.si/it/) (accessed on 10 February 2025), Strunjan Cliffs Park (https://parkstrunjan.si/it/) (accessed on 10 February 2025), and Sečovlje Salt Pans Park (https://www.kpss.si/si/intro) (accessed on 10 February 2025).

2.2. Sampling Strategy and Analysis

The phytoplankton were sampled between 2001 and 2022 in the vicinity of the mussel farms located in three distinct areas. From east to west, the first site is situated in the Bay of Muggia (area = 16 ha), followed by the largest site, extending from Grignano to the Bay of Sistiana (area = 150 ha), and, finally, the third site is located in the Gulf of Panzano, off the coast of Duino (area = 37 ha). As a first step, four macro-areas were identified on the basis of their geomorphological, physico-chemical, and meteorological differences: Muggia (A1), Grignano–Santa Croce (A2), Aurisina–Sistiana (A3), and Duino (A4). Specifically, A1 is located in the southernmost part of the Italian sector of the GoT and is generally only minimally influenced by freshwater inflow. Its characteristics, regarding sea currents, weather, climatic conditions, and water column structure, differ significantly from those of A4, which is situated near the mouth of the Isonzo river, a continuous source of freshwater that decreases the salinity and supplies the area with nutrients. A2 and A3 display intermediate characteristics between these two extremes [35].

Ancillary physico-chemical parameters, including temperature, salinity, and the concentration of chlorophyll *a*, as well as nutrient concentrations, were measured at three sampling sites, namely Duino, Grignano, and Muggia, corresponding to areas A4, A2 + A3, and A1, respectively. The physico-chemical parameters were measured in situ using a CTD multiparameter probe (Idronaut Ocean Seven 316). It is important to note that salinity was measured according to the Practical Salinity Scale, and, as required by the Joint Panel on Oceanographic Tables and Standards, is reported without units or symbols.

Water samples for the nutrient analysis were collected in pre-cleaned (HCl) polyethylene bottles (V = 100 mL). The dissolved inorganic species, including ammonium (N-NH_4^+), nitrite (N-NO_2^-), nitrate (N-NO_3^-), phosphorous (P-PO_4^{3-}), and silicate (Si-SiO_2), were analysed, following filtration (GF/F filters, porosity = 0.45 μM). The total nitrogen (TN) and total phosphorous (TP) were analysed, after persulphate oxidation [36]. Final determinations were carried out using a segmented flow autoanalyser (Autoanalyser II AAII and Analytical QuAAttro, SEAL), following the use of modified Grasshoff methods [37]. Method accuracy was verified for each analytical batch, using certified reference standards and sample spikes (Inorganic Ventures standard solutions and MOOS-2, NRC). Analytical performance was verified by proficiency tests (PTs) organised by the European network of PT providers (QUASIMEME programmes AQ1 and AQ2). The calculated limit of quantification (LOQ) was as follows: 0.025 μM for N-NO_2^-, 0.71 μM for N-NO_3^- and TN, 0.05 μM for N-NH_4^+, 0.01 μM for P-PO_4^{3-}, and 0,75 μM for Si-SiO_2.

Phytoplankton was collected at fortnightly intervals, except in the event of adverse weather conditions or legal restrictions on mussel harvesting, from ten sites belonging to the four study areas. Notably, three different sampling techniques were used. From 2001 to 2005, a Niskin bottle was used; from 2006 to 2016, sampling was conducted using a phytoplankton net; and, most recently, in compliance with European standards (UNI EN 15972:2012), a Lund tube was used to integrate the entire water column. In all cases, the water samples were collected in dark glass bottles (V = 500 mL), fixed with Lugol's iodine solution (0.5–1.0%) [38], and stored in the dark at T = 4 °C, until the final determination, which was performed according to the Utermöhl method [39].

In this study, rainfall was considered a proxy variable influencing the drivers of primary production and, consequently, the presence of toxic phytoplankton. The rainfall data were obtained from OSMER FVG (Friuli–Venezia Giulia Regional Meteorological Observatory; https://www.osmer.fvg.it/clima.php?ln=) (accessed on 20 January 2025) and were expressed as millimetres of accumulated precipitation, calculated on a decadal basis. Given the high meteorological variability in the GoT, data from multiple rainfall stations located near the mussel farming areas were selected as reference points. Specifically, rainfall data from Fossalon were used for areas A2, A3, and A4, while data from the Trieste–Molo Fratelli Bandiera station were used for area A1.

A corresponding map of the sampling sites is presented in Figure 1.

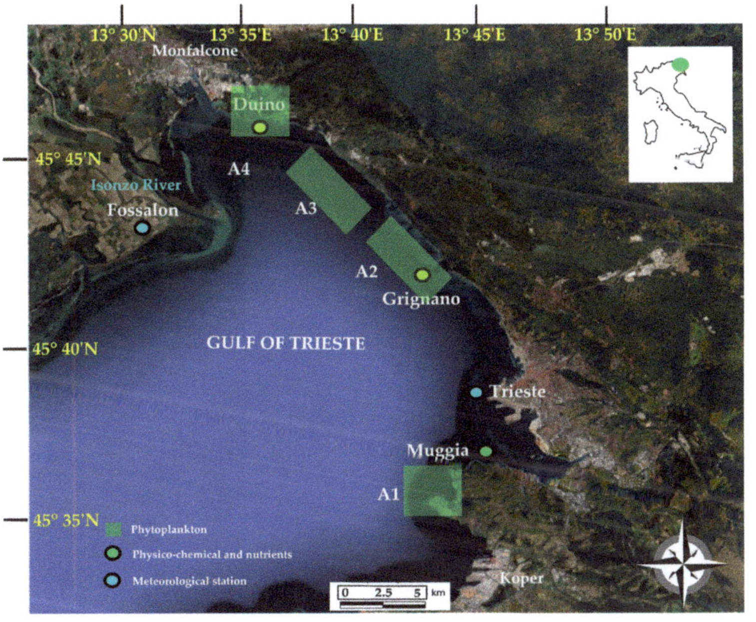

Figure 1. Map of sampling sites in the Gulf of Trieste (modified from Google Earth®).

2.3. Database Setting and Statistical Analysis

The final dataset comprised approximately 10,900 measurements. Statistical analyses were performed using the following software packages: Excel TM®, PAST 2.17 (Paleontological STatistics) [40], ODV (Ocean Data View, 5.0.0) [41], and R 1.4 (https://www.r-project.org/) (accessed on 10 January 2025).

The spatial distribution and seasonal trends were illustrated using histograms (mean ± standard error). The Shapiro–Wilk test was applied to test data normality [42], while the non-parametric Kruskal–Wallis H test (K-W) [43] was used to evaluate significant differences in the median values. The presence of outliers was assessed using Grubbs' test [44]. Spearman's rank correlation coefficients (r) were calculated to examine the monotonic relationships between the variables [45], and the Mann–Kendall trend test (M–K) was employed to assess positive or negative temporal trends in the selected parameters [46]. The significance level for all the tests was set at $p < 0.05$.

The phytoplankton database was compiled from samples collected over a 20-year period, using different sampling techniques. As a result, preliminary assumptions were required prior to the statistical analysis. The selected phytoplankton species belong to four genera: *Alexandrium, Dinophysis, Lingulodinium,* and *Pseudo-nitzschia*. The first three genera are members of the class Dinophyceae, whereas *Pseudo-nitzschia* is classified within the class Bacillariophyceae [2]. In this study, only genus-level identification was performed during the initial sampling period; species-level identification of toxic taxa became a more prominent focus in subsequent years. Therefore, in order to ensure consistency across the full dataset, the analysis was conducted at the genus level. Considering the variability in sampling methodologies, particularly those involving the Niskin bottle, the values corresponding to different sampling depths were averaged, and phytoplankton counts below 120 cells L^{-1} were excluded. Moreover, each sampling method was statistically analysed separately to account for methodological differences.

3. Results

3.1. Physico-Chemical Parameters

The water temperature exhibited a wide range of values (3.6–28.6 °C). On average, the highest values were recorded in Duino (mean ± standard error = 16.7 ± 0.1 °C), while the lowest were observed in Grignano (16.0 ± 0.1 °C) (Figure 2). The salinity had an average value of 37.0 ± 0.01, with the lowest value measured in Duino (8.7) and the highest in Muggia (38.8). The chlorophyll *a* concentrations ranged from <0.20 to 7.36 μg L^{-1} (mean = 0.94 ± 0.01 μg L^{-1}), with the highest concentration observed in Grignano. A significant difference in terms of the spatial distribution was observed for all the parameters (K-W test: $p = 1.4 \times 10^{-11}$ for temperature, 0 for salinity, and 6.5×10^{-243} for chlorophyll *a*). Notably, numerous positive outliers were detected for the chlorophyll *a* concentration, while negative outliers were observed for salinity.

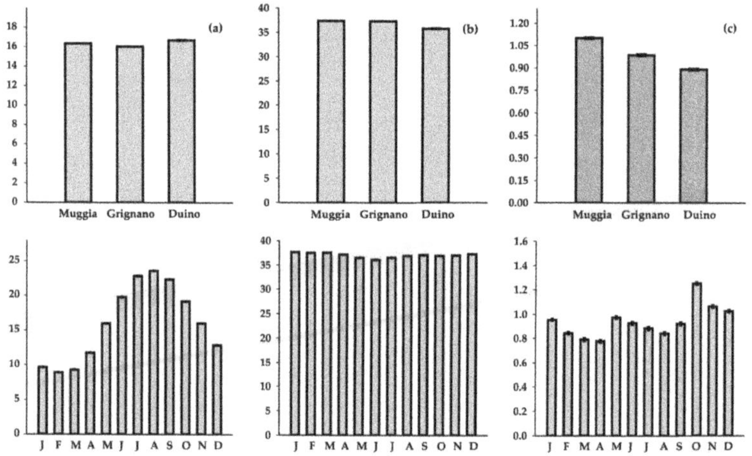

Figure 2. Spatial variability and monthly trends of physico-chemical parameters (mean ± standard error): (**a**) water temperature (°C), (**b**) salinity, and (**c**) chlorophyll *a* concentration (μg L^{-1}).

A pronounced and statistically significant monthly variability was observed for all the parameters (K-W: $p = 0$ for all the variables). On average, temperature exhibited marked seasonal variability, with the minimum value occurring in February (8.93 ± 0.03 °C) and the maximum in August (23.56 ± 0.05 °C). Salinity showed less variation, ranging from 36.11 ± 0.05 in June to 37.64 ± 0.02 in January, with negative outliers distributed homogeneously throughout the year. The chlorophyll *a* concentrations were characterised by enrichments in late spring (0.97 ± 0.01 μg L^{-1} in May) and early autumn (1.25 ± 0.01 μg L^{-1} in October).

The time series analysis revealed significant positive trends for both the temperature and salinity (Table S1). Specifically, the temperature increased between +0.06 °C $year^{-1}$ and +0.07 °C $year^{-1}$ depending on the site (average = +0.06 °C $year^{-1}$) ($p = 4 \times 10^{-32}$). The salinity also exhibited a positive trend (average = +0.01 per year) ($p = 2.5 \times 10^{-8}$), and the highest value was found in Duino (+0.03 per year). In contrast, the chlorophyll *a* concentration remained quite stable (+0.00 μg L^{-1}), except for a significant increase found in Duino (+0.01 μg L^{-1}; $p = 6.3 \times 10^{-5}$) (Figure 3).

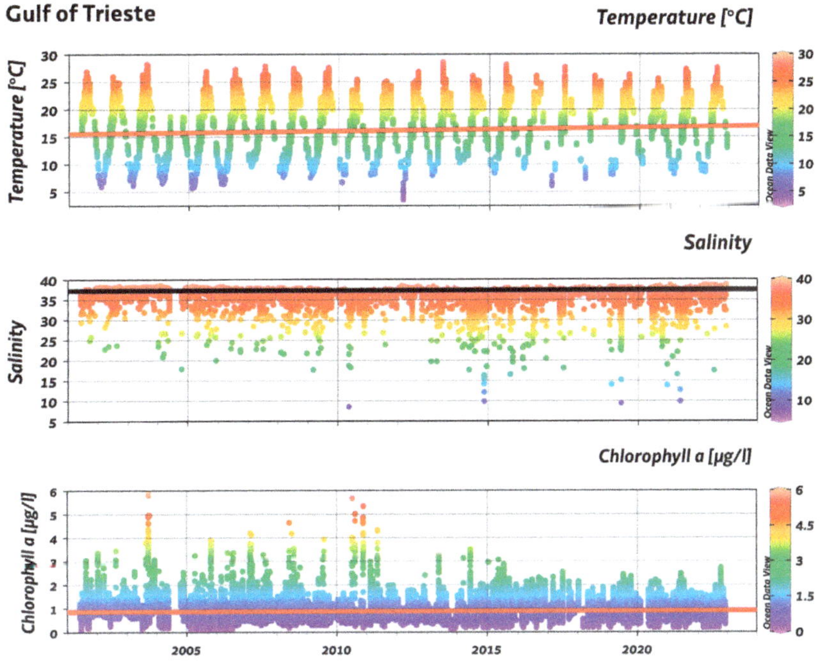

Figure 3. Time series analysis of temperature, salinity, and chlorophyll *a* concentration.

3.2. Nutrients

The results of the spatial distribution and monthly variability of the nutrients are reported in Figures 4 and 5. Concentrations of N-NO_3^- + N-NO_2^- reached up to 130.93 μM in Duino, which also exhibited the highest mean concentration (19.00 ± 1.16 μM), which is one order of magnitude higher than the mean values recorded in Muggia and Grignano. The ammonium (N-NH_4^+) concentrations ranged, on average, from 1.47 ± 0.08 in Duino to 1.05 ± 0.08 in Grignano. The spatial distribution of TN mirrored that of N-NO_3^- + N-NO_2^-, with an average concentration of 18.74 ± 0.62 μM. Several measurements of P-PO_4^{3-} and TP were the LOQ (0.01 μM). Nonetheless, P-PO_4^{3-} was most enriched in Duino (0.14 ± 0.01 μM), while the TP concentrations were fivefold higher, ranging from 0.53 ± 0.04 to 0.60 ± 0.05 in Grignano and Duino, respectively. The distribution of Si-SiO_2 followed a similar pattern to the other nutrients, with the highest mean concentration recorded in Duino (17.21 ± 1.19 μM) and the lowest in Muggia (3.74 ± 0.21 μM). All the nutrients, with the exception of TP, exhibited significant spatial variability (K-W: $p = 8.1 \times 10^{-59}$, 3.8×10^{-13}, 5.5×10^{-48}, 6.1×10^{-7}, 0.41, and 8.7×10^{-45} for N-NO_3^- + N-NO_2^-, N-NH_4^+, TN, P-PO_4^{3-}, TP, and Si-SiO_2, respectively), along with the presence of numerous positive outliers.

The monthly variability patterns revealed typical nutrient enrichments during late spring/early summer and early autumn, consistent across all the nutrients considered. However, no statistically significant differences were detected, with the exception of Si-SiO_2 (K-W: $p = 0.4377$, 0.0821, 0.332, 0.4774, 0.0831, and 5.3×10^{-5}, for N-NO_3^- + N-NO_2^-, N-NH_4^+, TN, P-PO_4^{3-}, TP, and Si-SiO_2, respectively). Overall, all the nutrient species exhibited numerous positive outliers.

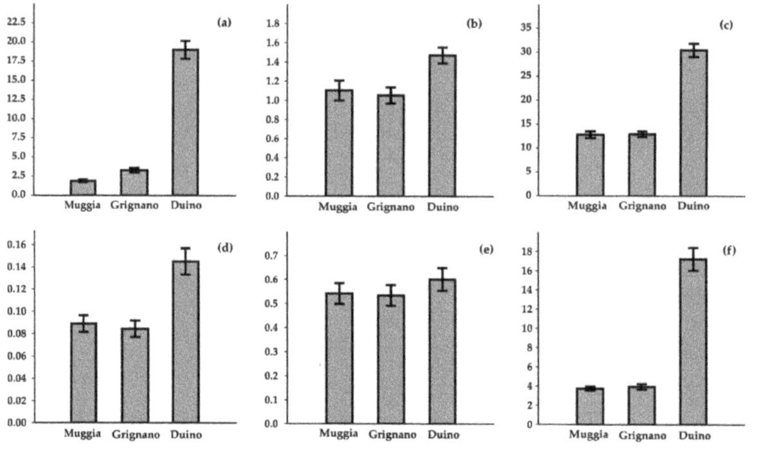

Figure 4. Spatial variability of nutrient concentrations, expressed as $\mu M\ l^{-1}$: (**a**) $N\text{-}NO_3^-$ + $N\text{-}NO_2^-$, (**b**) $N\text{-}NH_4^+$, (**c**) TN, (**d**) $P\text{-}PO_4^{3-}$, (**e**) TP, and (**f**) $Si\text{-}SiO_2$.

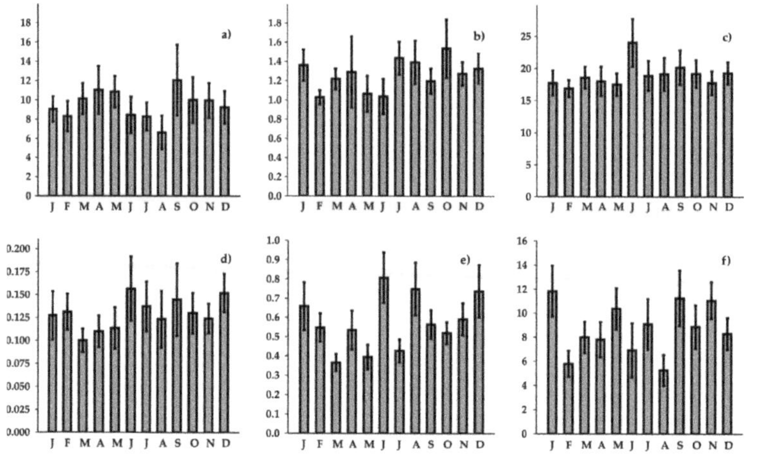

Figure 5. Monthly trends in nutrient concentrations, expressed as μM: (**a**) $N\text{-}NO_3^-$ + $N\text{-}NO_2^-$, (**b**) $N\text{-}NH_4^+$, (**c**) TN, (**d**) $P\text{-}PO_4^{3-}$, (**e**) TP, and (**f**) $Si\text{-}SiO_2$.

The elevated concentrations of N compounds relative to P resulted in a marked deviation from the optimal nutrient ratio for phytoplankton growth, as defined by the Redfield ratio (DIN/PO_4^{3-}), which is typically 16:1 [47]. This imbalance was observed in all of the investigated sites. The mean ratio was 227 (minimum value 1, maximum value 6385). The highest average ratio was recorded in Duino (418), and the lowest in Grignano (58).

The monthly time series analysis revealed both positive and negative trends for the nutrient species. Ammonium ($N\text{-}NH_4^+$) exhibited a significant increase in late summer and early autumn months. The nitrate ($N\text{-}NO_3^-$), total nitrogen (TN), and silicate ($Si\text{-}SiO_2$) concentration showed significant positive trends across all the months considered. In contrast, the patterns in terms of phosphorus species were quite different. Orthophosphate ($P\text{-}PO_4^{3-}$) showed no significant trends, except for a positive trend observed in November. Total phosphorus (TP), on the other hand, exhibited a consistent negative trend throughout the year, with the exception of March, April, and December, during which no significant trends were detected (Table S2).

3.3. Toxic Phytoplankton

Among the toxic phytoplankton genera, *Pseudo-nitzschia* was the most frequently detected, being present in 68.1% of the collected samples, followed by *Dinophysis* (22.7%), while *Lingulodinium* was found in only 8.3% of the samples. On average, *Pseudo-nitzschia* also exhibited the highest abundance (9060 ± 673 cells L^{-1}), with a cell density approximately two orders of magnitude higher than that of *Lingulodinium* (12 ± 2 cell L^{-1}). The spatial distribution of the toxic genera at the investigated sites is shown in Figure 6.

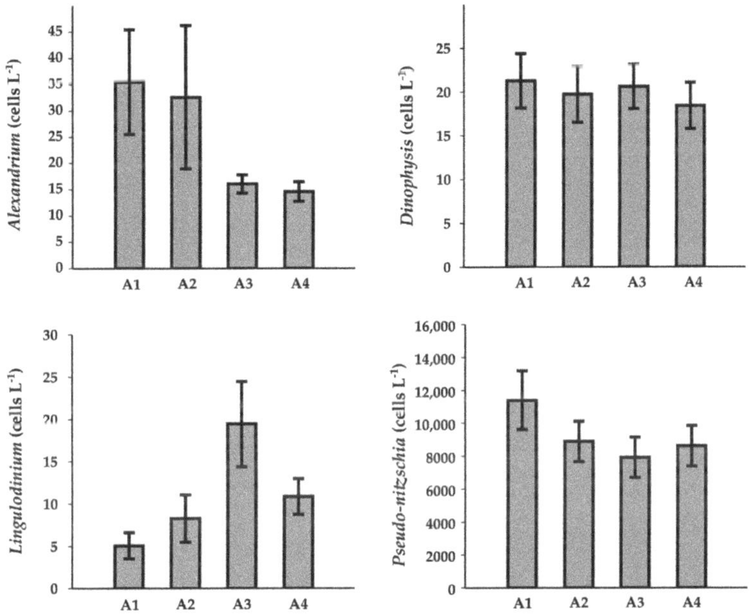

Figure 6. Spatial variability of toxic phytoplankton in the four selected areas.

It is noteworthy that site A1 exhibited higher concentrations of all the genera compared to the other sites. In particular, *Pseudo-nitzschia* (mean = 11,403 ± 1787 cells L^{-1}) was predominant, exceeding the abundance of *Alexandrium* (35 ± 10 cells L^{-1}) and *Dinophysis* (21 ± 3 cells L^{-1}). Conversely, *Lingulodinium* showed the highest abundance in site A3 (19 ± 5 cells L^{-1}). Despite these spatial differences, the Kruskal–Wallis test did not indicate any statistically significant differences in the distribution of the genera among the four sites (p = 0.392, 0.064, 0.091, and 0.5302 for *Alexandrium*, *Dinophysis*, *Lingulodinium*, and *Pseudo-nitzschia*, respectively). The same result was obtained when considering the sampling methods independently (Table S3).

The analysis of monthly variability revealed that high abundances of the genera were primarily found in early spring and late summer/early autumn, a pattern consistent across all the genera. However, some specific differences were reported. *Alexandrium* exhibited a progressive increase in its monthly abundance from winter to spring, reaching a maximum of 110 ± 39 cells L^{-1}, followed by a decline in summer and a secondary, less pronounced, peak in early autumn (8 ± 3 cells L^{-1}). The monthly patterns of *Dinophysis* and *Pseudo-nitzschia* were notably similar, both displaying two distinct peaks: the first, in late winter/early spring (8 ± 3 and 14,114 ± 2736 cells L^{-1}, respectively), and, the second, in early autumn (71 ± 10 and 24,454 ± 3770 cells L^{-1}, respectively). *Lingulodinium* was almost absent in winter, with peak abundances occurring in late spring and late summer (13 ± 3 and 44 ± 13 cells L^{-1}, respectively) (Figure 7).

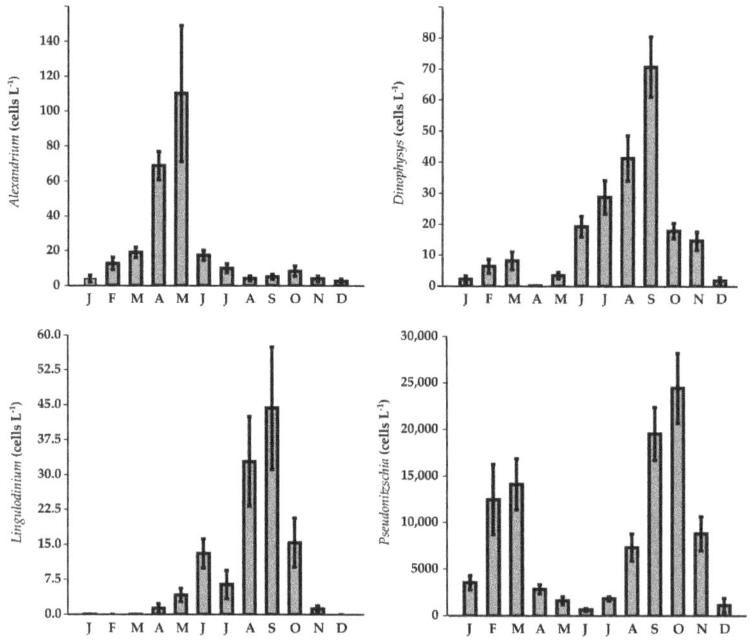

Figure 7. Monthly abundance of toxic phytoplankton genera.

The analysis of the trends in cell abundance for the four microalgal genera in relation to the different sampling techniques revealed a significant decrease over time exclusively for the genus *Lingulodinium* when sampled using a Lund tube. In contrast, no significant increasing or decreasing trends were observed over the years for the other genera, regardless of the sampling method applied (Table S4).

3.4. *Analysis of Precipitation*

Precipitation analysis was conducted by considering the accumulated rainfall (mm), calculated as the sum of daily precipitation occurring during the ten days preceding each sampling event. The maximum value was recorded in October 2015 in Fossalon (177 mm), while in Trieste, the highest value was observed in November 2019 (150 mm). On average, Fossalon experienced greater rainfall impact, with a mean value of 28 ± 34 mm (median = 15 mm), compared to Trieste (22 ± 24 mm). A statistically significant difference between the two sites was identified (K-W: $p = 0.031$).

3.5. *Relationships Between Variables*

The occurrence of significant correlations among the investigated parameters, considering the entire study period and the different sampling methods, is presented in Figure 8. Temperature was generally positively correlated with the abundance of toxic phytoplankton at all sites, except for the genus *Alexandrium* in Muggia. Conversely, temperature showed a negative correlation with salinity, except in Duino. It is noteworthy that the different phytoplankton genera were generally positively correlated with one another. Regarding nutrients, a general positive correlation was observed at all the sites, except for TP, along with a consistent inverse correlation with salinity. The concentration of some nutrient species, particularly nitrogen and silicate, was positively associated with the presence of phytoplankton, whereas TP exhibited a negative correlation, except in the case of

Pseudo-nitzschia. $P\text{-}PO_4^{3-}$ was positively correlated with the abundance of *Lingulodinium* in Muggia and, with *Dinophysis*, when considering the entire study area.

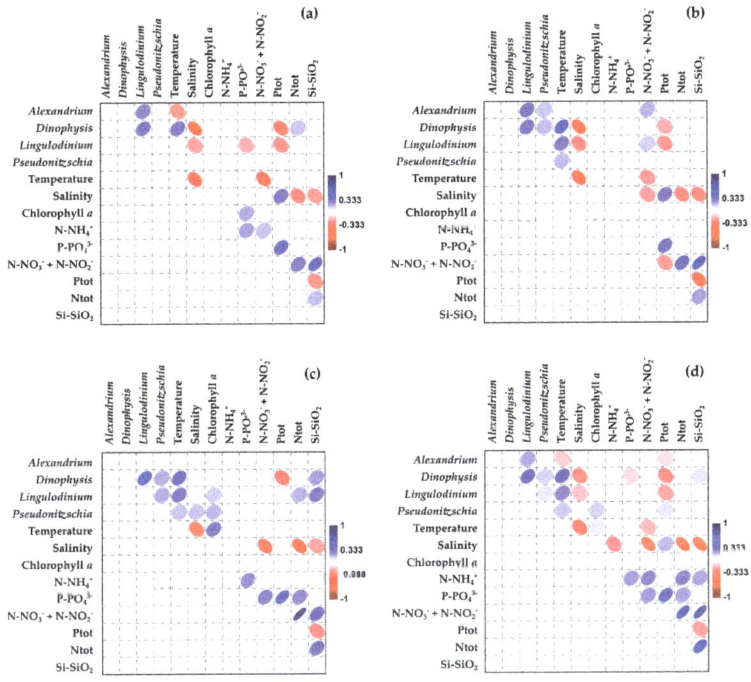

Figure 8. Spearman's correlation between physico-chemical parameters, nutrients, and the abundance of toxic phytoplankton: (**a**) Muggia, (**b**) Grignano, (**c**) Duino, and (**d**) entire study area. The colour of the shapes is proportional to the correlation coefficient (r).

With regard to phytoplankton abundance and their relationship with rainfall, a positive correlation was observed for the genera, *Dinophysis* and *Lingulodinium*. This association is graphically illustrated by superimposing the trends in the cell abundance and rainfall data, using the two-year period, 2015–2016, as an example. Notably, the seasonal peaks in these two genera substantially coincide with the periods of increased rainfall (Figure 9). In contrast, no correlation was found between rainfall and the genera, *Alexandrium* and *Pseudo-nitzschia*.

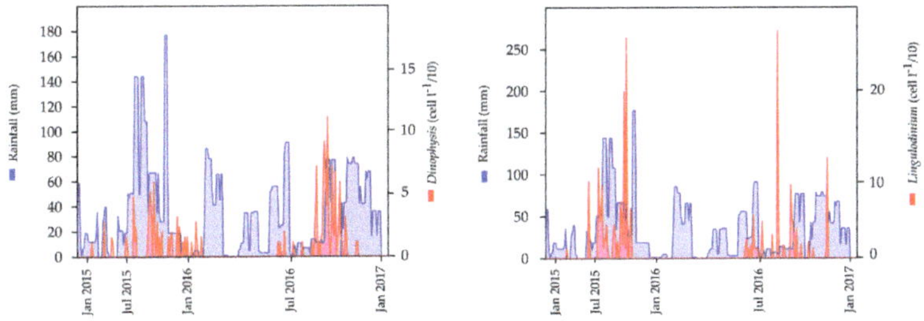

Figure 9. Trends in rainfall and cell abundance for the genera *Dinophysis* and *Lingulodinium* during the 2015–2016 period.

4. Discussion

4.1. Physico-Chemical Parameters and Nutrients

The presence of toxic phytoplankton represents a well-known risk for mussel farming, due to its potential impact on human health [1] and the associated economic losses [48]. These species are a permanent component of the phytoplankton community in both the Mediterranean and Adriatic Sea [49–52]. However, a comprehensive, long-term investigation of the abundance of toxic phytoplankton and its relationship with physico-chemical and meteorological parameters in the Gulf of Trieste (GoT) has yet to be conducted. This is relevant given the variability and long-term trends observed in phytoplankton communities as a consequence of climate change [53–56]. For instance, this issue is exemplified by the significant increase in air temperature recorded in Europe (+0.9 °C from 1901 to 2005, with a trend of +0.4 °C decade^{-1}) [57], as well as in the Mediterranean Sea [58].

In this study, the spatial and temporal variability typical of ecological systems [59] was confirmed by the occurrence of numerous anomalies across all the investigated parameters. The water column exhibited a significant increase in temperature of 0.06 °C per year over the period 2001–2022, which is comparable to the findings by Malačič et al. [35] for the GoT, where an increase of approximately 0.1 °C per year in the surface temperature was reported in spring and summer, during the period 1991–2003. A significant temperature gradient was observed, in contrast to the findings by Cossarini et al. [60], who reported that seasonal variability predominates over spatial variability in the northern Adriatic. This is likely attributable to the differing influence of freshwater inputs (e.g., Isonzo river, Karst springs, and minor Slovenian watercourses) across the GoT area [61], which also contributes to the salinity gradient and its year-round fluctuation. The observed increase in salinity was lower than that calculated by Malačič et al. [35], ranging from 0.1 to 0.3 per year, but it should be noted that their values refer to surface waters. The increase in salinity, particularly during the spring–summer period, is possibly related to a reduction in river discharge [62,63], as observed in Duino, which exhibited the lowest average salinity, but the highest increase during the study period.

The chlorophyll *a* levels indicate an oligotrophic condition, consistent with previous findings by Mozetič et al. [64], who reported similar results, based on field analysis and satellite data collected over the period 1970–2007. High chlorophyll *a* concentrations were rare, with the 90th percentile reaching only 1.57 μg L^{-1}, as also observed by Lipizer et al. [65]. The trophic status was further supported by the application of the trophic index (TRIX) [66], which ranged from 2.09 to 5.92 (average 3.74 ± 0.62), corresponding to conditions classified from a high to a good state [67]. The time series analysis revealed no significant trends overall, except for a slight increase observed in Duino, similar to findings in the Venice Lagoon and the northern Adriatic, reported by Mozetič et al. and Bernard-Aubry et al. [64,68], respectively. Cozzi et al. [69] reported substantial fluctuations over a longer time series (1986–2018), partially overlapping with the present study. Notably, the highest chlorophyll *a* values were recorded before 1992, followed by a marked decrease from 1997 to 2009, and a subsequent shift toward oligotrophic conditions up to 2018, consistent with the levels observed in this investigation.

The processes influencing nutrient dynamics in the GoT have been extensively described in the literature, including riverine inputs [29,61,70,71], urban discharges [24,72–74], atmospheric deposition [75], and benthic fluxes [76]. In this study, the spatial distribution of nutrients, particularly nitrogen species and silicates, and, to a lesser extent, phosphorus, highlights the key role of the Isonzo river in transporting agricultural runoff from inland soils, both in dissolved and particulate forms. This is supported by the significant positive correlations observed among the nutrient concentrations, along with the negative correlation with salinity. Seasonal patterns showed the lowest nutrient levels during summer and a marked increase in autumn, consistent with trends reported in previous studies involving shorter time series (1999–2006; 1999–2010) [60,65]. Overall, the anomalies in terms of the nutrient concentrations are widespread as consequence of the forces acting in the GoT (i.e., dilution from fluvial inputs, mixing of water column, winds), which are common in coastal areas of the Mediterranean [32,77–80]. The Mediterranean Sea is recognised as a phosphorous-limited system [81–83] and this is also observed in regard to the GoT [65], as confirmed by this long-term survey. The regulatory reduction of polyphosphates in detergents (G.U. no. 178, 1983) has had significant effects since the early 2000s [84,85], a trend that persists and is reflected in the observed decrease in the total phosphorus (TP) concentrations. In contrast, nitrogen species have shown a substantial increase, supporting the findings reported by Giani et al. [34].

4.2. Toxic Phytoplankton

The genera *Alexandrium*, *Dinophysis*, *Lingulodinium*, and *Pseudo-nitzschia* are typical components of the phytoplankton community in the northern Adriatic Sea [86–91], and were periodically detected throughout the two decades assessed in this study, exhibiting genus-level differences consistent with previous findings [4,72–74]. *Alexandrium* typically occurred in late winter and spring, primarily represented by *Alexandrium minutum* in the southern sector of the GoT [92]. *Dinophysis* was generally observed between June and October; however, in recent years, its presence has extended into the late autumn–winter period. Francé and Mozetič [4] reported that *Dinophysis sacculus* is typical of the late spring/early summer period, while *D. caudata* and *D. fortii* are more prevalent in late summer and autumn. Similarly, Ninčević-Gladan et al. [93] described the seasonal occurrence of *D. acuminata* and *D. sacculus* in spring, *D. caudata* and *D. fortii* in summer, and, finally, *D. tripos* in winter in the central Adriatic. As previously noted [89], *Lingulodinium* was generally present at low abundances. In the present study, both net and tube sampling proved more effective in detecting this genus [38], allowing for the identification of abundance peaks in May and October, an observation not previously documented. Finally, *Pseudo-nitzschia* was more abundant in spring and autumn (net and tube collections), consistent with the findings from both the Lim Channel (Istria, Croatia) and the GoT [92,94].

The metabolism of dinoflagellates and diatoms differs markedly, mixotrophic versus obligate autotrophic, respectively, resulting in distinct environmental requirements. Diatoms, such as *Pseudo-nitzschia*, rely on specific light and nutrient conditions. In the GoT, population peaks are clearly observed when the photoperiod favours photosynthetic activity and nutrient availability is enhanced by rainfall, supporting bloom development [95,96]. In contrast, dinoflagellates exhibit the ability to survive and proliferate during summer, when nutrient concentrations are low, as well as in late autumn, despite the reduced occurrence of photoperiods [88,97–101]. Overall, a significant positive correlation was found among the abundances of the studied genera, suggesting the existence of shared environmental triggers promoting population growth.

No statistically significant increase or decrease in phytoplankton abundance was observed over the study period, with the sole exception of *Lingulodinium*, which showed a decreasing trend when sampled using the tube method. This is particularly noteworthy, as *Lingulodinium* is considered a rare genus. Our findings are consistent with those reported by Zingone et al. [101] for the Adriatic Sea, who found no specific trends in the abundance or frequency of toxic species. However, the same authors did observe a sustained increase in species richness, marked by the continuous emergence of other toxic species.

4.3. Correlation of Toxic Phytoplankton with Physico-Chemical Parameters, Nutrients, and Rainfall

The diversity and dynamics of phytoplankton populations are influenced by a wide range of complex processes, involving both abiotic and biotic factors. Among the biotic drivers, predation pressure on primary producers and viral infections have been identified as key contributors to the structuring of the phytoplankton community [90].

In general, the trophic state of a coastal system tends to favour the presence of toxic phytoplankton species, as nutrients promote their growth [102]. However, considerable variability exists depending on the site, time, and species, which can lead to deviations from this general pattern. As expected, temperature showed a positive effect on phytoplankton growth, whereas salinity was inversely correlated, reflecting the role of freshwater as the primary source of nutrients in the GoT. The abundance of *Dinophysis* was negatively correlated with TP, while TN and Si-SiO_2 were positively associated. Due to its mixotrophic metabolism, it may be hypothesised that *Dinophysis* benefit indirectly from nutrient enrichment via the stimulation of its prey, rather than through the direct uptake of bioavailable nutrients [103–105].

Lingulodinium also exhibited a negative correlation with TP, consistent with the findings by Domingues and Lima [106], who reported an inverse relationship between the proliferation of *Lingulodinium polyedra* and the phosphorus concentration in the Ria Formosa Lagoon (Portugal). Conversely, positive correlations were observed between *Lingulodinium* and both nitrogen species and Si-SiO_2.

Pseudo-nitzschia showed a positive correlation with TP, and several studies have also highlighted associations with both organic and inorganic nitrogen in various environments [107,108], but not with silicate, which is essential for frustule formation. However, no clear correlation was observed with silicate, despite its essential role in frustule formation. A lack of correlation between silicate concentrations and diatom abundance may occur when silicate is in excess and no longer a limiting factor, allowing other environmental variables to play a more influential role [109,110]. Additionally, in oligotrophic conditions, limited nutrient availability, particularly phosphorus, can constrain diatom growth, even when the amount of silicate is sufficient [111].

Finally, *Alexandrium* was positively correlated with dissolved nitrogen at the Grignano site only. These results support the hypothesis that toxic microalgae often exhibit low affinity for the direct uptake of inorganic nitrogen and phosphorus and frequently rely on mixotrophic strategies [96,110,112]. Some authors have also suggested that these microalgae are capable of adapting to limited phosphorus and nitrogen concentrations [113].

A significant positive correlation between the toxic phytoplankton abundance and decadal cumulative rainfall was observed for the genera *Dinophysis* and *Lingulodinium*, whereas no such relationship was found for *Alexandrium* and *Pseudo-nitzschia*. Ninčević-Gladan et al. [93] reported that among the *Dinophysis* species generally associated with high temperatures, such as *D. caudate, D. fortii, D. acuta*, and *D. tripos,* they are instead linked to elevated precipitation. In the northern Adriatic, intense rainfall events combined with high temperatures have been shown to trigger the proliferation of *D. fortii* during summer and late summer [93], with a frequency greater than that observed in the southern Adriatic basin [114]. The same authors also reported a correlation between precipitation and the genus *Pseudo-nitzschia*, particularly during the colder months; however, this was not confirmed in the present study.

Similarly, *Alexandrium* has been linked to freshwater inputs and conditions involving high water column stability [93], which are typically associated with prolonged water residence times and elevated nutrient availability [115]. Dinoflagellates are well-adapted to relatively calm, stratified water column conditions. Their ability for active movement allows them to remain within the euphotic zone and to access deeper, nutrient-rich layers when surface nutrients are depleted [116]. Nevertheless, Weise et al. [117] observed a positive correlation between *Alexandrium tamarense* blooms and precipitation events, along with other environmental factors, such as temperature, salinity, river discharge, and wind. They concluded that bloom development depends largely on the persistence of favourable conditions.

Phytoplankton distribution is influenced by biological interactions, such as zooplankton grazing, and is further modulated by hydrodynamic factors, including local circulation patterns and prevailing currents. Previous studies conducted in the Gulf of Trieste (GoT) have generally reported a negligible impact of zooplankton grazing on nutrient-enriched phytoplankton assemblages [118,119]. Unfortunately, no supporting data were available in this study, such as measurements of the current intensity and prevailing current direction, thereby limiting a more detailed interpretation.

4.4. Toxic Phytoplankton and Sampling Techniques

The use of different sampling techniques posed a challenge in terms of the comparability of phytoplankton abundance across different time periods. To address this limitation, the data were initially treated separately. Nonetheless, this apparent constraint provided an opportunity to evaluate and compare the efficiency of the three sampling methods employed.

The Niskin bottle method presents limitations in regard to representing low-density phytoplankton species. Additionally, some species exhibit vertical migration within the water column, as reported by Francé and Mozetič, who demonstrated the ability of *Dinophysis sacculus* to migrate from surface layers to the thermocline. As a result, sampling with the Niskin bottle may underestimate the actual phytoplankton abundance.

Conversely, the Niskin method proved to be the most suitable for assessing correlations between phytoplankton abundance and the physico-chemical parameters and nutrients, whereas the net and tube methods were less effective in this context. The phytoplankton net offers the advantage of concentrating the samples, as evidenced by the successful detection of *Lingulodinium*, a genus typically found at very low densities. The tube method represents an improvement, as it integrates the entire water column, thereby enabling a more accurate qualitative and quantitative representation of the phytoplankton community.

5. Conclusions and Future Perspectives

The relationship between phytoplankton and the climate is a highly topical issue, as noted by Henson et al. [55], who emphasised that climate change may alter phytoplankton diversity by accelerating community turnover, thereby increasing instability and reducing ecosystem resilience. This presents a considerable challenge for the development of reliable long-term predictive models, particularly concerning the behaviour of potentially toxic dinoflagellates, such as *Dinophysis*. As a predominantly opportunistic genus, its environmental interactions during periods of widespread toxicity remain difficult to interpret [120].

One major concern relates to the unpredictability of mussel trade bans, which have varied significantly over time. For instance, in the Slovenian sector of the GoT, no clear annual pattern of trade bans was observed prior to 2012, with restrictions occurring throughout the year, although more frequently in summer and autumn. In contrast, from 2013 onwards, trade bans consistently occurred during late summer and autumn (1999–2019 period) [14]. The authors concluded that the observed interannual variability in toxicity and phytoplankton dynamics, potentially exacerbated by future climatic and anthropogenic stressors, underscores the need for more efficient and adaptive surveillance and monitoring strategies.

To date, following the issuance of EU Regulation (EU, 2019/627), in Article 61, namely "Sampling plans", it is reported that: "Sampling plans to check for the presence of toxin-producing plankton in the water in classified production and relaying areas and for marine biotoxins in live bivalve molluscs shall take particular account of possible variations in the presence of plankton containing marine biotoxins". The planned actions can be summarised as follows: periodic sampling to monitor changes in the composition and distribution of toxic phytoplankton, followed by intensified sampling and periodic toxicity tests on live bivalve molluscs. During harvesting periods, toxin analysis must be performed weekly, unless a risk assessment indicates a very low probability of toxic episodes. If the assessment suggests that weekly sampling is insufficient, the frequency must be increased. If the monitoring results indicate that health standards are not being met or that there is a risk to human health, the competent authorities must close the affected classified production or relaying area. Reopening is allowed only when toxin levels fall below regulatory thresholds and phytoplankton trends are considered.

This study suggests that rainfall and salinity are valuable proxies for developing early warning models to predict conditions affecting mussel farming. Incorporating nutrient data may improve model accuracy. Validating these relationships could support the integration of weather forecasts with trophic status and phytoplankton seasonality, enabling proactive monitoring. This approach aligns with the study by Boivin-Rioux et al. [121], which successfully used variations in salinity and temperature to predict the occurrence of the potentially toxic species, *Alexandrium catenella*. More recently, explainable machine learning techniques were applied to forecast mussel toxicity in the Slovenian sector of the GoT, using a 28-year dataset of toxic phytoplankton records from mussel-farming areas [122]. Salinity, river discharge, and precipitation emerged as the strongest predictors of diarrhetic shellfish poisoning (DSP) toxin levels exceeding regulatory thresholds.

Building on the findings in this study, future research should pursue more integrated and predictive approaches to better understand the dynamics of toxic phytoplankton in coastal environments. In particular, the following directions are recommended:

- Implementing continuous, in situ measurements (e.g., automated sensors for monitoring the temperature, salinity, chlorophyll *a* concentration, and turbidity) would enable the real-time detection of environmental shifts that precede bloom events;
- Developing and calibrating hydrodynamic models integrated with phytoplankton dynamics can improve forecasts on bloom formation, spread, and persistence. The adoption of machine learning algorithms, particularly explainable models, should be further explored to enhance transparency and stakeholder confidence in early warning systems;
- Long-term, cross-border collaboration in regard to data collection and the harmonisation of sampling protocols will be critical to ensure comparability and broader applicability of predictive tools;
- Interdisciplinary studies that include economic risk assessments, especially for the aquaculture sector, can support evidence-based decision making and adaptive management under future climate scenarios.

These perspectives aim to bridge scientific research with practical applications, offering tools to anticipate and mitigate the impacts of harmful algal blooms in the context of increasing environmental variability.

Supplementary Materials: The following supporting information can be downloaded at: https://www.mdpi.com/article/10.3390/environments12050152/s1, Table S1: Univariate statistics of physico-chemical parameters. Table S2: Monthly temporal trends in nutrient concentrations. Table S3: Statistical test of significant differences in phytoplankton abundance by sampling area and collection method. Table S4: Temporal trends in phytoplankton abundance by genus and collection method.

Author Contributions: Conceptualisation, L.T., O.B., N.B. and A.A.; methodology, L.T. and O.B.; software, L.T. and A.A.; validation, L.T., O.B., A.A. and M.C.; investigation, O.B. and A.A.; data curation, L.T., O.B., N.B., M.C. and A.A.; writing—original draft preparation, A.A.; writing—review and editing, L.T., O.B., N.B., M.C. and A.A. All authors have read and agreed to the published version of the manuscript.

Funding: This research received no external funding.

Data Availability Statement: The data presented in this study are partially openly available via the following website: https://www.dati.friuliveneziagiulia.it/Ambiente/Acqua-Acque-di-classificazione-Superficiali-marino/qcsf-bwk5/about_data, accessed on 3 March 2025. Further requests should be submitted to urp@arpa.fvg.it and to the corresponding author.

Acknowledgments: The authors kindly acknowledge Bruno Zanolin for his support in the collection of phytoplankton data, as well as our colleagues from the microbiology laboratory for their analysis of toxic phytoplankton. We also extend our sincere thanks to all the staff at ARPA FVG involved in the sampling operations. The valuable suggestions provided by the two anonymous reviewers are gratefully acknowledged and have significantly contributed to the improvement of the manuscript.

Conflicts of Interest: The authors declare that there are no conflicts of interest.

References

1. Lassus, P.; Chomérat, N.; Hess, P.; Nézan, E. *Toxic and Harmful Microalgue of the World Ocean/Micro-Algues Toxiques et Nuisibles de l'océan Mondial*; Manuals and guides/IOC; International Society for the Study of Harmful Algae: Copenhagen, Denmark, 2016; ISBN 978-87-990827-6-6.
2. Hasle, G.R.; Syvertsen, E.E.; Steidinger, K.A.; Tangen, K.; Tomas, C.R. *Identifying Marine Diatoms and Dinoflagellates*; Elsevier: Amsterdam, The Netherlands, 1996.
3. Valbi, E.; Ricci, F.; Capellacci, S.; Casabianca, S.; Scardi, M.; Penna, A. A Model Predicting the PSP Toxic Dinoflagellate Alexandrium Minutum Occurrence in the Coastal Waters of the NW Adriatic Sea. *Sci. Rep.* **2019**, *9*, 4166. [CrossRef] [PubMed]
4. France, J.; Mozetič, P. Ecological Characterization of Toxic Phytoplankton Species (*Dinophysis* Spp., Dinophyceae) in Slovenian Mariculture Areas (Gulf of Trieste, Adriatic Sea) and the Implications for Monitoring. *Mar. Pollut. Bull.* **2006**, *52*, 1504–1516. [CrossRef]
5. Paz, B.; Daranas, A.H.; Norte, M.; Riobó, P.; Franco, J.M.; Fernández, J.J. Yessotoxins, a Group of Marine Polyether Toxins: An Overview. *Mar. Drugs* **2008**, *6*, 73–102. [CrossRef] [PubMed]
6. Bates, S.S.; Hubbard, K.A.; Lundholm, N.; Montresor, M.; Leaw, C.P. *Pseudo-Nitzschia*, Nitzschia, and Domoic Acid: New Research since 2011. *Harmful Algae* **2018**, *79*, 3–43. [CrossRef] [PubMed]
7. Oh, J.-W.; Pushparaj, S.S.C.; Muthu, M.; Gopal, J. Review of Harmful Algal Blooms (HABs) Causing Marine Fish Kills: Toxicity and Mitigation. *Plants* **2023**, *12*, 3936. [CrossRef]
8. Alvarez, S.; Brown, C.E.; Garcia Diaz, M.; O'Leary, H.; Solís, D. Non-Linear Impacts of Harmful Algae Blooms on the Coastal Tourism Economy. *J. Environ. Manag.* **2024**, *351*, 119811. [CrossRef]
9. Denis, L. Clearance Rate Responses of Mediterranean Mussels, *Mytilus Galloprovincialis*, to Variations in the Flow, Water Temperature, Food Quality and Quantity. *Aquat. Living Resour.* **1999**, *12*, 279–288. [CrossRef]
10. Curpan, A.-S.; Impellitteri, F.; Plavan, G.; Ciobica, A.; Faggio, C. Review: *Mytilus Galloprovincialis*: An Essential, Low-Cost Model Organism for the Impact of Xenobiotics on Oxidative Stress and Public Health. *Comp. Biochem. Physiol. Part C Toxicol. Pharmacol.* **2022**, *256*, 109302. [CrossRef]
11. Melaku Canu, D.; Solidoro, C. Socio-Economic Analysis and Stakeholder Involvement: Mussel-Farming in the Gulf of Trieste. *Mar. Policy* **2014**, *43*, 55–62. [CrossRef]
12. Solidoro, C.; Del Negro, P.; Mosetti, R.; De Walderstein, W.; Gomes Ferreira, J.; Zentilin, A.; Bricelj, A.; Beran, A.; Libralato, S.; Melaku Canu, D.; et al. *Sostenibilità Della Mitilicoltura Triestina*; Istituto Nazionale di Oceanografia e di Geofisica Sperimentale: Borgo Grotta Gigante, Italy, 2010.
13. European Commission. Commission Implementing Regulation (EU) 2019/627 of 15 March 2019 Laying down Uniform Practical Arrangements for the Performance of Official Controls on Products of Animal Origin Intended for Human Consumption in Accordance with Regulation (EU) 2017/625 of the European Parliament and of the Council and Amending Commission Regulation (EC) No 2074/2005 as Regards Official Controls. *Off. J. Eur. Union* **2019**, *131*, 51–100.

14. Henigman, U.; Mozetič, P.; Francé, J.; Knific, T.; Vadnjal, S.; Dolenc, J.; Kirbiš, A.; Biasizzo, M. Okadaic Acid as a Major Problem for the Seafood Safety (*Mytilus Galloprovincialis*) and the Dynamics of Toxic Phytoplankton in the Slovenian Coastal Sea (Gulf of Trieste, Adriatic Sea). *Harmful Algae* **2024**, *135*, 102632. [CrossRef] [PubMed]
15. Karydis, M. Toxic Phytoplankton in Eutrophic Regional Seas: An Overview. *Glob. NEST J.* **2023**, *25*, 178–211. [CrossRef]
16. Glibert, P.M. Harmful Algae at the Complex Nexus of Eutrophication and Climate Change. *Harmful Algae* **2020**, *91*, 101583. [CrossRef]
17. Gobler, C.J. Climate Change and Harmful Algal Blooms: Insights and Perspective. *Harmful Algae* **2020**, *91*, 101731. [CrossRef] [PubMed]
18. Heneash, A.M.M.; Tadrose, H.R.Z.; Hussein, M.M.A.; Hamdona, S.K.; Abdel-Aziz, N.; Gharib, S.M. Potential Effects of Abiotic Factors on the Abundance and Distribution of the Plankton in the Western Harbour, South-Eastern Mediterranean Sea, Egypt. *Oceanologia* **2015**, *57*, 61–70. [CrossRef]
19. Drakulovic, D.; Gvozdenovic, S.; Joksimovic, D.; Mandic, M.; Pestoric, B. Toxic and Potentially Toxic Phytoplankton in the Mussel and Fish Farms in the Transitional Area of Montenegrin Coast (South-Eastern Adriatic Sea). *Turk. J. Fish. Aquat. Sci.* **2017**, *17*, 885–900. [CrossRef]
20. Tamvakis, A.; Tsirtsis, G.; Karydis, M.; Patsidis, K.; Kokkoris, G.D. Drivers of Harmful Algal Blooms in Coastal Areas of Eastern Mediterranean: A Machine Learning Methodological Approach. *Math. Biosci. Eng.* **2021**, *18*, 6484–6505. [CrossRef]
21. Tsikoti, C.; Genitsaris, S. Review of Harmful Algal Blooms in the Coastal Mediterranean Sea, with a Focus on Greek Waters. *Diversity* **2021**, *13*, 396. [CrossRef]
22. Karydis, M.; Kitsiou, D. Eutrophication and Environmental Policy in the Mediterranean Sea: A Review. *Environ. Monit. Assess.* **2012**, *184*, 4931–4984. [CrossRef]
23. Olivotti, R.; Faganeli, J.; Malej, A. Eutrophication of Coastal Waters–Gulf of Trieste. *Water Sci. Technol.* **1986**, *18*, 303–316. [CrossRef]
24. Cozzi, S.; Mistaro, A.; Sparnocchia, S.; Colugnati, L.; Bajt, O.; Toniatti, L. Anthropogenic Loads and Biogeochemical Role of Urea in the Gulf of Trieste. *Sci. Total Environ.* **2014**, *493*, 271–281. [CrossRef] [PubMed]
25. Malej, A.; Malacic, V. Factors Affecting Bottom Layer Oxygen Depletion in the Gulf of Trieste(Adriatic Sea). *Ann. An. Istrske Mediter. Stud. (Hist. Nat.)* **1995**, *6*, 33–42.
26. Comici, C.; Bussani, A. Analysis of the River Isonzo Discharge (1998–2005). *Boll. Geofis. Teor. Appl.* **2007**, *48*, 435–454.
27. Cozzi, S.; Giani, M. River Water and Nutrient Discharges in the Northern Adriatic Sea: Current Importance and Long Term Changes. *Cont. Shelf Res.* **2011**, *31*, 1881–1893. [CrossRef]
28. Turk, V.; Mozetic, P.; Malej, A. Overview of Eutrophication-Related Events and Other Irregular Episodes in Slovenian Sea (Gulf of Trieste, Adriatic Sea). In *Annales: Series Historia Naturalis*; Scientific and Research Center of the Republic of Slovenia: Koper, Slovenia, 2007; Volume 17, p. 197.
29. Malej, A.; Mozetic, P.; Malacic, V.; Terzic, S.; Ahel, M. Phytoplankton Responses to Freshwater Inputs in a Small Semi-Enclosed Gulf (Gulf of Trieste, Adriatic Sea). *Mar. Ecol. Prog. Ser.* **1995**, *120*, 111–121. [CrossRef]
30. Stravisi, F. Some Characteristics of the Circulation in the Gulf of Trieste. *Thalass. Jugosl.* **1983**, *19*, 355–363.
31. Stravisi, F. The Vertical Structure Annual Cycle of the Mass Field Parameters in the Gulf of Trieste. *Boll. Oceanogr. Teor. Appl.* **1983**, *3*, 239–250.
32. Boldrin, A.; Carniel, S.; Giani, M.; Marini, M.; Bernardi Aubry, F.; Campanelli, A.; Grilli, F.; Russo, A. Effects of Bora Wind on Physical and Biogeochemical Properties of Stratified Waters in the Northern Adriatic. *J. Geophys. Res. Ocean.* **2009**, *114*, C08S92. [CrossRef]
33. Pagano, M.; Fernetti, M.; Busetti, M.; Ghribi, M.; Camerlenghi, A. Multicriteria GIS-Based Analysis for the Evaluation of the Vulnerability of the Marine Environment in the Gulf of Trieste (North-Eastern Adriatic Sea) for Sustainable Blue Economy and Maritime Spatial Planning. *People Nat.* **2023**, *5*, 2006–2025. [CrossRef]
34. Giani, M.; Djakovac, T.; Degobbis, D.; Cozzi, S.; Solidoro, C.; Umani, S.F. Recent Changes in the Marine Ecosystems of the Northern Adriatic Sea. *Estuar. Coast. Shelf Sci.* **2012**, *115*, 1–13. [CrossRef]
35. Malačič, V.; Celio, M.; Čermelj, B.; Bussani, A.; Comici, C. Interannual Evolution of Seasonal Thermohaline Properties in the Gulf of Trieste (Northern Adriatic) 1991–2003. *J. Geophys. Res.* **2006**, *111*, 2005JC003267. [CrossRef]
36. Valderrama, J.C. The Simultaneous Analysis of Total Nitrogen and Total Phosphorus in Natural Waters. *Mar. Chem.* **1981**, *10*, 109–122. [CrossRef]
37. Grasshoff, K.; Kremling, K.; Ehrhardt, M. *Methods of Seawater Analysis*; John Wiley & Sons: Hoboken, NJ, USA, 2009.
38. Zingone, A.; Totti, C.; Sarno, D.; Cabrini, M.; Caroppo, C.; Giacobbe, M.; Lugliè, A.; Nuccio, C.; Socal, G. CAPITOLO 21. FITOPLANCTON: METODICHE DI ANALISI QUALI-QUAN-TITATIVA. In *Metodologie di Studio del Plancton Marino*; ISPRA: Rome, Italy, 2010.
39. Uthermöhl, H. On the Perfecting of Quantitative Phytoplankton Method. *Int. Assoc. Theor. Appl. Limnol. Commun.* **1958**, *9*, 1–38.

40. Hammer, O.; Harper, D.A.T.; Ryan, P.D. PAST: Paleontological Statistics Software Package for Education and Data Analysis. *Palaeontol. Electron.* **2001**, *4*, 1–9.
41. Schlitzer, R.; Anderson, R.; Dodas, E.; Lohan, M.; Geibert, W.; Tagliabue, A.; Schlitzer, R.; Cullen, J.T.; Janssen, D.J.; Velazquez, S.; et al. UVicSPACE: Research & Learning Repository. *Chem. Geol.* **2018**, *493*, 210–223.
42. Shapiro, S.S.; Wilk, M.B. An Analysis of Variance Test for Normality (Complete Samples). *Biometrika* **1965**, *52*, 591–611. [CrossRef]
43. Zar, J.H. *Biostatistical Analysis*; Pearson Education India: Noida, India, 1999.
44. Grubbs, F.E. *Sample Criteria for Testing Outlying Observations*; University of Michigan: Ann Arbor, MI, USA, 1949.
45. Press, W.; Teukolsky, S.; Vetterling, W.; Flannery, B. *Numerical Recipes in C*, 2nd ed.; Cambridge University Press: New York, NY, USA, 1992.
46. Gilbert, R.O. *Statistical Methods for Environmental Pollution Monitoring*; John Wiley & Sons: Hoboken, NJ, USA, 1987.
47. Redfield, A.C. The Biological Control of Chemical Factors in the Environment. *Sci. Prog.* **1960**, *11*, 150–170.
48. Ragkos, A.; Skordos, D.; Koutouzidou, G.; Giantsis, I.A.; Delis, G.; Theodoridis, A. Socioeconomic Appraisal of an Early Prevention System against Toxic Conditions in Mussel Aquaculture. *Animals* **2022**, *12*, 2832. [CrossRef]
49. Socal, G.; Boldrin, A.; Bianchi, F.; Civitarese, G.; De Lazzari, A.; Rabitti, S.; Totti, C.; Turchetto, M.M. Nutrient, Particulate Matter and Phytoplankton Variability in the Photic Layer of the Otranto Strait. *J. Mar. Syst.* **1999**, *20*, 381–398. [CrossRef]
50. Orsini, L.; Sarno, D.; Procaccini, G.; Poletti, R.; Dahlmann, J.; Montresor, M. Toxic *Pseudo-Nitzschia Multistriata* (Bacillariophyceae) from the Gulf of Naples: Morphology, Toxin Analysis and Phylogenetic Relationships with Other Pseudo-Nitzschia Species. *Eur. J. Phycol.* **2002**, *37*, 247–257. [CrossRef]
51. Quiroga, I. Pseudo-Nitzschia Blooms in the Bay of Banyuls-Sur-Mer, Northwestern Mediterranean Sea. *Diatom Res.* **2006**, *21*, 91–104. [CrossRef]
52. Bosak, S.; Burić, Z.; Djakovac, T.; Viličić, D. Seasonal Distribution of Plankton Diatoms in Lim Bay, Northeastern Adriatic Sea. *Acta Bot. Croat.* **2009**, *68*, 351–365.
53. Wells, M.L.; Trainer, V.L.; Smayda, T.J.; Karlson, B.S.O.; Trick, C.G.; Kudela, R.M.; Ishikawa, A.; Bernard, S.; Wulff, A.; Anderson, D.M.; et al. Harmful Algal Blooms and Climate Change: Learning from the Past and Present to Forecast the Future. *Harmful Algae* **2015**, *49*, 68–93. [CrossRef] [PubMed]
54. Botana, L.M. Toxicological Perspective on Climate Change: Aquatic Toxins. *Chem. Res. Toxicol.* **2016**, *29*, 619–625. [CrossRef]
55. Henson, S.A.; Cael, B.B.; Allen, S.R.; Dutkiewicz, S. Future Phytoplankton Diversity in a Changing Climate. *Nat. Commun.* **2021**, *12*, 5372. [CrossRef]
56. Edullantes, B.; Low-Decarie, E.; Steinke, M.; Cameron, T. Comparison of Thermal Traits between Non-Toxic and Potentially Toxic Marine Phytoplankton: Implications to Their Responses to Ocean Warming. *J. Exp. Mar. Biol. Ecol.* **2023**, *562*, 151883. [CrossRef]
57. Alcamo, J.; Olesen, J.E. *Life in Europe Under Climate Change*; John Wiley & Sons: Hoboken, NJ, USA, 2012.
58. Marullo, S.; Artale, V.; Santoleri, R. The SST Multidecadal Variability in the Atlantic–Mediterranean Region and Its Relation to AMO. *J. Clim.* **2011**, *24*, 4385–4401. [CrossRef]
59. Hastings, A. Timescales, Dynamics, and Ecological Understanding. *Ecology* **2010**, *91*, 3471–3480. [CrossRef]
60. Cossarini, G.; Solidoro, C.; Umani, S.F. Dynamics of Biogeochemical Properties in Temperate Coastal Areas of Freshwater Influence: Lessons from the Northern Adriatic Sea (Gulf of Trieste). *Estuar. Coast. Shelf Sci.* **2012**, *115*, 63–74. [CrossRef]
61. Cozzi, S.; Falconi, C.; Comici, C.; Čermelj, B.; Kovac, N.; Turk, V.; Giani, M. Recent Evolution of River Discharges in the Gulf of Trieste and Their Potential Response to Climate Changes and Anthropogenic Pressure. *Estuar. Coast. Shelf Sci.* **2012**, *115*, 14–24. [CrossRef]
62. Zanchettin, D.; Traverso, P.; Tomasino, M. Po River Discharges: A Preliminary Analysis of a 200-Year Time Series. *Clim. Change* **2008**, *89*, 411–433. [CrossRef]
63. Solidoro, C.; Bastianini, M.; Bandelj, V.; Codermatz, R.; Cossarini, G.; Melaku Canu, D.; Ravagnan, E.; Salon, S.; Trevisani, S. Current State, Scales of Variability, and Trends of Biogeochemical Properties in the Northern Adriatic Sea. *J. Geophys. Res. Ocean.* **2009**, *114*, C07S91. [CrossRef]
64. Mozetič, P.; Solidoro, C.; Cossarini, G.; Socal, G.; Precali, R.; Francé, J.; Bianchi, F.; De Vittor, C.; Smodlaka, N.; Fonda Umani, S. Recent Trends Towards Oligotrophication of the Northern Adriatic: Evidence from Chlorophyll a Time Series. *Estuaries Coasts* **2010**, *33*, 362–375. [CrossRef]
65. Lipizer, M.; De Vittor, C.; Falconi, C.; Comici, C.; Tamberlich, F.; Giani, M. Effects of Intense Physical and Biological Forcing Factors on CNP Pools in Coastal Waters (Gulf of Trieste, Northern Adriatic Sea). *Estuar. Coast. Shelf Sci.* **2012**, *115*, 40–50. [CrossRef]
66. Vollenweider, R.; Giovanardi, F.; Montanari, G.; Rinaldi, A. Characterization of the Trophic Conditions of Marine Coastal Waters with Special Reference to the NW Adriatic Sea: Proposal for a Trophic Scale, Turbidity and Generalized Water Quality Index. *Environmetrics Off. J. Int. Environmetrics Soc.* **1998**, *9*, 329–357. [CrossRef]
67. Pettine, M.; Casentini, B.; Fazi, S.; Giovanardi, F.; Pagnotta, R. A Revisitation of TRIX for Trophic Status Assessment in the Light of the European Water Framework Directive: Application to Italian Coastal Waters. *Mar. Pollut. Bull.* **2007**, *54*, 1413–1426. [CrossRef]

68. Aubry, F.B.; Cossarini, G.; Acri, F.; Bastianini, M.; Bianchi, F.; Camatti, E.; De Lazzari, A.; Pugnetti, A.; Solidoro, C.; Socal, G. Plankton Communities in the Northern Adriatic Sea: Patterns and Changes over the Last 30 Years. *Estuar. Coast. Shelf Sci.* **2012**, *115*, 125–137. [CrossRef]
69. Cozzi, S.; Cabrini, M.; Kralj, M.; De Vittor, C.; Celio, M.; Giani, M. Climatic and Anthropogenic Impacts on Environmental Conditions and Phytoplankton Community in the Gulf of Trieste (Northern Adriatic Sea). *Water* **2020**, *12*, 2652. [CrossRef]
70. Cantoni, C.; Cozzi, S.; Pecchiar, I.; Cabrini, M.; Mozetič, P.; Catalano, G.; Umani, S.F. Short-Term Variability of Primary Production and Inorganic Nitrogen Uptake Related to the Environmental Conditions in a Shallow Coastal Area (Gulf of Trieste, N Adriatic Sea). *Oceanol. Acta* **2003**, *26*, 565–575. [CrossRef]
71. Mozetič, P.; Francé, J.; Kogovšek, T.; Talaber, I.; Malej, A. Plankton Trends and Community Changes in a Coastal Sea (Northern Adriatic): Bottom-up vs. Top-down Control in Relation to Environmental Drivers. *Estuar. Coast. Shelf Sci.* **2012**, *115*, 138–148. [CrossRef]
72. Cozzi, S.; Reisenhofer, E.; Di Monte, L.; Cantoni, C.; Adami, G. Effect of Environmental Forcing on the Fate of Nutrients, Dissolved Organic Matter and Heavy Metals Released by a Coastal Wastewater Pipeline. *Chem. Ecol.* **2008**, *24*, 87–107. [CrossRef]
73. Mozetič, P.; Malačič, V.; Turk, V. A Case Study of Sewage Discharge in the Shallow Coastal Area of the Northern Adriatic Sea (Gulf of Trieste). *Mar. Ecol.* **2008**, *29*, 483–494. [CrossRef]
74. Scroccaro, I.; Ostoich, M.; Umgiesser, G.; De Pascalis, F.; Colugnati, L.; Mattassi, G.; Vazzoler, M.; Cuomo, M. Submarine Wastewater Discharges: Dispersion Modelling in the Northern Adriatic Sea. *Environ. Sci. Pollut. Res.* **2010**, *17*, 844–855. [CrossRef] [PubMed]
75. Malej, A.; MozetiČ, P.; MalaČČ, V.; Turk, V. Response of Summer Phytoplankton to Episodic Meteorological Events (Gulf of Trieste, Adriatic Sea). *Mar. Ecol.* **1997**, *18*, 273–288. [CrossRef]
76. Faganeli, J.; Ogrinc, N. Oxic–Anoxic Transition of Benthic Fluxes from the Coastal Marine Environment (Gulf of Trieste, Northern Adriatic Sea). *Mar. Freshw. Res.* **2009**, *60*, 700–711. [CrossRef]
77. Duarte, C.M.; Agustı, S.; Kennedy, H.; Vaqué, D. The Mediterranean Climate as a Template for Mediterranean Marine Ecosystems: The Example of the Northeast Spanish Littoral. *Prog. Oceanogr.* **1999**, *44*, 245–270. [CrossRef]
78. Gremare, A.; Amouroux, J.; Cauwet, G.; Charles, F.; Courties, C.; de Bovee, F.; Dinet, A.; Devenon, J.-L.; Durrieu de Madron, X.; Ferre, B.; et al. Effets d'une Forte Tempête Hivernale Sur Les Variables Physiques et Biologiques à Une Station Côtière Méditerranéenne. *Oceanol. Acta* **2003**, *26*, 407–419.
79. De Vittor, C.; Paoli, A.; Umani, S.F. Dissolved Organic Carbon Variability in a Shallow Coastal Marine System (Gulf of Trieste, Northern Adriatic Sea). *Estuar. Coast. Shelf Sci.* **2008**, *78*, 280–290. [CrossRef]
80. Guadayol, Ò.; Peters, F.; Marrasé, C.; Gasol, J.M.; Roldán, C.; Berdalet, E.; Massana, R.; Sabata, A. Episodic Meteorological and Nutrient-Load Events as Drivers of Coastal Planktonic Ecosystem Dynamics: A Time-Series Analysis. *Mar. Ecol. Prog. Ser.* **2009**, *381*, 139–155. [CrossRef]
81. Lipizer, M.; Cozzi, S.; Catalano, G.; Falconi, C. Seasonal Fluctuations of DIN/DIP and DON/DOP Ratio in the Northern Adriatic Sea. *Ann. Dell'istituto Super. Di Sanità* **1999**, *35*, 383–388.
82. Gismondi, M.; Giani, M.; Savelli, F.; Boldrin, A.; Rabitti, S. Particulate Organic Matter in the Northern and Central Adriatic. *Chem. Ecol.* **2002**, *18*, 27–38. [CrossRef]
83. Lucea, A.; Duarte, C.M.; Agustí, S.; Kennedy, H. Nutrient Dynamics and Ecosystem Metabolism in the Bay of Blanes (NW Mediterranean). *Biogeochemistry* **2005**, *73*, 303–323. [CrossRef]
84. Degobbis, D.; Precali, R.; Ivancic, I.; Smodlaka, N.; Fuks, D.; Kveder, S. Long-Term Changes in the Northern Adriatic Ecosystem Related to Anthropogenic Eutrophication. *Int. J. Environ. Pollut.* **2000**, *13*, 495–533. [CrossRef]
85. Degobbis, D.; Precali, R.; Ferrari, C.R.; Djakovac, T.; Rinaldi, A.; Ivančić, I.; Gismondi, M.; Smodlaka, N. Changes in Nutrient Concentrations and Ratios during Mucilage Events in the Period 1999–2002. *Sci. Total Environ.* **2005**, *353*, 103–114. [CrossRef]
86. Mozetic, P.; Umani, S.F.; Kamburska, L. Plankton Variability in the Gulf of Trieste (Northern Adriatic). *Archo Oceanogr. Limnol.* **2002**, *23*, 7–19.
87. Viličić, D.; Marasović, I.; Mioković, D. Checklist of Phytoplankton in the Eastern Adriatic Sea. *Acta Bot. Croat.* **2002**, *61*, 57–91.
88. Cabrini, M.; Fornasaro, D.; Cossarini, G.; Lipizer, M.; Virgilio, D. Phytoplankton Temporal Changes in a Coastal Northern Adriatic Site during the Last 25 Years. *Estuar. Coast. Shelf Sci.* **2012**, *115*, 113–124. [CrossRef]
89. Pistocchi, R.; Guerrini, F.; Pezzolesi, L.; Riccardi, M.; Vanucci, S.; Ciminiello, P.; Dell'Aversano, C.; Forino, M.; Fattorusso, E.; Tartaglione, L.; et al. Toxin Levels and Profiles in Microalgae from the North-Western Adriatic Sea—15 Years of Studies on Cultured Species. *Mar. Drugs* **2012**, *10*, 140–162. [CrossRef]
90. Cerino, F.; Fornasaro, D.; Kralj, M.; Giani, M.; Cabrini, M. Phytoplankton Temporal Dynamics in the Coastal Waters of the North-Eastern Adriatic Sea (Mediterranean Sea) from 2010 to 2017. *Nat. Conserv.* **2019**, *34*, 343–372. [CrossRef]
91. Totti, C.; Romagnoli, T.; Accoroni, S.; Coluccelli, A.; Pellegrini, M.; Campanelli, A.; Grilli, F.; Marini, M. Phytoplankton Communities in the Northwestern Adriatic Sea: Interdecadal Variability over a 30-Years Period (1988–2016) and Relationships with Meteoclimatic Drivers. *J. Mar. Syst.* **2019**, *193*, 137–153. [CrossRef]

92. Vascotto, I.; Mozetič, P.; Francé, J. Phytoplankton Time-Series in a LTER Site of the Adriatic Sea: Methodological Approach to Decipher Community Structure and Indicative Taxa. *Water* **2021**, *13*, 2045. [CrossRef]
93. Gladan, Ž.N.; Matić, F.; Arapov, J.; Skejić, S.; Bužančić, M.; Bakrač, A.; Straka, M.; Dekneudt, Q.; Grbec, B.; Garber, R.; et al. The Relationship between Toxic Phytoplankton Species Occurrence and Environmental and Meteorological Factors along the Eastern Adriatic Coast. *Harmful Algae* **2020**, *92*, 101745. [CrossRef] [PubMed]
94. Ljubešić, Z.; Bosak, S.; Viličić, D.; Borojević, K.K.; Marić, D.; Godrijan, J.; Ujević, I.; Peharec, P.; Đakovac, T. Ecology and Taxonomy of Potentially Toxic *Pseudo-Nitzschia* Species in Lim Bay (North-Eastern Adriatic Sea). *Harmful Algae* **2011**, *10*, 713–722. [CrossRef]
95. Round, F.E.; Crawford, R.M.; Mann, D.G. *Diatoms: Biology and Morphology of the Genera*; Cambridge University Press: Cambridge, UK, 1990.
96. Burkholder, J.M.; Glibert, P.M.; Skelton, H.M. Mixotrophy, a Major Mode of Nutrition for Harmful Algal Species in Eutrophic Waters. *Harmful Algae* **2008**, *8*, 77–93. [CrossRef]
97. Fonda, S.; Franco, P.; Ghirardelli, E.; Malej, A. Outline of Oceanography and the Plankton of the Adriatic Sea. In *Marine Eutrophication and Population Dynamics, Proceedings of the 25th EMBS, İstanbul, Turkey, 18–20 August 1992*; Colombo, G., Ferrari, I., Ceccarelli, V.U., Rossi, R., Eds.; Olsen & Olsen: Fredensborg, Denmark, 1992.
98. Godrijan, J.; Marić, D.; Tomažić, I.; Precali, R.; Pfannkuchen, M. Seasonal Phytoplankton Dynamics in the Coastal Waters of the North-Eastern Adriatic Sea. *J. Sea Res.* **2013**, *77*, 32–44. [CrossRef]
99. Brush, M.J.; Mozetič, P.; Francé, J.; Aubry, F.B.; Djakovac, T.; Faganeli, J.; Harris, L.A.; Niesen, M. Phytoplankton Dynamics in a Changing Environment. In *Coastal Ecosystems in Transition: A Comparative Analysis of the Northern Adriatic and Chesapeake Bay*; American Geophysical Union: Washington, DC, USA, 2020; pp. 49–74.
100. Cabrini, M.; Cok, S.; Pecchiar, I. Dinamica e Struttura Del Fitoplancton Nella Fascia Costiera Del Golfo Di Trieste. *Biol. Mar. Mediterr.* **2000**, *7*, 850–853.
101. Zingone, A.; Escalera, L.; Aligizaki, K.; Fernández-Tejedor, M.; Ismael, A.; Montresor, M.; Mozetič, P.; Taş, S.; Totti, C. Toxic Marine Microalgae and Noxious Blooms in the Mediterranean Sea: A Contribution to the Global HAB Status Report. *Harmful Algae* **2021**, *102*, 101843. [CrossRef]
102. Anderson, D.M.; Glibert, P.M.; Burkholder, J.M. Harmful Algal Blooms and Eutrophication: Nutrient Sources, Composition, and Consequences. *Estuaries* **2002**, *25*, 704–726. [CrossRef]
103. Hattenrath-Lehmann, T.K.; Marcoval, M.A.; Mittlesdorf, H.; Goleski, J.A.; Wang, Z.; Haynes, B.; Morton, S.L.; Gobler, C.J. Nitrogenous Nutrients Promote the Growth and Toxicity of *Dinophysis Acuminata* during Estuarine Bloom Events. *PLoS ONE* **2015**, *10*, e0124148. [CrossRef]
104. Tong, M.; Smith, J.L.; Kulis, D.M.; Anderson, D.M. Role of Dissolved Nitrate and Phosphate in Isolates of Mesodinium Rubrum and Toxin-Producing *Dinophysis Acuminata*. *Aquat. Microb. Ecol.* **2015**, *75*, 169–185. [CrossRef]
105. García-Portela, M.; Reguera, B.; Gago, J.; Le Gac, M.; Rodríguez, F. Uptake of Inorganic and Organic Nitrogen Sources by *Dinophysis Acuminata* and *D. Acuta*. *Microorganisms* **2020**, *8*, 187. [CrossRef] [PubMed]
106. Domingues, R.B.; Lima, M.J. Unusual Red Tide of the Dinoflagellate *Lingulodinium Polyedra* during an Upwelling Event off the Algarve Coast (SW Iberia). *Reg. Stud. Mar. Sci.* **2023**, *63*, 102998. [CrossRef]
107. Caroppo, C.; Congestri, R.; Bracchini, L.; Albertano, P. On the Presence of *Pseudo-Nitzschia Calliantha Lundholm*, Moestrup et Hasle and *Pseudo-Nitzschia Delicatissima* (Cleve) Heiden in the Southern Adriatic Sea (Mediterranean Sea, Italy). *J. Plankton Res.* **2005**, *27*, 763–774. [CrossRef]
108. Sahraoui, I.; Grami, B.; Bates, S.S.; Bouchouicha, D.; Chikhaoui, M.A.; Mabrouk, H.H.; Hlaili, A.S. Response of Potentially Toxic *Pseudo-Nitzschia* (Bacillariophyceae) Populations and Domoic Acid to Environmental Conditions in a Eutrophied, SW Mediterranean Coastal Lagoon (Tunisia). *Estuar. Coast. Shelf Sci.* **2012**, *102*, 95–104. [CrossRef]
109. Brzezinski, M.A. The Si: C: N Ratio of Marine Diatoms: Interspecific Variability and the Effect of Some Environmental Variables 1. *J. Phycol.* **1985**, *21*, 347–357. [CrossRef]
110. Smayda, T.J. Harmful Algal Blooms: Their Ecophysiology and General Relevance to Phytoplankton Blooms in the Sea. *Limnol. Oceanogr.* **1997**, *42*, 1137–1153. [CrossRef]
111. Krom, M.; Herut, B.; Mantoura, R. Nutrient Budget for the Eastern Mediterranean: Implications for Phosphorus Limitation. *Limnol. Oceanogr.* **2004**, *49*, 1582–1592. [CrossRef]
112. Smayda, T.J. Turbulence, Watermass Stratification and Harmful Algal Blooms: An Alternative View and Frontal Zones as "Pelagic Seed Banks". *Harmful Algae* **2002**, *1*, 95–112. [CrossRef]
113. Klemm, K.; Cembella, A.; Clarke, D.; Cusack, C.; Arneborg, L.; Karlson, B.; Liu, Y.; Naustvoll, L.; Siano, R.; Gran-Stadniczeñko, S.; et al. Apparent Biogeographical Trends in Alexandrium Blooms for Northern Europe: Identifying Links to Climate Change and Effective Adaptive Actions. *Harmful Algae* **2022**, *119*, 102335. [CrossRef]
114. Ninčević-Gladan, Ž.; Skejić, S.; Bužančić, M.; Marasović, I.; Arapov, J.; Ujević, I.; Bojanić, N.; Grbec, B.; Kušpilić, G.; Vidjak, O. Seasonal Variability in *Dinophysis* Spp. Abundances and Diarrhetic Shellfish Poisoning Outbreaks along the Eastern Adriatic Coast. *Bot. Mar.* **2008**, *51*, 449–463. [CrossRef]

115. Spatharis, S.; Danielidis, D.B.; Tsirtsis, G. Recurrent *Pseudo-Nitzschia Calliantha* (Bacillariophyceae) and Alexandrium Insuetum (Dinophyceae) Winter Blooms Induced by Agricultural Runoff. *Harmful Algae* **2007**, *6*, 811–822. [CrossRef]
116. Margalef, R. Life-Forms of Phytoplankton as Survival Alternatives in an Unstable Environment. *Oceanol. Acta* **1978**, *1*, 493–509.
117. Weise, A.M.; Levasseur, M.; Saucier, F.J.; Senneville, S.; Bonneau, E.; Roy, S.; Sauvé, G.; Michaud, S.; Fauchot, J. The Link between Precipitation, River Runoff, and Blooms of the Toxic Dinoflagellate *Alexandrium Tamarense* in the St. Lawrence. *Can. J. Fish. Aquat. Sci.* **2002**, *59*, 464–473. [CrossRef]
118. Turner, J.T.; Tester, P.A.; Lincoln, J.A.; Carlsson, P.; Granéli, E. Effects of N: P: Si Ratios and Zooplankton Grazing on Phytoplankton Communities in the Northern Adriatic Sea. III. Zooplankton Populations and Grazing. *Aquat. Microb. Ecol.* **1999**, *18*, 67–75. [CrossRef]
119. Mozetic, P.; Lipej, L. Phytoplankton-zooplankton Trophic Interactions along the Salinity Gradient (Gulf of Trieste). *Rapp. De La Comm. Int. Pour L'exploration Sci. De La Mer Méditerranée* **1998**, *35*, 468–469.
120. Bill, B.D.; Moore, S.K.; Hay, L.R.; Anderson, D.M.; Trainer, V.L. Effects of Temperature and Salinity on the Growth of *Alexandrium* (Dinophyceae) Isolates from the Salish Sea. *J. Phycol.* **2016**, *52*, 230–238. [CrossRef]
121. Boivin-Rioux, A.; Starr, M.; Chassé, J.; Scarratt, M.; Perrie, W.; Long, Z. Predicting the Effects of Climate Change on the Occurrence of the Toxic Dinoflagellate *Alexandrium Catenella* Along Canada's East Coast. *Front. Mar. Sci.* **2021**, *7*, 608021. [CrossRef]
122. Marzidovšek, M.; Francé, J.; Podpečan, V.; Vadnjal, S.; Dolenc, J.; Mozetič, P. Explainable Machine Learning for Predicting Diarrhetic Shellfish Poisoning Events in the Adriatic Sea Using Long-Term Monitoring Data. *Harmful Algae* **2024**, *139*, 102728. [CrossRef]

environments

MDPI

Article

Integration of Earth Observation and Field-Based Monitoring for Morphodynamic Characterisation of Tropical Beach Ecosystems

James Murphy [1,*], Jonathan E. Higham [1], Andrew J. Plater [1], Kasey E. Clark [1,2] and Rachel Collin [2]

1 Department of Geography and Planning, School of Environmental Sciences, University of Liverpool, Liverpool L69 7ZX, UK; j.e.higham@liverpool.ac.uk (J.E.H.); gg07@liverpool.ac.uk (A.J.P.)
2 Smithsonian Tropical Research Institute, Luis Clement Avenue, Bldg. 401 Tupper, Ancon 0801, Panama; collinr@si.edu
* Correspondence: jfmurphy@liverpool.ac.uk

Abstract: Coastal erosion poses a significant threat to small tropical island regions, where coastal tourism and infrastructure play vital economic roles. However, the processes affecting tropical beaches, particularly in Central America, remain underexplored due to a lack of data on waves and atmospheric conditions. We propose a novel approach that utilises low-cost smartphone and satellite imagery to characterise beach ecosystems, where typically expensive and technologically intensive monitoring strategies are impractical and background data are scarce. As a test of its performance under real conditions, we apply this approach to four contrasting beaches in the low-lying islands of the Bocas del Toro Archipelago, Panama. We employ Earth Observation data and field-based monitoring to enhance understanding of beach erosion. Optical flow tracking velocimetry (OFTV) is applied to smartphone camera footage to provide a quantitative metric of wave characteristics during the high wave energy season. These data are combined with satellite-derived shoreline change data and additional field data on beach profiles and grain size. The results reveal distinct patterns of accretion and erosion across the study sites determined by wave climate, beach morphology, and grain size. Accreting beaches are generally characterised by longer wave periods, more consistent wave velocities, and finer, positively skewed sediments indicative of swell-dominated conditions and dissipative beach profiles. Conversely, more erosive sites are associated with shorter wave periods, more variable wave velocities, coarser and better-sorted sediments, and a shorter, steeper beach profile. Seasonal erosion during the high-energy wave season (January–April) and subsequent recovery were observed at most sites. This work demonstrates how foundational data for evidence-based coastal management can be generated in remote locations that lack essential baseline data.

Keywords: coastal erosion monitoring; tropical beach dynamics; wave velocity; optical flow; low cost; remote sensing; Caribbean

1. Introduction

Coastal erosion poses a significant threat to communities worldwide, where cities, vital infrastructure, and tourist economies are often located adjacent to the coast [1–4]. This presents a particular challenge in tropical and remote areas, where limited funding, expertise, and available data can hinder the sustainable management of marine and coastal assets [5]. Threats to beach ecosystems from coastal erosion will become even more pressing with climate change due to sea-level rise and changes in the magnitude and frequency of storms and hurricanes [6]. Coastal zone management aimed at mitigating these pressures necessitates robust data on contemporary and historical trends, which are often lacking and can be costly and resource-intensive to acquire [7].

Significant portions of the Caribbean coastline in Latin America remain under-researched, and the absence of data on wave and climatic conditions presents challenges for effective coastal management [8,9]. Nevertheless, local reports have highlighted increasing rates of erosion along the sandy shorelines of southern Costa Rica, near the Panamanian border [10,11]. The straight beaches along this coastline, like numerous others on the Caribbean coast, are particularly susceptible to concentrated wave energy [12], which is often linked with erosional behaviour [13]. While qualitative monitoring of tropical beach ecosystems has provided significant insights into local beach conditions, these efforts are generally focused on areas that are particularly vulnerable to extreme weather events, such as tropical cyclones [14].

Sandy shores throughout the Caribbean are highly valued for recreation [15], yet they are increasingly vulnerable to erosion due to their inherent geomorphological variability and growing anthropogenic pressure. Many of these coastlines are closely linked with coral reef systems, which not only enhance their aesthetic appeal for tourism but also provide critical ecosystem services, such as wave energy attenuation and sediment stabilisation, thereby contributing to beach resilience [16]. These are characteristic of perched beaches, which are coastal environments where natural, shallow-submerged features like coral reefs function similarly to submerged breakwaters, dissipating wave energy and offering protection to the adjacent shoreline.

Reef systems can also introduce temporal and spatial variability in coastal processes, driving localised patterns of beach erosion and accretion [17]. In particular, the presence of reef flats influences wave dynamics by increasing the surface roughness, altering wave transformation and energy dissipation across the nearshore zone [18]. The interconnected nature of these reef–beach systems, therefore, adds complexity to understanding local coastal morphology, as reef health directly influences shoreline stability [19,20].

To understand local coastal dynamics, it is essential to characterise wave behaviour accurately. A common approach involves using wave buoys; however, there are a limited number of buoys with long records in the Caribbean [21], and they often require specialised teams for regular calibration, maintenance, and data collection, necessitating robust maintenance plans and the availability of spare parts, which limits their dissemination in resource-limited settings. [22,23]. To address these challenges, researchers have explored a variety of cost-effective strategies for wave monitoring, including open-source and DIY wave loggers and gauges made with commercially available tools [24–26]; low-cost wave measurement devices such as GNSS buoys [27,28] and spotter buoys [29,30] and land-based vertical seismometers [31]. While these techniques offer an accurate and cost-effective means of collecting wave dynamics, their implementation is particularly beneficial in the short term, sometimes requiring specialised and dedicated teams for maintenance and calibration.

In addition to hardware-based approaches, numerical wave models have become an important resource for characterising wave conditions, particularly in areas lacking consistent in situ data. Global and regional models, such as WaveWatch III, and operational products from platforms like the Copernicus Marine Environment Monitoring Service (CMEMS), provide publicly available, continuously updated data on a variety of spatial and temporal scales. These model outputs have been widely adopted by the scientific community and have been validated against in situ measurements around the globe [32,33] demonstrating their utility for estimating wave parameters.

Optical video footage has proven effective in estimating nearshore bathymetry and hydrodynamic processes [34–36] This effectiveness stems from the capability for feature tracking that video footage offers. While particle image velocimetry (PIV) and particle-/feature-tracking velocimetry (PTV) have been extensively employed to quantify motion from sequences of images or videos [37–40], considerable opportunities remain for optical flow velocimetry (OF) techniques, which have shown notable advantages over PIV methods [41–43]. These data can be collected easily and consistently using dedicated video camera systems or even a mobile phone camera.

This paper addresses this data resource challenge through the development of a novel multi-platform methodology tested in real-world conditions through the case study of Bocas del Toro Archipelago in Panama. The study aims to integrate in situ measurements with freely available online Earth Observation (EO) data to create an accessible, comprehensive toolkit for quantifying and comparing beach morphodynamic characteristics. We present an approach that integrates beach slope, sediment size, and smartphone video footage collected during the high-energy wave season with beach width data derived from satellite imagery [44]. Using an optical flow tracking velocimetry (OFTV) algorithm, we extract wave velocities from video footage, enabling the analysis of wave speeds and periods across four study sites. By linking the wave variability, sediment characteristics, and beach slope, we quantify the processes driving long-term accretion and erosion trends observed in satellite data. Finally, we assess the effectiveness of this multi-platform toolkit in comparing coastal processes across different environments, providing new insight into the factors influencing beach response.

2. Materials and Methods

2.1. Data and Study Sites

The beaches of the Bocas del Toro Archipelago, located on the Caribbean coast of Panama, are particularly vulnerable to coastal erosion, with some areas of the coastal road being notably at risk due to their proximity to the shoreline. Bocas del Toro province is one of the poorest in Panama, where significant economic activity stems from ocean-based tourism and banana plantations on the nearby mainland's coastal plain [45]. In Bocas del Toro province, about 95% of all economic activity is linked to the tourism sector [46], which continues to grow, especially with the rise of accessible digital infrastructure, such as travel applications, for international tourists [47] Thus, understanding the coastal dynamics within the archipelago is vital. Of immediate concern is the effect of coastal erosion on essential and culturally important infrastructure, such as the airport, reservoir, hospital, and roads—much of which lies directly adjacent to exposed beaches.

Four sites were selected for a real-world test of the multi-platform approach, from February to April 2024, known as the dry season, which typically experiences high wave energy. The sites were chosen based on a combination of local knowledge and accessibility (Figure 1).

We characterised the beaches using shoreline change analysis conducted with CoastSat [44]. The selected sites provide a representative cross-section of beaches in the Bocas del Toro Archipelago, as they exhibit differences in wave climate, geology, and sediment type.

Primary data collection involved beach profile measurements (slope and beach width), sediment sampling for grain-size analysis, and video footage to document wave dynamics. At each beach, multiple transects were established to capture the spatial variability. Transects were spaced as evenly as possible along the shoreline, subject to accessibility constraints. The number of transects per beach was as follows: Playa Bluff—4; Playa Istmito—3; Playa Carenero—3; and Playa Wizard—3.

Little to no secondary data are available regarding the morphologies of the selected sites and many beaches in the surrounding archipelago, and the hydrodynamics of the archipelago were not comprehensively monitored at the time of this study. This data gap presents both a challenge and an opportunity; while it highlights the need for an accessible monitoring toolkit with easy-to-deploy methods and data collection techniques, it means that baseline information on historical coastal evolution is largely absent. Consequently, long-term monitoring and management strategies will need to be developed in the absence of historical context, limiting the ability to detect longer-term hydrological trends.

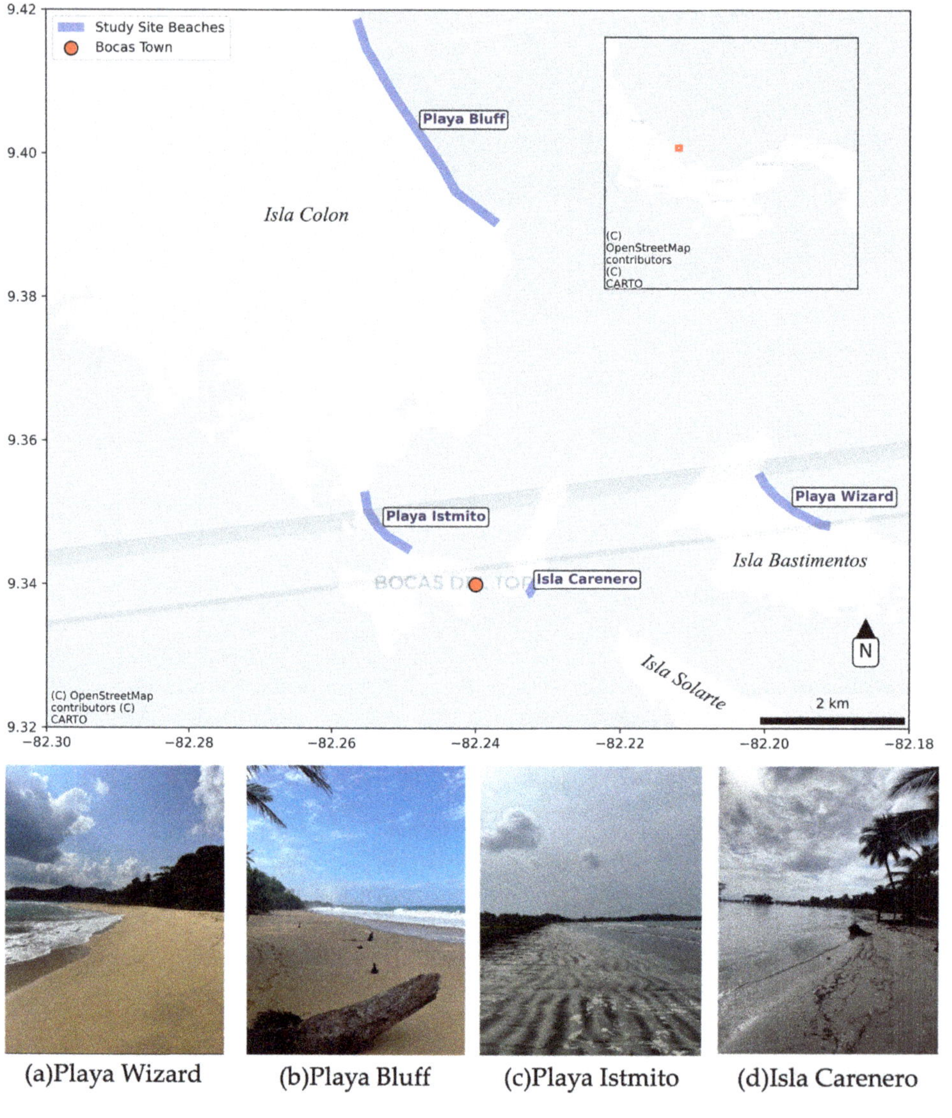

Figure 1. (**above**) The study sites within the Bocas del Toro Archipelago; (**inset**) the greater Bocas del Toro province, with the coordinates configured in EPSG:4326; (**below**) in situ photographs of the study sites, featuring (**a**) Playa Wizard, (**b**) Playa Bluff, (**c**) Playa Istmito, and (**d**) Isla Carenero. The images were taken by the author during field visits from February to March 2024.

In the absence of local historical wave data, regional models become essential. The nearest available NOAA wave buoy is the Western Caribbean wave buoy [48], located approximately 800 km from the archipelago. Consequently, the wave climate can be validated using a regional wave model. The WaveWatch III regional wave model was used in this case, as covered in Section 3.2.2.

Of the four study sites, Playa Wizard (also known as Playa Primera, see Figure 1a) measures approximately 1.2 km in length and is located on the western end of Isla Bastimentos, facing north–east towards the Caribbean Sea. Like the Bluff, this site is characterised by higher-energy waves and strong riptides, particularly during the dry season (January to April). Playa Bluff (Figure 1b) is situated in northern Isla Colon; at 4 km, it ranks as one of the largest beaches in the archipelago. The beach faces north–east towards the Caribbean Sea and is immediately backed by a rainforest that covers much of northern Isla Colón, separated by Bluff Beach Road—a road that follows the coastline from the central narrow strip of Isla Colón. This beach is a popular destination for tourism and surfing, hosting the Bocas Invitacional surfing competition at this location. Playa Istmito (Figure 1c) is positioned in the La Cabana neighbourhood in central Isla Colon. This beach is just over 1.3 km long and lies on a narrow strip of land connecting the island's northern part to Bocas Town in the south. Although facing northeast, it is in a bay sheltered by northern Isla Colon to the north and Isla Carenero to the east. Isla Carenero (Figure 1d) is an island east of Bocas Town on Isla Colon and is one of the smallest of the main islands within the Bocas del Toro Archipelago. It serves as a hub for recreational activities, boasting several restaurants and bars along the island's eastern beaches. This side of Isla Carenero experiences low wave energy due to its position between Isla Bastimentos to the east and Isla Solarte to the southeast. The beaches here are narrow, spanning only a few metres.

2.2. Primary Data Collection

2.2.1. Optical Flow Tracking Velocimetry (OFTV)

Optical flow tracking velocimetry (OFTV) [49] is especially suitable for monitoring dynamic wave fields because of its capability to handle the varying lighting conditions between the whitewater of breaking waves and the surrounding water. In contrast to traditional methods, such as particle image velocimetry (PIV) [37], which depends on seeded particles for tracking, OFTV tracks natural-intensity features within the video sequence directly, providing a non-intrusive and efficient approach to wave analysis [50].

This method was considered optimal for the study due to the challenges of installing permanent wave monitoring equipment in remote tropical regions. The region's climate and environmental conditions are highly damaging to stationary equipment, and the logistical difficulties of setting up and maintaining large instruments across scattered islands and beaches further complicate traditional monitoring approaches. The OFTV method provides a cost-effective alternative, requiring only an optical camera or smartphone to capture footage, making it particularly advantageous for use in these remote coastal settings.

Video footage of the wave environment was collected with an iPhone 12 Mini at a resolution of 1920×1080 and a video quality of 1080p. The phone was secured atop a tripod at a height of 5′1″. Each recording featured at least ten full wave periods to provide a robust dataset for analysis (see Appendix A Table A1 for a full breakdown of the survey times by site). For each beach transect, two videos were recorded, as follows: one near the waterline and another farther inland to provide a broader view of the incoming waves (Figure 2a). At Carenero Beach, where waves break very close to the shore, only a single field of view was recorded due to spatial constraints. Footage was taken weekly or as often as feasible. At times, certain transects could not be collected due to unexpected events, such as recreational surfing events, dangerous conditions, and algae growth that obstructed the installation of rectification markers.

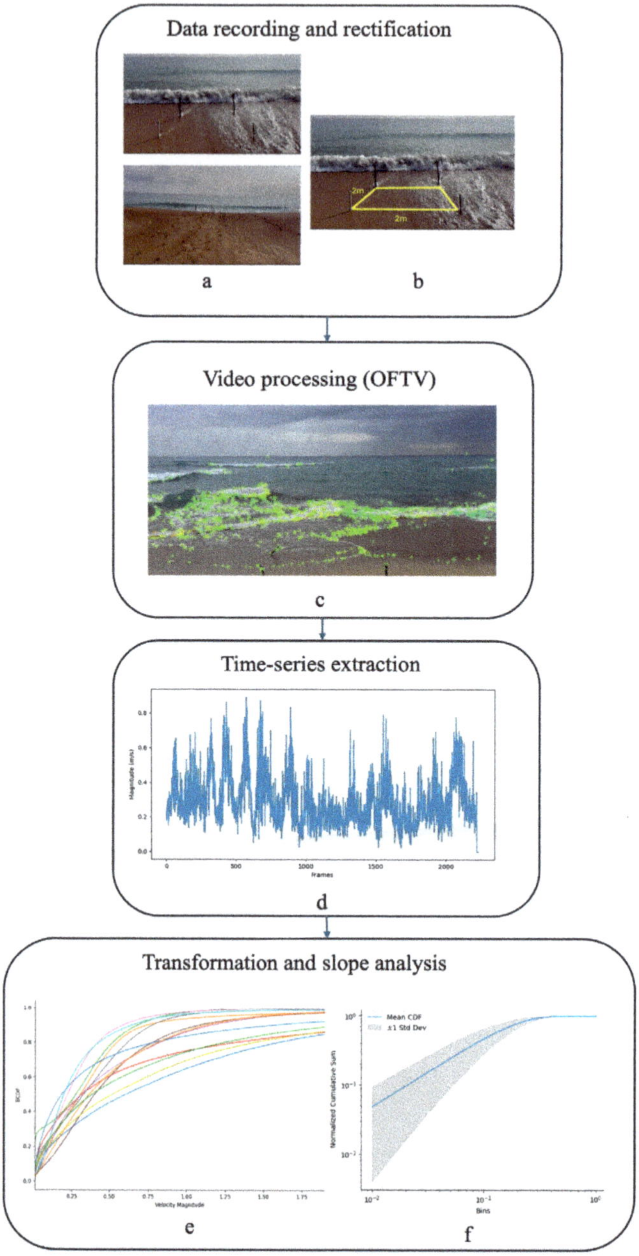

Figure 2. Overview of the OFTV processing pipeline detailing: (**a**) raw video footage from two fields of view with rectification markers; (**b**) rectification zone detailing a known distance for further processing; (**c**) feature tracking using OFTV; (**d**) deriving wave crest velocities into a time-series; (**e**) conversion of wave velocities into a CDF; and (**f**) log scale transformation and calculation of the mean and standard deviation of the linearised CDF.

All videos incorporated rectification markers within the frame to ensure accurate wave velocity calculations. These markers provided a known reference scale for the OFTV algorithm, enabling real-world distances to be derived from pixel measurements (Figure 2b). The pixel-to-distance conversion was calibrated by integrating four points of a known distance apart, ensuring precise and comparable wave speed calculations across all study sites.

Wave crest tracking was conducted using the Lucas–Kanade optical flow algorithm [51], which estimates motion by solving for the pixel displacement between consecutive video frames. A binary mask was applied to each frame, focusing on isolating regions of significant wave activity and ensuring that only regions of actual wave crests were retained for further analysis. Subsequently, a nearest-neighbour distance criterion was employed to track them across frames, ensuring continuous and reliable measurement of the wave motion dynamics (Figure 2c). The tracked horizontal positions of the wave crests were recorded as time-series data, capturing their displacement over time (i.e., their velocity, Figure 2d). These values served as reference points for the wave spectral analysis. To compare the variability in wave energy across the four beaches, velocity data were aggregated into representative distributions for the entire study period. These were then transformed into cumulative density functions (CDFs), supporting quantitative comparisons across sites through a simple metric linked to energy consistency (Figure 2e). These CDFs were log-scaled to linearise the naturally curved CDFs, enabling a slope calculation (Figure 2f). The slope of the linearised CDF, calculated as the gradient (dy/dx), offered a quantitative measure of the wave velocity variability; a lower gradient reflects more consistent, uniform wave velocities, while a steeper slope indicates a higher variability. For each beach, the mean slope across all transects was established, along with the standard deviation, to account for the variability in velocity distributions.

A wave period analysis was also conducted using Welch's power spectrum method [52]. This technique decomposed wave motion into its constituent frequencies, allowing for the identification of the dominant wave frequency. The peak of the power spectral density function was identified as the dominant wave frequency, which was then converted into the wave period by taking the inverse of this frequency. This process enabled in-depth examination of the wave dynamics, allowing for a refined understanding of how the wave energy is distributed across the different study sites. To assess the accuracy and applicability of the OFTV-derived wave periods, we conducted a comparative analysis against predictions from the NOAA/NCEP WaveWatch III (WWIII) model—a widely used, third-generation wave model [53]. For this comparison, we selected the WWIII grid point at 9°30′00.0″ N 82°00′00.0″ W, located approximately 27 km northeast of Isla Bastimentos. This location was chosen as it is the closest model point to Playa Wizard, one of the study sites facing north–east towards the Caribbean Sea.

2.2.2. Sediment Sampling and Beach Profile

Sediment was collected from the low- and high-tide points at each transect location at each beach. For a quantitative assessment of the beach grain size, fifty sediment grains for each site were examined using a Leica MS5 microscope, and the grain size was converted into millimetres (mm). The grain sizes in mm were converted into the Krumbein phi scale (see [54]) for a statistical analysis of the sediment size featuring sorting, skewness, and kurtosis (for the equations, see [55]).

To obtain beach profile data, the beach width was measured from the vegetation line, which was typically rainforest, to the point of low tide along the waterline, for every transect on every beach. The beach slope was measured at every two metres (every one metre for Carenero, as the beach width was significantly smaller) in degrees using a TruPulse

200 laser rangefinder. The average profile across the transects was then calculated to provide a representative slope for each beach.

2.3. Secondary Data Collection

CoastSat and Tidal Correction

A time-series analysis of the satellite images provided a longer-term context for the primary data collection to determine underlying trends in coastal erosion and accretion. This analysis was conducted using CoastSat [44], an open-source toolkit that extracts shoreline positions from satellite imagery over a specified time period. CoastSat uses a pre-trained machine learning algorithm to identify the water/land boundary and corrects this by employing data from nearby tide gauges. The limitations of using optical satellite data for shoreline monitoring are often exacerbated in areas with significant cloud cover, such as tropical regions [56]. Although the algorithm used by CoastSat incorporates a cloud filter, the prevalence of extensive cloud cover in the Bocas del Toro region rendered several images unusable for shoreline detection, resulting in a reduced temporal resolution of the available imagery.

To fill these data gaps as much as possible, various optical satellites with overlapping datasets—Landsat-7 (30 m resolution) and Sentinel-2 (10 m resolution)—were employed in this case. A total of 114 and 338 images were analysed for Landsat-7 and Sentinel-2, respectively. For a detailed list of the satellite observation dates, platforms, and their respective spatial resolutions, refer to Appendix A Table A2. The differing resolutions of these satellite platforms can influence shoreline precision; although CoastSat incorporates adjustments to normalise these differences across datasets, the Sentinel-2 imagery generally allowed for a more accurate shoreline delineation, particularly where the beach width was smaller. Visual verification was conducted image by image to ensure consistency across datasets and correct for any anomalies.

Limited satellite imagery, mainly due to cloud cover, hindered the coverage of the region prior to the launch of Sentinel-2A in 2016. As a result, the CoastSat analysis conducted here begins in 2016, providing a higher temporal resolution. Shorelines were extracted from satellite imagery between 1 January 2016 and 13 December 2023 to align with available tide gauge data.

Tide gauge data [57] used for correction were taken from a standard mechanical float-type tide gauge in Limon, Costa Rica (10.00 N, 83.02 W). Tidal data were collected hourly and available consistently from October 2009 until December 2023, with some data missing between late 2021 and early 2022. Following the tidal correction process, outliers in the datapoints are removed through an Otsu thresholding filter. In addition, further outliers were removed using a z-score method [58], which removed data points that sat outside a given deviation from the mean. Following this filtering process, the number of datapoints that remained for analysis for Playa Bluff, Playa Wizard, Playa Carenero, and Playa Istmito were 72, 146, 114, and 120, respectively.

The annual rate of change for each beach is calculated by first extracting shoreline positions using CoastSat and then deriving cross-shore distances from a fixed baseline to each shoreline. These distances were averaged annually, and the differences between consecutive years were used to compute rates of change, along with their standard deviations. To account for the natural high seasonal variability of shorelines [59], cross-shore distances were analysed not only over the entire study period but also monthly. This approach facilitated an examination of both long-term trends and short-term fluctuations arising from the shoreline extraction.

3. Results

3.1. CoastSat Shoreline Analysis

Throughout the entire study period, as shown in Figure 3, Playa Bluff and Playa Istmito exhibited a general increase in cross-shore distance, particularly in recent years. However, the data for Playa Bluff demonstrated considerable variability, especially in 2023, indicating that while long-term accretion occurs, significant short-term fluctuations are also evident.

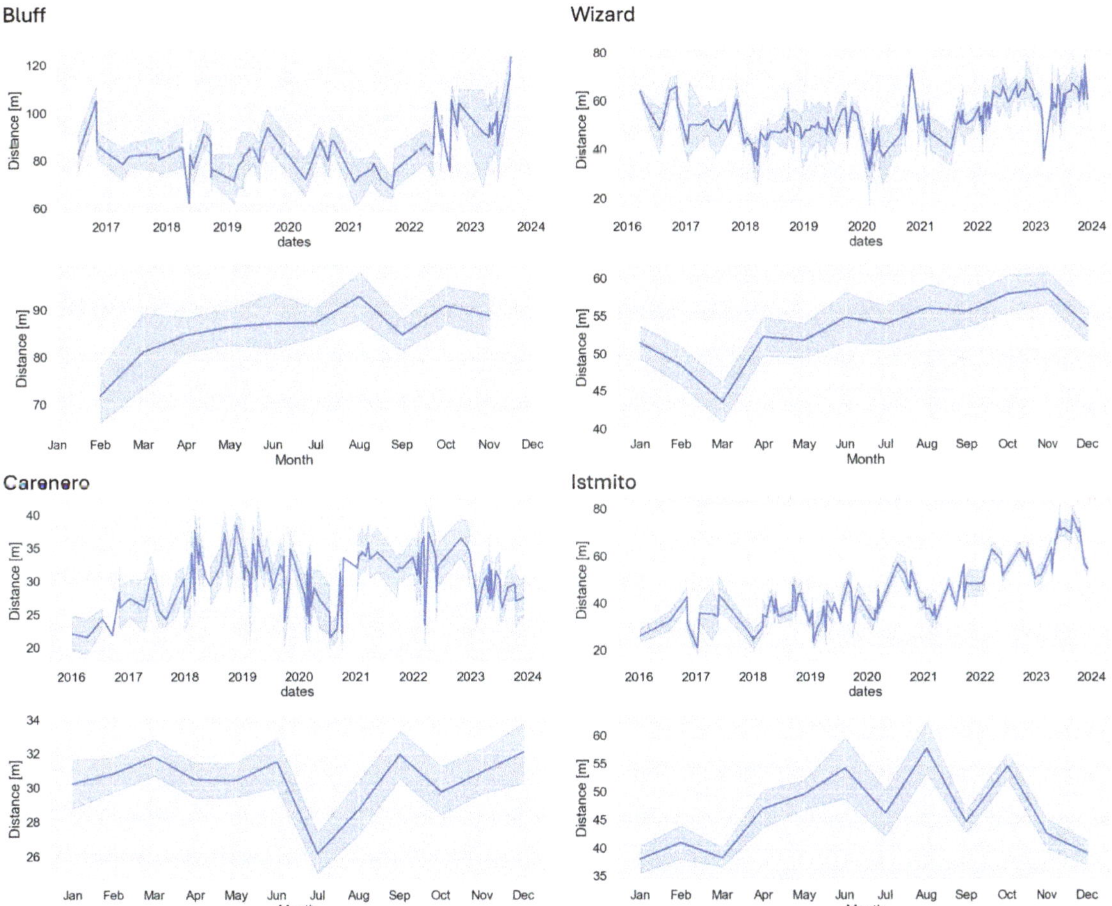

Figure 3. The temporal dynamics in cross-shore distances for each site calculated by the CoastSat algorithm. Temporal dynamics are calculated over the total study period (**top**) and monthly averages (**bottom**). The lines represent the mean values for each given time interval, with the shaded area representing the 95th confidence interval for this central tendency. Note: No imagery was available for Playa Bluff for the month of January.

In contrast, no discernible predominant pattern could be detected at Carenero and Playa Wizard over the long term, indicating that these areas exhibit more complex seasonal and interannual processes. For Playa Wizard, Playa Bluff, and Playa Istmito, there is a significant reduction in cross-shore distances during the high wave season (January to April), followed by a general recovery over the subsequent months. Across all sites, the greatest variability in shoreline patterns was observed outside the high wave season, suggesting that while consistent high-energy waves cause immediate and noticeable shoreline changes, other factors—such as tidal currents, fluctuations in sediment supply, and anthropogenic activity, such as coastal developments or boat traffic—likely contribute to variability throughout the rest of the year.

Carenero, however, deviates from the general seasonal pattern displayed by the other study sites. At this location, the cross-shore distance appears to significantly decrease between June and July, followed by subsequent recovery outside this period. This could be attributed to the unique characteristics of Carenero's protected coastline or other seasonal phenomena that differ from those affecting the more exposed study sites.

Wizard Beach experienced a significant reduction in cross-shore distance in March, which appears to be unique among the study sites. Notably, personal observations indicated that the geomorphology of Playa Wizard was actively changing throughout the study period, distinctly more so than the other sites.

A notable limitation is the potential for error in the shoreline positions extracted for the beaches of Carenero. Given their small size, even a sub-pixel accuracy error (~10 m) could affect the measurements. This potential error must be considered when interpreting the results, as it can likely account for the additional variability within the shoreline position and, subsequently, the cross-sectional distance. More broadly, the use of both Landsat (30 m resolution) and Sentinel-2 (10 m resolution) imagery likely introduced spatial inconsistencies in the shoreline detection. Although CoastSat applies sub-pixel edge detection, the difference in the native resolution between the sensors affects the accuracy of the change detection, particularly in sites with narrow or rapidly changing shorelines. Nonetheless, combining the two sensors extends the coverage of the region, and all imagery was manually screened to ensure quality and comparability across time. This trade-off is addressed further in Section 4.1

3.2. *Integrated Monitoring of Wave Dynamics and Beach Morphology*

3.2.1. OFTV-Derived Wave Characteristics

Figure 4 presents the OFTV-derived velocities (velocity magnitude) over time and the power spectral density for a single video, providing an example of wave energy distribution and flow dynamics at a specific stage in space and time. The site selected for this example was Carenero, chosen for simplicity, as the active wave zone was closer to the shore than at other sites. The velocity magnitude reveals a sinusoidal pattern, where peaks in wave speed indicate incoming waves. This pattern tends to be relatively repetitive over time, suggesting a more stable flow of incoming waves. For this transect, the calculated velocities remained below 1 m/s, representing a relatively low-energy wave climate. Observing the power spectra, the predominant wave energy occurs at a frequency of approximately 0.25 Hz, indicating a wave period of about 4 s. This spectral profile was a benchmark for exploring how the wave energy and flow patterns differed across the study sites. The characteristic frequency distributions and flow intensities were captured by averaging the data for each beach.

3.2.2. Data Verification Against Wave Model Output

Figure 5 compares the peak wave periods from the WWIII model and those derived from the OFTV algorithm based on the footage collected at Playa Wizard. Generally, the two data sources align, suggesting that OFTV is reliable for capturing the dominant wave periods under various conditions. Some differences are noted between the wave model output and the OFTV-derived wave periods. For instance, on 8 March 2024, the OFTV-derived wave period showed an increase from previous measurements, whereas the WWIII wave period showed a decrease from this point. Deviations between the wave periods calculated by these two methods may be explained by phenomena such as the following: (1) the WWIII estimates are provided hourly, while the OFTV-derived periods are calculated instantaneously in situ; and (2) the offshore nature of the singular WWIII grid point likely does not capture local environmental factors (shoaling, refraction, reflection, and interference) influencing wave conditions. OFTV, being directly observational, will likely better capture these local variances, whereas the model provides a larger scale, offshore perspective.

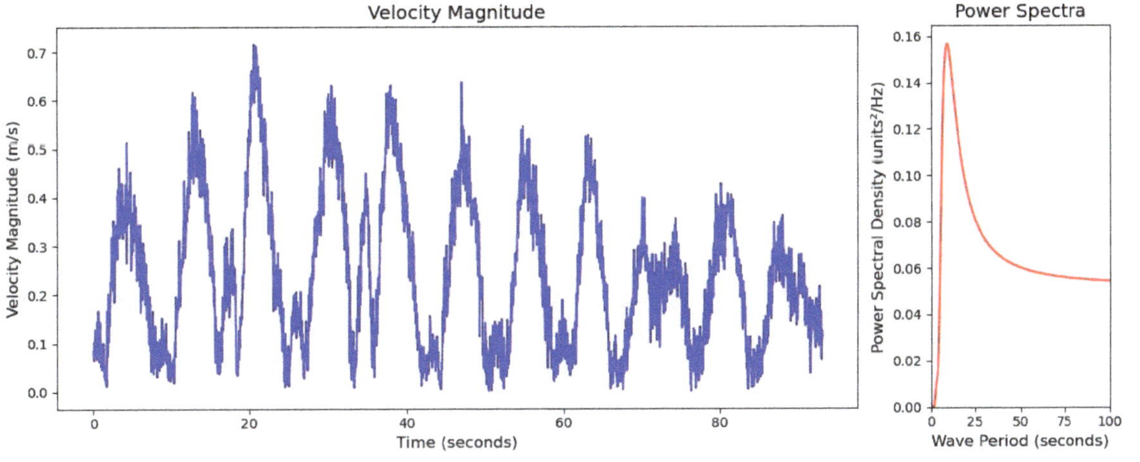

Figure 4. The OFTV-derived wave characteristics for a given transect at Isla Carenero, displaying the average wave velocity over the video (**left**) and the power spectra density estimate calculated by use of the Welch equation (**right**).

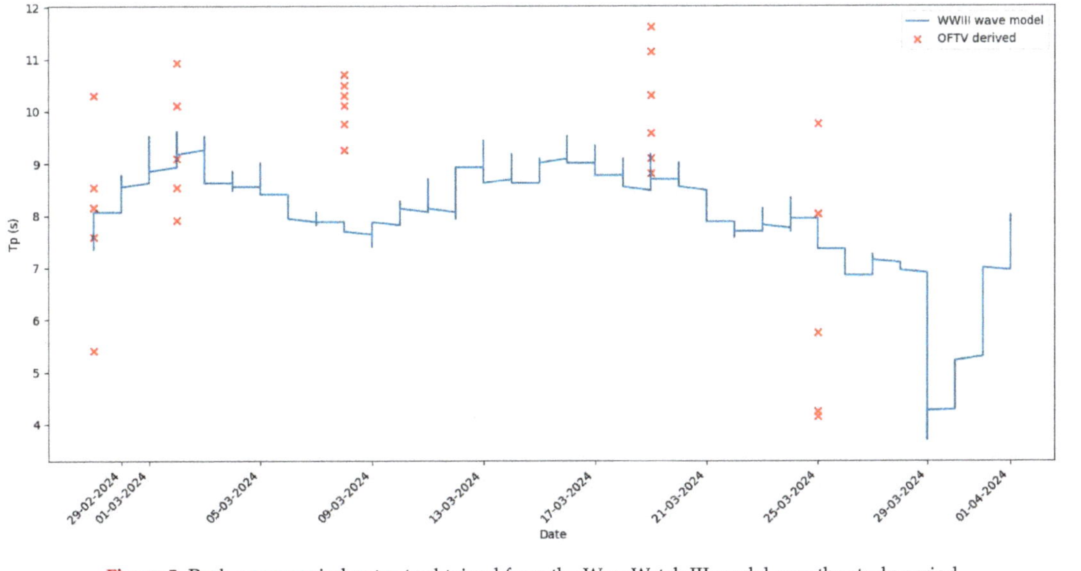

Figure 5. Peak-wave-period outputs obtained from the WaveWatch III model over the study period. Crosses indicate the OFTV-derived wave period from the in situ footage obtained from Playa Wizard.

3.2.3. Summary of Beach Characteristics

Observing Table 1, the beaches are arranged from top to bottom according to the overall accretive/erosive pattern of the change derived from the CoastSat beach width analysis, considering both the calculated rate of change (average ± SD) and the time-series of the change. The satellite data for the four studied beaches reveal a gradient from accretionary to erosional behaviours. In the table, a positive rate of change denotes accretion, while a negative rate indicates erosion. Istmito showed the most significant positive shoreline accretion, sustaining a consistent trend throughout the study period, followed by Wizard and Bluff, both displaying moderate accretion. Carenero exhibited a nearly neutral shoreline change over time, with a slight retreat in the final year of the dataset.

Table 1. A summary of beach characteristics for each study site reflects the changes in the shoreline data observed from the satellite imagery, ranging from more accretive conditions (Istmito) to more recent sediment loss (Carenero), as illustrated in Figure 3. Profile charts in the first column show cross-shore topography (X-axis: cross-shore distance in meters; Y-axis: height from the back-beach in meters). Characteristics are categorised into slope, wave features via OFTV, sediment analysis, and satellite observations (from 2016 to 2024).

	Site Name	OFTV Wave Characteristics		Sediment Characteristics					Satellite
		Wave Velocity Gradient	Wave Period (s)	Mean (mm)	Median (mm)	Sorting (Φ)	Skewness (Φ)	Kurtosis (Φ)	Avg RoC (m/y)
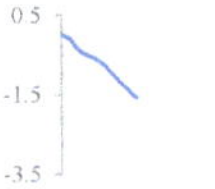	Istmito	6.6 ± 3.5	6.67	0.16	0.15	1.058	0.274	0.859	5.2 ± 5.8
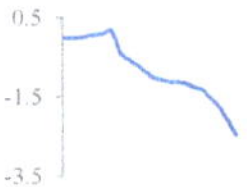	Wizard	4.8 ± 2.2	10.00	0.40	0.38	0.433	0.047	0.941	0.2 ± 7.8
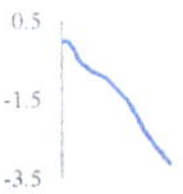	Bluff	4.5 ± 2.4	4.76	0.41	0.40	0.476	−0.282	0.657	0.2 ± 8.3
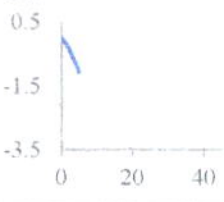	Carenero	3.6 ± 2.4	4.00	0.53	0.41	0.304	−0.677	0.402	0.8 ± 4.5

Sedimentological analysis further supports the shoreline changes derived from satellite observations, revealing systematic differences in the skewness and, to some extent, the mean and sorting across the four beaches. The transition from Istmito to Carenero is a reliable sedimentological indicator of a stable, accretive beach compared to a less stable, recently eroding one. Table 1 also indicates a general systematic decrease in sorting and an increase in grain size from top to bottom. Thus, the more accretionary trend is characterised by a combination of decreasing sorting, positive skew, and finer grain size. Following the classification of [56] kurtosis offers further insight into Bluff and Carenero. These locations exhibit platykurtic distributions alongside well-sorted, negatively skewed, and coarser grain sizes. Such a distribution suggests active sediment transport and redistribution under high-energy conditions, typical of dynamic, erosive coastal environments. Notably, Wizard and Bluff display similar trends in their satellite analysis and grain size statistics, with slight differences in kurtosis and skewness.

The wave velocity analysis shows consistent patterns with sediment characteristics and shoreline trends, highlighting the hydrodynamic forces shaping these environments. A general shift from shoreline progradation to recent retreat corresponds with a systematic decrease in the wave velocity gradient, suggesting that higher rates of beach accretion are linked to more consistent wave velocities. Although wave period data only partially align with this pattern, longer wave periods are observed at the accreting sites (Istmito and Wizard). In contrast, shorter periods are noted at Bluff and Carenero, likely due to localised wave modification near the shore. This inconsistency in the alignment of the wave period may arise from limitations in the temporal sampling of the OFTV data, which might not fully capture the variability of wave conditions. These findings imply that shorter-period waves and more variable wave velocities are less conducive to beach accretion.

Further insights arise from the beach slope data, which reflect each beach's capacity to dissipate wave energy and reinforce observed trends in sediment and shoreline data. Istmito and Wizard have shallower beach slopes, likely promoting wave energy attenuation and supporting accretion. In contrast, Bluff's slightly steeper profile aligns with its transition to more moderate erosion, while Carenero—with the steepest slope and narrowest beach width—experiences minimal wave attenuation, resulting in stronger erosional forces and sediment loss, as indicated by satellite observations and the negative skew in particle size distribution.

4. Discussion

4.1. Efficacy of the Multi-Platform Toolkit for Beach Characterisation

Integrating optical flow tracking velocimetry (OFTV) with the satellite-derived shoreline data, sedimentological measurements, and morphological characteristics provides a powerful multi-faceted approach to beach characterisation. This combination enables a cheap, low-cost depth of analysis that bridges observational data with explanatory insights, revealing potential causal mechanisms within coastal dynamics.

Using OFTV as a direct measure of the wave velocity, coupled with satellite-based shoreline trends from CoastSat, clarifies seasonal and long-term shoreline dynamics. The satellite data alone determine clear accretion and erosion patterns in time and space. Adding the local wave velocity analysis allows us to interpret these trends regarding the hydrodynamic forces driving them. For example, as the results indicate, accretion at sites such as Istmito and Wizard is associated with more consistent wave velocities, while the variable velocities—observed at Carenero and Bluff—correlate with erosional trends. This finding suggests that beaches with more stable hydrodynamic regimes are better able to retain sediments and promote accretion, while those subjected to variable velocities experience higher erosional pressures. However, the extent to which these relationships indicate causality rather than correlation requires further scrutiny. For instance, external factors such as nearshore geomorphology, human intervention, or episodic weather events may also play significant roles and are not fully captured within the current dataset. Future research might include human activity mapping or 3D modelling of shoreline structures using drones to better account for these influences.

OFTV-derived wave characteristics also enhance our understanding of sedimentological attributes such as grain size, skewness, and sorting. Coarser, negatively skewed sediments at Bluff and Carenero correspond with variable wave velocities, suggesting that dynamic wave forces contribute to the ongoing sediment redistribution observed at these sites. Therefore, OFTV offers an indirect yet meaningful perspective on sediment transport mechanisms, which are otherwise difficult or costly to observe directly, particularly in remote tropical locations.

Beach slope measurements add another layer to this characterisation, linking sediment retention with the beaches' capacity to dissipate wave energy. The relatively shallow slopes at Istmito and Wizard appear to dissipate incoming wave energy, reinforcing accretionary trends. At the same time, the steeper profiles of Bluff and Carenero are less effective at attenuating wave forces, enhancing erosion potential. By correlating OFTV wave data with these morphological profiles, we can identify mechanisms that help explain why certain beaches are more susceptible to erosion or conducive to accretion.

Fundamentally, this integrated multi-platform toolkit also assists in overcoming the limitations inherent to each data type. For instance, the seasonal and interannual shoreline fluctuations derived from CoastSat are susceptible to short-term anomalies, particularly on smaller beaches such as Carenero, where the pixel resolution can introduce variability. However, when combined with the OFTV-derived wave patterns, these fluctuations can be better contextualised, distinguishing between anomalous short-term events and sustained morphodynamic responses. The use of OFTV has been effectively employed to track features and their velocities from high-speed video [49,60–62]. Unlike satellite observations, which capture longer-term trends, OFTV enables the direct observation of fine-scale fluctuations in wave energy that significantly influence coastal sediment transport and erosion processes. Furthermore, this framework provides a comparative baseline against predictive wave models, such as WWIII, offering a localised validation tool that accounts for nearshore environmental variability often overlooked by offshore model data.

One methodological consideration in interpreting the satellite-derived shoreline positions is the combined use of Landsat-7 and Sentinel-2 imagery within the CoastSat framework. While CoastSat applies sub-pixel edge detection techniques to improve shoreline delineation, the differing native resolutions between these sensors can introduce differences in positional accuracy that can affect change detection, particularly in narrow or highly dynamic beach environments [44]. This is particularly relevant for sites like Carenero, where the scale of the beach width approaches the resolution threshold, and small shifts in shoreline position may fall within the margin of spatial uncertainty [63]. Nonetheless, combining these datasets was necessary to ensure a greater spatial and temporal coverage of the Bocas del Toro region, especially since this area experiences significant cloud cover, a common limitation of shoreline delineation in tropical regions [56]. To minimise inconsistencies related to resolution, all images were manually reviewed prior to shoreline extraction.

In this context, Sentinel-2 imagery offered more detailed shoreline delineation in recent years, while Landsat-7 offered valuable long-term continuity. Previous studies have quantified the impact of the resolution on the shoreline extraction accuracy. [64] found that the mean positional error for Landsat-derived shorelines was reduced from around 12.9 ± 15.3 m to 3.75 ± 7.0 m, using sub-pixel and georeferencing corrections, as utilised in CoastSat. Sentinel-2, with its finer native resolution, typically produces lower error margins. While variability under 10–15 m should be interpreted cautiously, the observed seasonal and interannual patterns remained consistent across the sensors and were further corroborated by OFTV-derived wave dynamics.

In sum, combining OFTV, sediment, morphological, and satellite-based shoreline data transitions beach characterisation from simple description to explanation. This new integrated approach captures both direct and indirect interactions between hydrodynamic forces and sediment transport, setting a new standard for multi-dimensional coastal analysis. Such comprehensive insights have broad implications for coastal resilience planning, where understanding the nuanced drivers of erosion and accretion can inform targeted interventions for beach preservation and adaptation.

4.2. Interpreting Patterns and Underlying Coastal Dynamics

The integrated dataset provides compelling insight into the processes shaping the accretive and, to some extent, erosive behaviour of the different beaches. By examining the results across the sediment, wave, and shoreline data, distinct patterns emerge that suggest underlying morphodynamic mechanisms at play. The more notable accretion observed as Istmito is associated with more consistent wave velocities, as indicated by the steeper wave velocity gradient derived from the OFTV data—phenomena also observed by [65] in various beaches across the globe. Additionally, wave period analysis provides further evidence supporting these dynamics. Beaches with higher accretionary rates, such as Istmito, generally display longer wave periods, which are typically associated with dominant swell waves generated offshore [66,67]. These longer wave periods promote sediment deposition by allowing finer sediments to settle, resulting in positively skewed grain size distributions and poorer sorting. However, Istmito's location within a bay suggests that its naturally sheltered conditions may influence its wave climate relative to more open beaches, potentially amplifying its accretive characteristics [68,69].

In contrast, less accretive sites characterised by a shallower wave velocity gradient, suggest greater variability in wave forcing. This process results in the deposition of coarser, better-sorted beach sediments with a more symmetrical grain size distribution. Indeed, finer sediments may be transported offshore, leading to negative skewness. Further, a more erosive shoreline trend is associated with variability in wave velocity coupled with a reduced wave period, which is indicative of locally generated waves. This pattern aligns with the OFTV-derived shorter wave periods observed at Carenero, where increased wave frequency likely contributes to intensified sediment transport and erosion, taking finer sediment offshore while leaving behind coarser, well-sorted beach deposits [70,71].

The sediment analysis results complement the wave velocity and wave period measurements findings and provide insight into local morphodynamic interactions. For example, the sediment profile at Istmito displays finer, well-sorted grains with a positive skew, indicative of low-energy conditions conducive to accretion and a dissipative beach profile [72]. Conversely, the sediment statistics at Carenero are characteristic of more erosional conditions, with the steeper, shorter profile being more reflective [72].

The seasonal trends in wave climate also play an important role. At Playa Wizard and Playa Bluff, seasonal wave variations lead to cyclic patterns of erosion and recovery. During the high-energy wave season (January to April), wave forces lead to erosion, followed by sediment deposition and shoreline recovery in the subsequent months when wave energy reduces. However, Carenero deviates from this pattern, significantly reducing cross-shore distance during June and July when other beaches recover. While the specific conditions pertaining to this anomaly are not explored here, seasonal differences in sediment transport are likely the result of localised phenomena such as offshore reef formations, seagrasses, or trade winds [73,74]. Future research should seek to integrate these observations with bathymetric and reef topography data, where available, to better understand their influence on the nearshore wave climate.

5. Conclusions

This study highlights the utility of combining satellite-based shoreline data, sedimentological and morphological analyses, and wave velocity variability and wave period derived from OFTV into a multi-platform toolkit for characterising coastal morphodynamics across four diverse beach ecosystems during the high-energy wave season. Long-term trends and short-term fluctuations in shoreline positions reveal complex interactions between hydrodynamics, sedimentology, and beach morphology.

Several key insights emerged by combining these easy-to-use methods and easy-to-collect datasets into a toolkit. Beaches that exhibit accretionary behaviour, such as Istmito, have more consistent wave velocities; longer wave periods; finer, poorly sorted, positively skewed sediments; and shallow, wider beach profiles. These characteristics indicate swell-dominated conditions operating across a dissipative beach profile. In contrast, Isla Carenero shows an erosional trend, characterised by variable wave velocities and shorter wave periods, coarser, negatively skewed, and well-sorted sediments. These properties suggest more locally influenced wave hydrodynamics at work over a more reflective beach profile.

Field testing of this new multi-platform toolkit in real-world conditions demonstrates significant advantages in capturing multiscale dynamics, bridging the gap between fine-scale local conditions and broader, long-term shoreline changes. Although there are limitations, such as sub-pixel errors in the satellite-derived shoreline trends on smaller beaches, integrating these datasets fosters a deeper understanding of coastal morphodynamics, particularly for remote tropical beaches. The development of this practical, low-cost toolkit for a comprehensive assessment of local beach dynamics offers a valuable and affordable foundation for targeted, evidence-based management interventions in dynamic coastal systems. Its adaptable, scalable design also makes it particularly well suited for replication in other developing nations facing similar logistical and environmental challenges.

Author Contributions: Conceptualisation, J.M., A.J.P., R.C., J.E.H. and K.E.C.; methodology, J.M., J.E.H. and A.J.P.; software, J.M. and J.E.H.; validation, J.M. and J.E.H.; formal analysis, J.M., J.E.H. and A.J.P.; investigation, J.M., J.E.H. and A.J.P.; resources, R.C.; data curation, J.M. and J.E.H.; writing—original draft preparation, J.M.; writing—review and editing, J.M., J.E.H., A.J.P., R.C. and K.E.C.; visualisation, J.M. and J.E.H.; supervision, J.E.H., A.J.P. and K.E.C.; project administration, J.M., K.E.C. and R.C.; funding acquisition, J.E.H., A.J.P., R.C. and K.E.C. All authors have read and agreed to the published version of the manuscript.

Funding: This research was funded by the Adrienne Arsht Community-Based Resilience Solutions Initiative at the Smithsonian Institution. 2024.

Data Availability Statement: Dataset available upon request from the authors.

Conflicts of Interest: The authors declare no conflicts of interest.

Appendix A

Table A1. The date, time, and duration of each video recording. Subtables represent (a) Playa Bluff, (b) Playa Istmito, (c) Playa Wizard, and (d) Isla Carenero.

	(a)	
	Playa Bluff	
Date	**Time**	**Duration (s)**
29 February 2024	10:04	126
29 February 2024	10:06	124
29 February 2024	10:17	103
29 February 2024	10:19	109
29 February 2024	09:55	100
29 February 2024	09:59	127
29 February 2024	10:32	121
29 February 2024	10:36	126
7 March 2024	09:50	119
7 March 2024	09:59	122
7 March 2024	10:25	110

Table A1. *Cont.*

7 March 2024	10:29	145
7 March 2024	11:04	104
7 March 2024	11:09	136
13 March 2024	10:04	100
13 March 2024	10:07	71
13 March 2024	10:27	92
13 March 2024	10:29	93
13 March 2024	10:48	92
13 March 2024	10:50	98
13 March 2024	11:05	91
13 March 2024	11:07	92
22 March 2024	08:38	92
22 March 2024	08:40	93
22 March 2024	09:17	92
22 March 2024	09:19	92
22 March 2024	09:54	93
22 March 2024	09:57	92
22 March 2024	10:17	92
22 March 2024	10:19	93
28 March 2024	08:13	97
28 March 2024	08:17	94
28 March 2024	08:36	93
28 March 2024	08:38	99
28 March 2024	08:59	95
28 March 2024	09:01	101
28 March 2024	09:15	95
	(b)	
	Playa Istmito	
Date	**Time**	**Duration (s)**
28 March 2024	09:21	93
1 March 2024	11:15	106
1 March 2024	11:24	77
1 March 2024	11:27	121
11 March 2024	16:15	93
11 March 2024	16:19	91
11 March 2024	16:21	93
11 March 2024	16:29	93
11 March 2024	16:31	96
11 March 2024	16:43	92
11 March 2024	16:45	93
18 March 2024	16:01	91
18 March 2024	16:06	91
18 March 2024	16:27	91

Table A1. *Cont.*

18 March 2024	16:30	94
18 March 2024	16:52	94
18 March 2024	16:54	91
29 March 2024	16:56	105
29 March 2024	16:59	94
29 March 2024	17:16	91
	(c)	
	Playa Wizard	
Date	**Time**	**Duration (s)**
29 March 2024	17:18	92
28 February 2024	13:49	100
28 February 2024	13:56	122
28 February 2024	14:09	91
28 February 2024	14:14	105
28 February 2024	14:23	82
28 February 2024	14:27	103
8 March 2024	09:30	93
8 March 2024	09:31	102
8 March 2024	09:42	101
8 March 2024	09:44	94
8 March 2024	09:53	120
8 March 2024	09:59	102
19 March 2024	09:57	92
19 March 2024	09:59	99
19 March 2024	10:09	93
19 March 2024	10:11	92
19 March 2024	10:26	92
19 March 2024	10:28	92
25 March 2024	09:26	92
25 March 2024	09:28	90
25 March 2024	09:59	91
25 March 2024	10:01	92
25 March 2024	10:34	93
25 March 2024	10:36	121
2 April 2024	12:44	106
2 April 2024	12:48	86
2 April 2024	13:14	91
2 April 2024	13:16	92
2 April 2024	13:29	91
	(d)	
	Isla Carenero	
Date	**Time**	**Duration (s)**
2 April 2024	13:31	91
30 January 2024	09:09	22

Table A1. *Cont.*

27 February 2024	15:46	88
27 February 2024	15:57	53
27 February 2024	16:05	31
12 March 2024	12:32	74
12 March 2024	12:41	94
12 March 2024	12:53	74
21 March 2024	10:30	92
21 March 2024	10:53	94
21 March 2024	11:06	92
27 March 2024	08:47	92
27 March 2024	09:00	93
27 March 2024	09:05	92

Table A2. Every satellite observation used for the CoastSat analysis. S2 = Sentinel-2; L7 = Landsat-7.

Site	Date	Cloud Cover	Satellite	Resolution (m)
BLUFF	2016-07-19 15:51:12+00:00	0.11052476	L7	30
BLUFF	2016-09-21 15:50:58+00:00	0	L7	30
BLUFF	2016-11-04 15:55:23+00:00	0	S2	10
BLUFF	2016-11-08 15:51:14+00:00	0	L7	30
BLUFF	2017-04-13 15:57:55+00:00	0	S2	10
BLUFF	2017-05-19 15:50:44+00:00	0	L7	30
BLUFF	2017-05-23 15:58:50+00:00	0	S2	10
BLUFF	2017-09-08 15:51:14+00:00	0	L7	30
BLUFF	2017-11-11 15:50:50+00:00	0	L7	30
BLUFF	2017-11-14 15:58:56+00:00	0.05404932	S2	10
BLUFF	2018-04-08 15:55:57+00:00	0	S2	10
BLUFF	2018-05-06 15:48:59+00:00	0.15043509	L7	30
BLUFF	2018-05-18 15:57:58+00:00	0.17213233	S2	10
BLUFF	2018-05-23 15:56:34+00:00	0	S2	10
BLUFF	2018-06-12 15:55:23+00:00	0	S2	10
BLUFF	2018-08-16 15:55:24+00:00	0	S2	10
BLUFF	2018-09-10 15:55:16+00:00	0.00083375	S2	10
BLUFF	2018-09-11 15:46:37+00:00	0.01755119	L7	30
BLUFF	2018-09-25 15:55:50+00:00	0.0130723	S2	10
BLUFF	2018-09-27 15:46:15+00:00	0	L7	30
BLUFF	2018-09-30 15:57:15+00:00	0	S2	10
BLUFF	2018-10-15 15:55:22+00:00	0.0012043	S2	10
BLUFF	2019-02-12 16:00:42+00:00	0	S2	10
BLUFF	2019-02-17 16:00:45+00:00	0.19530221	S2	10
BLUFF	2019-03-09 16:00:43+00:00	0.02988101	S2	10
BLUFF	2019-04-03 16:00:46+00:00	0	S2	10
BLUFF	2019-04-13 16:00:48+00:00	0.30771369	S2	10
BLUFF	2019-05-23 16:00:49+00:00	0	S2	10
BLUFF	2019-06-02 16:00:48+00:00	0.00605237	S2	10

Table A2. *Cont.*

Site	Date	Cloud Cover	Satellite	Resolution (m)
BLUFF	2019-07-12 16:00:51+00:00	0.00241889	S2	10
BLUFF	2019-07-17 16:00:54+00:00	0	S2	10
BLUFF	2019-08-31 16:00:46+00:00	0	S2	10
BLUFF	2020-04-17 16:00:46+00:00	0.00205863	S2	10
BLUFF	2020-07-14 15:18:32+00:00	0.01264976	L7	30
BLUFF	2020-07-16 16:00:50+00:00	0	S2	10
BLUFF	2020-08-20 16:00:49+00:00	0.25062788	S2	10
BLUFF	2020-08-31 15:15:06+00:00	0	L7	30
BLUFF	2020-09-19 16:00:47+00:00	0	S2	10
BLUFF	2020-10-04 16:00:51+00:00	0.03706563	S2	10
BLUFF	2020-10-09 16:00:49+00:00	0	S2	10
BLUFF	2021-02-06 16:00:45+00:00	0	S2	10
BLUFF	2021-03-03 16:00:45+00:00	0	S2	10
BLUFF	2021-05-17 16:00:47+00:00	0.00169837	S2	10
BLUFF	2021-06-01 16:00:46+00:00	0.48851285	S2	10
BLUFF	2021-07-16 16:00:47+00:00	0.00313941	S2	10
BLUFF	2021-09-03 14:46:42+00:00	0	L7	30
BLUFF	2021-09-19 14:45:41+00:00	0	L7	30
BLUFF	2021-10-05 14:43:52+00:00	0	L7	30
BLUFF	2022-04-07 16:00:46+00:00	0.15902915	S2	10
BLUFF	2022-05-22 16:00:46+00:00	0.18941453	S2	10
BLUFF	2022-06-06 16:00:53+00:00	0	S2	10
BLUFF	2022-07-06 16:00:58+00:00	0.10995142	S2	10
BLUFF	2022-07-16 14:16:43+00:00	0.25490196	L7	30
BLUFF	2022-07-16 16:00:58+00:00	0	S2	10
BLUFF	2022-08-30 16:00:46+00:00	0.44499341	S2	10
BLUFF	2022-09-09 16:00:47+00:00	0.0313941	S2	10
BLUFF	2022-10-04 16:00:52+00:00	0.03369977	S2	10
BLUFF	2022-10-14 16:00:50+00:00	0.1197402	S2	10
BLUFF	2022-10-24 16:00:48+00:00	0	S2	10
BLUFF	2023-03-18 16:00:49+00:00	0.00428195	S2	10
BLUFF	2023-04-12 16:00:46+00:00	0	S2	10
BLUFF	2023-04-22 16:00:44+00:00	0.0002985	S2	10
BLUFF	2023-04-27 16:00:47+00:00	0	S2	10
BLUFF	2023-05-12 16:00:45+00:00	0.01034461	S2	10
BLUFF	2023-05-17 16:00:50+00:00	0.00062788	S2	10
BLUFF	2023-05-27 16:00:50+00:00	0	S2	10
BLUFF	2023-06-11 16:00:49+00:00	0	S2	10
BLUFF	2023-06-12 13:37:30+00:00	0	L7	30
BLUFF	2023-07-06 16:00:51+00:00	0	S2	10
BLUFF	2023-07-09 13:37:32+00:00	0	L7	30
BLUFF	2023-08-25 16:00:53+00:00	0	S2	10
BLUFF	2023-08-30 16:00:51+00:00	0.08399209	S2	10
CARENERO	2016-01-09 15:49:47+00:00	0	L7	30

Table A2. *Cont.*

Site	Date	Cloud Cover	Satellite	Resolution (m)
CARENERO	2016-04-14 15:50:29+00:00	0.05726634	L7	30
CARENERO	2016-07-19 15:51:12+00:00	0	L7	30
CARENERO	2016-09-21 15:51:22+00:00	0	L7	30
CARENERO	2016-11-04 15:55:23+00:00	0	S2	10
CARENERO	2016-11-08 15:50:50+00:00	0	L7	30
CARENERO	2017-01-11 15:50:25+00:00	0.01381065	L7	30
CARENERO	2017-04-13 15:57:55+00:00	0	S2	10
CARENERO	2017-05-23 15:58:50+00:00	0	S2	10
CARENERO	2017-07-17 15:55:25+00:00	0	S2	10
CARENERO	2017-08-31 15:59:09+00:00	0.01699219	S2	10
CARENERO	2017-11-11 15:51:14+00:00	0	L7	30
CARENERO	2017-12-24 15:55:13+00:00	0	S2	10
CARENERO	2017-12-29 16:00:33+00:00	0	S2	10
CARENERO	2018-02-02 15:58:36+00:00	0	S2	10
CARENERO	2018-02-12 15:57:10+00:00	0	S2	10
CARENERO	2018-03-03 15:49:34+00:00	0	L7	30
CARENERO	2018-03-04 15:56:09+00:00	0	S2	10
CARENERO	2018-03-29 15:55:24+00:00	0	S2	10
CARENERO	2018-04-03 16:00:35+00:00	0.01523438	S2	10
CARENERO	2018-04-08 15:55:57+00:00	0	S2	10
CARENERO	2018-04-28 15:55:27+00:00	0	S2	10
CARENERO	2018-05-06 15:48:59+00:00	0	L7	30
CARENERO	2018-06-12 15:55:23+00:00	0	S2	10
CARENERO	2018-08-16 15:55:24+00:00	0	S2	10
CARENERO	2018-09-10 15:55:16+00:00	0	S2	10
CARENERO	2018-09-27 15:46:15+00:00	0	L7	30
CARENERO	2018-10-10 15:56:19+00:00	0	S2	10
CARENERO	2018-11-04 15:55:21+00:00	0	S2	10
CARENERO	2018-11-24 16:00:38+00:00	0	S2	10
CARENERO	2018-12-29 16:00:42+00:00	0	S2	10
CARENERO	2019-01-13 16:00:41+00:00	0	S2	10
CARENERO	2019-01-18 16:00:45+00:00	0	S2	10
CARENERO	2019-02-12 16:00:42+00:00	0	S2	10
CARENERO	2019-02-18 15:41:57+00:00	0.20694018	L7	30
CARENERO	2019-03-09 16:00:43+00:00	0	S2	10
CARENERO	2019-04-03 16:00:46+00:00	0	S2	10
CARENERO	2019-04-08 16:00:49+00:00	0.0078125	S2	10
CARENERO	2019-05-18 16:00:54+00:00	0	S2	10
CARENERO	2019-05-23 16:00:49+00:00	0	S2	10
CARENERO	2019-06-07 16:00:53+00:00	0	S2	10
CARENERO	2019-07-17 16:00:54+00:00	0	S2	10
CARENERO	2019-08-31 16:00:46+00:00	0	S2	10
CARENERO	2019-09-20 16:00:45+00:00	0	S2	10
CARENERO	2019-09-25 16:00:45+00:00	0	S2	10

Table A2. *Cont.*

Site	Date	Cloud Cover	Satellite	Resolution (m)
CARENERO	2019-09-30 15:33:26+00:00	0	L7	30
CARENERO	2019-11-01 15:32:28+00:00	0.10612691	L7	30
CARENERO	2019-11-04 16:00:46+00:00	0	S2	10
CARENERO	2019-11-24 16:00:42+00:00	0	S2	10
CARENERO	2020-02-07 16:00:37+00:00	0	S2	10
CARENERO	2020-02-21 15:26:31+00:00	0	L7	30
CARENERO	2020-02-22 16:00:42+00:00	0	S2	10
CARENERO	2020-03-08 15:26:03+00:00	0	L7	30
CARENERO	2020-03-08 16:00:41+00:00	0	S2	10
CARENERO	2020-03-13 16:00:43+00:00	0	S2	10
CARENERO	2020-03-18 16:00:42+00:00	0	S2	10
CARENERO	2020-04-07 16:00:43+00:00	0	S2	10
CARENERO	2020-05-12 16:00:44+00:00	0.0656901	S2	10
CARENERO	2020-07-14 15:18:08+00:00	0	L7	30
CARENERO	2020-07-16 16:00:50+00:00	0.02604167	S2	10
CARENERO	2020-08-31 15:15:30+00:00	0	L7	30
CARENERO	2020-09-19 16:00:47+00:00	0	S2	10
CARENERO	2020-10-02 15:13:24+00:00	0	L7	30
CARENERO	2020-10-04 16:00:51+00:00	0	S2	10
CARENERO	2020-10-09 16:00:49+00:00	0	S2	10
CARENERO	2021-01-07 16:00:45+00:00	0	S2	10
CARENERO	2021-01-17 16:00:45+00:00	0	S2	10
CARENERO	2021-01-27 16:00:45+00:00	0	S2	10
CARENERO	2021-02-06 16:00:45+00:00	0	S2	10
CARENERO	2021-03-03 16:00:45+00:00	0	S2	10
CARENERO	2021-03-18 16:00:45+00:00	0	S2	10
CARENERO	2021-03-28 16:00:44+00:00	0	S2	10
CARENERO	2021-05-17 16:00:47+00:00	0.27552083	S2	10
CARENERO	2021-09-19 14:45:41+00:00	0	L7	30
CARENERO	2021-09-29 16:00:50+00:00	0	S2	10
CARENERO	2021-10-21 14:42:51+00:00	0	L7	30
CARENERO	2021-12-28 16:00:47+00:00	0	S2	10
CARENERO	2022-01-25 14:33:14+00:00	0	L7	30
CARENERO	2022-02-01 16:00:40+00:00	0	S2	10
CARENERO	2022-02-06 16:00:45+00:00	0	S2	10
CARENERO	2022-02-16 16:00:46+00:00	0.03990885	S2	10
CARENERO	2022-02-21 16:00:40+00:00	0	S2	10
CARENERO	2022-03-08 16:00:49+00:00	0	S2	10
CARENERO	2022-03-14 14:28:28+00:00	0	L7	30
CARENERO	2022-04-07 16:00:46+00:00	0	S2	10
CARENERO	2022-05-22 16:00:46+00:00	0	S2	10
CARENERO	2022-05-26 14:19:29+00:00	0	L7	30
CARENERO	2022-06-06 16:00:53+00:00	0	S2	10
CARENERO	2022-08-30 16:00:46+00:00	0.00214844	S2	10

Table A2. *Cont.*

Site	Date	Cloud Cover	Satellite	Resolution (m)
CARENERO	2022-10-29 16:00:44+00:00	0	S2	10
CARENERO	2022-12-08 16:00:42+00:00	0	S2	10
CARENERO	2022-12-28 16:00:45+00:00	0	S2	10
CARENERO	2023-01-07 16:00:43+00:00	0	S2	10
CARENERO	2023-02-01 16:00:44+00:00	0	S2	10
CARENERO	2023-02-11 16:00:44+00:00	0	S2	10
CARENERO	2023-03-18 16:00:49+00:00	0.43307292	S2	10
CARENERO	2023-04-12 16:00:46+00:00	0	S2	10
CARENERO	2023-04-17 16:00:47+00:00	0	S2	10
CARENERO	2023-04-22 16:00:44+00:00	0	S2	10
CARENERO	2023-04-27 16:00:47+00:00	0	S2	10
CARENERO	2023-04-29 13:43:00+00:00	0	L7	30
CARENERO	2023-05-04 13:46:58+00:00	0	L7	30
CARENERO	2023-05-12 16:00:45+00:00	0	S2	10
CARENERO	2023-05-17 16:00:50+00:00	0	S2	10
CARENERO	2023-05-27 16:00:50+00:00	0	S2	10
CARENERO	2023-06-11 16:00:49+00:00	0	S2	10
CARENERO	2023-07-06 16:00:51+00:00	0.02415365	S2	10
CARENERO	2023-07-09 13:37:32+00:00	0	L7	30
CARENERO	2023-07-21 16:00:50+00:00	0	S2	10
CARENERO	2023-08-25 16:00:53+00:00	0	S2	10
CARENERO	2023-10-09 16:00:45+00:00	0	S2	10
CARENERO	2023-10-14 16:00:48+00:00	0.1750651	S2	10
CARENERO	2023-11-13 16:00:46+00:00	0	S2	10
CARENERO	2023-12-03 16:00:42+00:00	0	S2	10
ISTMITO	2016-01-09 15:50:11+00:00	0.03811944	L7	30
ISTMITO	2016-07-19 15:50:48+00:00	0	L7	30
ISTMITO	2016-11-04 15:55:23+00:00	0	S2	10
ISTMITO	2016-11-08 15:50:50+00:00	0	L7	30
ISTMITO	2016-12-24 15:59:57+00:00	0	S2	10
ISTMITO	2017-01-11 15:50:49+00:00	0	L7	30
ISTMITO	2017-01-23 15:55:18+00:00	0	S2	10
ISTMITO	2017-04-13 15:57:55+00:00	0.05828841	S2	10
ISTMITO	2017-05-19 15:51:07+00:00	0	L7	30
ISTMITO	2017-05-23 15:58:50+00:00	0	S2	10
ISTMITO	2017-07-17 15:55:25+00:00	0	S2	10
ISTMITO	2017-08-31 15:59:09+00:00	0	S2	10
ISTMITO	2017-11-11 15:50:50+00:00	0	L7	30
ISTMITO	2017-12-24 15:55:13+00:00	0.00269542	S2	10
ISTMITO	2017-12-29 15:50:49+00:00	0	L7	30
ISTMITO	2017-12-29 16:00:33+00:00	0	S2	10
ISTMITO	2018-02-02 15:58:36+00:00	0	S2	10
ISTMITO	2018-02-12 15:57:10+00:00	0	S2	10
ISTMITO	2018-03-03 15:49:34+00:00	0	L7	30

Table A2. *Cont.*

Site	Date	Cloud Cover	Satellite	Resolution (m)
ISTMITO	2018-03-04 15:56:09+00:00	0	S2	10
ISTMITO	2018-03-29 15:55:24+00:00	0.00269542	S2	10
ISTMITO	2018-04-08 15:55:57+00:00	0	S2	10
ISTMITO	2018-04-28 15:55:27+00:00	0	S2	10
ISTMITO	2018-05-18 15:57:58+00:00	0	S2	10
ISTMITO	2018-05-22 15:48:20+00:00	0.25753769	L7	30
ISTMITO	2018-05-23 15:56:34+00:00	0	S2	10
ISTMITO	2018-06-12 15:55:23+00:00	0	S2	10
ISTMITO	2018-08-16 15:55:24+00:00	0	S2	10
ISTMITO	2018-09-10 15:55:16+00:00	0	S2	10
ISTMITO	2018-09-11 15:46:14+00:00	0	L7	30
ISTMITO	2018-09-25 15:55:50+00:00	0	S2	10
ISTMITO	2018-09-27 15:45:51+00:00	0.281652	L7	30
ISTMITO	2018-10-10 15:56:19+00:00	0	S2	10
ISTMITO	2018-10-15 15:55:22+00:00	0.00516622	S2	10
ISTMITO	2018-11-04 15:55:21+00:00	0.06738544	S2	10
ISTMITO	2018-11-09 16:00:43+00:00	0.02021563	S2	10
ISTMITO	2018-11-24 16:00:38+00:00	0	S2	10
ISTMITO	2018-12-19 16:00:40+00:00	0	S2	10
ISTMITO	2018-12-24 16:00:38+00:00	0	S2	10
ISTMITO	2019-01-13 16:00:41+00:00	0	S2	10
ISTMITO	2019-01-18 16:00:45+00:00	0	S2	10
ISTMITO	2019-02-12 16:00:42+00:00	0	S2	10
ISTMITO	2019-03-09 16:00:43+00:00	0	S2	10
ISTMITO	2019-03-24 16:00:44+00:00	0	S2	10
ISTMITO	2019-04-08 16:00:49+00:00	0	S2	10
ISTMITO	2019-04-13 16:00:48+00:00	0	S2	10
ISTMITO	2019-05-18 16:00:54+00:00	0	S2	10
ISTMITO	2019-05-23 16:00:49+00:00	0	S2	10
ISTMITO	2019-06-02 16:00:48+00:00	0	S2	10
ISTMITO	2019-07-12 15:36:48+00:00	0.25826972	L7	30
ISTMITO	2019-07-17 16:00:54+00:00	0	S2	10
ISTMITO	2019-08-31 16:00:46+00:00	0	S2	10
ISTMITO	2019-09-20 16:00:45+00:00	0	S2	10
ISTMITO	2019-09-25 16:00:45+00:00	0	S2	10
ISTMITO	2019-09-30 15:33:50+00:00	0.16372796	L7	30
ISTMITO	2019-11-01 15:32:05+00:00	0	L7	30
ISTMITO	2019-11-04 16:00:46+00:00	0	S2	10
ISTMITO	2020-02-07 16:00:37+00:00	0	S2	10
ISTMITO	2020-02-21 15:26:54+00:00	0	L7	30
ISTMITO	2020-02-22 16:00:42+00:00	0	S2	10
ISTMITO	2020-03-08 15:25:39+00:00	0	L7	30
ISTMITO	2020-03-13 16:00:43+00:00	0	S2	10
ISTMITO	2020-03-18 16:00:42+00:00	0.02493261	S2	10
ISTMITO	2020-07-14 15:18:08+00:00	0	L7	30

Table A2. *Cont.*

Site	Date	Cloud Cover	Satellite	Resolution (m)
ISTMITO	2020-08-31 15:15:30+00:00	0	L7	30
ISTMITO	2020-09-19 16:00:47+00:00	0	S2	10
ISTMITO	2020-10-02 15:13:00+00:00	0	L7	30
ISTMITO	2020-10-04 16:00:51+00:00	0	S2	10
ISTMITO	2020-10-09 16:00:49+00:00	0	S2	10
ISTMITO	2020-11-08 16:00:47+00:00	0	S2	10
ISTMITO	2020-12-03 16:00:44+00:00	0	S2	10
ISTMITO	2020-12-05 15:09:03+00:00	0	L7	30
ISTMITO	2020-12-18 16:00:41+00:00	0	S2	10
ISTMITO	2021-01-06 15:06:43+00:00	0	L7	30
ISTMITO	2021-01-07 16:00:45+00:00	0	S2	10
ISTMITO	2021-01-27 16:00:45+00:00	0	S2	10
ISTMITO	2021-02-06 16:00:45+00:00	0	S2	10
ISTMITO	2021-02-26 16:00:44+00:00	0	S2	10
ISTMITO	2021-03-03 16:00:45+00:00	0	S2	10
ISTMITO	2021-03-11 15:01:34+00:00	0	L7	30
ISTMITO	2021-03-18 16:00:45+00:00	0	S2	10
ISTMITO	2021-06-15 14:53:59+00:00	0	L7	30
ISTMITO	2021-08-02 14:49:54+00:00	0	L7	30
ISTMITO	2021-09-03 14:46:42+00:00	0	L7	30
ISTMITO	2021-09-19 14:45:41+00:00	0	L7	30
ISTMITO	2021-09-19 16:00:48+00:00	0	S2	10
ISTMITO	2021-09-29 16:00:50+00:00	0	S2	10
ISTMITO	2021-10-05 14:44:16+00:00	0	L7	30
ISTMITO	2021-10-21 14:42:27+00:00	0	L7	30
ISTMITO	2022-01-12 16:00:41+00:00	0	S2	10
ISTMITO	2022-01-25 14:33:14+00:00	0	L7	30
ISTMITO	2022-01-27 16:00:46+00:00	0	S2	10
ISTMITO	2022-02-01 16:00:40+00:00	0	S2	10
ISTMITO	2022-03-14 14:28:28+00:00	0	L7	30
ISTMITO	2022-05-12 16:00:44+00:00	0.12297844	S2	10
ISTMITO	2022-05-22 16:00:46+00:00	0	S2	10
ISTMITO	2022-06-06 16:00:53+00:00	0	S2	10
ISTMITO	2022-09-09 16:00:47+00:00	0	S2	10
ISTMITO	2022-10-24 16:00:48+00:00	0.06738544	S2	10
ISTMITO	2022-10-29 16:00:44+00:00	0	S2	10
ISTMITO	2022-12-08 16:00:42+00:00	0	S2	10
ISTMITO	2022-12-23 16:00:43+00:00	0	S2	10
ISTMITO	2022-12-28 16:00:45+00:00	0	S2	10
ISTMITO	2023-02-01 16:00:44+00:00	0	S2	10
ISTMITO	2023-02-11 16:00:44+00:00	0	S2	10
ISTMITO	2023-04-12 16:00:46+00:00	0	S2	10
ISTMITO	2023-04-17 16:00:47+00:00	0.0791779	S2	10
ISTMITO	2023-04-27 16:00:47+00:00	0.00516622	S2	10
ISTMITO	2023-05-12 16:00:45+00:00	0	S2	10

Table A2. *Cont.*

Site	Date	Cloud Cover	Satellite	Resolution (m)
ISTMITO	2023-05-17 16:00:50+00:00	0	S2	10
ISTMITO	2023-05-27 16:00:50+00:00	0	S2	10
ISTMITO	2023-06-11 16:00:49+00:00	0	S2	10
ISTMITO	2023-06-12 13:37:30+00:00	0	L7	30
ISTMITO	2023-07-09 13:37:32+00:00	0	L7	30
ISTMITO	2023-08-25 16:00:53+00:00	0	S2	10
ISTMITO	2023-08-30 16:00:51+00:00	0	S2	10
ISTMITO	2023-10-09 16:00:45+00:00	0	S2	10
ISTMITO	2023-10-14 16:00:48+00:00	0	S2	10
ISTMITO	2023-11-13 16:00:46+00:00	0	S2	10
ISTMITO	2023-12-03 16:00:42+00:00	0	S2	10
WIZARD	2016-03-19 15:57:10+00:00	0.2044108	S2	10
WIZARD	2016-04-14 15:50:53+00:00	0	L7	30
WIZARD	2016-07-19 15:50:48+00:00	0.27461904	L7	30
WIZARD	2016-08-04 15:51:15+00:00	0	L7	30
WIZARD	2016-09-21 15:51:22+00:00	0	L7	30
WIZARD	2016-11-04 15:55:23+00:00	0	S2	10
WIZARD	2016-11-08 15:50:50+00:00	0	L7	30
WIZARD	2017-01-11 15:50:25+00:00	0	L7	30
WIZARD	2017-01-23 15:55:18+00:00	0.0055135	S2	10
WIZARD	2017-04-13 15:57:55+00:00	0	S2	10
WIZARD	2017-05-19 15:50:44+00:00	0	L7	30
WIZARD	2017-05-23 15:58:50+00:00	0	S2	10
WIZARD	2017-06-04 15:50:48+00:00	0.28098897	L7	30
WIZARD	2017-07-17 15:55:25+00:00	0.00038402	S2	10
WIZARD	2017-08-31 15:59:09+00:00	0.06067588	S2	10
WIZARD	2017-09-20 15:55:21+00:00	0.46938776	S2	10
WIZARD	2017-11-11 15:50:50+00:00	0	L7	30
WIZARD	2017-12-24 15:55:13+00:00	0	S2	10
WIZARD	2017-12-29 15:50:26+00:00	0.0163688	L7	30
WIZARD	2018-01-13 16:00:02+00:00	0.29731731	S2	10
WIZARD	2018-02-12 15:57:10+00:00	0	S2	10
WIZARD	2018-03-03 15:49:34+00:00	0	L7	30
WIZARD	2018-03-04 15:56:09+00:00	0	S2	10
WIZARD	2018-03-19 15:49:22+00:00	0	L7	30
WIZARD	2018-03-19 15:55:22+00:00	0.00290761	S2	10
WIZARD	2018-03-29 15:55:24+00:00	0	S2	10
WIZARD	2018-04-08 15:55:57+00:00	0	S2	10
WIZARD	2018-05-06 15:48:59+00:00	0.14407129	L7	30
WIZARD	2018-05-18 15:57:58+00:00	0	S2	10
WIZARD	2018-05-22 15:48:20+00:00	0.03667938	L7	30
WIZARD	2018-05-23 15:56:34+00:00	0	S2	10
WIZARD	2018-06-12 15:55:23+00:00	0	S2	10
WIZARD	2018-08-16 15:55:24+00:00	0	S2	10
WIZARD	2018-09-10 15:55:16+00:00	0	S2	10

Table A2. *Cont.*

Site	Date	Cloud Cover	Satellite	Resolution (m)
WIZARD	2018-09-25 15:55:50+00:00	0	S2	10
WIZARD	2018-09-27 15:45:51+00:00	0	L7	30
WIZARD	2018-10-10 15:56:19+00:00	0	S2	10
WIZARD	2018-10-15 15:55:22+00:00	0.19541365	S2	10
WIZARD	2018-11-04 15:55:21+00:00	0	S2	10
WIZARD	2018-11-09 16:00:43+00:00	0.01280996	S2	10
WIZARD	2018-11-24 16:00:38+00:00	0	S2	10
WIZARD	2018-12-14 16:00:37+00:00	0.0696456	S2	10
WIZARD	2018-12-19 16:00:40+00:00	0	S2	10
WIZARD	2019-01-01 15:43:26+00:00	0	L7	30
WIZARD	2019-01-13 16:00:41+00:00	0	S2	10
WIZARD	2019-01-18 16:00:45+00:00	0	S2	10
WIZARD	2019-02-12 16:00:42+00:00	0	S2	10
WIZARD	2019-02-18 15:41:57+00:00	0.06611433	L7	30
WIZARD	2019-03-09 16:00:43+00:00	0	S2	10
WIZARD	2019-04-03 16:00:46+00:00	0	S2	10
WIZARD	2019-04-08 16:00:49+00:00	0.03895106	S2	10
WIZARD	2019-05-13 16:00:49+00:00	0.04495831	S2	10
WIZARD	2019-05-18 16:00:54+00:00	0.24094799	S2	10
WIZARD	2019-05-23 16:00:49+00:00	0	S2	10
WIZARD	2019-06-02 16:00:48+00:00	0	S2	10
WIZARD	2019-06-07 16:00:53+00:00	0	S2	10
WIZARD	2019-07-12 15:37:12+00:00	0.08455608	L7	30
WIZARD	2019-07-17 16:00:54+00:00	0	S2	10
WIZARD	2019-08-31 16:00:46+00:00	0	S2	10
WIZARD	2019-09-20 16:00:45+00:00	0.00175554	S2	10
WIZARD	2019-09-25 16:00:45+00:00	0	S2	10
WIZARD	2019-09-30 15:33:26+00:00	0	L7	30
WIZARD	2019-11-01 15:32:28+00:00	0.35059159	L7	30
WIZARD	2019-11-04 16:00:46+00:00	0	S2	10
WIZARD	2019-12-03 15:30:34+00:00	0.27018176	L7	30
WIZARD	2020-02-07 16:00:37+00:00	0	S2	10
WIZARD	2020-02-21 15:26:54+00:00	0.16339211	L7	30
WIZARD	2020-02-22 16:00:42+00:00	0	S2	10
WIZARD	2020-03-03 16:00:43+00:00	0	S2	10
WIZARD	2020-03-08 15:25:39+00:00	0	L7	30
WIZARD	2020-03-08 16:00:41+00:00	0	S2	10
WIZARD	2020-03-13 16:00:43+00:00	0	S2	10
WIZARD	2020-03-18 16:00:42+00:00	0	S2	10
WIZARD	2020-04-17 16:00:46+00:00	0	S2	10
WIZARD	2020-04-22 16:00:40+00:00	0.03214834	S2	10
WIZARD	2020-05-12 16:00:44+00:00	0.04641211	S2	10
WIZARD	2020-07-14 15:18:08+00:00	0	L7	30
WIZARD	2020-08-31 15:15:30+00:00	0	L7	30
WIZARD	2020-09-19 16:00:47+00:00	0	S2	10

Table A2. *Cont.*

Site	Date	Cloud Cover	Satellite	Resolution (m)
WIZARD	2020-10-02 15:13:00+00:00	0.00433876	L7	30
WIZARD	2020-10-09 16:00:49+00:00	0.0057878	S2	10
WIZARD	2020-11-08 16:00:47+00:00	0	S2	10
WIZARD	2020-11-23 16:00:47+00:00	0	S2	10
WIZARD	2020-12-18 16:00:41+00:00	0.01837832	S2	10
WIZARD	2021-01-06 15:06:20+00:00	0	L7	30
WIZARD	2021-01-17 16:00:45+00:00	0	S2	10
WIZARD	2021-01-27 16:00:45+00:00	0	S2	10
WIZARD	2021-02-06 16:00:45+00:00	0.00131666	S2	10
WIZARD	2021-02-21 16:00:45+00:00	0	S2	10
WIZARD	2021-03-03 16:00:45+00:00	0	S2	10
WIZARD	2021-07-16 16:00:47+00:00	0	S2	10
WIZARD	2021-09-03 14:47:06+00:00	0	L7	30
WIZARD	2021-09-19 14:45:41+00:00	0	L7	30
WIZARD	2021-09-29 16:00:50+00:00	0.06152622	S2	10
WIZARD	2021-10-05 14:43:52+00:00	0	L7	30
WIZARD	2021-10-21 14:42:27+00:00	0	L7	30
WIZARD	2021-12-03 16:00:41+00:00	0	S2	10
WIZARD	2022-01-09 14:34:46+00:00	0.06424628	L7	30
WIZARD	2022-01-12 16:00:41+00:00	0	S2	10
WIZARD	2022-01-25 14:33:14+00:00	0	L7	30
WIZARD	2022-02-01 16:00:40+00:00	0	S2	10
WIZARD	2022-02-06 16:00:45+00:00	0	S2	10
WIZARD	2022-02-16 16:00:46+00:00	0	S2	10
WIZARD	2022-02-21 16:00:40+00:00	0	S2	10
WIZARD	2022-03-08 16:00:49+00:00	0	S2	10
WIZARD	2022-03-14 14:28:28+00:00	0	L7	30
WIZARD	2022-04-07 16:00:46+00:00	0.0322032	S2	10
WIZARD	2022-05-12 16:00:44+00:00	0	S2	10
WIZARD	2022-05-22 16:00:46+00:00	0.03969168	S2	10
WIZARD	2022-06-06 16:00:53+00:00	0.04682357	S2	10
WIZARD	2022-07-06 16:00:58+00:00	0.00065833	S2	10
WIZARD	2022-07-16 16:00:58+00:00	0	S2	10
WIZARD	2022-08-31 14:06:57+00:00	0.01111729	L7	30
WIZARD	2022-09-05 14:12:17+00:00	0.05200819	L7	30
WIZARD	2022-09-09 16:00:47+00:00	0.04158438	S2	10
WIZARD	2022-10-04 16:00:52+00:00	0	S2	10
WIZARD	2022-10-24 16:00:48+00:00	0	S2	10
WIZARD	2022-10-29 16:00:44+00:00	0	S2	10
WIZARD	2022-12-08 16:00:42+00:00	0	S2	10
WIZARD	2022-12-23 16:00:43+00:00	0	S2	10
WIZARD	2022-12-28 16:00:45+00:00	0	S2	10
WIZARD	2023-01-07 16:00:43+00:00	0.03280667	S2	10
WIZARD	2023-02-01 16:00:44+00:00	0	S2	10
WIZARD	2023-02-11 16:00:44+00:00	0	S2	10

Table A2. *Cont.*

Site	Date	Cloud Cover	Satellite	Resolution (m)
WIZARD	2023-02-21 16:00:43+00:00	0.14897411	S2	10
WIZARD	2023-04-12 16:00:46+00:00	0	S2	10
WIZARD	2023-04-17 16:00:47+00:00	0.17599298	S2	10
WIZARD	2023-04-22 16:00:44+00:00	0	S2	10
WIZARD	2023-04-27 16:00:47+00:00	0	S2	10
WIZARD	2023-05-12 16:00:45+00:00	0	S2	10
WIZARD	2023-05-17 16:00:50+00:00	0.00817424	S2	10
WIZARD	2023-05-27 16:00:50+00:00	0	S2	10
WIZARD	2023-06-11 16:00:49+00:00	0	S2	10
WIZARD	2023-06-12 13:37:30+00:00	0	L7	30
WIZARD	2023-07-06 16:00:51+00:00	0	S2	10
WIZARD	2023-07-09 13:37:32+00:00	0	L7	30
WIZARD	2023-07-21 16:00:50+00:00	0	S2	10
WIZARD	2023-08-25 16:00:53+00:00	0.01017665	S2	10
WIZARD	2023-08-30 16:00:51+00:00	0	S2	10
WIZARD	2023-09-29 16:00:48+00:00	0.05392802	S2	10
WIZARD	2023-10-09 16:00:45+00:00	0	S2	10
WIZARD	2023-10-14 16:00:48+00:00	0	S2	10
WIZARD	2023-10-29 16:00:45+00:00	0	S2	10
WIZARD	2023-11-13 16:00:46+00:00	0	S2	10
WIZARD	2023-11-16 13:21:29+00:00	0	L7	30
WIZARD	2023-12-03 16:00:42+00:00	0	S2	10

References

1. Bigio, A.G. *Cities and Climate Change. Building Safer Cities: The Future of Disaster Risk*; World Bank: Washington DC, USA, 2003; pp. 91–99.
2. Aerts, J.; Botzen, W.; Bowman, M.; Dircke, P.; Ward, P. *Climate Adaptation and Flood Risk in Coastal Cities*; Routledge: London, UK, 2013.
3. Oppenheimer, M.; Glavovic, B.C.; Hinkel, J.; van de Wal, R.; Magnan, A.K.; Abd-Elgawad, A.; Cai, R.; Cifuentes-Jara, M.; DeConto, R.M.; Ghosh, T.; et al. Sea Level Rise and Implications for Low-Lying Islands, Coasts and Communities. In *IPCC Special Report on the Ocean and Cryosphere in a Changing Climate*; Pörtner, H.-O., Roberts, D.C., Masson-Delmotte, V., Zhai, P., Tignor, M., Poloczanska, E., Mintenbeck, K., Alegría, A., Nicolai, M., Okem, A., et al., Eds.; Intergovernmental Panel on Climate Change: Geneva, Switzerland, 2019.
4. Pal, I.; Kumar, A.; Mukhopadhyay, A. Risks to coastal critical infrastructure from climate change. *Annu. Rev. Environ. Resour.* **2023**, *48*, 681–712. [CrossRef]
5. Alder, J. Have tropical marine protected areas worked? An initial analysis of their success. *Coast. Manag.* **1996**, *24*, 97–114. [CrossRef]
6. Griggs, G.; Reguero, B.G. Coastal Adaptation to Climate Change and Sea-Level Rise. *Water* **2021**, *13*, 2151. [CrossRef]
7. Silva, R.; Martínez, M.L.; Hesp, P.A.; Catalan, P.; Osorio, A.F.; Martell, R.; Fossati, M.; Miot da Silva, G.; Mariño-Tapia, I.; Pereira, P.; et al. Present and Future Challenges of Coastal Erosion in Latin America. *J. Coast. Res.* **2014**, *71*, 1–16. [CrossRef]
8. Reguero, B.G.; Méndez, F.J.; and Losada, I.J. Variability of multivariate wave climate in Latin America and the Caribbean. *Glob. Planet. Change* **2013**, *100*, 70–84. [CrossRef]
9. Vallarino Castillo, R.; Negro Valdecantos, V.; del Campo, J.M. Understanding the impact of hydrodynamics on coastal erosion in Latin America: A systematic review. *Front. Environ. Sci.* **2023**, *11*, 1267402. [CrossRef]
10. Lizano, O.G. Beach erosion of Costa Rica, including Coco island. *Intersedes* **2013**, *14*, 06–27.
11. Barrantes-Castillo, G.; Arozarena-Llopis, I.; Sandoval-Murillo, L.F.; Valverde-Calderón, J.F. Critical beaches due to coastal erosion in the Caribbean south of Costa Rica during the period 2005–2016. *Rev. Geogr. De Am. Cent.* **2020**, *64*, 75–102. [CrossRef]

12. Fernandez, G.B.; Bulhões, E.; da Rocha, T.B. Impacts of severe storm occurred in April 2010 along Rio de Janeiro coast, Brazil. *J. Coast. Res.* **2011**, *64*, 1850–1854.
13. Speranski, N.; Calliari, L. Bathymetric lenses and localized coastal erosion in southern Brazil. *J. Coast. Res.* **2001**, *SI34*, 209–215.
14. Dolique, F.; Jeanson, M.; Besson, J. A monitoring network for assessing the impact of extreme marine meteorological events on tropical beaches. *J. Coast. Res.* **2007**, *50*, 77–81. [CrossRef]
15. Harris, L.R.; Defeo, O. Sandy shore ecosystem services, ecological infrastructure, and bundles: New insights and perspectives. *Ecosyst. Serv.* **2022**, *57*, 101477. [CrossRef]
16. Barbier, E.B.; Hacker, S.D.; Kennedy, C.; Koch, E.W.; Stier, A.C.; Silliman, B.R. The value of estuarine and coastal ecosystem services. *Ecol. Monogr.* **2011**, *81*, 169–193. [CrossRef]
17. Gallop, S.L.; Bosserelle, C.; Eliot, I.; Pattiaratchi, C.B. The influence of coastal reefs on spatial variability in seasonal sand fluxes. *Mar. Geol.* **2013**, *344*, 132–143. [CrossRef]
18. Quataert, E.; Storlazzi, C.; Van Rooijen, A.; Cheriton, O.; Van Dongeren, A. The influence of coral reefs and climate change on wave-driven flooding of tropical coastlines. *Geophys. Res. Lett.* **2015**, *42*, 6407–6415. [CrossRef]
19. Guannel, G.; Arkema, K.; Ruggiero, P.; Verutes, G. The power of three: Coral reefs, seagrasses and mangroves protect coastal regions and increase their resilience. *PLoS ONE* **2016**, *11*, e0158094. [CrossRef]
20. de Almeida, L.R.; Ávila-Mosqueda, S.V.; Mendoza, E.; van Tussenbroek, B.I.; Silva, R. Beyond the Beach: Multi-Parameter Interpretation of Shore Deterioration in a Caribbean Reef System. *Diversity* **2024**, *16*, 266. [CrossRef]
21. Morales-Márquez, V.; Cáceres-Euse, A.; Hernández-Carrasco, I.; Molcard, A.; Orfila, A. Extreme waves in the Caribbean Sea: Spatial regionalization and long-term analysis. *Front. Mar. Sci.* **2023**, *10*, 1294189. [CrossRef]
22. Morang, A.; Larson, R.; Gorman, L. Monitoring the coastal environment; part I: Waves and currents. *J. Coast. Res.* **1997**, *13*, 111–133.
23. Tendero-Peiró, C.; Asensio-Montesinos, F.; Anfuso, G.; Corbí, H. The Coastal Scenery of São Miguel Island, Azores Archipelago: Implications for Coastal Management. *J. Mar. Sci. Eng.* **2024**, *12*, 707. [CrossRef]
24. Lyman, T.P.; Elsmore, K.; Gaylord, B.; Byrnes, J.E.; Miller, L.P. Open Wave Height Logger: An open source pressure sensor data logger for wave measurement. *Limnol. Oceanogr. Methods* **2020**, *18*, 335–345. [CrossRef]
25. Temple, N.A.; Webb, B.M.; Sparks, E.L.; Linhoss, A.C. Low-Cost Pressure Gauges for Measuring Water Waves. *J. Coast. Res.* **2020**, *36*, 661–667. [CrossRef]
26. Godfrey, S.; Cooper, J.R.; Plater, A.J. Roving multiple camera array with Structure-from-Motion for coastal monitoring. *J. Mar. Sci. Eng.* **2023**, *11*, 591. [CrossRef]
27. Knight, P.J.; Bird, C.O.; Sinclair, A.; Plater, A.J. A low-cost GNSS buoy platform for measuring coastal sea levels. *Ocean. Eng.* **2020**, *203*, 107198. [CrossRef]
28. Knight, P.J.; Bird, C.O.; Sinclair, A.; Higham, J.; Plater, A.J. Beach deployment of a low-cost GNSS buoy for determining sea-level and wave characteristics. *Geosciences* **2021**, *11*, 494. [CrossRef]
29. Raghukumar, K.; Chang, G.; Spada, F.; Jones, C.; Janssen, T.; Gans, A. Performance characteristics of "Spotter," a newly developed real-time wave measurement buoy. *J. Atmos. Ocean. Technol.* **2019**, *36*, 1127–1141. [CrossRef]
30. Lancaster, O.; Cossu, R.; Boulay, S.; Hunter, S.; Baldock, T.E. Comparative wave measurements at a wave energy site with a recently developed low-cost wave buoy (Spotter), ADCP, and pressure loggers. *J. Atmos. Ocean. Technol.* **2021**, *38*, 1019–1033. [CrossRef]
31. Zopf, D.O.; Creech, H.C.; Quinn, W.H. The wavemeter: A land-based system for measuring nearshore ocean waves. *MTS J.* **1976**, *10*. Available online: https://ir.library.oregonstate.edu/concern/technical_reports/jm214w120 (accessed on 7 June 2025).
32. Ravdas, M.; Zacharioudaki, A.; Korres, G. Implementation and validation of a new operational wave forecasting system of the Mediterranean Monitoring and Forecasting Centre in the framework of the Copernicus Marine Environment Monitoring Service. *Nat. Hazards Earth Syst. Sci.* **2018**, *18*, 2675–2695. [CrossRef]
33. Brus, S.R.; Wolfram, P.J.; Van Roekel, L.P.; Meixner, J.D. Unstructured global to coastal wave modeling for the Energy Exascale Earth System Model using WAVEWATCH III version 6.07. *Geosci. Model Dev.* **2021**, *14*, 2917–2938. [CrossRef]
34. Lippmann, T.C.; Holman, R.A. Phase speed and angle of breaking waves measured with video techniques. In *Coastal Sediments*; ASCE: Reston, VA, USA, 1991; pp. 542–556.
35. Bos, C. Wave Characteristics Derived from Video. Master's Thesis, Delft University of Technology, Delft, The Netherlands, 2006.
36. Thomson, J.; Jessup, A.T. A Fourier-based method for the distribution of breaking crests from video observations. *J. Atmos. Ocean. Technol.* **2009**, *26*, 1663–1671. [CrossRef]
37. Adrian, R.J.; Westerweel, J. *Particle Image Velocimetry*; Cambridge University Press: Cambridge, UK, 2011.
38. Brevis, W.; Niño, Y.; Jirka, G.H. Integrating cross-correlation and relaxation algorithms for particle tracking velocimetry. *Exp. Fluids* **2011**, *50*, 135–147. [CrossRef]
39. Umeyama, M. Theoretical considerations, flow visualization and pressure measurements for rear-end collisions of two unequal solitary waves. *J. Math. Fluid Mech.* **2019**, *21*, 1–23. [CrossRef]

40. Abu Rowin, W.; Kevin, X.; Bhirawa, T.; Lee, W.J.; Philip, J.; Marusic, I.; Monty, J. High-resolution PIV measurement over wind-generated waves. *Exp. Fluids* **2024**, *65*, 77. [CrossRef]
41. Dérian, P.; Almar, R. Wavelet-based optical flow estimation of instant surface currents from shore-based and UAV videos. *IEEE Trans. Geosci. Remote Sens.* **2017**, *55*, 5790–5797. [CrossRef]
42. Oertel, M.; Suefke, F. Two-dimensional dam-break wave analysis: Particle image velocimetry versus optical flow. *J. Hydraul. Res.* **2020**, *58*, 326–334. [CrossRef]
43. Li, J.; Kong, X.; Yang, Y.; Yang, Z.; Hu, J. Optical flow based measurement of flow field in wave-structure interaction. *Ocean. Eng.* **2022**, *263*, 112336. [CrossRef]
44. Vos, K.; Splinter, K.D.; Harley, M.D.; Simmons, J.A.; Turner, I.L. CoastSat: A Google Earth Engine-Enabled Python Toolkit to Extract Shorelines from Publicly Available Satellite Imagery. *Environ. Model. Softw.* **2019**, *122*, 104528. [CrossRef]
45. ATP. *Plan Maestro de Turismo Sostenible*; Autoridad de Tourismo de Panama: Panama City, Panama, 2020. Available online: https://www.atp.gob.pa/Plan_Maestro_de_Turismo_Sostenible_2020-2025.pdf (accessed on 8 May 2023).
46. Klytchnikova, I.; Dorosh, P. Tourism sector in Panama: Regional economic impacts and the potential to benefit the poor. *Nat. Resour. Forum.* **2013**, *37*, 70–79. [CrossRef]
47. Pleasant, T.; Spalding, A. Development and dependency in the periphery: From bananas to tourism in Bocas del Toro, Panama. *World Dev. Perspect.* **2021**, *24*, 10. [CrossRef]
48. NBDC (National Data Buoy Center). Station 42057—Western Caribbean Buoy. 2025. Available online: https://www.ndbc.noaa.gov/station_page.php?station=42057 (accessed on 19 March 2025).
49. Fullmer, W.D.; Higham, J.E.; LaMarche, C.Q.; Issangya, A.; Cocco, R.; Hrenya, C.M. Comparison of velocimetry methods for horizontal air jets in a semicircular fluidized bed of Geldart Group D particles. *Powder Technol.* **2020**, *359*, 323–330. [CrossRef]
50. Higham, J.E.; Shahnam, M.; Vaidheeswaran, A. Anomalous diffusion in a bench-scale pulsed fluidized bed. *Phys. Rev. E* **2021**, *103*, 043103. [CrossRef]
51. Lucas, B.D.; Kanade, T. An iterative image registration technique with an application to stereo vision. In Proceedings of the IJCAI'81: 7th International Joint Conference on Artificial Intelligence, Vancouver, BC, Canada, August 1981; Volume 2, pp. 674–679.
52. Welch, P. The use of fast Fourier transform for the estimation of power spectra: A method based on time averaging over short, modified periodograms. *IEEE Trans. Audio Electroacoust.* **1967**, *15*, 70–73. [CrossRef]
53. The WAVEWATCH III Development Group (WW3DG). *User Manual and System Documentation of WAVEWATCH III*; Version 6.07. Tech. Note 333; NOAA/NWS/NCEP/MMAB: College Park, MD, USA, 2019; 465p.
54. Krumbein, W.C. Size frequency distributions of sediments and the normal phi curve. *J. Sediment. Res.* **1938**, *8*, 84–90. [CrossRef]
55. Folk, R.L.; Ward, W.C. Brazos River bar [Texas]; a study in the significance of grain size parameters. *J. Sediment. Res.* **1957**, *27*, 3–26. [CrossRef]
56. Whitcraft, A.K.; Vermote, E.F.; Becker-Reshef, I.; Justice, C.O. Cloud cover throughout the agricultural growing season: Impacts on passive optical earth observations. *Remote Sens. Environ.* **2015**, *156*, 438–447. [CrossRef]
57. Permanent Service for Mean Sea Level (PSMSL). Tide Gauge Data. 2024. Available online: https://psmsl.org/data/obtaining/stations/2191.php (accessed on 14 November 2024).
58. Barnett, V. *Outliers in Statistical Data*; John Wiley & Sons: Hoboken, NJ, USA, 1994; Volume 2, pp. 705–708.
59. McAllister, E.; Payo, A.; Novellino, A.; Dolphin, T.; Medina-Lopez, E. Multispectral satellite imagery and machine learning for the extraction of shoreline indicators. *Coast. Eng.* **2022**, *174*, 104102. [CrossRef]
60. Higham, J.E.; Shahnam, M.; Vaidheeswaran, A. Using a proper orthogonal decomposition to elucidate features in granular flows. *Granul. Matter* **2020**, *22*, 86. [CrossRef]
61. Higham, J.E.; Isaac, O.S.; Rigby, S.E. Optical flow tracking velocimetry of near-field explosions. *Meas. Sci. Technol.* **2022**, *33*, 047001. [CrossRef]
62. Weber, J.; Higham, J.E.; Musser, J.; Fullmer, W.D. Critical analysis of velocimetry methods for particulate flows from synthetic data. *Chem. Eng. J.* **2021**, *415*, 129032. [CrossRef]
63. Luijendijk, A.; Hagenaars, G.; Ranasinghe, R.; Baart, F.; Donchyts, G.; Aarninkhof, S. The State of the World's Beaches. *Sci. Rep.* **2018**, *8*, 6641. [CrossRef] [PubMed]
64. Pardo-Pascual, J.E.; Sánchez-García, E.; Almonacid-Caballer, J.; Palomar-Vázquez, J.M.; Priego De Los Santos, E.; Fernández-Sarría, A.; Balaguer-Beser, Á. Assessing the accuracy of automatically extracted shorelines on microtidal beaches from Landsat 7, Landsat 8 and Sentinel-2 imagery. *Remote Sens.* **2018**, *10*, 326. [CrossRef]
65. Preston, J.; Hurst, M.D.; Mudd, S.M.; Goodwin, G.C.; Newton, A.J.; Dugmore, A.J. Sediment accumulation in embayments controlled by bathymetric slope and wave energy: Implications for beach formation and persistence. *Earth Surf. Process. Landf.* **2018**, *43*, 2421–2434. [CrossRef]
66. Masselink, G.; Hughes, M.; Knight, J. *Introduction to Coastal Processes and Geomorphology*; Routledge: London, UK, 2014.

67. Komar, P.D. Beach processes and erosion—An introduction. In *Handbook of Coastal Processes and Erosion*; CRC Press: Boca Raton, FL, USA, 2018; pp. 1–20.
68. Shimozono, T. Long wave propagation and run-up in converging bays. *J. Fluid Mech.* **2016**, *798*, 457–484. [CrossRef]
69. Vila-Concejo, A.; Gallop, S.L.; Largier, J.L. Sandy beaches in estuaries and bays. In *Sandy Beach Morphodynamics*; Elsevier: Amsterdam, The Netherlands, 2020; pp. 343–362.
70. Kemp, P.H. The relationship between wave action and beach profile characteristics. *Coast. Eng. Proc.* **1960**, *7*, 14. [CrossRef]
71. Nielsen, P. *Coastal Bottom Boundary Layers and Sediment Transport*; World Scientific: River Edge, NJ, USA, 1992.
72. Wright, L.D.; Short, A.D. Morphodynamic variability of surf zones and beaches: A synthesis. *Mar. Geol.* **1984**, *56*, 93–118. [CrossRef]
73. Kohsiek, L.H.M.; Hulsbergen, C.H.; Terwindt, J.H.J. Beach erosion along the west coast of Aruba, Netherlands Antilles. *J. Coast. Res.* **1987**, *3*, 37–53.
74. Ortega, A.Y.; Díaz, L.J.O.; Cueto, J.E. Assessment and management of coastal erosion in the marine protected area of the Rosario Island archipelago (Colombian Caribbean). *Ocean. Coast. Manag.* **2023**, *239*, 106605. [CrossRef]

environments

MDPI

Article

Two Decades of CARICOMP Mangrove Monitoring (1992–2013) Reveal Variability in Tree Structure and Productivity of *Rhizophora mangle* Across the Wider Caribbean

Björn Kjerfve [1,2,*], Hazel A. Oxenford [3], Rachel Collin [4], Inácio Abreu Pestana [1], Jimena Samper-Villarreal [5], Israel Medina-Gómez [6], Jorge Cortés [5], Struan R. Smith [7], Karen Koltes [8], Ilka C. Feller [9], Carolina Bastidas [10], Rahanna Juman [11], Francisco X. Geraldes [12], Alessandro Filippo [13], Ramon Varela [14], Croy McCoy [15], Jaime Garzón-Ferreira [16,†], Jaime Polanía [17], Juan C. Capelo [18] and John Ogden [19,‡]

1 Departamento de Geoquímica, Universidade Federal Fluminense, Niterói 24210-141, RJ, Brazil; inaciopestana@id.uff.br
2 School of Earth, Ocean & Environment, University of South Carolina, Columbia, SC 29208, USA
3 Centre for Resource Management and Environmental Studies (CERMES), University of the West Indies, Cave Hill, Bridgetown BB11000, Barbados; hazel.oxenford@cavehill.uwi.edu
4 Smithsonian Tropical Research Institute, Balboa Ancón 0843-03092, Panama; collinr@si.edu
5 Centro de Investigación en Ciencias del Mar y Limnología (CIMAR), Ciudad de la Investigación, Universidad de Costa Rica, San José 11501-2060, Costa Rica; jimena.sampervillarreal@ucr.ac.cr (J.S.-V.); jorge.cortes@ucr.ac.cr (J.C.)
6 Escuela Nacional de Estudios Superiores (ENES-UNAM), Universidad Nacional Autónoma de México, Unidad Mérida, Mérida 97205, Mexico; imedgomez@gmail.com
7 Natural History Museum and the Bermuda Aquarium Museum and Zoo and the Bermuda Institute for Ocean Sciences, Hamilton HM 834, Bermuda; srsmith@gov.bm
8 U. S. Department of the Interior, Washington, DC 20240, USA; karenkoltes@yahoo.com
9 Smithsonian Environmental Research Center, Edgewater, MD 21037, USA; felleri@si.edu
10 Sea Grant College Program, Massachusetts Institute of Technology, Cambridge, MA 02139, USA; bastidas@mit.edu
11 Institute of Marine Affairs, Hilltop Lane, Chaguaramas, Trinidad, Trinidad and Tobago; rajuman@ima.gov.tt
12 Centro de Investigaciones Biología Marina, Universidad Autónoma de Santo Domingo, Santa Domingo 10100, Dominican Republic; fgeraldes0@gmail.com
13 Departamento de Oceanografia Física e Meteorologia, FAOC, Universidade do Estado do Rio de Janeiro, Rio de Janeiro 20550-900, RJ, Brazil; afilippo@uerj.br
14 Fundación La Salle de Ciencias Naturales, Caracas 1010, Venezuela; varelaallegue@hotmail.com
15 Cayman Islands Environmental Centre, Grand Cayman Island KY1-1002, Cayman Islands; croy.mccoy@gmail.com
16 INVEMAR, Santa Marta 470003, Colombia; anisotremus@gmail.com
17 Área Curricular de Bosques y Conservación, Universidad Nacional de Colombia Sede Medellín, Medellín 050034, Colombia; jhpolaniav@unal.edu.co
18 Museo Marino de Margarita Dr. Fernando Cervigón, Boulevard el Paseo, Boca del Río, Margarita Island 6304, Venezuela; capelosky@gmail.com
19 University of South Florida, St. Petersburg, FL, USA; verdnan23@gmail.com
* Correspondence: bkjerfve@id.uff.br
† Current address: Sage Ridge School, 2515 Crossbow Ct., Reno, NV 89511, USA.
‡ Co-author John Ogden passed prior to publication of this manuscript.

Abstract: The Caribbean Coastal Marine Productivity (CARICOMP) program was conceptualized in 1985 to monitor coral reefs, seagrass beds, and mangrove forests at multiple sites across the wider Caribbean. Mangrove monitoring was focused on the dominant Caribbean species, red mangrove (*Rhizophora mangle*). Forest structure and productivity were monitored at 21 sites (18 countries) across different geomorphological settings, from tropical to subtropical mainland and island systems. Here, we provide the key findings from the CARICOMP mangrove data collected, mostly from 1992 to 2013, to assess spatial and temporal variability across the region. Red mangrove above-ground biomass averaged 190 t ha^{-1} (far higher than previously reported) but ranged widely across sites from 33 to 590 t ha^{-1}, equating to an average above-ground 'blue carbon' of 84 t ha^{-1} (range 15–260 t ha^{-1}). Tree density averaged 3237 trees ha^{-1}, tree basal area averaged 19.7 m^2 ha^{-1}, tree height averaged 6.1 ± 2.8 m, and seedling density varied from 1.2 to 74 seedlings m^{-2} across the sites. Among the environmental factors that influence mangroves, local temperature and rainfall explained 48% of the variability in measured tree structure parameters. Annual litterfall, as a proxy for productivity, measured on average 1.24 ± 0.70 kg m^{-2} yr^{-1}, with 60% of the total litterfall composed of leaves. Litterfall varied seasonally by 42%. No relationship was apparent between litterfall and seasonal ocean–atmosphere climate indices (ONI and AMM). With exception of the three most southwesterly CARICOMP sites, hurricanes and tropical storms impacted the mangrove sites repeatedly, resulting in considerable damage. A direct strike by a category-4 hurricane in 1998 in Dominican Republic killed 67% of the red mangrove trees, lowered above-ground biomass by 91%, basal area by 89%, litterfall by 63%, and resulted in the subsequent growth of many tall and thin saplings, totally changing the structure of the forest ecosystem in the first few years after the hurricane. In comparing mangrove systems, major differences may be explained by time elapsed since the last destructive event (hurricane) affecting each site. This highlights the fact that despite an increasing focus on preserving these valuable ecosystems, they are still highly vulnerable to natural hazards and likely to face a poor outcome under ongoing climate change.

Keywords: mangroves; Caribbean; CARICOMP; monitoring; productivity; structure; hurricane; coastal blue carbon

1. Introduction

Mangrove forests are marine wetlands that cover narrow vegetated strips along the coasts in the tropics and subtropics. Globally, mangroves cover some 147,000 km^2 [1]. They are highly productive and structurally and functionally biodiverse ecosystems [2] that provide diverse ecological services as well as storm protection to coasts. Mangroves provide habitat for birds and terrestrial and marine flora and fauna, including a number of endangered species; serve as nursery areas for commercially important fish and shellfish species; yield honey and tannin; support commercial harvest of land crabs; but are also exploited unsustainably, contributing to their decline globally [3,4].

Together with seagrass meadows and salt marshes, mangrove forests are recognized among the globally important 'blue carbon ecosystems'. Despite their critical role in climate mitigation (through carbon sequestration) and adaptation (through coastal protection), mangroves are among the most threatened habitats anywhere [5–7]. These transitional ecosystems have accumulated large carbon stocks (particularly in organic rich soils as well as in above-ground biomass) and continue to serve actively as key carbon sinks. However, when degraded or destroyed, they no longer sequester carbon at the same rate even with warmer temperatures and increased atmospheric CO_2 levels [8] and instead become a source of atmospheric CO_2 [9]. Protection and restoration of mangrove ecosystems thus has an important role in climate change mitigation [7].

Mangroves have been declining globally for several decades due to both direct and indirect anthropogenic impacts. For example, 35% of mangroves worldwide were lost in just two decades from the late 1970s to the late 1990s [10,11]. However, conservation and protection efforts appear to have been effective at slowing the global rate of net loss of mangroves to 3.4% over the last two decades (up to 2020) [1].

Mangrove losses have been largely driven by direct anthropogenic impacts such as conversion to agricultural lands and aquaculture ponds; urban expansion and shoreline alterations for tourism, marinas and ports; hydrological alterations; and over-harvesting for timber and charcoal [12]. Indirect processes such as shoreline erosion, sea level rise, compaction, subsidence, vegetation loss due to storms, altered sediment supply, and pollution have also contributed to mangrove losses [13,14]. With regard to climate change, mangrove ecosystems are mainly impacted by sea level rise and storm events [15,16]. Since the 1990's, global sea level rise has accelerated to the current rate of 3.2 ± 0.4 mm yr^{-1} and is likely to increase still further [16,17]. The resulting coastal squeeze, the increasing limits placed on the distribution of coastal habitats by rising sea level along the seaward margin, and changes in land use on the landward margin, are a threat to coastal wetlands worldwide. Mangrove systems, which are susceptible to flooding from increased inundation, can migrate inland. However, when potential for landward migration is limited by human occupation or by steep topography, mangroves drown where sediment accretion is not rapid enough to keep pace with relative sea level rise [18].

The combined effect of recurring mangrove inundation and the length of inundation (hydroperiod) help control the growth rate of salt-tolerant mangrove plants, with both occasional hypersaline conditions during droughts and freshwater pulses balancing the interstitial water salinity [19]. The hydroperiod reflects oscillations of the hydraulic water table, governed by drought, rainfall/runoff, and tidal inundation [20,21]. As the climate and relative sea level change, the response of the interstitial salinity will prove critical to the growth response of the mangrove trees, as will the supply of terrigenous sediments.

In the wider Caribbean, both mainland and island mangrove systems have experienced the same stressors and exhibited the same responses as mangroves elsewhere [10,22–24], with reported losses of 24% in total area in the 25 years between 1980 and 2005 [23]. Caribbean-wide degradation of mangroves, together with that of coral reefs and seagrasses, translates into economic losses on the order of billions of US dollars [23,25], as these ecosystems support many key ecosystem services that underpin the livelihoods of coastal populations and the blue economies on which many Caribbean nations rely [26].

To better understand long-term changes in the ecologically and economically important mangrove systems in the wider Caribbean, the Caribbean Coastal Marine Productivity (CARICOMP) program was proposed in 1985 as a regional monitoring program [27–29]. The first of its kind, CARICOMP was a regional network of local scientists and environmental managers who planned and executed standardized protocols across geographically variable mangrove, reef, and seagrass systems.

The objective of the mangrove monitoring program was to assess differences among sites, determine seasonal patterns of productivity, and identify changes over time [29] with a focus on *R. mangle*, the dominant mangrove species in the wider Caribbean. Here, we assess the CARICOMP mangrove data collected during the 22 years between 1992 and 2013 (when the program was sunsetted), and a few years beyond for some sites where data collection continued, in order to reveal spatial and temporal variability across the region, investigate potential drivers of change across protected mangrove sites, and provide insights for future monitoring of these valuable coastal forests. We highlight differences and changes among the sites with more than six years of observations. At the initiation of this research effort, the concept of 'blue carbon' had not yet been coined, and carbon sequestration in mangrove soils was not a monitoring goal at that time. Our results, however, shed light on regional variation in above-ground carbon stocks.

2. Materials and Methods

2.1. Mangrove Sites

In 1992, the CARICOMP program began monitoring mangroves at 21 sites in 18 countries across the wider Caribbean [30,31]. The mangrove sites were located in the insular Caribbean and on mainland coasts of South, Central, and North America, and ranged from Laguna de Celestún (México) in the west, to Barbados in the east, and from Bermuda in the north, to Bocas del Toro (Panamá) in the south (Figure 1; Table 1). The physical setting of the mangrove sites at most of the participating CARICOMP partner institutions, descriptions of the community types, estimates of total areas, freshwater inputs, and exposure to wave action and/or currents are described in [32].

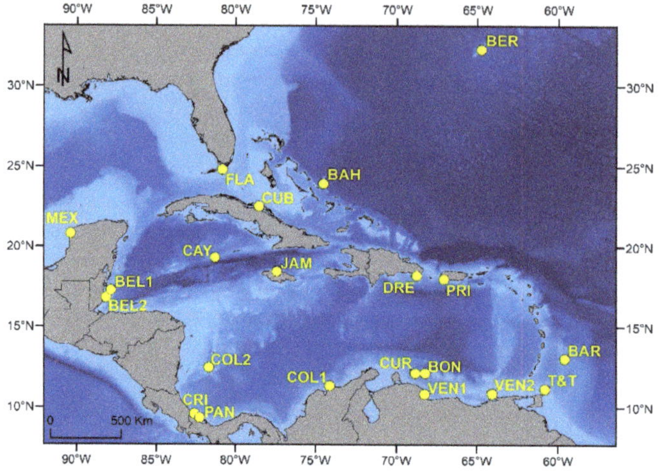

Figure 1. Location of the 21 CARICOMP mangrove sites with acronyms listed in Table 1.

Table 1. Location of the 21 CARICOMP sites with records of tree structure (S) and productivity (P). Y* indicates limited observations.

Acronym	Site	Country	S	P	Lat N°	Long W°
BAH	Blackwood Bay, San Salvador	Bahamas	Y	Y	23.9541	−74.5277
BAR	Graeme Hall Swamp	Barbados	Y	Y	13.0717	−59.5770
BEL1	Calabash Cay, Turneffe Atoll	Belize	Y	Y	17.2827	−87.8156
BEL2	Twin Cays (near Carrie Bow Cay)	Belize	Y	N	16.8302	−88.0992
BER	Hungry Bay	Bermuda	Y	Y	32.2906	−64.7591
BON	Lac Bay	Bonaire, N.A.	Y	N	12.1167	−68.2282
CAY	Central Swamp, North Sound	Cayman Islands	Y	Y	19.3211	−81.2801
COL1	Ensenada de Chengue	Colombia	Y	Y	11.3166	−74.1278
COL2	Hooker Bay, San Andrés	Colombia	Y	Y	12.5611	−81.7045
CRI	Laguna Gandoca	Costa Rica	Y	Y	9.5893	−82.5983
CUB	Punta la Jaula, Cayo Coco	Cuba	Y	Y	22.5264	−78.5196
CUR	St. Jorisbaai	Curaçao, N.A.	Y	Y	12.1199	−68.8275
DRE	Canal Punta Mangle al Catuano	Dominican Republic	Y	Y	18.2034	−68.7533
FLA	Long Key, Florida	USA	Y*	N	24.8209	−80.8112
JAM	Discovery Bay	Jamaica	Y	Y	18.4697	−77.4150
MEX	Laguna de Celestún	Mexico	Y	N	20.8223	−90.3838
PAN	Isla de Colón, Bocas del Toro	Panama	Y	Y	9.3520	−82.2590
PRI	Pitahaya, La Parguera	Puerto Rico	Y	Y	17.9569	−67.0946
T&T	Bon Accord Lagoon, Tobago	Trinidad and Tobago	Y	Y	11.1660	−60.8300
VEN1	Tumba Cuatro, Morrocoy	Venezuela	Y	Y	10.8352	−68.2613
VEN2	Punta de Mangle, Isla de Margarita	Venezuela	Y	Y	10.8640	−64.0560

The CARICOMP mangrove effort focused on monitoring tree structure and productivity within plots of fringing red mangroves, *Rhizophora mangle*, with minimal anthropogenic disturbances, particularly from coastal alteration and pollution [32]. However, the plan to collect data in 'pristine locations' to establish a 'natural baseline' without human disturbance proved rather futile. Although some sites were less disturbed than others, no site was pristine [28]. At each mangrove site, a team of local scientists, managers and/or students established between 2–6 'fixed' plots in the mangrove forest fringing the water's edge (within 0–30 m, and located a haphazard distance apart), where red mangroves are typically the most common mangrove tree species.

Monitoring was conducted to record tree structure and productivity for the duration of the study according to the CARICOMP methods manual [31], that was based on general methods for measurement of mangrove ecosystem structure and function described by Lugo and Snedaker [33], Pool et al. [34], and Snedaker and Snedaker [35]. To define local patterns of change, the plan was to monitor mangrove tree structure and productivity parameters annually.

2.2. Tree Structure Measurements

Mangrove tree structure was assessed in fixed 10 × 10 m replicate plots at each site. Within each plot, all mangrove trees with a trunk diameter > 2.5 cm were identified, numbered, and permanently marked around the trunk at breast height for the duration of the study, and the X and Y coordinates of all trees within each plot recorded. The species of each tree was noted, and standard forestry measurements were made on the marked trees, including trunk circumference at breast height, and height to the top of the canopy [31]. *R. mangle* circumference was measured immediately above the first prop-root. This trunk circumference (C) was then used to calculate the diameter at breast height (DBH) for each tree, where DBH = C/π. Circumference measurements are robust measures, whereas tree height (H) measurements are challenging for tall trees (Note #S1).

Tree structure measurements were carried out mostly between 1992 and 2010. They were initially scheduled to be repeated every year but were in fact only measured 1–3 times at most sites, although 18 times at Graeme Hall Swamp (BAR), and 13 times at Ensenada de Chengue (COL1), with sampling still continuing at these sites. At several sites, the trees in the designated plots were not sampled in the same year, rather one plot in one year, then another plot the next year, or two plots the next year. The intention was to monitor three plots at each site, but some sites established only two plots, while additional plots were added at several other sites and included in the analyses. An additional two plots at Lac Bay (BON), including only *Avicennia germinans* trees, and one plot at Graeme Hall Swamp (BAR), comprising only *Laguncularia racemosa* trees, are reported in the results but not included in the analyses.

The mean number of trees per hectare (for each species) was based on the sum of all trees in all plots at each site over the years monitored. Due to turn-over (recruitment/growth and mortality of individual trees and storm damage) the number of trees varied over time by as much as 10 trees per site. A mean number of 2470 live *R. mangle* trees were measured across the 21 sites over the duration of sampling, in addition to 233 black, *A. germinans*, 115 white, *L. racemosa*, and 1 buttonwood, *Conocarpus erectus* trees (Note #S2, Table S2A).

"Samples", or tree-measures, refer to the total number of circumference and/or height measurements on individual trees at each site. "Sets", or repeat measures, on the trees at a site is a key variable used in lieu of time. It is calculated as the number of samples divided by the mean number of trees at each site for each species. The number of sets is not necessarily an integer due to the fact that some plots were sometimes omitted, and the number of trees sampled varied between sets, especially for sites with long records. It proved necessary to use "set" (not time) because of sampling variability, e.g., the measurements at the three plots at a site sometimes required two years to complete.

Based on the circumference measurements (cm), the basal area (BA) of individual trees was calculated as $(\pi/4) \times (\mathrm{DBH}^2)$. The overall mean BA ($m^2\ ha^{-1}$) of the stand of trees of the same species at each site was calculated as the sum of all individual basal areas divided by sets × plots × 100, where 100 is a scale factor to render the results as m^2 ha^{-1}. Similarly, overall mean tree density (D, trees ha^{-1}) for each species and site was calculated as the overall number of samples divided by sets × plots × 100, where 100 again is a scale factor to render the density as number of trees per hectare for each species and site.

Above-ground biomass (AGB) is the amount of standing dry organic matter per unit area at any one time. We adopted the power relationship (Note #S3):

$$AGB = a \times DBH^b$$

where a and b are regression coefficients.

We selected "best-fit" relationships generated for each mangrove species (Table 2) and applied these to estimate biomass for each tree in the CARICOMP mangrove data set. The mean AGB (t ha^{-1}) for each species and site was then calculated as individual tree biomass and summed for all trees and divided by sets × plots × 10, where 10 is a scale factor to render the results in metric tons per hectare. To present these AGB values as above-ground carbon (AGC, t ha^{-1}), we used the balanced conversion factors of 44.1% for *R. mangle* and 44.0% for *L. racemosa* and *A. schaueriana* (the latter as a proxy for *A. germinans*) following Rodrigues et al. [36] (Note #S3).

Table 2. Summary of allometric relationships we adopted between AGB (kg) and DBH (cm), for three mangrove species; a and b are dimensional regression coefficients, n is the number of trees felled, and R^2 represents the variability of AGB explained by DBH.

Species	N	R^2	a	B	Study Location	References
Rhizophora mangle	74	0.99	0.2537	2.4048	NE Brazil	[37]
Avicennia germinans	25 (DBH > 4 cm)	0.97	0.1400	2.4000	French Guiana	[38]
Laguncularia racemosa	35	0.92	0.1442	2.3250	SE Brazil	[39]

2.3. Seedling Measurements

Within each of the tree structure plots, up to five 1 × 1 m sub-plots were established haphazardly. In the sub-plots at six sites, all live rooted seedlings and tree saplings (DBH < 2.5 cm diameter) were tagged, identified, mapped, and their heights measured [31]. At some sites, all *R. mangle* seedlings were harvested in an additional sub-plot (outside but adjacent to a main plot) at five sites, measured for height and individually weighed to allow for the calculation of wet weight for mangrove seedlings based on seedling height using the following power expression:

$$W = a \times \exp(H)^b$$

where W is wet weight (g) and H is seedling height (cm).

R. mangle plants taller than 194 cm with DBH < 2.5 cm are referred to as saplings and not included in this calculation. Seedling counts were used to determine seedling density.

2.4. Productivity/Litterfall Measurements

An index of productivity is taken as the sum of all plant material (litterfall) produced during a fixed time and was measured at 17 of the 21 CARICOMP sites. To estimate mangrove productivity, anywhere from 5 to 10 litterfall traps (each 0.25 m^2) were placed within each of the monitoring plots. Each trap was secured to trunks, branches, or prop-roots of red mangrove trees approximately 1 m above the ground and above the highest high tide [31]. The traps were fitted with a 1.5 mm mesh to collect litterfall from the mangrove canopy. Litterfall was collected from the traps monthly at most sites during the first year; four times during the second year, corresponding to summer, winter, wet season, and dry season; and subsequently, every other year during maximum and minimum litter production periods. Most traps (72%) were emptied of litterfall at monthly intervals (i.e., every 27–32 days). After collecting, litterfall samples were dried, and each of the plant components weighed separately, including leaves, flowers, fruits (propagules for *R. mangle*), bracts, wood, and miscellaneous [31]. Simultaneously with the first litter collection, loose surface litter on the forest floor just below each of the traps was collected, washed of sediment and salt, and dried and weighed just like the trap litter samples. This measure represented the standing crop of litter but was not implemented at most sites

2.5. Physical Environmental Parameters

Adjacent to the mangrove sites, repeated measurements of water temperature and salinity (using a handheld thermometer and refractometer) were made at irregular intervals [31,40]. Interstitial salinity (i.e., the salinity of the sub-surface water penetrating the sediments surrounding the mangrove roots) was measured whenever possible and surface salinity measurements were taken if plots were inundated with water. Tidal water level measurements were made at some of the sites but with poor consistency (Note #S4).

Early in the program, each research site made weather observations (max/min air temperature and rainfall), water temperature and turbidity, and salinity [41], but with extensive data gaps. Although a sub-set of the CARICOMP sites were equipped with semi-automated weather stations several years later, reasonably complete weather data were retrieved only from Carrie Bow Cay (3 km east of BEL2) with data recorded 1995–1996, and 2000–2013 [42], and Bocas del Toro (PAN) 2003–2013 [43]. However, we obtained annual rainfall time series from an additional eight CARICOMP mangrove sites from national meteorological services, with rain gauges located anywhere from 0.2 to 28 km from the sites. We considered the variability of the Ocean Niño Index (ONI) and Atlantic Meridional Mode (AMM) climate indices [44–46]. We also summarized the tracks and categories of tropical cyclones with the potential to impact the CARICOMP mangrove sites (NOAA Historical Hurricane Tracks; accessed 8 April 2024). We considered a mangrove site potentially impacted by hurricanes and tropical storms when the track of a tropical cyclone came within 80 km of a mangrove site [47].

3. Results

3.1. Environmental Setting

The CARICOMP mangrove sites, including mainland and island systems, covered a wide range of environmental settings (Table 3), including highly productive estuarine environments, coastal areas, low-rainfall island settings, and over-wash barrier islands; and highly variable geomorphology. The biophysical typology of the CARICOMP mangrove sites, according to the geomorphic and sedimentary setting [48], indicates that 12 of the sites can be classified as lagoonal, and nine sites as open coast (Table 3). The astronomical tidal range is less than 1 m at all sites (Note #S4). The biophysical topology classification works better for the Caribbean mangrove sites than the ecomorphodynamics approach [49,50] due to lack of monitoring of sedimentation rates and nutrient fluxes. The areas occupied by Caribbean mangroves vary widely among the sites from 1 ha (JAM) to 3000 or more ha (MEX and CAY). The substrate is calcareous silt and sand at seven sites, mostly peat at eight sites, terrigenous silt and sand at five sites, and limestone substrate at JAM. Annual rainfall varies from 383 mm yr^{-1} (VEN2) to 3428 mm/yr (PAN), with 12 sites experiencing annual rainfall between 1000 and 1873 mm yr^{-1}. A total of 124 hurricanes and 107 tropical storms either entered or formed in the wider Caribbean during the CARICOMP monitoring program (https://oceanservice.noaa.gov/news/historical-hurricanes/, accessed on

2 May 2025). Some tropical cyclones formed within the Caribbean, as was the case with the very destructive category-5 Hurricane Mitch in 1998 [51,52]. BER experienced 15 tropical cyclones during the 22-year monitoring program, while COL1, PAN, and CRI, located south of the cyclone tracks, experienced none.

Table 3. Summary of site parameters. Those representing biophysical topology, area, substrate and anthropogenic impact are synthesized from individual site descriptions given in Kjerfve [32] and updated with personal knowledge of the authors. Annual rainfall data represent averaged records from national meteorological stations where available or were extracted from site descriptions given in Kjerfve [32]. The number of tropical cyclones (hurricanes/tropical storms) with the eye within 80 km of a mangrove site during the CARICOMP monitoring program are also shown. Anthropogenic impacts: minimal = site virtually undisturbed; moderate = some nutrient input, visitation, and/or harvesting; extensive = alteration of drainage, high nutrient inputs, and adjacent infrastructure (housing, hotels, and roads).

Site	Type	Mangrove Area (ha)	Soil/Substrate	Rainfall (mm/Year)	Anthropogenic Impact	Tropical Cyclones
BAH	Open coast	~5	Calcareous	1000	Moderate	8
BAR	Lagoon	~10	Terrigenous/limestone	1291	Extensive	7
BEL1	Open coast	~5	Peat	1800	Minimal	9
BEL2	Open coast	~10	Peat	1873	Moderate	6
BER	Open coast	2.43	Peat	1464	Moderate	15
BON	Lagoon	~100	Calcareous	491	Minimal	3
CAY	Lagoon	3430	Peat	1493	Minimal	6
COL1	Lagoon	9	Terrigenous	400	Minimal	0
COL2	Lagoon	~50	Calcareous	2150	Moderate	5
CRI	Lagoon	250	Terrigenous	3321	Moderate	0
CUB	Open coast	~100	Peat/calcareous	1076	Minimal	12
CUR	Lagoon	~10	Calcareous	570	Moderate	3
DRE	Open coast	1500	Calcareous	1000	Moderate	4
FLA	Lagoon	~50	Calcareous	1059	Moderate	6
JAM	Open coast	~1	Limestone	1105	Minimal	4
MEX	Lagoon	>3000	Terrigenous	760	Minimal	1
PAN	Lagoon	~50	Peat	3428	Minimal	0
PRI	Open coast	996	Terrigenous	873	Moderate	5
T&T	Lagoon	80	Peat/limestone	1534	Moderate	3
VEN1	Lagoon	>1000	Calcareous	1213	Minimal	1
VEN2	Open coast	16	Terrigenous	383	Moderate	1

3.2. Anthropogenic Impacts

Mangrove wetlands are legally protected in all countries with CARICOMP mangrove sites, and direct human impact is minimal-to-moderate at most of the mangrove sites (Table 3). However, this was not always the case, and until the 1980s, mangroves were often cut down for construction of roads, piers, hotels, housing projects, and mangrove wetlands were often channelized, but have since then mostly enjoyed protection as bird and wildlife reserves, national parks, in the cases of BAR, BER, BON, and CAY as RAMSAR sites, and in the case of DRE, as a UNESCO World Heritage site (Parque Nacional Cotubanamá).

Graeme Hall Swamp (BAR) is probably the site with the most ongoing anthropogenic impact. Whereas the 32 ha wetland, of which ~10 ha consists of mangroves, is protected, dense residential development borders the wetland on three sides, with nutrient-rich runoff reaching the swamp from the northeast. The channel to the ocean is regularly cleared, and a sluicegate, which used to control ocean water entering the wetland across the sandbar, is no longer functional and remains closed, making BAR the site with the lowest salinity (Figure 2). Sand accretion and widening of the beach have also contributed to the swamp becoming more isolated from tidal seawater influx. Other sites are indirectly impacted by anthropogenic activities. For example, the mangrove forest at Laguna Gandoca (CRI) has nutrient-rich agricultural runoff that floods the mangroves and deposits silt across the wetland [53,54].

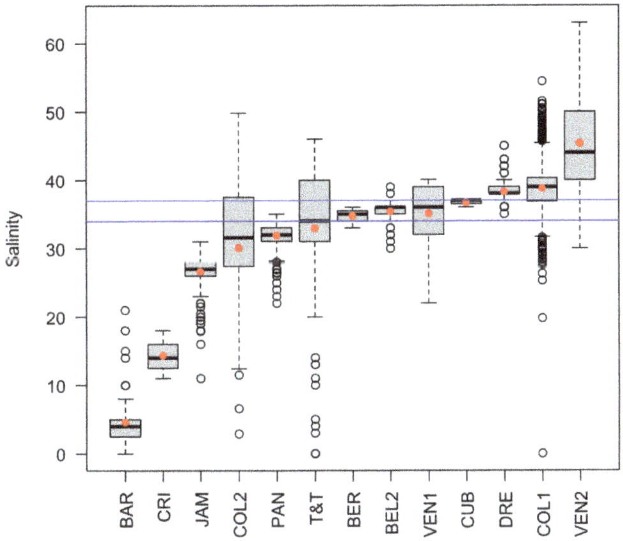

Figure 2. Salinity measurements at 13 CARICOMP sites ordered from lowest to highest salinity. Boxes represent upper and lower quartiles, and whiskers represent 1.5 IQR (interquartile range) above and below the limits of the box. As the medians (short black line inside each box) are almost the same as the means (red dots), the salinity data are approximately normally distributed. The two horizontal blue lines indicate the upper and lower limits of typical Caribbean Sea surface salinities.

Despite the protected status, clearing mangroves for wood and construction did occur at some sites. For example, at Twin Cays (BEL2), data collection had stopped by 2000 due to repeated theft of the litter traps, followed ultimately by illegal clear-cutting of the CARICOMP mangrove plots. At Punta de Mangle (VEN2), a site showing mangrove decline, plots had different longevity, and some of them were substituted in later years to avoid repeated theft of the litter traps, and to avoid data collection in heavily deteriorated areas.

3.3. Salinity

A total of 2467 mangrove salinity (salinity is non-dimensional, thus without units, but for practical purposes equal to ‰) measurements (Note #S5, Table S5A) were made at 13 sites (Figure 2). Most (81% of records) were for surface salinity whilst only 19% were interstitial measurements.

The overall mean salinity was 32.1 ± 10.3 with n = 13 (Note #S5, Table S5A). Only three of the site means were outside one standard deviation from the overall mean: Graeme Hall Swamp (BAR) with a mean salinity of 4.6 ± 3.6 with n = 84, Laguna Gandoca (CRI) with a mean salinity of 14.3 ± 3.5 and n = 3, and Punta de Mangle on Isla de Margarita (VEN2) with a mean salinity of 45.4 ± 6.0 with n = 227. Based on Aquarius satellite data [55], in situ measurements, and numerical modeling, sea surface salinity (SSS) in the Caribbean varies interannually by 0.5, ranging spatially from 34 to 37, mostly due to changes in the Amazon River plume [56], and the SSS in the southwestern Caribbean (Colombian Basin) is somewhat diluted by heavy regional rainfall and river runoff [57]. Four of the sites exhibited mean salinities within the 34–37 Caribbean SSS range. Six sites had lower mean salinities, BAR (4.6 ± 3.6), CRI (14.3 ± 3.5), JAM (26.6 ± 2.4), COL2 (30.1 ± 13.1), PAN (31.9 ± 1.9), and T&T (33.0 ± 9.6), and three sites had higher mean salinities, DRE (38.2 ± 1.3), COL1 (38.9 ± 4.1), and VEN2 (45.4 ± 6.0).

3.4. Mangrove Tree Structure

Tree structure of red mangroves varied widely among the sites. Repeat measurements of tree structure were made on 2470 red mangrove trees, summarized in Figure 3 (and Note #S6, Table S6A). There was considerable variability in the number of measurements among sites with a maximum number of 1238 measures on *R. mangle* at Ensenada de Chengue (COL1) but only 50 at Isla de Margarita (VEN2). Eight red mangrove trees exceeded 18.5 m in height (three at T&T, two at BAR, two at CRI, and one at VEN2), the two tallest trees reaching 20.4 and 20.0 m at T&T. Average tree heights exceeded 10 m only at BAR, CRI, and VEN2, but the average tree height (H) across all sites was more modest, 6.1 ± 2.8 m (Figure 3A, Note #S6, Table S6A). At ten sites, the trees were on average shorter than 5 m (BAH, BEL1, BEL2, CAY, COL2, CUB, CUR, MEX, PAN, and VEN1) (Figure 3A, Note #S6, Table S6A). Repeated tree circumference (C) measurements on the same tree typically varied by less than ±1 cm, yielding robust estimates of DBH (Figure 3B) and associated above-ground biomass (AGB). The average DBH measure for red mangroves across all sites was 9.0 ± 5.2 cm. The trees with the greatest DBH measured 42.8 cm at CRI and 41.7 cm at VEN2.

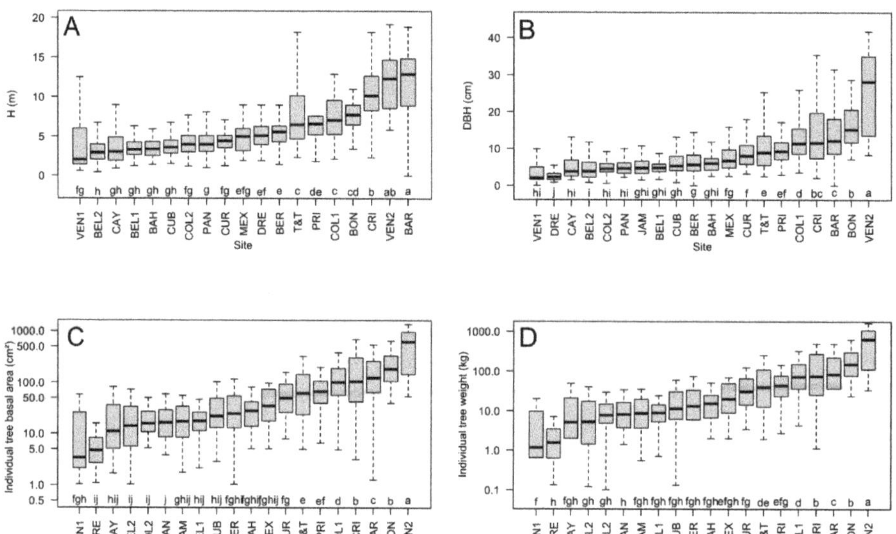

Figure 3. *R. mangle* tree structure parameters (**A**) tree height (H m), (**B**) tree diameter at breast height (DBH cm), (**C**) individual tree basal area (BA cm^2), and (**D**) individual tree above-ground weight (kg) (when summed equal to AGB) for the CARICOMP sites, ordered from lowest to highest. Each box corresponds to 50% of sample variation (i.e., it is the interquartile range—IQR) and the central bar inside the box is the median. The whiskers represent 1.5 × the IQR above and below the limits of the box, which correspond to the 3° and 1° quartile, respectively. Distance between y-axis values were log-transformed to optimize data visualization in plots (**C**,**D**). Letters within the plot indicate statistical differences between sites (α = 0.05), according to the Tukey's post hoc test [58] (these data are summarized in Note #S6, Table S6A).

The mean basal area (BA) varied significantly among sites from 4.1 m^2 ha^{-1} (BAH) to 44.8 m^2 ha^{-1} (VEN2), with an overall mean of 19.7 ± 11.1 m^2 ha^{-1} (Figure 3C, Note #S6,

Table S6A). Based on data from 74 plots across the 21 sites, red mangrove tree density (D) ranged from 733 trees ha^{-1} (BON) to 6030 trees ha^{-1} (PAN) (Note #S6, Table S6A). Five sites had mean D < 2000 trees ha^{-1} (BON, VEN2, BAH, CRI, and MEX); thirteen sites had mean D from 2000 to 5000 trees ha^{-1}, and three sites had mean D > 5000 tress ha^{-1} (PAN, COL2, BEL2) (Note #S6, Table S6A). *R. mangle* AGB varied across 20 sites from 32.9 t ha^{-1} (BAH) to 589.6 t ha^{-1} (VEN2) with a mean AGB across all sites of 189.9 ± 142.8 t ha^{-1}, equating to a mean AGC of 83.7 ± 63.0 t ha^{-1} (Note #S6, Table S6A). A significant trend line (Figure 4) indicates lower tree density for sites with greater AGB.

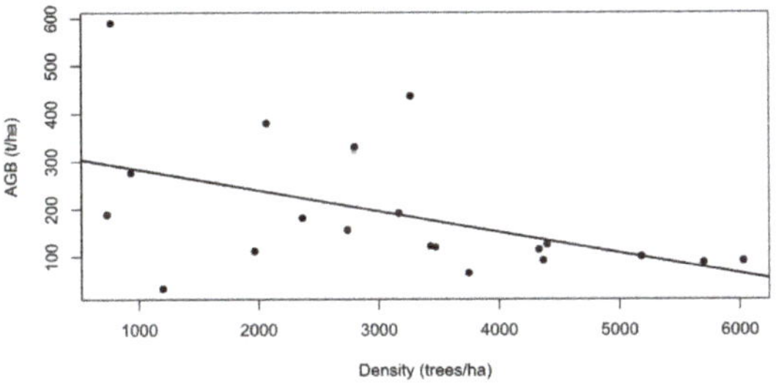

Figure 4. Regression of above-ground biomass (AGB t ha^{-1}) on tree density (D trees ha^{-1}) for 20 sites with $R^2 = 0.23$ and $p = 0.03$, where AGB = −0.0435 × D + 326.195.

3.5. Mangrove Seedlings

From 1998 to 2002, a total of 1228 *R. mangle* seedlings were measured for height and weight at five sites, BAR, BON, COL2, CRI, and DRE (Note #S7), and regressed as ln(W) on ln(H) (Figure 5), resulting in a highly significant relationship with $R^2 = 0.86$ (Figure 5). The tallest seedling measured 112 cm.

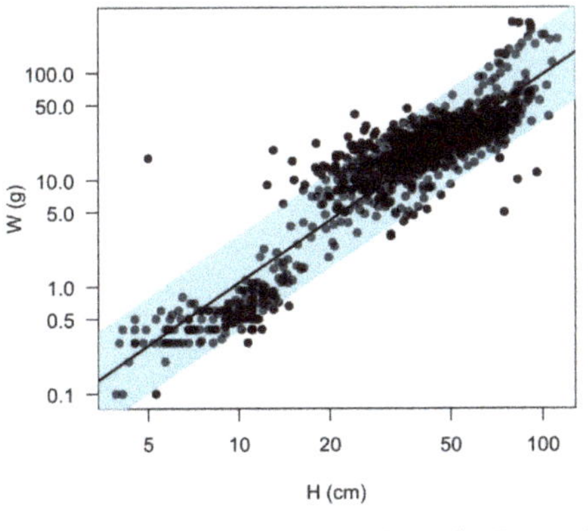

Figure 5. Regression of ln(W) on ln(H) for *R. mangle* seedlings, where the blue shading indicates the 95% confidence interval, ln(W) = 1.952 × ln(H) − 4.416, with n = 1228, $R^2 = 0.86$, $p < 0.0001$.

From 1992 to 2017, a total of 8440 *R. mangle* seedlings were counted and measured for height at six sites, BAR, BER, BON, CRI, DRE, and PAN (Note #S7, Table S7A). Data beyond 2013 were only measured at BAR. The comparisons of seedling biomass and density are plotted in Figure 6 (and summarized in Note #S7, Table S7A).

The average seedling height was 61 ± 29 cm with outliers, reaching 194 cm at BER, BAR, and PAN (Note #S7, Table S7A). The greatest density of seedlings was observed at BAR with an overall mean of 74.2 seedlings m^{-2}, whereas PAN and CRI both had very low seedling density, 1.2 and 1.7 seedlings m^{-2}, respectively. Seedling biomass was calculated for each site using the derived height vs. weight regression equation (Figure 5), although

23 seedlings at BER (1.7%), 112 seedlings at BAR (2.5%), and 182 seedlings (69.2%) were taller than 112 cm. BAR had by far the highest mean seedling biomass, 21.3 t ha^{-1} and CRI the lowest, 0.5 t ha^{-1} (Figure 6). Seedling biomass equaled 5.6% of tree AGB at BAR; 3.3% at DRE, 2.5% at BER, 2.4% at BON, and 2.1% at PAN, while at CRI, the seedling biomass measured only 0.2% of AGB.

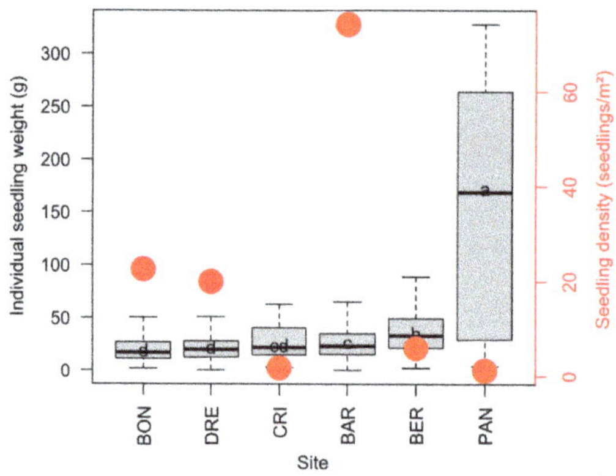

Figure 6. Individual *R. mangle* seedling biomass (box plots) and mean seedling density (red dots), ordered from lowest to highest seedling weights, for the six CARICOMP mangrove sites with recorded seedling height measurements (Note #S7, Table S7A). Letters within the boxes indicate statistical differences ($\alpha = 0.05$) according to the Tukey's post hoc test [58].

3.6. *Litterfall Productivity*

Mangrove productivity (i.e., litterfall) was measured at 17 of the 21 CARICOMP sites (Figure 7), with the first traps set in 1992, and the last included in this analysis set in 2013. Litterfall measurements were not made at FLA, MEX, BEL2, and JAM. There were a total of 23,957 trap collections from which red mangrove litter was measured and analyzed. During the 22 years of monitoring, red mangrove litter was collected during 729 month-long sets, each set normalized to 30 days. As traps measured 0.25 m^2, litterfall rates were normalized to m^{-2} (Note #S8, Tables S8A and S9A).

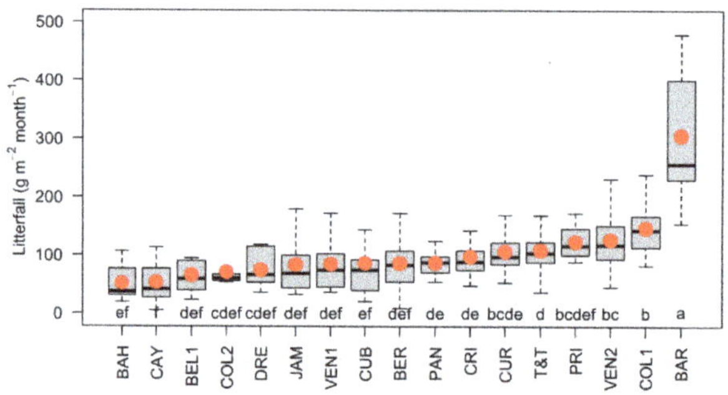

Figure 7. *R. mangle* litterfall (g m^{-2} month $^{-1}$ dry weight) (box plots) with mean values (red dots) for 17 CARICOMP sites ordered from lowest to highest. Letters below the boxes indicate statistical differences ($\alpha = 0.05$) according to the Tukey's post hoc test [58]. (Note #S8, Table S8A).

Across the 17 sites and all sampling periods, the mean productivity of *R. mangle* was 1.238 ± 0.695 kg m^{-2} yr^{-1} (dry weight) with leaves contributing 57.3% of the litterfall, fruits (propagules) 22.0%, bracts 5.6%, flowers 5.0%, and wood 7.1% (Note #S8, Table S8A). BAR proved to be an outlier, generating red mangrove litterfall at 3.660 kg m^{-2} yr^{-1}, exceeding the overall mean *R. mangle* litterfall for all sites by more than three standard deviations, but with all samples collected in 2005, an El Niño year [59]. CAY, BAH, COL2, and CUB generated the least *R. mangle* litterfall (Figure 7).

By averaging *R. mangle* litterfall monthly across the sites, the statistics are slightly different (1.264 ± 0.209 kg m^{-2} yr^{-1} dry weight; Note #S9, Table S9A). A distinct seasonal litterfall cycle is apparent (Figure 8). Total litterfall peaks in June, more than one standard deviation above the annual mean, mostly due to fruit litterfall. The lowest red mangrove litterfall occurs December–February, these three months having litterfall more than one standard deviation below the annual mean. The six measured plant components contribute differently to the overall litterfall. Leaves produce 59.6% of the total annual litterfall for red mangroves, fruits/propagules were the second most important litterfall component at 15.9%. The leaf litterfall was similarly high from March to November. Fruit litterfall is highest May–August, bract litterfall (7.0%) peaks in June and is lowest January–March, flower litterfall (6.6%) peaks August–October, and wood debris (7.8%) is highest in November (Figure 8).

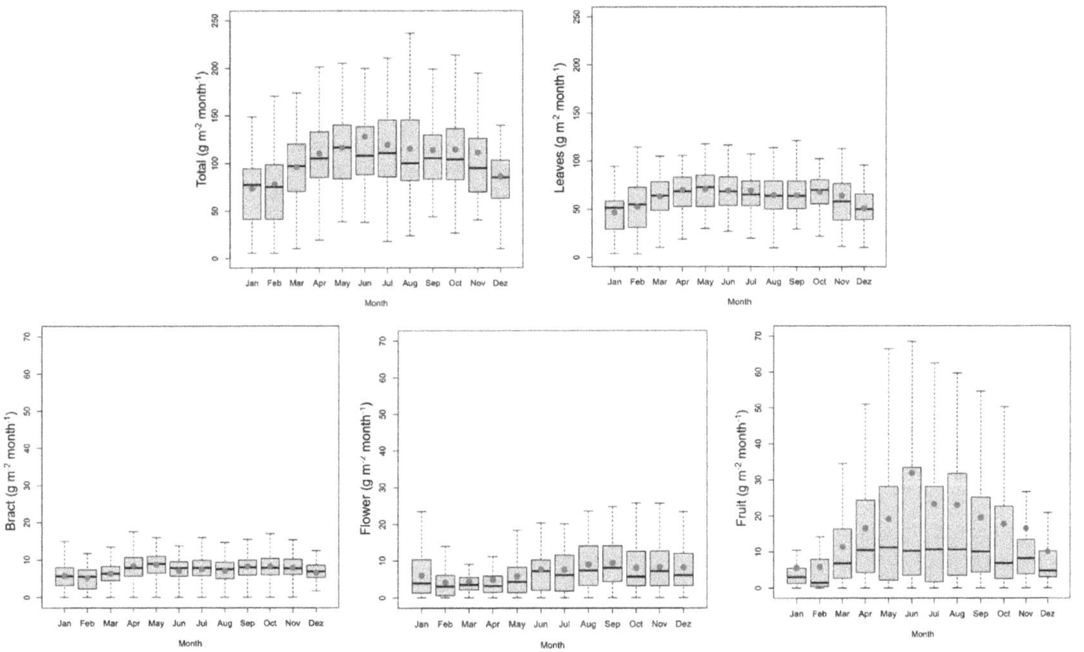

Figure 8. Seasonal litterfall (g m^{-2} $month^{-1}$ dry weight) shown as monthly means across sites separately for total litter (including wood and miscellaneous debris), leaves, bracts, flowers, and fruits (propagules). Horizontal lines in boxes represent the monthly medians, red dots represent the monthly means (Note #S9, Table S9A).

Only seven of the sites had sufficiently long records to allow exploration of the extent to which litterfall changed with time. Sites with litterfall measurements for at least 9 years (Note #S8, Table S8A) are BER 9 years, CAY 12 years, CRI 14 years, COL1 15 years, PAN 15 years, VEN2 16 years, and T&T 19 years. Since start and end years were different among the sites, we chose to examine change over time at individual sites rather than merging the data. The resulting linear regressions proved to be non-significant for CAY, CRI, and T&T, i.e., no temporal trend in litterfall at these sites for the duration of the CARICOMP monitoring. Two sites exhibited significant decline in litterfall with time, VEN2 with $R^2 = 0.66$ ($p = 0.0006$), and

COL1 with $R^2 = 0.29$ ($p = 0.02$); and two sites exhibited significant increase in litterfall with time, PAN with $R^2 = 0.27$ ($p = 0.04$); and BER with $R^2 = 0.40$ ($p = 0.09$) (Figure 9; Note #S10). The decline in total litterfall at VEN2 from 1992 to 2008 by approximately 50% is most remarkable.

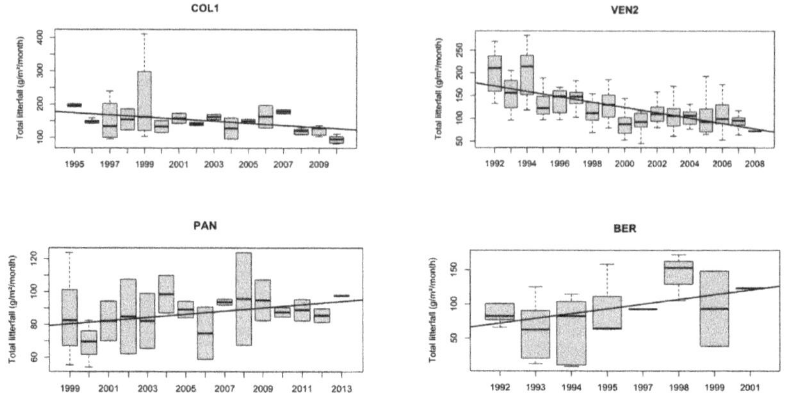

Figure 9. Time series box plots of *R. mangle* total litterfall (g m^{-1} $month^{-1}$ dry weight) for the four sites with a significant decreasing or increasing litterfall rate over time.

3.7. *Temperature and Rainfall*

Mean tree height (H), diameter at breast height (DBH), and tree density (D) are three independent tree structure parameters, whereas above-ground biomass (AGB) is calculated from DBH, and basal area (BA) is calculated from DBH and the number of trees. Considering *R. mangle*, we explored to what extent these independent tree structure parameters and litterfall productivity varied with latitude, which can be viewed as a proxy of sustained minimum annual temperature, irradiance, and length of day. Our null hypotheses (H_0) were that these parameters do not vary with latitude. By regressing separately each parameter on latitude, we found that DBH increased with decreasing latitude ($R^2 = 0.20$, $p = 0.045$, n = 20), indicating that 20% of the variability in DBH is explained by latitude, increasing towards the equator (Figure 10), likewise, H, AGB, and BA but not density (Note #S11). There was no significant trend of mean litterfall with latitude.

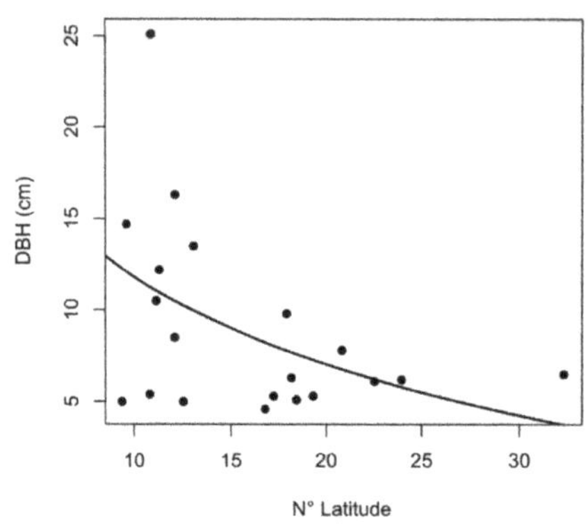

Figure 10. Relationship between DBH (cm) and latitude for 20 CARICOMP sites with $R^2 = 0.20$ and $p = 0.045$, with DBH (cm) = 27.647 − 6.878 × ln(°Lat).

Regression of DBH on annual rainfall measurements at 20 CARICOMP mangrove sites (Figure 11) yielded highly significant regression for DBH on rainfall with $R^2 = 0.27$, $p = 0.018$, with DBH (cm) = 4.95 − 0.40 × ln Rain (mm yr^{-1}), indicating that sites with higher mean DBH experiences lower mean rainfall. This was also the case for AGB, whereas regression of H, BA, and Density on rainfall were not significant (Note #S12). Overall, rainfall explains 27% of the variability in mean site DBH.

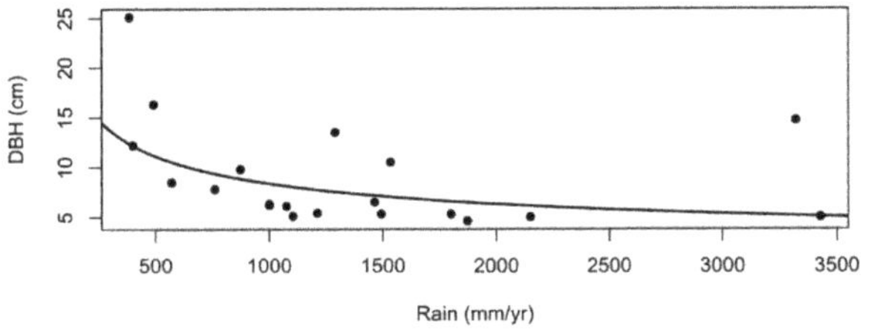

Figure 11. Relationship between DBH (cm) and rainfall (mm yr^{-1}), R^2 = 0.27 (p = 0.018) for n = 20, with ln (DBH) = 4.95 − 0.40 × ln(Rain).

Regressing total site litterfall on mean rainfall (Table 3) for 17 sites yielded no significant trend. However, considering separately the six sites with available annual rainfall data (Note #S12; Table S12A) and at least 35 sets of litterfall measurements (i.e., BER; CAY; CRI; PAN; T&T; and VEN2; Note #S8; Table S8A); litterfall significantly decreased with increasing rainfall at VEN2 (R^2 = 0.28; p < 0.0001) and CAY (R^2 = 0.15; p = 0.02). In contrast, litterfall significantly increased with increasing rainfall at BER (R^2 = 0.15; p = 0.02) and PAN (R^2 = 0.14; p = 0.01), whilst at CRI, T&T, and VEN1, there was no significant correlation between litterfall and rainfall.

Further, we explored the potential effects of two largely independent ocean–atmosphere climate indices (ONI and AMM), sometimes correlated with interannual climate variability in the wider Caribbean but found no relationship between either index and litterfall (Note #S13).

3.8. Hurricane Impact

Recording of tropical cyclone impacts at the CARICOMP mangrove sites was not conducted systematically, and recollections are limited to observations at only a few sites. DRE seemingly suffered the most significant hurricane damage of the CARICOMP mangrove sites during the monitoring program. At BER, ~25% biomass loss was recorded in 1999. Significant hurricane damage was also noted at CAY, and VEN2, albeit not quantified. Fortuitously, however, CARICOMP tree structure (Table 4) and litterfall (Table 5) measurements were made at DRE both before and after Hurricane George (Note #S14, Figure S14A) crossed the site as a category-4 storm in September 1998, permitting quantification of the hurricane impact at this site. Based on these measurements, only 33% of the trees survived the hurricane and the site mean AGB and BA were reduced to 8% and 6% of pre-hurricane values, respectively (Table 4). Likewise, litterfall in 2000 was only 37% of the litterfall recorded prior to H. George (Table 5).

In the two years subsequent to H. George, DRE became overgrown by tall, spindly saplings (mean H = 4.2 ± 1.2 m, mean DBH = 1.7 ± 0.5 cm). This may simply be typical forest succession (with one species) in what could be called "a race for the canopy". Using Santos et al.'s [37] allometric relationship to calculate sapling AGB (possibly an overestimate) indicates that the saplings would have boosted site plant AGB by 17% to a total 10.8 t ha^{-1}, and plant basal area by 56% to a total 1.4 m^2 ha^{-1} if included with the trees (Table 4). If the saplings were included in the tree density calculations, DRE tree density would have increased by 140% to a total of 3433 trees ha^{-1} subsequent to H. George (Table 4), but still far from the tree density prior to the hurricane. Ignoring that some of the saplings in 2000 probably were seedlings in 1996, the combined seedling and sapling above-ground biomass at DRE subsequent to H. George could have amounted to as much as 62% compared to the tree AGB at DRE (Table 4; Note #S7, Table S7A).

Table 4. *R. mangle* tree structure measurements at DRE before and after Hurricane George (1998). Saplings are trees with DBH < 2.5 cm, and H 195–610 cm.

Site	Year	Sets	Plots	Trees	Samples	Height Mean (m)	Height SD (m)	DBH Mean (cm)	DBH SD (cm)	Density (Trees/ha)	Basal Area (m^2/ha)	AGB (t/ha)
Trees	1996	1	3	131	131	7	2.6	6.3	2.6	4367	16	122.4
Saplings	1996	1	3	0	0	-	-	-	-	0	0	0
Trees	2000	1	3	43	43	6.4	1.4	3.5	0.8	1433	0.9	9.2
Saplings	2000	1	3	60	60	4.2	1.2	1.7	0.5	2000	0.5	1.6

Table 5. *R. mangle* litterfall measurements at DRE before and after Hurricane George (1998).

Start	End	Traps	Traps/Set	Sets	Bracts	Flowers	Fruits	Leaves	Misc.	Wood	Mean (g m^{-2} $month^{-1}$)	Annual (kg m^{-2} yr^{-1})
1997	1997	30	1	1	4.9	6.8	3.6	181.7	3.1	26.4	226.5	2.7
2000	2001	30	30	13	5.3	3.4	24.5	47.5	0.7	5.7	87.1	1.0

Hungry Bay, Bermuda (BER) was exposed to 15 tropical cyclones during the CARICOMP monitoring program. The most seaward plot at BER was severely impacted by swells from Hurricane Gert (category-2; 1999), losing about 25% of the biomass, and then essentially destroyed by category-3 Hurricane Fabian (2003). The mangrove area shrunk in response to coastal erosion, driven by sea level rise, and destruction of the protective peninsula, allowing storm waves to kill many trees in 2003. BER also suffered an absence of any mangrove seedlings in the two most seaward plots, indicating recruitment failure, and thus no capacity for recovery from hurricane tree loss. Hurricanes Igor (category-1, 2010), Faye (category-1, 2014), and Gonzalo (category-3, October 2014) caused additional damage to the Hungry Bay mangroves, with additional canopy loss, and destruction of the seaward mangrove margin. A substantive seawall was constructed in 2023 to close a gap created by Hurricane Fabian in the peninsula protecting what is left of the Hungry Bay mangroves. Some re-planting has been successful.

Category-4 Hurricane Ivan (2004) destroyed much of the tall mangrove forest throughout Central Swamp (CAY). The CAY mangroves receded so much that the leaf litter plots were no longer under the tree canopy in most areas. The forest is still recovering two decades later.

The eye of category-5 Hurricane Lenny (1999) only came within 600 km of Margarita Island (thus not included in Table 3), but nevertheless significantly changed the shore morphology at VEN2 by sedimentation from hurricane swells, and as a result, impairing tidal flooding of the Punta de Mangle (VEN2) mangroves. This caused the *R. mangle* trees to degenerate over the following years, with many trees dead by 2008. H. Lenny, albeit far away, caused decreased mangrove litter production, especially fruit production, which became minimal and irregular. By the end of data collection in 2008, many trees at VEN2 had died in the two permanent plots that were monitored for 16 years.

4. Discussion

Spanning more than 22 years across the wider Caribbean region, the CARICOMP program represents one of the geographically largest and longest comprehensive mangrove monitoring programs. As expected from such a diverse and long-time collaboration, there are variations in the number and frequency of data collection efforts among sites. However, the dataset was strengthened by the considerable effort put towards a standardized methodology. This effort included training, development, and methods shared within the network, which have been detailed in previous publications [27–31].

We made measurements on 2470 individual *R. mangle* trees and found substantial variations among sites in mean tree height (H, 3.1 to 11.9 m), mean diameter at breast height (DBH, 4.6 to 25.1 cm), mean basal area (BA, 4.1 to 44.8 m^2 ha^{-1}), tree density (D, 733 to 6030 trees ha^{-1}), and mean above-ground biomass (AGB, 33 to 590 t ha^{-1}; equating to above-ground carbon (AGC) of 15 to 260 t ha^{-1}). These mean values for H, DBH, and BA, are consistent with other studies in Latin America and the Caribbean [60]. However, our AGB for several sites (BAR, COL1, CRI, T&T, and VEN2) is higher than the previously reported AGB range in Latin America and the Caribbean that spans from 8 t ha^{-1} in Florida to 279 t ha^{-1} in Panama [60]. Overall, AGB recorded across the CARICOMP sites (overall mean 190 t ha^{-1}) is higher than previously reported for mangrove forests in North and Central America (mean 145 t ha^{-1}) but similar to values modelled for South America (mean 186 t ha^{-1}) [61]. Differences in allometric calculations of AGB based on DBH (and sometimes also H) could, in part, be responsible for these differences in AGB values reported in the literature.

Among potential drivers for the observed spatial variability in mangrove forest structure across the wider Caribbean, are variations in environmental conditions (Table 3) linked to the latitudinal differences, *inter alia*, among the CARICOMP sites that span 23 degrees of latitude from N 9° (CRI, PAN) to N 32° (BER). Chollett et al. [40] corroborated the wide variations in environmental setting across the CARICOMP sites. Our results indicate that latitude (as a proxy for temperature/radiation) explains 20% of the variability in DBH, with both DBH and H increasing towards the equator. Rainfall explains 27% of the variation in DBH, which decreases with increasing rainfall across the same 20 sites. However, we found no significant influence of the seasonal ocean–atmosphere climate indices (ONI and AMM) on mangrove tree structure. Further, we found no consistent influence of rainfall or climate variability on mangrove productivity (measured as litterfall) for sites with sufficient data to test, and no consistent change across sites over time.

Our results are in partial agreement with a predictive model developed by [61] using red mangrove AGB values from 102 points (merged into 52 locations) across the globe together with mean temperature and rainfall at varying temporal scales to predict red mangrove biomass. The model indicates that mangrove biomass will be higher at locations with higher temperature (in line with our findings) and higher rainfall (contrary to our findings). The true influence of rainfall on mangrove biomass across the CARICOMP sites could perhaps be masked by differences in their geomorphology (Table 3), runoff from local rivers, tidal flushing, and/or anthropogenic impacts. For example, in BAR, where extensive anthropogenic impacts have resulted in high nutrient input and low salinity conditions (Table 3, Figure 2), mangrove trees are tall with high AGB (Figure 3), and with the highest productivity (litterfall) of all CARICOMP mangrove sites (Figure 7). Likewise at CRI, which receives nutrient-rich agricultural runoff, the trees are particularly tall with high AGB (Figure 3), albeit with low litterfall.

Mangrove seedlings likewise exhibited great variability across the sites. At Graeme Hall Swamp (BAR), the seedling density was very high at a mean of 74 seedlings m^{-2}, and seedling biomass of 21.3 t ha^{-1}. The high seedling biomass is most likely due to the presence of large mature trees and zero tidal flushing at the Barbados site, which essentially is located in what is now an almost freshwater lake, and therefore, propagules are not flushed from the system but simply fall to the ground and propagate among enormous competition. At the other extreme, PAN and CRI exhibited extremely low seedling density, 1.2 and 1.7 seedlings m^{-2}, respectively. A few additional and separate plots at Bocas del Toro (PAN) were established, where seedlings were tagged but most of them died, partially due to trampling during illegal tree cutting and removal of logs. One plot was completely lost due to tree cutting and subsequent construction of a dock and the framing of a house. Further, the water across the mangrove plots was uncharacteristically deep, at least waist deep at high tide, hindering seedlings from becoming established. The almost constant inundation of the CARICOMP mangrove plots at PAN is probably the result of slumping during an earthquake with evidence of slumping across the entire Bocas del Toro mangrove system [62,63]. Since the PAN seedlings were on average more than twice as tall than those at CRI, PAN seedling biomass (1.9 t ha^{-1}) was greater than seedling biomass at CRI (0.5 t ha^{-1}). Across all sites, seedling biomass varied substantially, representing between 0.2% of AGB at CRI and 5.6% of AGB at BAR.

Additionally, variation may be linked to the fact that some of the CARICOMP mangrove sites are located on the mainland, while others are located on Caribbean islands with different hydrological characteristics. Interstitial salinity is particularly important to the health of mangroves [64–66] and is in turn affected by the availability of freshwater, which depends on both precipitation and river runoff. Laguna Gandoca (CRI) has high tree structure parameter values (Figure 2), and low median salinity (14.3) as a result of both the second highest rate of precipitation among the CARICOMP mangrove sites and discharge by Rio Gandoca. Graeme Hall Swamp (BAR), with high mangrove AGB, receives relatively modest precipitation but is fed by a large watershed and a sluicegate limits tidal exchange, resulting in the lowest mean salinity (4.6) among the CARICOMP mangrove sites. Similarly to mangroves, seagrass meadows at the same CARICOMP monitoring sites showed wide spatial variation in seagrass community characteristics and productivity that made it difficult to generalize about changes occurring across Caribbean seagrass meadows [67].

Within the CARICOMP timeframe, from 1992 to 2013, some locations were able to monitor the mangrove plots across all or many of the 22 years, while others were not, making assessment of changes over time across the region challenging. Mangrove forest structure at a few sites where sufficient data were available (BAR, BER, COL1, PAN, and T&T) underwent changes through time but these were not consistent across sites. For example, average tree height declined consistently over time in Graeme Hall Swamp (BAR) due to the loss of several large mature trees and continuous recruitment of many saplings (DBH > 2.5 cm) that were then included in the tree dataset at this site. Tree density declined in Chengue Bay (COL1), while in Bocas del Toro (PAN) and Bon Accord Lagoon (T&T), there was a marked increase in biomass. Decreases in water quality and increasing temperatures and human density were identified through anecdotal evidence of CARICOMP participants as potential stressors over time that varied significantly among sites [40]. For example, while mangrove forests are able to act as filters for nutrients and suspended particles, at excessive levels, these variables can negatively affect mangrove trees [68]. Hydrological changes, hypersalinity, hurricanes, logging, and sea level rise were all put forward as likely stressors at given sites, by participants, but were not specifically measured by the CARICOMP program, although well recognized as causative factors in mangrove mortality, e.g., [69,70]. At some sites, a phase shift in forest maturity as a result of recovery from a disturbance appeared to explain changes seen. These examples illustrate heterogeneity among the sites but also support the influence of recognized drivers of decline in mangroves such as sea level rise, promoting the retreat of fringing mangroves, as has been observed in Hungry Bay (BER) for more than three decades [15], and hurricane impact.

When the CARICOMP monitoring program was conceived in 1985, measuring hurricane impact was not an explicit objective. Four decades later, in analyzing the CARICOMP mangrove data, it is apparent that the major changes to *R. mangle* tree structure across the Caribbean were largely, directly or indirectly, the result of hurricanes. With exception of COL1, PAN, and CRI, all CARICOMP sites experienced hurricanes during the 22-year monitoring program. Some of the impacts were major, but we only have quantitative data for one site (DRE), and reliable qualitative observations for another three sites (BER, CAY, and VEN2), while, presumably, all other sites experienced destruction by tropical cyclones to undocumented degrees. The measurements of tree structure parameters and litterfall at DRE both prior and subsequent to Hurricane George in 1998, was a rare opportunity to measure the impact and destruction due to a hurricane. The storm destroyed all tall trees in Parque Nacional Cotubanamá, including Canal Punta Mangle al Catuano (DRE), and totally devastated the mangroves site. Similarly, *R. mangle* trees at BER, CAY, and VEN2 suffered severely from hurricane impacts. At BER, storm waves during Hurricane Fabian (2003) eroded completely at least one of the mangrove plots. At CAY, hurricane winds destroyed the foliage to the point that litter traps were no longer below the tree canopy. At VEN2, hurricane swells from far away restructured the beach to the point that the mangrove plots were no longer flooded by the tide, which caused trees to degenerate and die. Certainly, the decline in litterfall by 50% at VEN2 can also be attributed to the same effect. In all cases, the mangrove sites have not recovered two decades later.

5. Conclusions

Although mangroves in the wider Caribbean mostly cover much less area than mangroves in Southeast Asia, Africa, and South America, they typically have high tree density thus providing valuable ecosystem services that support key economic sectors (e.g., tourism and fisheries) in both mainland and island nations across the region [26]. These services include enhancement of both terrestrial and marine biodiversity through provision of unique marine forest habitat, and services that improve climate resilience such as shoreline protection and sequestration of substantial amounts of blue carbon.

We demonstrate high variability in red mangrove structure and productivity across wider Caribbean sites, less than half of which can be attributed to latitude and annual rainfall. We also reveal a lack of consistency among sites in long-term trends in mangrove productivity. These differences suggest that local-level drivers of forest structure, productivity, and change are important. Anecdotal and some quantitative evidence reported here suggest that hurricanes, abrupt sea level events, and local nutrient enrichment have been important drivers of change over the period 1992–2013. Furthermore, since this time inundation by pelagic sargassum is now causing widespread damage to fringing mangroves across the region [71]. This emphasizes that conservation and management strategies including blue carbon accounting cannot follow a 'one-size fits all' approach but need to be site-specific and supported by the monitoring of local environmental events.

Considering the future of mangrove monitoring, we note that the unique CARICOMP monitoring program was stopped after 22 years, as funding and regional coordination could not be sustained without the continued efforts of core founding members (most of whom are now retired or deceased). Important lessons learned from collecting and analyzing CARICOMP mangrove data include (1) field data collection in mangroves is time consuming, costly and even hazardous, and therefore difficult to sustain over the long-term; (2) standardized data collection is essential for building a shared regional database to compare across sites and monitor regional change, but is difficult over such varied physical environments; and (3) explicit protocols for monitoring abrupt events should be included to quantify major drivers of change in regional mangroves such as hurricanes and sea-level events.

Looking forward, many parameters can now be measured using remote-sensing technologies, which could reduce cost and increase efficiency and safety if adopted by future monitoring programs. Such technologies include drone and satellite imagery to assess canopy height, areal extent, tree defoliation and mortality, inundation by water and sargassum, and other measures [72–76]. Automated loggers to continuously record temperature, salinity, water level, and other parameters are also being used successfully [77].

Despite advances in remote sensing technologies, individual tree measurements are still required for monitoring changes in mangrove forest structure and productivity (including accurate rates of carbon sequestration and emissions), and these cannot yet be conducted remotely [72,75]. However, time and cost of manual field monitoring could be greatly reduced by simplifying the range of parameters collected (where tree number and DBH are the most important and robust measures for forest structure and AGC, and leaf litter for productivity).

Reviving and maintaining a regional mangrove monitoring program that can detect changes and reveal drivers of change in support of management will require a new injection of funding, regional commitment, and wise use of remote-sensing technology together with intelligently structured field data collection that improves efficiency and reduces costs. As intimated by Howard et al. [72] and Ávila-Acosta et al. [78], a focus on the role of mangroves in coastal blue carbon and climate adaptation and resilience could bring new opportunities for funding long-term regional monitoring.

Supplementary Materials: The following supporting information can be downloaded at: https://www.mdpi.com/article/10.3390/environments12120463/s1, including: Note #S1—Tree Structure Data [79,80]; Note #S2—Mean Number of Live Mangrove Trees; Note #S3—Calculating Above-Ground Biomass [81–84]; Note #S4—Tides [85–89]; Note #S5—Salinity; Note #S6—Results of Tree Structure Measurements; Note #S7—Mangrove Seedlings; Note #S8—Mangrove Litterfall by Site; Note #S9—Seasonal Litterfall; Note #S10—Regression of Litterfall on Time; Note #S11—Regressions of Tree Parameters on Latitude; Note #S12—Rainfall [90,91]; Note #S13—Climate Indices [92,93]; Note #S14—Hurricanes [94–99].

Author Contributions: Conceptualization: J.O., J.C., J.S.-V. and C.B. Writing and editing: B.K., H.A.O., I.C.F., J.S.-V., S.R.S. and R.V. Editing: R.C., J.C., K.K., R.J., C.M., F.X.G., J.G.-F., J.P., J.C.C., I.A.P., I.M.-G. and A.F. Data curation and analysis: B.K., I.M.-G., J.S.-V., A.F. and C.B. Statistical analysis: I.A.P. and B.K. Data monitoring: teams of local scientists, managers, and students at the 21 CARICOMP sites. Author John Ogden passed away prior to the publication of this manuscript. All authors have read and agreed to the published version of the manuscript.

Funding: Funding for the CARICOMP program was obtained from UNESCO, The John D. and Cathrine T. McArthur Foundation, Summit Fund, Henry Foundation, National Science Foundation, U.S. Department of State, U.S. Department of the Interior, and UNEP.

Data Availability Statement: All data are summarized in the tables in the Supplemental Information or by requesting the raw and unedited data from bkjerfve@id.uff.br in the form of some 60 Excel files and a large Access database file.

Acknowledgments: Marc Steyaert (deceased) of the UNESCO Coastal Marine Program (later, Coastal Regions and Small Islands) recognized the scientific opportunity and funded the initial workshops and much of the first decade of the CARICOMP program. The Steering Committee: E. Jordán-Dahlgren, J. Ogden (deceased), J. Garzón-Ferreira, I. Goodbody (deceased), B. Kjerfve, P. Penchaszadeh, W. Wiebe, J. Woodley (deceased), and J. Zieman (deceased) implemented the scientific monitoring programs, negotiated funding, and made administrative and logistical arrangements. Peter Bacon (deceased) of the University of the West Indies, St. Augustine campus, was the original architect of the mangrove monitoring program. The many Caribbean marine laboratories and institutions, which formally joined the CARICOMP program over the years, contributed facilities and personnel. The Site Directors worked diligently with local scientists, managers, and students to collect data and made our annual meetings educational, productive, and great fun. Many appear as co-authors of CARICOMP publications listed in the "References" section. The West Indies Laboratory of Fairleigh Dickinson University in St. Croix hosted CARICOMP from 1985 to 1987. The

Florida Institute of Oceanography (FIO) of the University of South Florida was the administrative center for the program from 1988 to 2010. We thank D. Milliken (deceased), S. Vargo, D. Haynes, R. Erdmann, and R. Walker for administration of grant and contract support and for facilitating meetings. The University of the West Indies, Centre for Marine Sciences (CMS) in Kingston provided operational and logistical support for the Data Management Center (later, Caribbean Coastal Data Centre—CCDC). We thank CMS Directors I. Goodbody, J. Woodley, G. Warner, D. Webber, and data managers D. Linton and M. Ford. UNESCO Coastal Marine Program, The John D. and Catherine T. MacArthur Foundation, Phase I (1991–1994), and Phase II (1995–1999), provided core funding for CARICOMP. Supplemental funding was provided by (1) Summit Fund; (2) Henry Foundation; (3) United States National Science Foundation; (4) Coral Reef Initiative, United States Department of State; (5) Coral Reef Program of the Office of Insular Affairs, United States Department of the Interior; and (6) UNEP Caribbean Environment Program. This is contribution number 1086 of the Caribbean Coral Reef Ecosystems Program (CCRE), Smithsonian Institution.

Conflicts of Interest: The authors declare no conflicts of interest and agree to this submission.

References

1. Bunting, P.; Rosenqvist, A.; Hilarides, L.; Lucas, R.M.; Thomas, N.; Tadono, T.; Worthington, T.A.; Spalding, M.; Murray, N.J.; Rebelo, L.-M. Global Mangrove Extent Change 1996–2020: Global Mangrove Watch Version 3.0. *Remote Sens.* **2022**, *14*, 3657. [CrossRef]
2. Rovai, A.S.; Twilley, R.R.; Castañeda-Moya, E.; Midway, S.R.; Friess, D.A.; Trettin, C.C.; Bukoski, J.J.; Stovall, A.E.L.; Pagliosa, P.R.; Fonseca, A.L.; et al. Macroecological patterns of forest structure and allometric scaling in mangrove forests. *Glob. Ecol. Biogeogr.* **2021**, *30*, 1000–1013. [CrossRef]
3. Vo, Q.T.; Kuenzer, C.; Vo, Q.M.; Moder, F.; Oppelt, N. Review of valuation methods for mangrove ecosystem services. *Ecol. Indic.* **2012**, *23*, 431–446. [CrossRef]
4. López-Angarita, J.; Roberts, C.M.; Tilley, A.; Hawkins, J.P.; Cooke, R.G. Mangroves and people: Lessons from a history of use and abuse in four Latin American countries. *For. Ecol. Manag.* **2016**, *368*, 151–162. [CrossRef]
5. Barbier, E.B.; Hacker, S.D.; Kennedy, C.; Koch, E.W.; Stier, A.C.; Silliman, B.R. The value of estuarine and coastal ecosystem services. *Ecol. Monogr.* **2011**, *81*, 169–193. [CrossRef]
6. Mcleod, E.; Chmura, G.L.; Bouillon, S.; Salm, R.; Björk, M.; Duarte, C.M.; Lovelock, C.E.; Schlesinger, W.H.; Silliman, B.R. A blueprint for blue carbon: Toward an improved understanding of the role of vegetated coastal habitats in sequestering CO_2. *Front. Ecol. Environ.* **2011**, *9*, 552–560. [CrossRef]
7. Duarte, C.M.; Atwood, T.B.; Kairo, J.G.; Kennedy, H.; Krause-Jensen, D.; Lovelock, C.E.; Serrano, O. *UNESCO Marine World Heritage: Custodians of the Globe's Blue Carbon Assets*; CLT-2021/WS/1; UNESCO: Paris, France, 2021; p. 15.
8. Farnsworth, E.J.; Ellison, A.M.; Gong, W.K. Elevated CO_2 alters anatomy, physiology, growth, and reproduction of red mangrove (*Rhizophora mangle* L.). *Oecologia* **1996**, *108*, 599–609. [CrossRef] [PubMed]
9. Adame, M.F.; Connolly, R.M.; Turschwell, M.P.; Lovelock, C.E.; Fatoyinbo, T.; Lagomasino, D.; Goldberg, L.A.; Holdorf, J.; Friess, D.A.; Sasmito, S.D.; et al. Future carbon emissions from global mangrove forest loss. *Glob. Change Biol.* **2021**, *27*, 2856–2866. [CrossRef]
10. Valiela, I.; Bowen, J.L.; York, J.K. Mangrove forests: One of the world's threatened major tropical environments. *BioScience* **2001**, *51*, 807–815. [CrossRef]
11. Feller, I.C.; Friess, D.A.; Krauss, K.W.; Lewis, R.R. The state of the world's mangroves in the 21st century under climate change. *Hydrobiologia* **2017**, *803*, 1–12. [CrossRef]
12. Goldberg, L.; Lagomasino, D.; Thomas, N.; Fatoyinbo, T. Global declines in human-driven mangrove loss. *Glob. Change Biol.* **2020**, *26*, 5844–5855. [CrossRef]
13. Murray, N.J.; Worthington, T.A.; Bunting, P.; Duce, S.; Hagger, V.; Lovelock, C.E.; Lucas, R.; Saunders, M.I.; Sheaves, M.; Spalding, M.; et al. High-resolution mapping of losses and gains of Earth's tidal wetlands. *Science* **2022**, *376*, 744–749. [CrossRef]
14. Zamboni, N.S.; Prudencio, M.D.C.; Amaro, V.E.; Fatima, A.D.M.M.; Verutes, G.M.; Carvalho, A.R. The protective role of mangroves in safeguarding coastal populations through hazard risk reduction: A case study in northeast Brazil. *J. Ocean Coast. Manag.* **2022**, *229*, 106353. [CrossRef]
15. Ellison, J.C. Mangrove retreat with rising sea-level, Bermuda. *Estuar. Coast. Shelf Sci.* **1993**, *37*, 75–87. [CrossRef]
16. Ward, R.D.; Lacerda, L.D. Responses of mangrove ecosystems to sea level change. In *Dynamic Sedimentary Environments of Mangrove Coasts*; Sidik., F., Friess, D.A., Eds.; Elsevier: Amsterdam, The Netherlands, 2021; pp. 235–253, ISBN 9780128164372. [CrossRef]

17. IPCC. *Climate Change 2021: The Physical Science Basis. Contribution of Working Group I to the Sixth Assessment Report of the Intergovernmental Panel on Climate Change*; Masson-Delmotte, V.P., Zhai, A., Pirani, S.L., Connors, C., Péan, S., Berger, N., Caud, Y., Chen, L., Goldfarb, M.I., Gomis, M., Eds.; Cambridge University Press: Cambridge, UK, 2021. Available online: https://www.ipcc.ch/report/ar6/wg1/downloads/ (accessed on 9 September 2024).
18. Alongi, D.M. Impact of global change on nutrient dynamics in mangrove forests. *Forests* **2018**, *9*, 596. [CrossRef]
19. Downs, C.; Krauss, K.; Kruse, S. Hydraulic redistribution in mangroves: Time-lapse electrical resistivity reveals diel patterns of subsurface salt mobilization consistent with exchange of water between trees and sediments. *ESS Open Arch.* **2022**. [CrossRef]
20. Cahoon, D.R.; Lynch, J.C. Vertical accretion and shallow subsidence in a mangrove forest of southwestern Florida, USA. *Mangroves Salt Marshes* **1997**, *1*, 173–186. [CrossRef]
21. Reef, R.; Lovelock, C.E. Regulation of water balance in mangroves. *Ann. Bot.* **2015**, *115*, 385–395. [CrossRef] [PubMed]
22. FAO. The World's Mangroves 1980–2005. In *FAO Forestry Paper*; Forest Resources Division, Food and Agriculture Organization of the United Nations: Rome, Italy, 2007; Volume 153, p. 77.
23. Polidoro, B.A.; Carpenter, K.E.; Collins, L.; Duke, N.C.; Ellison, A.M.; Ellison, J.C.; Farnsworth, E.J.; Fernando, E.S.; Kathiresan, K.; Koedam, N.E.; et al. The Loss of species: Mangrove extinction risk and geographic areas of global concern. *PLoS ONE* **2010**, *5*, e10095. [CrossRef] [PubMed]
24. Hamilton, S.E.; Casey, D. Creation of a high spatio-temporal resolution global database of continuous mangrove forest cover for the 21st century (CGMFC-21). *Glob. Ecol. Biogeogr.* **2016**, *25*, 729–738. [CrossRef]
25. CARSEA. Caribbean Sea Ecosystem Assessment. A contribution to the Millennium Ecosystem Assessment (MA). In *Caribbean Marine Studies*; Special Edition; Agard, J., Cropper, A., Garcia, K., Eds.; 2007; p. 85. Available online: https://www.millenniumassessment.org/documents_sga/CARSEA_2007.pdf (accessed on 1 November 2025).
26. Oxenford, H.A.; Mahon, R. The state of marine ecosystems that support Blue Economies in the Wider Caribbean. In *The Caribbean Blue Economy*; Clegg, P., Mahon, R., McConney, P., Oxenford, H.A., Eds.; Routledge: Oxford, UK, 2020.
27. Ogden, J.; Alcolado, P.; Alleng, G.; Bonair, K.; Laydoo, R.; Bone, D.; Klein, E.; Penchaszadeh, P.; Buchan, K.; Walker, S.; et al. Caribbean coastal marine productivity (CARICOMP): A research and monitoring network of marine laboratories, parks, and reserves. In Proceedings of the 8th International Coral Reef Symposium, Balboa, Panama, 24–29 June 1996; Volume 1, pp. 641–646.
28. Alcolado, P.M.; Alleng, G.; Bonair, K.; Bone, D.; Buchan, K.; Bush, P.G.; De Meyer, K.; Garcia, J.R.; Garzón-Ferreira, J.; Gayle, P.M.; et al. The Caribbean coastal marine productivity program (CARICOMP). *Bull. Mar. Sci.* **2001**, *69*, 819–829.
29. Cortés, J.; Oxenford, H.A.; van Tussenbroek, B.I.; Jordán-Dahlgren, E.; Cróquer, A.; Bastidas, C.; Ogden, J.C. The CARICOMP Network of Caribbean Marine Laboratories (1985–2007): History, key findings, and lessons learned. *Front. Mar. Sci.* **2019**, *5*, 1–18. [CrossRef]
30. Wiebe, W.J.; Gerace, D.T.; Flowers, L.; Johnson, L.; Ward, C.; Oxenford, H.; Parker, C.; Tschirky, J.; Smith, S.R.; Ellison, J.; et al. Structure and productivity of mangrove forests in the Caribbean region. In Proceedings of the 8th International Coral Reef Symposium, Balboa, Panama, 24–29 June 1996; Volume 1, pp. 669–672.
31. CARICOMP. *A Cooperative Research and Monitoring Network of Marine Laboratories, Parks, and Reserves. CARICOMP Methods Manual Levels 1 & 2: Manual of Methods for Mapping and Monitoring of Physical and Biological Parameters in the Coastal Zone of the Caribbean*; CARICOMP Data Management Centre, Centre for Marine Sciences, University of the West Indies: Kingston, Jamaica; Jamaica and Florida Institute of Oceanography, University of South Florida: St. Petersburg, PL, USA, 2001; p. 91.
32. Kjerfve, B. *CARICOMP: Caribbean Coral Reef, Seagrass and Mangrove Sites*; Coastal Region and Small Islands Papers 3; UNESCO: Paris, France, 1998; p. 347.
33. Lugo, A.E.; Snedaker, S.C. Properties of a mangrove forest in southern Florida. In Proceedings of the International Symposium on the Biology and Management of Mangroves, Honolulu, HI, USA, 8–11 October 1974; Volume 1, pp. 170–212.
34. Pool, J.; Snedaker, S.C.; Lugo, A. Structure of mangrove forests in Florida, Puerto Rico, Mexico, and Costa Rica. *Biotropica* **1977**, *9*, 195–212. [CrossRef]
35. Snedaker, S.C.; Snedaker, J.G. *The Mangrove Ecosystem: Research Methods*; UNESCO Monographs in Oceanographic Methodology: Paris, France, 1984; Volume 8, p. 251, ISBN 92-3-102181-8.
36. Rodrigues, D.P.; Hamacher, C.; Estrada, G.C.D.; Soares, M.L.G. Variability of carbon content in mangrove species: Effect of species, compartments and tidal frequency. *Aquat. Bot.* **2015**, *120*, 346–351. [CrossRef]
37. Santos, H.V.S.; Hollanda, F.S.R.; Santos, T.O.; Andrade, K.V.S.; Santana, M.B.S.; Estrada, G.C.D.; Soares, M.L.G. Allometric models for estimating the aboveground biomass of the mangrove *Rhizophora mangle*. *Braz. J. Oceanogr.* **2017**, *65*, 44–53. [CrossRef]
38. Fromard, F.; Puig, H.; Mougin, E.; Marty, G.; Betoulle, J.L.; Cadamuro, L. Structure, above-ground biomass and dynamics of mangrove ecosystems: New data from French Guiana. *Oecologia* **1998**, *115*, 39–53. [CrossRef]
39. Medeiros, T.C.C.; Sampaio, E.V.S.B. Allometry of above-ground biomasses in mangrove species in Itamaracá, Pernambuco, Brazil. *Wetl. Ecol. Manag.* **2008**, *16*, 323–330. [CrossRef]
40. Chollett, I.; Collin, R.; Bastidas, C.; Cróquer, A.; Gayle, P.M.H.; Jordán-Dahlgren, E.; Koltes, K.; Oxenford, H.; Rodriguez-Ramirez, A.; Weil, E.; et al. Widespread local chronic stressors in Caribbean coastal habitats. *PLoS ONE* **2017**, *12*, e0188564. [CrossRef]

41. Koltes, K.H.; Ruiz Renteira, F.; Kjerfve, B.; Smith, S.R.; Alleng, G.; Bonair, K.; Bone, D.; Buchan, K.; Bush, P.; DeMeyer, K.; et al. Meteorological and oceanographic characterization of coral reef, seagrass and mangrove habitats in the Wider Caribbean. In Proceedings of the 8th International Coral Reef Symposium, Balboa, Panama, 24–29 June 1996; Volume 1, pp. 657–662.
42. Koltes, K.H.; Opishinski, T.B. Patterns of water quality and movement in the vicinity of Carrie Bow Cay, Belize. *Smithson. Contrib. Mar. Sci.* **2009**, *38*, 379–390.
43. Paton, S. Monthly summary Bocas Station. *Smithson. Trop. Res. Inst. Dataset* **2019**. [CrossRef]
44. Giannini, A.; Kushnir, Y.; Cane, M.A. Interannual variability of Caribbean rainfall, ENSO, and the Atlantic Ocean. *J. Clim.* **2000**, *13*, 297–311. [CrossRef]
45. Giannini, A.; Kushnir, Y.; Cane, M.A. Seasonality in the impact of ENSO and the north Atlantic high on Caribbean rainfall. *Phys. Chem. Earth Part B Hydrol. Ocean. Atmos.* **2001**, *26*, 1464–1909. [CrossRef]
46. Martinez, C.; Kushnir, Y.; Goddard, L.; Ting, M. Interannual variability of the early and late-rainy seasons in the Caribbean. *Clim. Dyn.* **2020**, *55*, 1563–1583. [CrossRef]
47. Amaral, C.; Poulter, B.; Lagomasino, D.; Fatoyinbo, T.; Taillie, P.; Lizcano, G.; Canty, S.; Herrera Silveira, J.A.; Teutli-Hernández, C.; Cifuentes-Jara, M.; et al. Drivers of mangrove vulnerability and resilience to tropical cyclones in the North Atlantic Basin. *Sci. Total Environ.* **2023**, *898*, 165413. [CrossRef]
48. Worthington, T.A.; zu Ermgassen, P.S.E.; Friess, D.A.; Krauss, D.W.; Lovelock, C.E.; Thorley, J.; Tingey, R.; Woodroffe, C.D.; Bunting, P.; Cormier, N.; et al. A global biophysical typology of mangroves and its relevance for ecosystem structure and deforestation. *Sci. Rep.* **2020**, *10*, 14652. [CrossRef] [PubMed]
49. Lugo, A.E.; Snedaker, S.C. The ecology of mangroves. *Annu. Rev. Ecol. Syst.* **1974**, *5*, 39–64. [CrossRef]
50. Twilley, R.R.; Rivera-Monroy, V.H. Ecogeomorphic models of nutrient biogeochemistry for mangrove wetlands. In *Coastal Wetlands: An Integrated Ecosystem Approach*; Perillo, G.M.E., Wolanski, E., Cahoon, D.R., Brinson, M.M., Eds.; Elsevier: New York, NY, USA, 2009; pp. 641–684.
51. Sheng, J.; Wang, L.; Andréfouët, S.; Hu, C.; Hatcher, B.G.; Muller-Karger, F.E.; Kjerfve, B.; Heyman, W.D.; Yang, B. Upper ocean response of the Mesoamerican Barrier Reef System to Hurricane Mitch and coastal freshwater inputs: A study using Sea-viewing Wide Field-of-view Sensor (SeaWiFS) ocean color data and a nested-grid ocean circulation model. *J. Geophys. Res.* **2007**, *112*, C07016. [CrossRef]
52. Kjerfve, B.; McField, M.; Thattai, D.; Giró, A. Coral reef health in the Gulf of Honduras in relation to fluvial runoff, hurricanes, and fishing pressure. *Mar. Pollut. Bull.* **2021**, *172*, 112865. [CrossRef]
53. Coll, M.; Fonseca, A.C.; Cortés, J. El manglar y otras asociaciones vegetales de la laguna Gandoca, Limón, Costa Rica. *Rev. Biol. Trop.* **2001**, *49*, 321–329.
54. Coll, M.; Cortés, J.; Sauma, D. Características físico-químicas y determinación de plaguicidas en el agua de la laguna de Gandoca, Limón, Costa Rica. *Rev. Biol. Trop.* **2004**, *52*, 33–42.
55. NASA Aquarius Project. *Aquarius Sea Surface Salinity Products. Ver. 5.0*; PO. DAAC: Pasadena, CA, USA, 4 April 2017. [CrossRef]
56. Grodsky, S.A.; Johnson, B.K.; Carton, J.A.; Bryan, F.O. Interannual Caribbean salinity in satellite data and model simulations. *J. Geophys. Res. Ocean.* **2015**, *120*, 1375–1387. [CrossRef]
57. Beier, E.; Bernal, G.; Ruiz-Ochoa, M.; Barton, E.D. Freshwater exchanges and surface salinity in the Colombian basin, Caribbean Sea. *PLoS ONE* **2017**, *12*, e0182116. [CrossRef] [PubMed]
58. Hsu, J. *Multiple Comparisons: Theory and Methods*; Chapman and Hall/CRC: New York, NY, USA, 1996; p. 296. eBook. ISBN 9780429170874. [CrossRef]
59. Eakin, C.M.; Morgan, J.A.; Heron, S.F.; Smith, T.B.; Liu, G.; Alvarez-Filip, L.; Baca, B.; Bartels, E.; Bastidas, C.; Bouchon, C.; et al. Caribbean corals in crisis: Record thermal stress, bleaching, and mortality in 2005. *PLoS ONE* **2010**, *5*, e13969.
60. Lacerda, L.D.; Conde, J.E.; Bacon, P.R.; Alarcon, C.; D'Croz, L.; Kjerfve, B.; Polania, J.; Vannucci, M. Mangrove ecosystems of Latin America and the Caribbean: A summary. In *Conservation and Sustainable Utilization of Mangrove Forests in Latin America and Africa Regions. Part I—Latin America*; Lacerda, L.D., Ed.; ITTO/International Society for Mangrove Ecosystems: Okinawa, Japan, 1993; pp. 1–42.
61. Hutchison, J.; Manica, A.; Swetnam, R.; Balmford, A.; Spalding, M. Predicting global patterns in mangrove forest biomass. *Conserv. Lett.* **2013**, *7*, 233–240. [CrossRef]
62. Phillips, S.; Bustin, R.M.; Lowe, L.E. Earthquake-induced flooding of a tropical coastal peat swamp: A modern analogue for high-sulfur coals? *Geology* **1994**, *22*, 929–932. [CrossRef]
63. Suárez, G.; Pardo, M.; Domínguez, J.; Ponce, L.; Montero, W.; Boschini, I.; Rojas, W. The Limón, Costa Rica earthquake of 22 April 1991: Back arc thrusting and collisional tectonics in a subduction environment. *Tectonics* **1995**, *14*, 518–530. [CrossRef]
64. Cintrón-Molero, G. Mangroves of arid regions of Puerto Rico and the Caribbean. In *Towards the Rational Use of High Salinity Tolerant Plants. Tasks for Vegetation Science*; Lieth, H., Al Masoom, A.A., Eds.; Springer: Dordrecht, Germany, 1993; pp. 117–122. [CrossRef]

65. Soares, M.L.G.; Chaves, F.O.; Estrada, G.C.D.; Fernandez, V. Mangrove forests associated with salt flats: A case study from southeast Brazil. *Braz. J. Oceanogr.* **2017**, *65*, 102–115. [CrossRef]
66. Zenteno-Palma, C.A.; Olivos-Ortiz, A.; Álvarez, M.C.; Quijano-Scheggia, S.I.; Jiménez-Ramon, G.A. The relationship of sediment and interstitial water properties with mangrove health in a subtropical coastal lagoon of Mexico. In *Estuary*; Foneman, W., Ed.; IntechOpen: London, UK, 2018; pp. 99–115. [CrossRef]
67. van Tussenbroek, B.I.; Cortés, J.; Collin, R.; Fonseca, A.C.; Gayle, P.M.H.; Guzmán, H.M.; Jácome, G.E.; Juman, R.; Koltes, K.H.; Oxenford, H.A.; et al. Caribbean-wide, long-term study of seagrass beds reveals local variations, shifts in community structure and occasional collapse. *PLoS ONE* **2014**, *9*, e98377. [CrossRef]
68. Kathiresan, K.; Bingham, B.L. Biology of mangroves and mangrove ecosystems. *Adv. Mar. Biol.* **2001**, *40*, 81–251. [CrossRef]
69. Jiménez, J.A.; Soto, R. Patrones regionales en la estructura y composición florística de los manglares de la Costa Pacífica de Costa Rica. *Rev. Biológica Trop.* **1985**, *33*, 25–37.
70. Jiménez, J.A. Mangrove forests of the Pacific coast of Central America. In *Physiological Ecology: Coastal Plant Communities of Latin America*; Seeliger, U., Ed.; Academic Press: Oxford, UK, 1992; pp. 259–267, ISBN 9780080925677. [CrossRef]
71. Rodríguez-Martínez, R.E.; Torres-Conde, E.G.; Rosellón-Druker, J.; Cabanillas-Terán, N.; Jáuregui-Haza, U. The Great Atlantic Sargassum Belt: Impacts on the Central and Western Caribbean—A review. *Harmful Algae* **2025**, *144*, 102838. [CrossRef]
72. Howard, J.; Hoyt, S.; Isensee, K.; Telszewski, M.; Pidgeon, E. *Coastal Blue Carbon: Methods for Assessing Carbon Stocks and Emissions Factors in Mangroves, Tidal Salt Marshes, and Seagrasses*; International Union for Conservation of Nature (IUCN): Arlington, VA, USA, 2014; p. 181. Available online: https://unesdoc.unesco.org/ark:/48223/pf0000372868 (accessed on 5 October 2025).
73. Simard, M.; Fatoyinbo, L.; Smetanka, C.; Rivera-Monroy, V.H.; Castañeda, E.; Thomas, N.; Stocken, T.V. Mangrove canopy height globally related to precipitation, temperature and cyclone frequency. *Nat. Geosci.* **2019**, *12*, 40–45. [CrossRef]
74. Simard, M.; Fatoyinbo, L.; Thomas, N.M.; Stovall, A.E.; Parra, A.; Barenblitt, A.; Bunting, P.; Hajnsek, I. A new global mangrove height map with a 12 meter spatial resolution. *Sci. Data* **2025**, *12*, 15. [CrossRef] [PubMed]
75. Jones, A.R.; Raja Segaran, R.; Clarke, K.D.; Waycott, M.; Goh, W.S.; Gillanders, B.M. Estimating mangrove tree biomass and carbon content: A comparison of forest inventory techniques and drone imagery. *Front. Mar. Sci.* **2020**, *6*, 784. [CrossRef]
76. Hernández, W.J.; Morell, J.M.; Armstrong, R.A. Using high-resolution satellite imagery to assess the impact of Sargassum inundation on coastal areas. *Remote Sens. Lett.* **2021**, *13*, 24–34. [CrossRef]
77. Dani, S.; Castaño-Isaza, J.; Lee, S.; Edwards, P. *Mangroves Monitoring and Evaluation Manual for Jamaica*; National Environment and Planning Agency (NEPA): Kingston, Jamaica, 2010; p. 18. Available online: https://documents1.worldbank.org/curated/en/965411613137686768/pdf/Mangroves-Monitoring-and-Evaluation-Manual-for-Jamaica.pdf (accessed on 29 September 2025).
78. Ávila-Acosta, C.R.; Domínguez-Domínguez, M.; Vázquez-Navarrete, C.J.; Acosta-Pech, R.G.; Martínez-Zurimendi, P. Above-ground biomass and carbon storage in mangrove forests in southeastern Mexico. *Resources* **2024**, *13*, 41. [CrossRef]
79. R Core Team. *R: A Language and Environment for Statistical Computing*; R Foundation for Statistical Computing: Vienna, Austria, 2024. Available online: https://www.R-project.org (accessed on 20 September 2025).
80. Posit Team. *RStudio: Integrated Development Environment for R*; Posit Software: Boston, MA, USA, 2024. Available online: http://www.posit.co// (accessed on 20 September 2025).
81. Golley, F.; Odum, H.T.; Wilson, R.F. The structure and metabolism of a Puerto Rican red mangrove forest in May. *Ecology* **1962**, *43*, 9–19. [CrossRef]
82. Clough, B.F.; Scott, K. Allometric relationships for estimating above-ground biomass in six mangrove species. *For. Ecol. Manag.* **1989**, *27*, 117–127. [CrossRef]
83. Cintrón, G.; Schaeffer Novelli, Y. Methods for studying mangrove structures. In *The Mangrove Ecosystem: Research Methods*; Snedaker, S.C., Snedaker, J.G., Eds.; UNESCO: Paris, France, 1984; pp. 91–113, ISBN 92-3-102181-8.
84. Ong, J.E.; Gong, W.K.; Wong, C.H. Biomass allometry and partitioning of the mangrove, *Rhizophora apiculata*. *For. Ecol. Manag.* **2004**, *188*, 395–408. [CrossRef]
85. Kjerfve, B. Tides of the Caribbean Sea. *J. Geophys. Res.* **1981**, *86*, 4243–4247. [CrossRef]
86. Torres, R.R.; Tsimplis, M.N. Tides and long-term modulations in the Caribbean Sea. *J. Geophys. Res.* **2011**, *116*, C10022. [CrossRef]
87. Krauss, K.W.; McKee, K.L.; Lovelock, C.E.; Cahoon, D.R.; Saintilan, N.; Reef, R.; Chen, L. How mangrove forests adjust to rising sea level. *New Phytol.* **2014**, *202*, 19–34. [CrossRef]
88. Ward, R.D.; Lacerda, L.D.; Silva Cerqueira, A.; Hugo, V.; Hernandez, O.C. Vertical accretion rates of mangroves in northeast Brazil: Implications for future responses and management. *Estuar. Coast. Shelf Sci.* **2023**, *289*, 108382. [CrossRef]
89. Woodroffe, C.D.; Rogers, K.; McKee, K.L.; Lovelock, C.E.; Mendelssohn, I.A.; Saintilan, N. Mangrove sedimentation and response to relative sea-level rise. *Annu. Rev. Mar. Sci.* **2016**, *8*, 243–266. [CrossRef]
90. Funk, C.; Peterson, P.; Landsfeld, M.; Pedreros, D.; Verdin, J.; Shukla, S.; Husak, G.; Rowland, J.; Harrison, L.; Hoell, A.; et al. The climate hazards infrared precipitation with stations—A new environmental record for monitoring extremes. *Sci. Data* **2015**, *2*, 150066. [CrossRef]

91. Touma, D.; Martinez, C.; National Center for Atmospheric Research. *The Climate Data Guide: CHIRPS: Climate Hazards InfraRed Precipitation with Station Data (Version 2)*; Nacional Center for Atmospheric Research: Boulder, CO, USA, 2023. Available online: https://climatedataguide.ucar.edu/climate-data/chirps-climate-hazards-infrared-precipitation-station-data-version-2 (accessed on 16 September 2024).
92. Bretherton, C.S.; Smith, C.; Wallace, J.M. An intercomparison of methods for finding coupled patterns in climate data. *J. Clim.* **1992**, *5*, 541–560. [CrossRef]
93. Chiang, C.H.; Vimont, D.J. Analogous Pacific and Atlantic Meridional Modes of Tropical Atmosphere-Ocean Variability. *J. Clim.* **2004**, *17*, 4143–4158. [CrossRef]
94. Smith, T.J., III; Anderson, G.; Balentine, K.; Tiling, G.; Ward, G.; Whelan, K. Cumulative impacts of hurricanes on Florida mangrove ecosystems: Sediment deposition, storm surges and vegetation. *Wetlands* **2009**, *29*, 24–34. [CrossRef]
95. Castañeda-Moya, E.; Rivera-Monroy, V.H.; Chambers, R.M.; Zhao, X.; Lamb-Wotton, L.; Gorsky, A.; Gaiser, E.E.; Troxler, T.G.; Kominoski, J.S.; Hiatt, M. Hurricanes fertilize mangrove forests in the Gulf of Mexico (Florida Everglades, USA). *Proc. Natl. Acad. Sci. USA* **2020**, *117*, 4831–4841. [CrossRef]
96. Imbert, D. Hurricane disturbance and forest dynamics in east Caribbean mangroves. *Ecosphere* **2018**, *9*, e02231. [CrossRef]
97. Branoff, B.L. Mangrove disturbance and response following the 2017 hurricane season in Puerto Rico. *Estuaries Coasts* **2020**, *43*, 1248–1262. [CrossRef]
98. Kauffman, J.B.; Cole, T.G. Micronesian mangrove forest structure and tree responses to a severe typhoon. *Wetlands* **2010**, *30*, 1077–1084. [CrossRef]
99. Han, X.; Feng, L.; Hu, C.; Kramer, P. Hurricane-Induced changes in the Everglades National Park mangrove forest: Landsat observations between 1985 and 2017. *J. Geophys. Res. Biogeosci.* **2018**, *123*, 3470–3488. [CrossRef]

environments

MDPI

Article

Threat of Alien Species to Native Biodiversity in Mangroves near Latin America's Largest Port: Pathways for Technological Innovation and Strengthening of Regulations

Sidnei Aranha [1,*], Felipe Rakauskas [2], Leonardo Ferreira da Silva [3], Caio Fernando Fontana [4] and Maurício Lamano Ferreira [2]

[1] Port of Santos. Environmental Management Superintendence, Santos 11015-900, Brazil
[2] Department of Basic and Environmental Sciences, University of São Paulo, Lorena 12602-810, Brazil; feliperakauskas@usp.br (F.R.)
[3] Department of Architecture and Urban Planning, University Presbyterian Mackenzie, São Paulo 01302-907, Brazil; leonardoferreira.silva@mackenzista.com.br
[4] Department of Bioproducts and Bioprocesses, Federal University of São Paulo, Santos 11013-000, Brazil; caio.fernando@unifesp.br
* Correspondence: sidneiaranhagja@gmail.com

Abstract: Mangrove forests are biodiverse and highly productive coastal ecosystems, fundamental to fisheries and tourism. However, they are severely threatened by human activities and invasive species, particularly in port areas such as the Port of Santos, necessitating effective environmental management. This study aimed to analyze the risks of biological invasion in mangrove ecosystems stemming from port activities, with a focus on the Port of Santos (PS), Brazil. To achieve this, we conducted a bibliometric review using the Web of Science and Scopus databases, analyzed vessel traffic flows arriving at the PS over 14 years (from 2010 to 2024), and discussed alternatives to address the challenge of biological invasion. The review revealed a significant gap in the scientific literature, as few studies (9.9%, n = 71) address the intersection of maritime transport, invasive species, and mangroves in Latin American contexts. The intense and constant flow of international vessels into the Port of Santos, totaling 15,193 arrivals from more than 200 ports worldwide between 2010 and 2024, poses a persistent threat of biological invasion. This high-volume connectivity, with several foreign hubs exceeding 300 departures in the period, reinforces the role of ships as vectors transporting exotic species in ballast water and through hull fouling. This can destabilize local ecosystems and cause significant socioeconomic losses unless control measures, mediated by effective policies, regulations, and technologies, are implemented in the short term. A spatiotemporal analysis of vessel traffic flows over a 14-year period revealed persistent high-risk corridors for bioinvasion, directly linking maritime activity patterns to the threat level for adjacent mangrove ecosystems. The data indicate a substantial challenge for the PS, yet one with a high potential for resolution in the medium term, contingent upon investment in technology and regulation.

Keywords: biological invasion; alien species; invasive species; mangrove ecosystems; port management; coastal cities

1. Introduction

Mangroves are highly productive ecosystems, occupying a transitional zone between terrestrial and aquatic environments and covering approximately 150,000 km^2 in tropical and subtropical regions [1]. These areas serve as vital biodiversity hotspots, harboring

 https://doi.org/10.3390/environments12120483

significant biological richness, including dozens of tree species and hundreds of fish and arthropod groups, with many taxa being endemic (e.g., *Pelliciera* and *Laguncularia*) [2,3]. However, over 50% of global mangrove cover has been severely impacted since the 1980s due to human activities such as deforestation, aquaculture, and climate change [4], directly affecting local economies.

Coastal tourism, including ecotourism in mangrove areas, encompasses a variety of recreational and economic activities developed in the coastal zone and adjacent waters, including hospitality infrastructure, dining, marinas, and support services [5]. Within the framework of the Blue Economy, this sector aims to reconcile socioeconomic development with environmental conservation, promoting inclusive and sustainable growth in coastal and marine ecosystems [6]. Various coastal environments worldwide possess preserved ecosystems [7] that attract tourism. Concurrently, mangroves play a crucial role in coastal fisheries, sustaining catches of bivalves, shrimp, and crabs [8], often conducted on a small scale, as defined by Teh et al. [9] as artisanal, low-technology activities geared toward local markets.

Brazil, for instance, is home to 25% of the world's mangroves, which underpin part of the local economies. However, these environments face threats from invasive species, especially near port regions [10]. This occurs because these ecosystems provide sufficient conditions and resources for the establishment and population explosion of opportunistic species introduced through various means, including ballast water. Ships transport goods and strengthen the global economy but also act as vectors for species dispersal via hulls, propellers, rudders, and ballast tanks [11]. Most marine invasions occur via ballast water or biofouling, accounting for 60% to 90% of such events [12], with their impacts reflected in economic losses (~5% of the global economy), disrupted fisheries, hull fouling, and infrastructure damage [13].

In southeastern Brazil, certain species are dominant in mangrove formation. For example, *Rhizophora mangle*, known as the red mangrove, is notable for its adventitious roots, which contribute to substrate consolidation and reduced coastal erosion. *Avicennia schaueriana*, the black mangrove, features pneumatophore, root structures that emerge from the soil to facilitate gas exchange in saturated substrates. *Laguncularia racemosa*, which forms white mangroves, occurs in less flooded zones, and its roots also contribute to terrain stabilization [14]. However, the recent occurrence of an invasive alien species has been reported near the Port of Santos (PS). The species *Sonneratia apetala*, commonly known as the apple mangrove, is a typical mangrove tree species belonging to the family Lythraceae. Native to Southeast Asia, it is found in countries such as India, Sri Lanka, and Indonesia [15]. With an average height exceeding 12 m, individuals of this species exhibit aerial roots (pneumatophores) that anchor in muddy substrates, along with oval, glossy leaves and exuberant yellow flowers that stand out in the landscape. This species has buoyant fruits and water-adapted seeds, which facilitate its natural expansion [16].

Biological invasions, propelled by global maritime commerce, constitute a paramount threat to global biodiversity and economic stability [11]. Invasive alien species can precipitate the collapse of native ecosystems through mechanisms such as rampant competition, novel predation, or the introduction of pathogens, while concurrently inflicting severe socioeconomic detriment on aquaculture, port infrastructure, and fisheries [12]. Within this context, the Port of Santos, a pivotal maritime hub in South America, occupies a critical frontline. The relentless influx of international vessels transforms each ship into a potential vector, ferrying non-indigenous organisms in ballast water and on hull fouling. Consequently, the pivotal inquiry is not whether the native system will be impacted, but which invasive species pose the most imminent peril and through what mechanisms the primary introduction pathway, i.e., shipping traffic, can be effectively mitigated [17].

A robust assessment of bioinvasion risks in maritime hubs must be grounded in a detailed characterization of the primary vector—the vessels themselves. This study therefore incorporates a comprehensive spatiotemporal analysis of shipping traffic destined for the Port of Santos, spanning from 2010 to 2024. Mapping the intensity and distribution of maritime routes provides the critical foundation for identifying persistent high-risk corridors and temporal shifts in invasion pressure [18]. This empirical groundwork is indispensable for linking theoretical models of species introduction with the tangible, dynamic patterns of naval movement that define the threat landscape.

The PS is the largest and most important port complex in Brazil and Latin America, ranking among the largest globally. With specialized terminals, it handles solid bulk (sugar, coffee, soy), liquid bulk, containers, and general cargo, accounting for 25% of Brazilian foreign trade and leading globally in the export of sugar, orange juice, and coffee. Its economic influence encompasses 67% of the national GDP, particularly in the Southeast and Central-West states. Ranked as the 39th largest container port and the 35th in tonnage, the PS has established itself as the busiest in the region and recently broke its historical record, handling over 17 million tons [19,20].

This growth was driven by increased exports, especially after the imposition of 50% tariffs by the U.S. on Brazilian products, which accelerated shipments to Europe and North America. Given its undeniable importance to the Brazilian economy and global exports, the PS faces severe challenges from the impacts caused by vessels and urgently requires attention to the control, monitoring, and eradication of alien species. Considering the existence of efficient technologies and regulatory mechanisms for preventing severe environmental impacts, such as biological invasion, this work aimed to present the opportunities and challenges of Brazil's largest logistics center regarding its activities and transport flows, as well as the potential for technological use to mitigate environmental conflicts. Thus, this study proposed to conduct a literature review, understand the flows of vessels arriving at and departing from the PS, and evaluate the technologies used in port management. These results will indicate viable pathways for the PS, also serving as a reference for other Brazilian and South American ports facing similar environmental challenges.

2. Materials and Methods

To present the opportunities and challenges of the Port of Santos regarding existing environmental conflicts, this study opted to present the state of the art of the scientific literature on environmental conflicts in the PS region through a bibliographic review. Subsequently, an analysis of vessel flows at the largest port in Latin America was conducted, followed by an approach to regulations and technologies that could help overcome such conflicts. The methodological descriptions of these central points of the manuscript follow.

2.1. Bibliographic Review Analysis

To ground the literature review and identify relevant studies on invasive species in mangrove ecosystems, with a focus on maritime transport vectors and the Latin America/Brazil region, systematic searches were conducted in the Web of Science (WoS) and Scopus databases. The searches were performed on 13 July 2025, using specific search strings, as detailed below. The choice of these two platforms is due to the fact that they are databases with strict editorial policies, which provide articles from journals that have undergone rigorous peer review with greater security, although there are good works published and available in other databases.

The first search strategy aimed to identify a general overview of non-native species in mangroves. The search string used was: ("alien species" OR "invasive species" OR "non-native species" OR "exotic species") AND (mangrove* OR "mangrove ecosystem*").

This initial search retrieved 27 articles in the Web of Science database and 44 articles in Scopus, totaling 71 publications addressing the relationship between invasive species and mangroves more broadly. This result demonstrates a considerable body of scientific literature dedicated to the study of biological invasions in mangrove ecosystems, indicating the scientific community's recognition of the importance and relevance of this theme.

Subsequently, to refine the scope of the review and focus on the specific aspects of maritime transport and the geographic region of interest, a second, more restrictive search string was applied: ("alien species" OR "invasive species" OR "non-native species" OR "exotic species") AND (mangrove* OR "mangrove ecosystem*") AND (port* OR "maritime transport*" OR "shipping" OR "vessel*") AND ("Latin America*" OR Brazil*). This refined search yielded a significantly smaller number of articles, highlighting an important gap in the scientific literature. In Web of Science, only 5 articles were found, while Scopus identified only 2 articles, totaling 7 publications that specifically address the intersection of invasive species in mangroves, maritime transport vectors, and the Latin America or Brazil region.

The drastic reduction in the number of publications, from 71 to just 7 articles, in the specific search reveals a significant knowledge gap in the scientific literature. This scarcity of studies suggests that the intersection of invasive species in mangroves, specifically associated with maritime transport vectors, and with a geographic focus on Latin America or Brazil, represents an underexplored area of research, despite its evident ecological and economic relevance.

2.2. Vessel Logistics Flows and Biological Invasion Risks

This analysis sought to understand the patterns of international vessel movement destined exclusively for the Port of Santos (Brazil) between 2010 and 2024. The data used consist of consolidated annual records of port departures by terminal and country of origin. It is important to note that for the year 2024, the records only cover the period up to March 21, which naturally results in underrepresentation compared to other years in the series.

All analytical steps were conducted in the R language (version 4.4.0), using the dplyr, tidyr, and ggplot2 packages, ensuring reproducibility, traceability, and methodological rigor. Initially, fields related to annual departure volumes (variables named SAIDAS_2010 to SAIDAS_2024) were automatically identified and treated for missing values, which were replaced by zero, following recommended best practices for time series with sporadic gaps. From this standardized base, a series of quantitative analyses were structured, focusing on three complementary dimensions of logistics dynamics: temporal evolution, interannual stability, and spatial distribution by country of origin.

The annual evolution of total global departures to Santos revealed marked fluctuations, allowing for the identification of contraction or recovery trends throughout the series. This aggregated visualization serves as a background for contextualizing the impacts of logistical, sanitary, and economic events on the international flows converging to the main Brazilian port.

In the second axis of analysis, international ports were classified based on the coefficient of variation (CV%), a statistical metric that expresses the relative instability of the series by relating the standard deviation to the mean of annual departures. Ports with a lower CV% were considered the most stable, while those with higher CV% were classified as more volatile. To ensure the statistical consistency of comparisons, a minimum criterion was adopted: a mean greater than one departure per year and, for volatile ports, at least eight years of operation recorded in the series. This approach allowed for distinguishing terminals with regular and predictable behavior from those subject to sharp fluctuations, often associated with external shocks, extreme seasonality, or dependence on specific logistics chains.

The accumulated departures by country of origin were aggregated, identifying the ten territories with the highest number of recorded departures for Santos in the period. This spatial dimension allows for an understanding of the relative centrality of certain economies within the logistics supply structure with Brazil, besides signaling consolidated trade relations, productive specializations, or competitive port advantages.

All graphs were generated with typographic standardization in Times New Roman 12 pt, scales adjusted to the behavior of each variable, and an interpretive focus on communicating patterns relevant for decision-making in logistics planning and public policy formulation. The adopted approach allows for an integrated reading of Brazil's maritime connectivity, providing subsidies for understanding the resilience, vulnerability, and concentration of international routes destined for the Port of Santos over recent decades.

2.3. Cartographic Elaboration of the Spatial Distribution of Port Departures (2010–2024)

The construction of thematic maps was carried out using QGIS software version 3.34.9-Prizren, an open-source geoprocessing platform widely used in spatial research for its stability, flexibility, and integration with complex databases [21,22]. The data used were the same as those processed in the statistical analyses described previously, containing consolidated annual records of port departures between 2010 and 2024.

The database was initially treated in the R environment, including the georeferencing of ports based on their geographic coordinates. Subsequently, the data were imported into QGIS in the WGS 84 coordinate reference system (EPSG:4326), suitable for representing global-scale phenomena with broad compatibility for analysis and web mapping tools [23].

To graphically represent the intensity of departures, graduated proportional circle symbology was chosen, based on the total departures per port in each time frame. Six thematic maps were produced: one map of accumulated totals (2010–2024) and five triennial maps covering the intervals: 2010–2012, 2013–2015, 2016–2018, 2019–2021, and 2022–2024.

The classification of values was defined based on manually adjusted quantile intervals, prioritizing ranges with rounded limits to favor readability (e.g., 1–100, 101–200, 201–300, 301+). The sizing of the circles was calibrated to preserve the visual hierarchy between classes without cartographic distortion, respecting the final layout scale of 1:250,000,000.

Each map was overlaid on a raster base layer (OpenStreetMap/ESRI), allowing for geographic contextualization and visual recognition of coastlines and regions of interest. The visual composition was finalized in the QGIS Print Composer, with the insertion of legends, graphic scales, data sources, and standardized cartographic elements.

The division into triennial periods enabled the observation of the evolution of maritime routes with analytical granularity, highlighting strategic shifts, logistics concentration, and reconfigurations induced by events such as economic crises, the pandemic, and regulatory changes. This approach is supported by spatial analysis studies in maritime transport that advocate for the use of intermediate time windows to capture patterns of change and stability [24].

3. Results and Discussion

3.1. Bibliographic Review Analysis

Within the context of this investigation, a comparative bibliometric analysis was conducted using the Web of Science (WoS) and Scopus databases, aiming to understand the global research landscape on invasive species in mangrove ecosystems, particularly those associated with maritime transport. This dual approach allows for a more robust assessment of scientific output, considering the different indexing policies and geographical coverage of the databases [25].

3.1.1. Distribution, Collaboration, and Impact in Studies on Ballast Water and Port Bioinvasion

This result seeks to describe some findings about the indexers selected in two databases (i.e., WoS and SCOPUS). The main objective is not to compare the results of both databases, but rather to complement them, although there are other studies on the subject in other indexers. The quantitative analysis of the geographical distribution of scientific output reveals distinct patterns between the two databases analyzed. In the WoS, the United States emerges as the absolute leader with 10 studies published in the analyzed period, followed by the People's Republic of China (8 studies), Australia (4 studies), and Brazil and Germany, both with 3 studies each. This distribution reflects the traditional scientific hegemony of North America and Europe in high-impact factor journals indexed by WoS (Figure 1A). In contrast, the Scopus database shows the People's Republic of China in the lead with 17 studies, followed by the United States (16 studies), Brazil (7 studies), Australia (5 studies), and Hong Kong (4 studies). This inversion in leadership between China and the United States may be attributed to Scopus's more inclusive indexing policies, which have significantly expanded their coverage of Asian journals over the last two decades. The greater Chinese representation in Scopus also reflects the exponential growth of Chinese scientific output and its increasing participation in regional and specialized journals (Figure 1B).

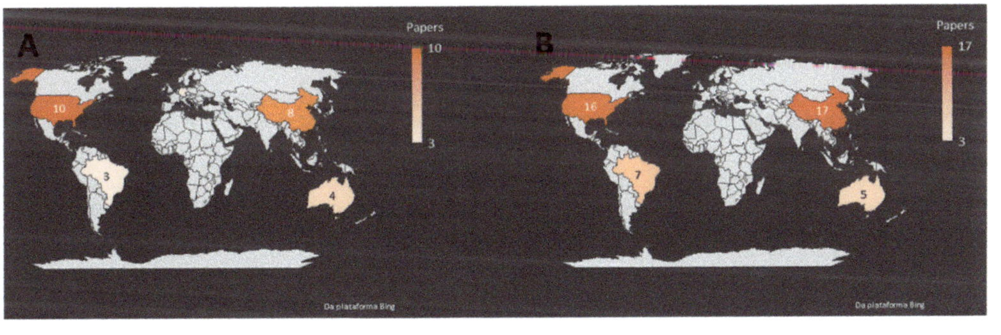

Figure 1. Geographical distribution of manuscripts in the selected databases. In (**A**), results from the WoS database. In (**B**), results from the Scopus database.

Brazil, occupying the third position in WoS and fourth in Scopus, presents a modest scientific output when compared to the extent of its coastline and the importance of its mangrove ecosystems. This underrepresentation is particularly significant considering that Brazil harbors approximately 25% of the world's mangroves [10] and possesses some of the busiest port complexes in Latin America, including the Port of Santos, which handles over 130 million tons of cargo annually [26].

Despite the diversity of countries represented in the authorship of the studies, the analysis reveals a pronounced geographical concentration of investigations in two main regions: North America/Caribbean and the Indo-Malay/Western Pacific region. This concentration is not fortuitous but reflects specific characteristics that make these regions particularly suitable for studies of marine biological invasion.

The Caribbean and Gulf of Mexico region has historically been recognized as a natural laboratory for studies of invasive alien species, particularly in mangroves, due to a confluence of favorable factors: intense international maritime traffic, connectivity between different biogeographical provinces, the presence of diverse and well-studied coastal ecosystems, and an extensively documented history of species introductions [27]. Additionally, the presence of consolidated research institutions and long-term monitoring programs in this region facilitates the detection and study of invasive species.

Similarly, the Indo-Pacific region, with emphasis on the coastal waters of China, Australia, and Southeast Asia, concentrates studies due to its exceptional native biological diversity, intense port and aquaculture activity, and well-established government monitoring programs for invasive alien species [28]. China, in particular, has substantially invested in research on biological invasions in response to the growing economic impacts of invasive species on its aquaculture and coastal systems.

The bibliometric analysis reveals a critical gap in the representation of South America, and particularly the Brazilian coast, in the international scientific literature. This underrepresentation is especially concerning when contrasted with the ecological and economic importance of Brazilian coastal ecosystems. Brazil possesses the world's second-largest extent of mangroves, behind only Indonesia [29], and its ports are among the busiest globally in terms of cargo volume.

This gap represents not only a research opportunity but an urgent necessity for scientific development in this field of knowledge. The absence of robust studies on biological invasions in Brazilian mangroves limits environmental management capacity and may result in significant undetected or inadequately managed ecological and economic impacts. In the specific context of the Port of Santos, this gap is particularly relevant, considering its position as the main entry point for goods in the country and its proximity to extensive mangrove areas.

The Scopus database presents a more diversified and dynamic network structure, with greater participation of Asian researchers and an emerging connectivity with research groups from Latin America. This difference suggests that Scopus is capturing emerging collaborations and regional perspectives that have not yet achieved full representation in the high-impact literature indexed by WoS.

Emerging themes identified in the literature include the integration of climate change and biological invasions, the growing use of genomic tools for tracking origins and dispersal routes, and the development of predictive models based on navigation data and environmental variables. These emerging themes are particularly relevant for the proposed research at the Port of Santos, suggesting innovative methodological approaches that could be applied in the Brazilian context.

The analysis of publication quality, based on the journals of publication and citation metrics, reveals that studies are predominantly concentrated in high-impact scientific journals. Marine Ecology Progress Series, Biological Invasions, Aquatic Invasions, and Marine Pollution Bulletin are positioned as the main publication vehicles, indicating that the theme has gained recognition in prestigious international journals.

In an interesting approach, the analysis of keyword co-occurrence networks offers a multifaceted understanding of the conceptual and thematic panorama that structures the literature on invasive species in mangrove ecosystems, as addressed in the present review (Figure 2A,B). These visualizations, generated from bibliometric data, mirror the trends and research foci outlined in the text, revealing the intrinsic interconnection between core concepts and specific areas of study.

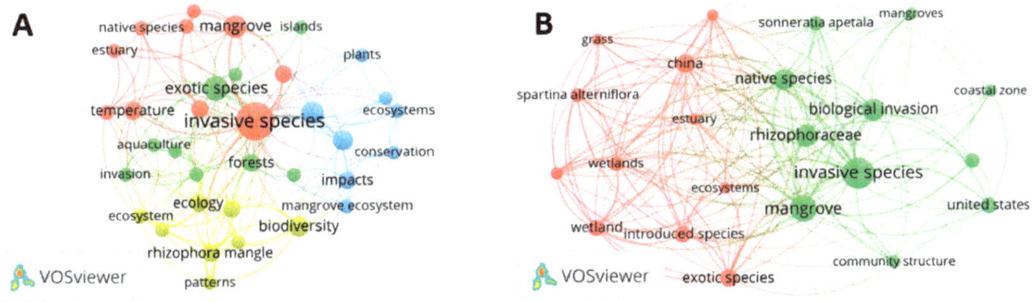

Figure 2. Keyword analysis of the articles selected in the bibliographic review: (**A**) WoS and (**B**) Scopus.

As is widely known in scientific literature, the prominence of the keywords "invasive species" and "exotic species" as central nodes in both networks unequivocally confirms the position of these themes as the epicenter of scientific investigation. This conceptual centrality reflects the consolidation of biological invasions as one of the main global ecological concerns, particularly in the context of coastal ecosystems [12,16,27].

The structural centrality of 'invasive species' and 'exotic species' within both co-occurrence networks suggests that these concepts are not only foundational but also act as key conceptual bridges, integrating disparate research threads. This pattern reflects the broader scientific consensus that prioritizes biological invasions as a leading threat to the integrity of coastal ecosystems globally.

The strong co-occurrence with terms such as "mangroves", "mangrove ecosystem", "estuary", and "wetlands" evidences the scientific recognition of these environments as systems particularly vulnerable and prioritized for invasion research. This conceptual configuration aligns perfectly with the findings of the literature review, which establishes mangroves as ecosystems highly sensitive to anthropogenic disturbances and introductions of exotic species.

The thematic structure revealed by the co-occurrence analysis manifests through two main axes that delineate the central concerns of the scientific community. The first axis focuses on the driving factors and contexts of biological invasions, evidenced by the co-occurrence of terms such as "temperature", "aquaculture", "invasion", and the dichotomy "exotic species"/"native species". This configuration suggests an investigative approach that recognizes the multifactorial nature of invasive processes, incorporating both anthropogenic elements and abiotic environmental factors.

The second axis focuses on ecosystem impacts and conservation strategies, materialized in the interconnection between "ecosystems", "conservation", "impacts", "biodiversity", and "mangrove ecosystem". This thematic duality reflects the paradigmatic evolution of research on biological invasions, which has transitioned from a predominantly descriptive approach to a perspective oriented toward conservation and adaptive management.

The analysis reveals significant conceptual gaps that have direct implications for the development of research in the Brazilian context. Although the literature review extensively emphasizes ballast water and biofouling as primary maritime transport vectors, these terms do not emerge with equivalent prominence in the keyword networks. This discrepancy suggests that transport vectors are often subsumed under broader concepts like "biological invasion" or that bibliographic indexing tends to privilege the species and their impacts over the specific mechanisms of introduction. Similarly, while the review documents in detail invasions by bivalves such as *Magallana gigas* and *Mytilopsis sallei*, these taxa are not proportionally represented in the networks, indicating a possible underrepresentation of invasive invertebrates in the indexed literature or a tendency to use more generic terms for these groups.

The keyword analysis, while reinforcing the geographical gap in the literature concerning the Port of Santos and biological invasions, is inherently limited by the scope of the bibliometric review. It must be acknowledged that high-quality research on this topic may well be published in journals outside the Web of Science and Scopus indices; however, such publications were not captured by the methodology of this study. The absence of terms related to Brazil, South America, or specific ports in the region in the co-occurrence networks evidences the critical underrepresentation of regional research in the international literature. This gap represents not only a research opportunity but an urgent need for scientific development that can contribute significantly to global knowledge on biological invasions in tropical mangrove ecosystems.

It is worth noting that several studies have pointed to biological invasion also originating from encrustation, demonstrating the potential of this dispersal strategy by some groups of marine and estuarine fauna [30–32].

3.1.2. Biological Invasion, Taxonomy, Vectors, and Impacts

The analysis of the literature reveals a concentration of studies on specific taxonomic groups. Invasive bivalves, particularly *Crassostrea gigas* (*Magallana gigas*), *Mytilopsis sallei*, and other species of the family Dreissenidae, dominate the literature, representing approximately 40% of the analyzed studies. This predominance reflects both the ease of detection and study of these organisms and their significant economic impacts on aquaculture and port infrastructure.

Alien mangrove species, including *Sonneratia apetala*, *Laguncularia racemosa*, and other species, constitute the second most studied group, followed by invasive crustaceans such as *Carcinus maenas*. This taxonomic distribution suggests a research trend driven by direct economic impacts and ease of detection, potentially neglecting less conspicuous but ecologically important groups.

Regarding transport vectors, ballast water dominates the literature, being addressed in approximately 60% of the studies, followed by biofouling (25%) and aquaculture (15%). This distribution reflects both the ease of studying ballast water and its recognition as a primary vector for the introduction of marine species. However, an underrepresentation of studies on adhered sediments and transported debris is observed; vectors that may be particularly relevant for ports specialized in handling solid bulk.

Research in the WoS databases indicates that *Schinus terebinthifolius* (Brazilian peppertree), a species native to South America that has exhibited a pronounced invasion capacity in the mangrove ecosystems of Florida, epitomizes the role of allelopathic mechanisms in biological invasions. Research conducted by Donnelly et al. [29] demonstrated that this species synthesizes secondary metabolites that significantly inhibit the growth and development of native mangrove species, specifically *Avicennia germinans* (black mangrove) and *Rhizophora mangle* (red mangrove) [29].

Other empirical studies indicate that the release of allelochemicals from macerated fruits leads to elevated concentrations of phytotoxic compounds in the soil, not only decreasing the biomass of native flora but also precipitating considerable mortality among associated faunal organisms, as demonstrated by experiments with Artemia. The reduction in photosynthetic efficiency among native mangroves, attributable to leaf abscission induced by allelopathic agents, confers immediate competitive advantages to *Schinus terebinthifolia* seedlings, thus perpetuating the cycle of dominance by the invasive species [33].

The establishment of *Spartina alterniflora* in mangrove ecosystems in southern China constitutes a paradigmatic example of the inherent complexity of biogeochemical modifications promoted by alien species in recipient environments. Scientific work conducted by research groups led by Xia et al. [34] provides robust evidence of transformations in soil organic carbon (SOC) and total nitrogen (TN) reservoirs, with particular emphasis on modifications observed in the surface soil horizon, between 0–10 cm depth [35].

The significant contribution of *Spartina alterniflora* to the composition of surface organic carbon, ranging between 42.6% and 62.2%, associated with its interference in soil microbial communities and the increase in nitrous oxide (N_2O) emissions through the intensification of denitrification processes, demonstrates the magnitude of the functional alterations imposed by this invasion on ecological systems [36].

Notables are studies that identified a mechanism of ecological resistance mediated by the herbivorous behavior of native rodents (*Rattus losea*). This biotic interaction demonstrates an effective capacity to suppress the development and persistence of *S. alterniflora*, manifesting more pronouncedly in microenvironments characterized by the presence of understory. Such findings indicate that the preservation of the functionality of original trophic networks may represent an endogenous biological control approach, offering promising perspectives for management strategies based on natural ecological processes [37].

The species *Sonneratia apetala* constitutes an exemplary paradigm of the risks associated with the introduction of alien species for ecological restoration purposes. After more than two decades of extensive use in reforestation programs in China, this species has demonstrated the capacity to develop invasive characteristics that compromise the integrity of native mangrove ecosystems [34,35].

Investigations conducted by Chen et al. [33] revealed that *Sonneratia apetala* exhibits phenotypic plasticity to tidal flooding equivalent to that observed in its native congener *Sonneratia caseolaris*. This adaptive convergence indicates a significant overlap of ecological niche along the tidal gradient, conferring substantial competitive advantages to the alien species over autochthonous species. The capacity for similar physiological response to periodic flooding conditions suggests that *Sonneratia apetala* can efficiently occupy the same microhabitats used by native species, resulting in intensified interspecific competition [33].

The first documented record of *Sonneratia apetala* in South America, made by Eysink et al. [15] in the Cubatão region, in the metropolitan region of Santos (Southeastern Brazil), represents a biogeographic milestone of continental relevance. The location of the occurrence site approximately two km from the Port of Santos provides robust circumstantial evidence that the introduction occurred via ballast water from commercial vessels. This event exemplifies the fundamental role of anthropogenic vectors in facilitating the transcontinental dispersal of species, particularly in port contexts of high global connectivity [15].

The characterization of *Sonneratia apetala* as an invasive alien species is based on its high dispersal potential and demonstrated capacity to promote significant structural alterations in mangrove ecosystems. The identification of four preferential niches for colonization of "microhabitats" (mudflats), tidal channels, canopy gaps, and abandoned aquaculture nurseries, provides essential technical support for the development of early detection strategies and the implementation of rapid response protocols. These microhabitats represent critical points of ecological vulnerability, where preventive intervention can be most effective in containing invasive processes [35,38].

Also noteworthy is the formation of reefs by *Magallana gigas* (Pacific oyster) in the Port River-Barker Inlet Estuary, Australia, which constitutes a paradigmatic example of how invasive species can act as ecosystem engineers, promoting substantial structural modifications in the physical architecture of the habitat. Investigations conducted by Martin et al. [34] demonstrated that these biogenic structures exert a significant influence on the population dynamics of native species, evidencing positive associations between the presence of oyster reefs and increased densities of Avicennia marina seedlings [39].

This facilitative relationship suggests that the calcareous structures formed by *Magallana gigas* create microhabitats favorable for the establishment of native mangroves, possibly through the modification of microclimatic conditions, sediment stabilization, and alteration of local hydrodynamic patterns. The observed phenomenon illustrates how invasive species can, paradoxically, contribute to the restoration of native communities through ecological engineering processes, challenging traditional paradigms about the exclusively negative impacts of biological invasions [40].

However, the analysis of ecosystem effects reveals a complex duality of consequences. Although the reefs resulted in measurable increases in species richness and biomass of invertebrates and fish, they simultaneously facilitated the establishment of other exotic species, including *Carcinus maenas* (European green crab). This duality of effects, concomitant facilitation of native and exotic species, illustrates the inherent complexity of ecological interactions in invaded systems and evidences the difficulty in predicting the outcomes of management strategies [41].

The ambivalent nature of these impacts suggests that the assessment of invasive species as ecosystem engineers requires multidimensional approaches that consider both ecological benefits and costs. The capacity of *M. gigas* to simultaneously promote the recovery of native species and the invasion of other alien species demonstrates that the outcomes of biological invasions can be significantly more nuanced than traditionally assumed, requiring more sophisticated conceptual frameworks to guide management decisions.

The comparative bibliometric analysis reveals significant strategic implications for the development of research on biological invasions in the context of the Port of Santos. The identification of a critical geographical gap in the representation of the Brazilian coast strongly justifies the need and relevance of local studies, positioning the proposed research as a pioneering and necessary contribution to regional and global knowledge. However, the concentration of studies on specific vectors (ballast water) suggests methodological opportunities for investigating less studied but potentially relevant vectors in the context of the Port of Santos, such as adhered sediments and debris associated with solid bulk transport. This approach could reveal previously undocumented introduction routes and contribute to a more comprehensive understanding of invasion mechanisms.

Furthermore, the identified collaboration networks offer strategic opportunities for international partnerships that could significantly strengthen national research capacity and facilitate the transfer of methodologies and specialized knowledge to the Brazilian context.

3.2. *Spatial Analysis of Vessel Flows Destined for the Port of Santos*

Figure 3 presents the spatial distribution of international port departures destined for the Port of Santos, Brazil, from 2010 to 2024. The data were organized into five triennial periods (2010–2012, 2013–2015, 2016–2018, 2019–2021, and 2022–2024), in addition to an aggregated panel with the accumulated totals for the complete period. It is important to note that the data for the year 2024 correspond to departures recorded up to March 21, meaning the values for this triennium are still partial.

The cartographic symbology is based on graduated proportional circles, classified into volume ranges (1–100, 101–200, 201–300, and above 300 departures), representing the magnitude of flows by port of origin. All mapped departures have the Port of Santos as their exclusive destination, which figures as a structuring logistical hub of the South American Atlantic coast.

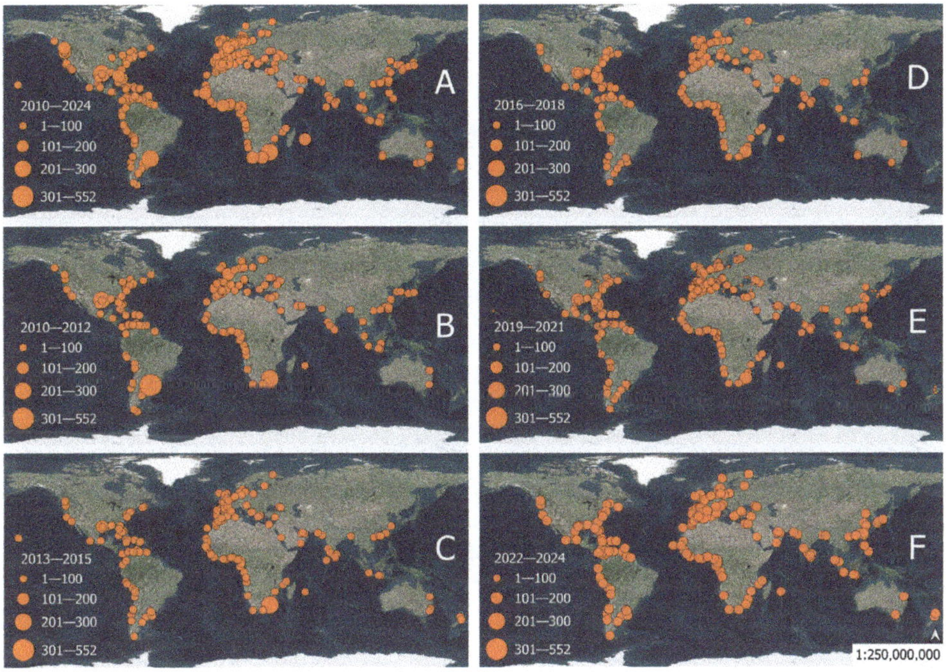

Figure 3. Geospatial distribution of vessels destined for the Port of Santos between 2010 and 2024. In (**A**), representation of the total period. In (**B**–**F**), different periods representing vessels destined for the Port of Santos. Scale 1:250,000,000.

Throughout the analyzed period, a robust articulation of the Port of Santos with ports from various regions of the globe is observed, delineating a transcontinental commercial connectivity network of great intensity. In the overall panel (2010–2024), dense logistical corridors originating in Western Europe stand out, particularly in countries such as Spain, Italy, Germany, Belgium, and The Netherlands, along with important connections with West Africa (notably Nigeria, Angola, and Ghana), South Asia (India and Sri Lanka), and the west coast of the American continent. Some African and European ports exceed 300 departures in the aggregated period, evidencing systematic links with the Brazilian port platform.

The segmented reading reveals structural changes over the triennia. Between 2010 and 2012, a total volume of 3474 departures is verified, with peaks in Latin American and African ports. This behavior reflects the intensification of flows linked to the commodity boom cycle, with strong insertion of Brazil in South-South trade and extractive chains. In the following triennium (2013–2015), the volume rises to 3734 departures, marked by a geographical expansion of the network, with the entry of new ports from the Middle East and South Asia, consistent with the densification of energy and mineral chains.

In the period from 2016 to 2018, with 3258 recorded departures, a certain spatial redistribution is observed, with a greater presence of flows originating from the northern hemisphere and a slight retraction in the African axis. This recomposition may be associated with geopolitical reconfigurations and the emergence of more selective and regional trade policies. In the pandemic triennium of 2019 to 2021, the total volume fell to 2925 departures, revealing the direct impact of COVID-19 on global supply chains, with logistical interruptions and rerouting of routes to ports considered more resilient. However, the persistence of the main departure hubs is noted, which maintain connectivity with Santos even in crisis scenarios.

In the most recent period (2022–2024), with data still partial, 1802 departures were recorded up to March 2024, signaling a gradual resumption of international flows. Despite the incompleteness of the data, the spatial configuration already points to the reactivation of traditional corridors, especially between the Mediterranean Europe, the Gulf of Guinea, and Southeast Asia. The persistence of large volumes in certain ports, even after sanitary, logistical, and geopolitical shocks, evidences their systemic robustness, reaffirming their centrality in long-distance maritime networks.

The multitemporal cartographic analysis presented here allows not only for the visualization of the hierarchy and density of flows destined for the Port of Santos but also for understanding the evolutionary dynamics of its global connection network. The articulation between structural stability and contextual adaptations of the flows reinforces the role of the Port of Santos as a strategic logistical node and offers valuable input for port planning, commercial risk modeling, and the formulation of integrated foreign trade and infrastructure policies [42,43] and draws attention to the need to create strategies that avoid or minimize the problem of biological invasion in coastal ecosystems. Brazilian reports have described some exotic species near by the Port of Santos, Southeastern Brazil [44,45].

The spatiotemporal analysis of maritime traffic elucidates the precise navigation corridors that function as persistent vectors of biological pressure on mangrove ecosystems. This finding underscores the imperative for risk management policies that transcend individual vessel inspection, targeting instead the consolidated dynamics of shipping routes. Consequently, naval movement data are elevated from mere metrics to a strategic foundation for pre-empting biological invasions.

3.3. Vessel Flows and Biological Invasion Risk

Figures 4A and 4B represent, respectively, the ten ports with the lowest and highest coefficient of variation (CV%) of international departures destined for the Port of Santos (Brazil) between 2010 and 2024. The CV, defined as the ratio between the standard deviation and the mean of annual departures, is a statistical measure sensitive to the relative instability of logistical flows and was therefore analyzed here. By isolating the proportional fluctuation of port movement, this metric allows for the capture of interannual operational patterns in a standardized manner, even in terminals with different volume scales.

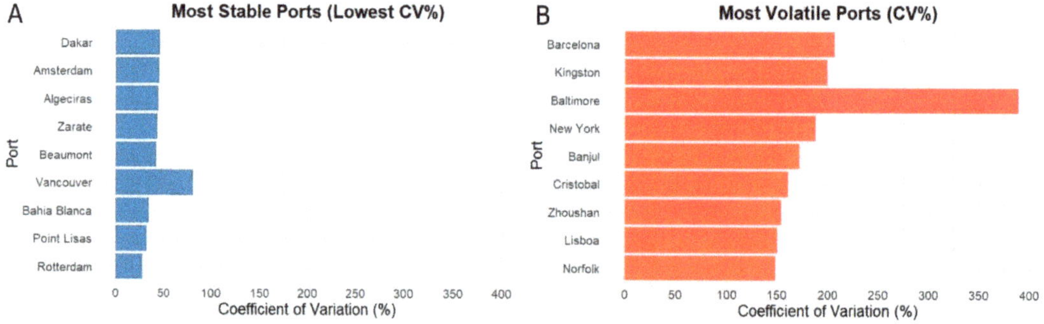

Figure 4. Representation of the most stable (**A**) and volatile Ports (**B**) (CV%).

Figure 4A presents the ten ports with the lowest CV%, constituting the terminals with the highest relative regularity in the annual movement of departures destined for the Port of Santos. The same criterion of statistical robustness was applied, requiring regular operation in at least eight of the fifteen years analyzed, with an annual mean greater than one departure. Notable ports include Amsterdam, Algeciras, Rotterdam, and Beaumont, all recognized logistical centers with consolidated infrastructure, high intermodal connectivity, and operational predictability. The performance of these terminals aligns with attributes such as advanced automation, long-term logistical contracts, and integration into mature trade routes, which reduces their sensitivity to external variations [46]. Zárate (Argentina) and Bahia Blanca (Argentina) reinforce the presence of South American terminals among the most stable, evidencing the existence of regular logistical corridors focused on the export of commodities, especially grains and derivatives, with recurring destinations to the main Brazilian port. The constancy of these flows appears to cushion the effects of external crises, even in economies marked by political or exchange rate instabilities.

The port of Dakar (Senegal) represents the only African case among the most stable. Its presence in this group may be related to the operation of regular cargoes, such as phosphates, ores, and agricultural products, destined for the Port of Santos, in a corridor that has been strengthening with Euro-African investments in logistical infrastructure and export agreements to South America. The case of Vancouver (Canada) is noteworthy; although it figures among the ten most stable, its CV% is noticeably higher than the others in the group, indicating a bimodal behavior or alternation between periods of high and low movement. This apparent contradiction may be linked to the seasonality of Canadian exports (timber, grains), impacts of forest fires, severe climatic restrictions, or labor interruptions. As discussed by Woo et al. [43], statistical stability should be interpreted cautiously, as seemingly regular patterns may mask cyclical behaviors or critical dependencies on specific markets. Point Lisas (Trinidad and Tobago) completes the group, suggesting that even small and medium-sized ports, when integrated into well-defined industrial chains (such as petrochemicals and fertilizers), can maintain relative stability in their flows destined for Brazil.

Figure 4B highlights the ten most unstable terminals, according to CV%. To avoid distortions caused by sporadic or one-off movements, only ports with recorded activity in at least eight of the fifteen years analyzed and an annual mean greater than one departure was included. This methodological filter ensures statistical consistency and comparability between cases. The port of Baltimore (USA) stands out with a CV close to 380%, ranking as the most volatile terminal in the sample. This magnitude of oscillation suggests the presence of significant operational shocks throughout the series, possibly linked to changes in vehicle export chains, port strikes, and fluctuations in US tariffs. Its instability directly impacts the USA-Brazil corridor, hindering logistical predictability in operations with Santos [42].

Other North American terminals, such as New York and Norfolk, complete the picture of logistical vulnerability in the Atlantic axis, reflecting the fragility of logistical centers even in high-capacity contexts. This volatility may have been aggravated by recent shocks—such as the COVID-19 pandemic—and by reconfigurations in international shipping [43]. In Europe, Barcelona and Lisbon also appear among the most volatile ports. This suggests the exposure of these terminals to changes in Mediterranean flows destined for Brazil, regional inter-port competition, and alternation between different types of cargo. In the case of Barcelona, the indirect effects of Catalan political-regional instability on port logistics may also have had an influence.

In Latin America and the Caribbean, the terminals of Cristóbal (Panama) and Kingston (Jamaica) stand out, whose volatility may be associated with high dependence on specific flows, such as transshipment containers and extractive exports destined for the Port of Santos, as well as political instabilities and customs reforms. These ports operate primarily as regional redistribution bases, making them particularly sensitive to reconfigurations in global maritime routes and fluctuations in intercontinental demand [47]. In Africa, the port of Banjul (The Gambia) illustrates the typical vulnerability of terminals in economies highly dependent on a limited number of commodities and markets. Its high volatility may be related to regional economic instability, fluctuations in primary exports, and limited local logistical resilience on routes destined for South America. The presence of Zhoushan (China) among the most volatile reflects specific oscillations linked to environmental control, restrictive policies, and alternation between regional and export operations, especially in cargoes directed to Brazil. Even within China, ports specialized in bulk and chemical products face regulatory regimes that impact the consistency of logistical flows.

These cases demonstrate that port volatility is not restricted to developing countries or low-capacity terminals but also affects relevant ports in central markets, especially when exposed to external shocks, severe seasonalities, or sectoral instabilities. The vulnerability in routes directed to the Port of Santos, Brazil's main entry point, represents a critical factor for the country's logistical security. The comparison between Figure 4A,B reveals that port logistical volatility is multifactorial and does not depend exclusively on the level of economic development or the scale of operation. Ports considered technically modern, such as Baltimore, Barcelona, and Vancouver, exhibit relevant fluctuation patterns, while regional terminals show notable stability.

The use of the coefficient of variation as a metric allowed, in this work, a comparative analysis of operational consistency, being instrumental for risk assessments, logistical insurance management, investment decisions, and the formulation of resilient port policies. Given the growing scenario of disruptive events such as pandemics, wars, and climate alterations, the capacity to predict and manage volatilities becomes a strategic differentiator for port governance in the 21st century [48].

Regarding the countries with the highest total number of port departures towards the Port of Santos, Figure 5 presents the ten most representative between 2010 and 2024. This focus allows for the identification of the main emitting poles of international maritime flows connected to the main Brazilian port terminal, highlighting those with the highest density of logistical connections throughout the historical series. It should be noted that the 2024 data only cover the period up to March 21, which may underestimate the final annual volume.

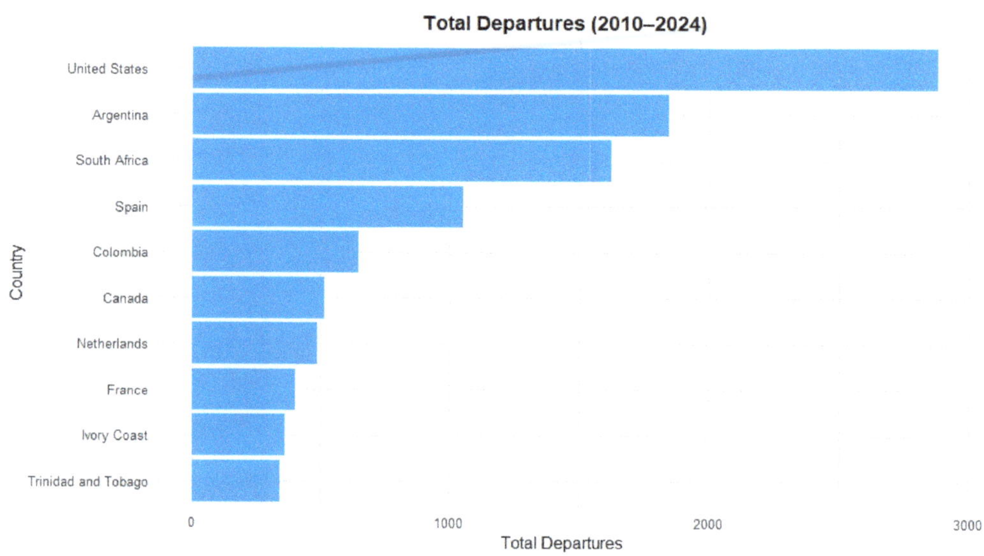

Figure 5. Total Departures to the Port of Santos (2010–2024).

The United States leads by a wide margin, totaling 2879 recorded departures in the period. This prominence reflects the intensity of commercial relations between Brazil and the USA, with emphasis on the transport of petroleum derivatives, chemicals, coal, grains, automotive parts, and other high-value-added cargoes. The consolidated presence of logistical terminals in the Gulf of Mexico and the US east coast, plus regular lines operated by major global carriers, contributes to this high inter-port connectivity. Argentina occupies the second position, with 1844 departures, evidencing the strong regional logistical integration in the Southern Cone. Geographical proximity, the shared use of international cabotage (treated as long haul), and productive complementarity explain this significant volume. Ports such as Zárate, Buenos Aires, and Bahía Blanca operate as relevant centers in the South American Atlantic axis.

The pronounced maritime connectivity identified in this study, particularly the high volume of departures from the United States (2879) and Argentina (1844), establishes direct and distinct pathways for biological invasions. Vessels originating from the logistical hubs of the US East Coast and the Gulf of Mexico represent a primary vector for the introduction of invasive macroalgae already prevalent in those waters, such as *Grateloupia turuturu* and *G. lanceolata*. These species have demonstrated their invasive potential by outcompeting native flora and altering community structures in the Northern Hemisphere; their dispersal to the Santos estuary via hull fouling poses a tangible and severe threat of analogous ecological disruption [49]. Concurrently, the intense regional integration with Argentine ports like Zárate and Bahía Blanca facilitates a separate, yet equally concerning, invasion corridor. This route has already proven functionally open, as evidenced by the significant transformations wrought in Argentine Patagonia by introduced species like the reef-builder *Ficopomatus enigmaticus* and the rapidly spreading kelp *Undaria pinnatifida*. The high frequency of arrivals from Argentina dramatically increases the probability of similar bioinvaders becoming established in Brazilian mangroves and estuaries, where they could trigger comparable regime shifts [50]. Therefore, these departure statistics must be interpreted not merely as trade figures, but as quantifiable proxies for specific and regionally contextualized bioinvasion risks.

In third place, South Africa stands out with 1620 departures, revealing the strategic importance of intercontinental South-South corridors. The routes between ports like Durban and Cape Town and the Port of Santos reflect the dynamism of exchanges involving mineral bulk, fertilizers, and vehicles, supported by bilateral trade agreements and robust logistical infrastructure. Spain, Colombia, and Canada appear next, comprising a group of countries with diversified export profiles. Spain acts as a logistical hub for cargo from the Mediterranean and North Africa; Colombia integrates regional circuits of energy and agricultural commodities; and Canada stands out in the supply of paper, pulp, fertilizers, and grains, with strong seasonality.

Countries such as France and The Netherlands, traditional commercial partners of Brazil, are present among the top ten emitters, sustaining regular transatlantic flows via ports like Le Havre and Rotterdam. These terminals concentrate containerized cargo, chemicals, and high-value-added manufactured goods. The inclusion of Côte d'Ivoire and Trinidad and Tobago is also noteworthy, representing connections with West Africa and the Caribbean, respectively. Although on a smaller scale, these countries play a strategic role in the circulation of commodities such as cocoa, aluminum, fuels, and fertilizers, contributing to the geographical diversification of maritime routes destined for Santos.

The integrated analysis of this empirical data and the scientific literature on biological invasions allows for a deeper understanding of the ecological risks associated with maritime traffic. As previously discussed by Sardain et al. [17] and Lohan et al. [51], the volume of port departures is directly proportional to the risk of accidental introduction of alien aquatic organisms, transported via ballast water, biofouling, or sediments adhered to vessels.

The leadership position of the United States in this port flow is consistent with its scientific relevance on the topic: it leads publications on invasive species in mangroves, with emphasis on Florida, where studies on *Schinus terebinthifolius* and *Crassostrea virginica* are concentrated [52,53]. Although Argentina does not have the same academic prominence, its strong logistical relationship with Brazil makes it a potential vector for invasive species, especially given the scarcity of detailed monitoring in many South American countries. Other relevant emitters, such as South Africa, Spain, and Canada, already feature in research on biological invasions, reinforcing the coherence between logistical connectivity and ecological risk. In the case of Trinidad and Tobago, the literature highlights its vulnerability to the introduction of species such as *Saccostrea cucullata* [51], confirming the strategic importance of the Caribbean region in the context of bioinvasions.

The international port system connected to the Port of Santos is characterized by a complex network of North–South, intra-Latin American, and transoceanic routes, involving countries from different biogeographical regions. This configuration, besides reflecting the centrality of Santos in the national economy, also reveals Brazil's exposure to diffuse environmental risks. The overlap between commercial flows and biological vectors reinforces the urgent need for public policies aimed at port environmental monitoring, especially to mitigate the introduction of alien species into sensitive ecosystems like Brazilian mangroves.

It is further highlighted that from 2010 to 2024, a significant volume of international port departures destined for the Port of Santos was observed, originating from over 200 locations distributed across all continents. The totality of the 15,193 departures recorded in this interval corresponds exclusively to inbound flows to Brazil, i.e., domestic maritime transits not being considered in this study. It is important to note that the data for the year 2024 comprises records only up to March 21, which is why the values for the most recent triennium (2022–2024) should be interpreted cautiously regarding their potential underreporting.

The multitemporal cartographic analysis of the data allowed for the identification of spatial and temporal patterns of direct connectivity between the Port of Santos and regions recognized as source zones of exotic species that currently threaten the integrity of Brazilian mangroves. Countries such as India, China, Indonesia, Saudi Arabia, Nigeria, Angola, the United States (with emphasis on Florida), Cuba, and other Caribbean nations presented, throughout the period, between 200 and over 1000 departures destined for the Port of Santos. This significant flow of connections, often recurring at specific intervals, coincides with the origin of invasive species such as *Crassostrea gigas*, *Saccostrea cucullata*, *Sonneratia apetala*, and *Mytilopsis sallei*, all already recorded in Brazilian estuaries with documented ecological impacts. It is important to highlight that maritime traffic as an indicator of biological invasion risk is not accurate. The two most important factors for ship-mediated species introduction identified in the literature are salinity [54,55] and the biogeographical region of the last ports of origin of the voyage, as shown in this study.

3.4. Normative and Technological Conflicts in Controlling Biological Invasion by Vessels at the Port of Santos

The Port of Santos is governed by various safety norms that include the control of invasive species, among which NORMAM-401/DPC stands out. This is an example of a norm issued by the Directorate of Ports and Coasts of Brazil, which establishes the rules for the management of ballast water on ships operating in Brazilian jurisdictional waters. This norm aims to implement the guidelines of the International Convention for the Control and Management of Ships' Ballast Water and Sediments (BWM Convention), to which Brazil is a signatory [56]. In this context, ships must follow a rigorous protocol that includes an International Ballast Water Management Certificate (BWMC), which ships must carry, issued after inspection; a Ballast Water Management Plan (BWMP), meaning each vessel must have an approved plan; a Ballast Water Record (BWR), where all ballast operations must be documented; and Ballast Water Treatment, which institutes the duty to meet the standards of Regulation D-2 of the BWM Convention, which limits viable organisms and harmful bacteria [57]. It should be considered that not only ballast water is responsible for biological invasion, but also fouling. Corrêa et al. [44] found six potentially invasive species coming from vessels (e.g., *Branchiomma luctuosum*, *Hydroides elegans*, *Bugula neritina*, *Styela plicata*, *Clavelina oblonga*, and *Botrylloides giganteus*), while Rangel-Pereira et al. [58] corroborates this statement.

At the local level, a very important regulation at the Port of Santos is NAP.SUMAS. OPR.023.2024, a technical norm issued by the Santos Port Authority that regulates the control and inspection of ballast water on moored ships [59]. Its main objective is to prevent environmental damage by ensuring vessels comply with the international rules established by the BWM Convention. This norm establishes procedures for verifying ship compliance, such as document analysis and, in some cases, ballast water sampling, ensuring there are no irregular discharges that could introduce invasive species or pollutants into the local ecosystem.

Some legal conflicts have placed the risk of biological invasion at the heart of highly sensitive ecological issues and further threaten the biogeographical problems in the coastal ecosystems of the largest port in Latin America, as shown in the literature review and the intensity of vessel flows arriving annually at the Port of Santos.

Recently, the National Waterway Transportation Agency (ANTAQ) considered that norm NAP.SUMAS.OPR.023.2024 encroached upon the competence of the Brazilian Maritime Authority (AMB), responsible for ballast water management in the country, and annulled it. However, after different judicial instances in the Brazilian scenario, it was understood that the APS acted within its attributions, supported by national legislation (Law 9.966/2000) that obliges port authorities to inspect and prevent pollution in their areas of jurisdiction, including ballast water control.

However, beyond normative issues, it is fundamental that instruments are developed and utilized for monitoring ballast water and the vessels arriving daily at the Port of Santos. The use of oceanographic buoys with sensors for continuous monitoring is being studied, addressing challenges and advantages in stability, movement, and instrument integration. This instrument provides balance parameters, materials, and inertial sensors, highlighting critical factors for platform efficiency [60].

In the second half of the last century, the U.S. Geological Survey (USGS) pioneered tests with satellite telemetry for the remote collection of hydrological data. The benefits included reduced operational costs and automatic data processing, though there were still needs for system expansion and greater efficiency. However, recently, advances in sensors, automation, and telemetry have revolutionized water quality monitoring, allowing for real-time analysis and more efficient mitigation strategies [61]. Modern systems transmit data via cellular or satellite to platforms accessible to the public, improving the response to events such as harmful algal blooms, which is associated with reduced costs and increased efficiency in providing crucial information for scientists and environmental managers.

Monitoring ballast water in ships faces significant challenges due to traditional verification methods, which rely mainly on salinity measurement and manual reports subject to errors and inconsistencies. Ballast water exchange (BWE) in open sea, although effective in reducing invasive organisms, depends on complex and costly inspections, often limited by a lack of infrastructure and technical resources. Furthermore, Ballast Water Reporting

Forms (BWRFs) frequently present flaws, such as incorrect geographical information or absence of effective exchange, compromising enforcement and increasing environmental risks [62].

This initiative aligns with a broader global network dedicated to addressing the challenge of biological invasions mediated by maritime vessels. To counter this pervasive issue, the GloBallast Partnerships Programme (a collaboration between the GEF, UNDP, and IMO) was instituted, employing a comprehensive and tripartite implementation strategy that functions across global, regional, and national spheres. The program fosters a multi-stakeholder partnership encompassing governments, industry, and non-governmental organizations, thereby promoting international coordination, the formulation of guidelines, the establishment of sectoral alliances, and the advancement of knowledge management. Its principal achievements encompass the implementation of national strategies, the enactment of legal reforms, and the promotion of public–private partnerships to pioneer effective technological solutions [63].

The project is operationalized across high-priority sub-regions, including the Caribbean and the Mediterranean, through a collaborative network of over 15 Lead Partnering Countries and more than 70 Partner Countries. The Lead Partnering Countries are committed to the adoption of National Ballast Water Management Strategies and pertinent legal reforms, simultaneously functioning as central hubs for regional capacity-building initiatives. The overarching objective is to support developing nations in the implementation of the Ballast Water Management Convention, thereby curtailing the transfer of harmful aquatic organisms, safeguarding marine biodiversity, and mitigating the profound ecological and economic detriments inflicted by these invasive species [63].

To overcome these limitations, innovative technologies, such as remote Ballast Water Reporting and Monitoring Systems (BWRMS), have been developed to collect real-time data directly from ballast tanks. These systems use integrated sensors to measure parameters like turbidity, pH, dissolved oxygen, and temperature, in addition to georeferencing each sample to ensure compliance with regulations. The automatic transmission of data via satellite eliminates the need for manual intervention, reducing fraud and providing greater transparency. This approach not only validates the effectiveness of water exchange but also monitors ballast quality, offering a more reliable and sustainable solution for environmental management in ports.

Some technological instruments that analyze parameters such as water salinity, pH, and temperature, combined with GPS and telemetry to transmit data in real-time to a control station on land, have been tested in the port region of Santos. The system aims to ensure compliance with Brazilian legislation, which requires water exchange at least 200 nautical miles from the coast, preventing the bioinvasion of alien species, a serious problem as demonstrated by the various cases cited in the bibliographic review of this study.

The technological solution is pioneering by integrating sensors, satellite communication, and GPS tracking into an automated system, eliminating the possibility of data manipulation by the crew. Brazilian developers, together with the Port of Santos, will present the technology to the International Maritime Organization (IMO), seeking its global adoption as a tool to combat not only biological pollution caused by ballast water, which moves over 100 million tons annually along the Brazilian coast, but also to minimize ecological risks from biological invasion in the vicinity of the Port of Santos. Currently, the Port of Santos adopts similar technologies, such as the Digital Twin system, to optimize operations and reduce environmental risks, demonstrating how innovation can mitigate the impacts of bioinvasion [64].

In this sense, the importance of artificial intelligence (AI) is highlighted, as it consolidates its role in strengthening environmental policies, especially in port contexts. In the port environment, characterized by its complexity and need for integration, AI enables digital networks capable of responding to environmental incidents, such as fuel spills or changes in maritime emissions, through real-time analysis and predictive simulations [65]. Furthermore, it enables the automation of regulatory reports, ensuring greater transparency and credibility for all involved agents, from public authorities to the local community.

It is emphasized that the use of AI in ports goes beyond the vision of a mere operational tool, as it is a vector of innovation that supports strategic decisions. Intelligent systems allow for predicting cargo flows, optimizing energy management, and incorporating renewable energy solutions into terminals. This transformation consolidates ports as intelligent organisms, where governance is anchored in verifiable data and the transparency demanded by investors and regulators [66]. The impact, therefore, is not limited to legal compliance: it strengthens competitiveness, reinforces institutional credibility, and aligns the port sector with global carbon neutrality goals.

Finally, it is worth noting that in addition to these technological tools, popular knowledge may also be an ally in the study and combat of alien species. Kousteni. et al., [66] mention that citizen science has proven to be a pivotal and complementary tool for monitoring marine Invasive Alien Species (IAS), which are among the greatest threats to oceanic biodiversity. By engaging the public in data collection, often facilitated by digital platforms and mobile applications, this approach enables an unprecedented spatial and temporal scale in the detection of introduced organisms. A comprehensive study in European waters substantiated the efficacy of this model, documenting 8513 records of alien fish and mollusks, which included eight first records for the region and three expansions of their known geographical distribution. The analysis revealed distinct distribution patterns, with a concentration of fish records in the Eastern Mediterranean and mollusks in the Northeast Atlantic, reflecting both biogeographical dynamics and the varying intensity of monitoring efforts among countries. Scientific validation is a critical pillar for data reliability; encouragingly, 88.5% of the records were validated by experts, although the availability of supporting imagery is an aspect that could be improved.

Citizen science, when integrated into projects with expert guidance and robust validation protocols, generates high-quality data. This information is indispensable for informing policies such as the EU Marine Strategy Framework Directive and the IAS Regulation, while simultaneously fostering public awareness, thus establishing itself as a vital component for the proactive management of marine biological invasions.

4. Conclusions

This study demonstrates that global maritime connectivity, especially at the Port of Santos, represents a potential critical vector for the introduction of invasive species into mangrove ecosystems. The bibliometric analysis revealed a scientific production concentrated in regions such as North America, Europe, and Asia, with significant underrepresentation of South America, despite the ecological and economic relevance of Brazilian mangroves. The predominance of studies on invasive bivalves (e.g., *Crassostrea gigas*, *Mytilopsis sallei*) and alien species (e.g., *Sonneratia apetala*) reflects the influence of vectors like ballast water and biofouling, while gaps persist regarding other dispersal mechanisms, such as adhered sediments. The overlap between high-traffic maritime routes and the biogeographic origin of invasive species already established in Brazil (e.g., *Saccostrea cucullata*, *Schinus terebinthifolius*) reinforces the urgency for more robust biosecurity policies.

The geospatial analysis of vessel flows evidenced that the Port of Santos maintains intense connections with ports in Asia, Europe, Africa, and the Americas, increasing the risk of bioinvasions. Periods such as 2013–2015 and 2016–2018 were notable for the increase in Asian and Caribbean routes, coinciding with records of invasive species in

Brazil. The logistical volatility in ports like Baltimore (USA) and Banjul (The Gambia) suggests that economic and geopolitical crises can alter traffic patterns, introducing new biological vectors. Furthermore, the stability of routes with countries like Argentina and South Africa indicates continuous risks of introducing species adapted to environments similar to Brazilian mangroves.

The normative and technological conflicts in ballast water control at the Port of Santos expose weaknesses in environmental governance. Although NORMAM-401/DPC and NAP.SUMAS.OPR.023.2024 seek to mitigate risks, the dispute of competences between regulatory agencies and the dependence on traditional inspection methods limit their effectiveness. Innovative solutions, such as automated sensors, satellite telemetry, and digital twin systems, offer opportunities for real-time monitoring, but their implementation on a global scale still faces technical and regulatory challenges. The harmonization between international norms (e.g., BWM Convention) and local policies is essential to reduce the accidental introduction of invasive species.

Considering these findings, it is recommended to: (1) prioritize research on less studied vectors (e.g., sediments, debris, hull fouling); (2) expand international scientific collaboration networks to strengthen monitoring in Brazil; and (3) adopt an integrated approach of predictive technologies (e.g., environmental DNA, ecological niche modeling) with port policies. The growing anthropogenic pressure on coastal ecosystems demands urgent actions to balance logistical development and conservation, positioning the Port of Santos as a strategic laboratory for innovations in marine biosecurity. This study not only fills critical knowledge gaps but also provides a methodological model applicable to other tropical ports, contributing to the global sustainability of threatened coastal ecosystems.

Moreover, AI should not be seen merely as an additional resource, but as a central element in the transformation of environmental and sustainable practices. Its application in port contexts demonstrates gains not only in efficiency and sustainability but also in the generation of significant economic value. By integrating technology, governance, and sustainability, ports position themselves as active agents in the transition towards a greener and more resilient global economy.

Author Contributions: Conceptualization, S.A. and C.F.F.; methodology, S.A., F.R., L.F.d.S. and M.L.F.; software, F.R. and L.F.d.S.; validation, S.A., C.F.F. and M.L.F.; formal analysis, S.A., F.R., L.F.d.S. and M.L.F.; investigation, S.A., F.R., L.F.d.S. and M.L.F.; resources, S.A. and C.F.F.; data curation, F.R. and L.F.d.S.; writing—original draft preparation, S.A., F.R., L.F.d.S. and M.L.F.; writing—review and editing, S.A., C.F.F. and M.L.F.; visualization, S.A. and C.F.F.; supervision, C.F.F.; project administration, S.A.; funding acquisition, C.F.F. All authors have read and agreed to the published version of the manuscript.

Funding: This study was funded by National Council for Scientific and Technological Development (CNPq) (Grant #307185/2023-0)—M.L.F. and also by Vanzolini Foundation (Grant 1).

Data Availability Statement: The original contributions presented in this study are included in the article. Further inquiries can be directed to the corresponding author(s).

Acknowledgments: The authors thank the Port of Santos for the data availability. Acknowledgments are extended to the Editor-in-Chief. Associate Editor, and anonymous reviewers.

Conflicts of Interest: The authors declare no conflicts of interest.

References

1. Leal, M.; Spalding, M.D. The State of the World's Mangroves 2022. Global Mangrove Alliance. Available online: https://www.mangrovealliance.org (accessed on 15 August 2025).
2. Kathiresan, K. Mangroves: Types and importance. In *Mangroves: Ecology, Biodiversity and Management*; Springer: Singapore, 2021; pp. 1–31.

3. Cheng, G.; Liao, H.; Li, X. Experiences and lessons learned from mangrove conservation in China. *Front. For. Glob. Change* **2025**, *8*, 1588710. [CrossRef]
4. Zheng, Y.; Takeuchi, W. Quantitative assessment and driving force analysis of mangrove forest changes in China from 1985 to 2018 by integrating optical and radar imagery. *ISPRS Int. J. Geo-Inf.* **2020**, *9*, 513. [CrossRef]
5. Belhabib, D.; Campredon, P.; Lazar, N.; Sumaila, U.R.; Baye, B.C.; Kane, E.A.; Pauly, D. Best for pleasure, not for business: Evaluating recreational marine fisheries in West Africa using unconventional sources of data. *Palgrave Commun.* **2016**, *2*, 15050. [CrossRef]
6. Das, J. Blue economy, blue growth, social equity and small-scale fisheries: A global and national level review. *Stud. Soc. Sci. Res.* **2023**, *4*, 38–82. [CrossRef]
7. Satia, B.P. An overview of the large marine ecosystem programs at work in Africa today. *Environ. Dev.* **2016**, *17*, 11–19. [CrossRef]
8. Diele, K.; Koch, V.; Abrunhosa, F.A.; de Farias Lima, J.; de Jesus de Brito Simith, D. The brachyuran crab community of the Caeté Estuary, North Brazil: Species richness, zonation and abundance. In *Mangrove Dynamics and Management in North Brazil*; Springer: Berlin/Heidelberg, Germany, 2010; pp. 251–263.
9. Teh, L.S.; Teh, L.C.; Sumaila, U.R. Quantifying the overlooked socio-economic contribution of small-scale fisheries in Sabah, Malaysia. *Fish. Res.* **2011**, *110*, 450–458. [CrossRef]
10. Liu, X.; Liao, J.; Shen, G.; Zhang, L.; Chen, B. Extraction of 10 m resolution global mangrove in 2022. *Remote Sens.* **2022**, *16*, 2723. [CrossRef]
11. Hewitt, C.L.; Gollasch, S.; Minchin, D. The vessel as a vector—Biofouling, ballast water and sediments. In *Biological Invasions in Marine Ecosystems*; Springer: Berlin/Heidelberg, Germany, 2009; pp. 117–131. [CrossRef]
12. Ruiz, G.M.; Fofonoff, P.W.; Steves, B.P.; Carlton, J.T. Invasion history and vector dynamics in coastal marine ecosystems: A North American perspective. *Aquat. Ecosyst. Health Manag.* **2015**, *18*, 299–311. [CrossRef]
13. Davidson, I.; Cahill, P.; Hinz, A.; Kluza, D.; Scianni, C.; Georgiades, E. A review of biofouling of ships' internal seawater systems. *Front. Mar. Sci.* **2021**, *8*, 761531. [CrossRef]
14. Schaeffer-Novelli, Y.; Coelho, C., Jr.; de Almeida, R.; Menezes, G.V.; Menghini, R.P.; Cunha-Lignon, M.; Cintrón-Molero, G. Manguezais na Ilha do Cardoso, Cananéia, Estado de São Paulo, Brasil: Uma narrativa histórica. *Hoehnea* **2025**, *52*, e262024. [CrossRef]
15. Eysink, G.G.; Hatamura, E.; Schaffer-Novelli, Y. First occurrence in mangroves of South America of the exotic species *Sonneratia apetala* Buch.-Ham. from the Indo-Malayan region. *Biota Neotrop.* **2023**, *23*, e20231575. [CrossRef]
16. Ren, H.; Lu, H.; Shen, W.; Huang, C.; Guo, Q.; Li, Z.A.; Jian, S. *Sonneratia apetala* Buch. Ham in the mangrove ecosystems of China: An invasive species or restoration species? *Ecol. Eng.* **2009**, *35*, 1243–1248. [CrossRef]
17. Sardain, A.; Sardain, E.; Leung, B. Global forecasts of shipping traffic and biological invasions to 2050. *Nat. Sustain.* **2019**, *2*, 274–282. [CrossRef]
18. Ojaveer, H.; Olenin, S.; Narščius, A.; Florin, A.B.; Ezhova, E.; Gollasch, S.; Jensen, K.; Lehtiniemi, M.; Minchin, D.; Normant-Saremba, M.; et al. Dynamics of biological invasions and pathways over time: A case study of a temperate coastal sea. *Biol. Invasions* **2017**, *19*, 799–813. [CrossRef]
19. Hilsdorf, W.D.C.; Nogueira Neto, M.D.S. Porto de Santos: Prospecção sobre as causas das dificuldades de acesso. *Gestão Produção* **2015**, *23*, 219–231. [CrossRef]
20. APS Autoridade Portuária de Santos. Relatório Integrado 2024. Available online: https://www.portodesantos.com.br/wp-content/uploads/Relatorio-integrado-2025-v-30-05-2025-.pdf (accessed on 8 September 2025).
21. Graser, A.; Dragaschnig, M.; Straub, M. QGIS and the processing toolbox in the context of spatial analysis education. *Trans. GIS* **2021**, *25*, 708–728. [CrossRef]
22. Sherrouse, B.C.; Clement, J.M.; Semmens, D.J. GIS tools for spatial analysis: The role of open-source platforms in environmental research. *Environ. Model. Softw.* **2020**, *134*, 104854. [CrossRef]
23. Hurni, L.; Kriz, K.; Cartwright, W. *Atlases: Trends in Cartography*; Springer: Berlin/Heidelberg, Germany, 2016.
24. Ducruet, C. *Maritime Networks: Spatial Structures and Time Dynamics*; Springer: Berlin/Heidelberg, Germany, 2015. [CrossRef]
25. Mongeon, P.; Paul-Hus, A. The journal coverage of Web of Science and Scopus: A comparative analysis. *Scientometrics* **2016**, *106*, 213–228. [CrossRef]
26. ANTAQ—Agência Nacional de Transportes Aquaviários. *Anuário Estatístico Portuário 2022*; ANTAQ: Brasília, Brazil, 2023.
27. Augusthy, S.; Nizam, A. The diversity, drivers, consequences and management of plant invasions in the mangrove ecosystems. *Sci. Total Environ.* **2024**, *945*, 173851. [CrossRef]
28. Lim, C.S.; Jolkifli, Z.H.; Jair, A.; Karim, N.; Wahab, R.A.; Desai, D.V.; Cho, K. An inter-site study of biofouling recruitment on static immersion panels in major ports of South East Asia and India. *ASEAN J. Sci. Technol. Dev.* **2024**, *41*, 167–176. [CrossRef]
29. Donnelly, M.J.; Green, D.M.; Walters, L.J. Allelopathic effects of fruits of the Brazilian pepper Schinus terebinthifolius on growth, leaf production and biomass of seedlings of the red mangrove Rhizophora mangle and the black mangrove Avicennia germinans. *J. Exp. Mar. Biol. Ecol.* **2008**, *364*, 149–156. [CrossRef]

30. Martinez-Laiz, G.; MacLeod, C.D.; Hesketh, A.V.; Konecny, C.A.; Ros, M.; Guerra-Garcia, J.M.; Harley, C.D. The journey of hull-fouling mobile invaders: Basibionts and boldness mediate dislodgement risk during transit. *Biofouling* **2022**, *38*, 837–851. [CrossRef] [PubMed]
31. Serrano, E.; Ribes, M.; Bolivar, M.; Coma, R. Fouling of a boat hull by the invasive zooxanthellate coral Oculina patagonica-could shipping be enhancing its unique large-scale spread? *Mediterr. Mar. Sci.* **2023**, *24*, 76–86. [CrossRef]
32. Gammon, M.; Floerl, O.; Happy, S.; Leonard, K.; Lass, H.; Walls, K.; Floerl, L.; Casanovas, P.; Goodwin, E.; Hucker, S.; et al. Policy, management, and the 'Level of Fouling' scale to transform marine invasion risk reduction from recreational boats. *J. Environ. Manag.* **2025**, *394*, 127493. [CrossRef]
33. Zuo, X.; Wang, Y.; Zhao, H.; Li, G.; Wang, Y.; Li, G.; Zhang, L. Allelopathic effects of *Amomum villosum* Lour. volatiles from different organs on selected plant species and soil microbiota. *Plants* **2022**, *11*, 3550. [CrossRef] [PubMed]
34. Xia, S.; Wang, W.; Song, Z.; Kuzyakov, Y.; Guo, L.; Zwieten, L.V.; Wang, H. Spartina alterniflora invasion controls organic carbon stocks in coastal marsh and mangrove soils across tropics and subtropics. *Glob. Change Biol.* **2020**, *27*, 1627–1644. [CrossRef]
35. Ren, H.; Jian, S.; Lu, H.; Zhang, Q.; Shen, W.; Han, W.; Guo, Q. Restoration of mangrove plantations and colonisation by native species in Leizhou bay, South China. *Ecol. Res.* **2008**, *23*, 401–407. [CrossRef]
36. Feng, J.; Zhou, J.; Wang, L.; Cui, X.; Ning, C.; Wu, H.; Lin, G. Effects of short-term invasion of Spartina alterniflora and the subsequent restoration of native mangroves on the soil organic carbon, nitrogen and phosphorus stock. *Chemosphere* **2017**, *184*, 774–783. [CrossRef]
37. Zhang, Y.; Meng, H.; Wang, Y.; He, Q. Herbivory enhances the resistance of mangrove forest to cordgrass invasion. *Ecology* **2018**, *99*, 1382–1390. [CrossRef]
38. Chen, B.; Zhen, L.; Yang, Z.; Liu, T.; Yang, S.; Mu, W.; Chen, J. miRNA–mRNA integrated analysis reveals candidate genes associated with salt stress response in halophytic *Sonneratia apetala*. *RNA Biol.* **2025**, *22*, 1–13. [CrossRef]
39. Martin, B.; Huveneers, C.; Reeves, S.; Baring, R. Mangrove-associated Pacific oysters (*Magallana gigas*) influence estuarine biodiversity. *NeoBiota* **2025**, *98*, 1–26. [CrossRef]
40. Halassy, M.; Batáry, P.; Csecserits, A.; Török, K.; Valkó, O. Meta-analysis identifies native priority as a mechanism that supports the restoration of invasion-resistant plant communities. *Commun. Biol.* **2023**, *6*, 1120. [CrossRef]
41. Garside, C.; Bishop, M. The distribution of the European shore crab, Carcinus maenas with respect to mangrove forests in southeastern Australia. *J. Exp. Mar. Biol. Ecol.* **2014**, *461*, 173–178. [CrossRef]
42. Notteboom, T.; Rodrigue, J.-P. *Port Economics, Management and Policy*; Routledge: London, UK, 2021. [CrossRef]
43. Woo, S.D.; Han, C.Y.; Lee, H.-C. Port resilience and performance evaluation using entropy–TOPSIS under COVID-19. *J. Transp. Geogr.* **2022**, *96*, 103192. [CrossRef]
44. Corrêa, T.R.; Calderon, E.; da Silva Lima, L.; de Freitas Delgado, J.; Neto, J.A.B.; de Carvalho Aguiar, V.M.; da Fonseco, E.M. Bioinvasion of benthic invertebrates in a eutrophicated tropical estuarine system in South America (Santos, Brazil). *Environ. Prot. Res.* **2023**, *3*, 38–55. [CrossRef]
45. Machado, A.L.; Diogo, B.L.; Santos, J.A. Espécies exóticas introduzidas no porto de Santos. *An. Do Congr. Bras. De Iniciação Científica* **2024**, *1*, 304–308.
46. Alavi-Borazjani, S.A.; Bengue, A.A.; Chkoniya, V.; Shafique, M.N. An Overview of Critical Success Factors for Digital Shipping Corridors: A Roadmap for Maritime Logistics Modernization. *Sustainability* **2025**, *17*, 5537. [CrossRef]
47. Pan, H.; Zhang, A. Port performance under global disruptions: A data-driven network perspective. *Transp. Res. Part E Logist. Transp. Rev.* **2022**, *160*, 102658. [CrossRef]
48. Rodrigue, J.-P.; Comtois, C.; Slack, B. *The Geography of Transport Systems*, 5th ed.; Routledge: London, UK, 2020. [CrossRef]
49. Aguilar-Rosas, L.E.; Núñez-Cebrero, F.; Aguilar-Rosas, C. Introduced marine macroalgae in the Port of Ensenada, Baja California, Mexico: Biological contamination. *Procedia Environ. Sci.* **2013**, *18*, 836–843. [CrossRef]
50. Orensanz, J.M.; Schwindt, E.; Pastorino, G.; Bortolus, A.; Casas, G.; Darrigran, G.; Elías, R.; Gappa, J.J.L.; Obenat, S.; Pascual, M.; et al. No longer the pristine confines of the world ocean: A survey of exotic marine species in the southwestern Atlantic. *Biol. Invasions* **2002**, *4*, 115–143. [CrossRef]
51. Lohan, K.M.; Hill-Spanik, K.M.; Torchin, M.E.; Strong, E.E.; Fleischer, R.C.; Ruiz, G.M. Molecular phylogenetics reveals first record and invasion of Saccostrea species in the Caribbean. *Mar. Biol.* **2015**, *162*, 231–243. [CrossRef]
52. Makowski, C.; Finkl, C.W. Invasive species within South Florida coastal ecosystems: An example of a marginalized environmental resource base. In *Impacts of Invasive Species on Coastal Environments*; Springer: Berlin/Heidelberg, Germany, 2019; pp. 1–60. [CrossRef]
53. Benson, G.; Donnelly, M.J.; Sacks, P.E.; Walters, L.J. Documenting loss and fragmentation of intertidal oyster (*Crassostrea virginica*) reefs in a subtropical estuary. *Environments* **2023**, *10*, 133. [CrossRef]
54. Paiva, F.; Barco, A.; Chen, Y.; Mirzajani, A.; Chan, F.T.; Lauringson, V.; Baltazar-Soares, M.; Zhan, A.; Bailey, S.A.; Javidpour, J.; et al. Is salinity an obstacle for biological invasions? *Glob. Change Biol.* **2018**, *24*, 2708–2720. [CrossRef]

55. Kalyvioti, G.; Galanidi, M.; Zenetos, A. Risk assessment to identify high-risk voyage origin ports and a watch list for NIS introduction in the Mediterranean with vessels: The case of Saronikos Gulf, Greece. *Manag. Biol. Invasions* **2024**, *15*, 337–370. [CrossRef]
56. Brito, E.P.; da Costa Fernandes, F.; Nunes, Z.M.P. Risk assessment of biological invasion via ballast water in an Amazonian river port. *Manag. Biol. Invasions* **2025**, *16*, 355–373. [CrossRef]
57. Marinha do Brasil. Brasil Segue Diretrizes Internacionais para o Controle de Água de Lastro. 2022. Available online: https://www.marinha.mil.br/economia-azul/noticias/brasil-segue-diretrizes-internacionais-para-o-controle-de-agua-de-lastro (accessed on 8 September 2025).
58. Rangel-Pereira, F.S.; Castro, M.C.T.; Scapolatempore, M.P.; Fernandes, L.D.A.; Coutinho, R.; Ruiz, G.M. Potential contribution of foreign-flagged recreational and craft vessels to the introduction and spread of non-indigenous species: A preliminary assessment for Brazil. *Mar. Pollut. Bull.* **2025**, *215*, 117841. [CrossRef]
59. APS Autoridade Portuária de Santos. NAP.SUMAS.OPR.023.2024, de 11 de abril de 2024: Norma para o Controle de Gestão e Conformidade da Água de Lastro de Navios e Embarcações em Atividade no Porto Organizado de Santos. 2024. Available online: https://www.portodesantos.com.br/ (accessed on 8 September 2025).
60. Fathurizqullah, M.; Malik, A.M.A.; Azman, S.; Ali, A.; Khairuddin, J.; Saad, K.A.M.; Zakaria, A. Environmental buoys for water quality assessment. In *Coastal Sustainability: Insights from Southeast Asia and Beyond*; Springer Nature: Cham, Switzerland, 2024; pp. 131–148.
61. Durieux, E.D.H.; Le Corre, K.; Marchessaux, G.; Veyssiere, D.; Etourneau, S.; Garrido, M. Movements and habitat use of the invasive blue crab Callinectes sapidus in a Mediterranean coastal lagoon. *bioRxiv* **2025**, 2025-08. Available online: https://www.biorxiv.org/content/10.1101/2025.08.13.663855v1 (accessed on 19 October 2025). [CrossRef]
62. Pereira, N.N.; Colombo, F.B.; Chávez, M.I.A.; Brinati, H.L.; Carreño, M.N.P. Challenges to implementing a ballast water remote monitoring system. *Ocean. Coast. Manag.* **2016**, *131*, 25–38. [CrossRef]
63. Globallast. The Invasive Aquatic Species. 2025. Available online: https://archive.iwlearn.net/globallast.imo.org/the-invasive-aquatic-species-2/index.html (accessed on 19 October 2025).
64. Dinh, G.H.; Pham, H.T.; Nguyen, L.C.; Dang, H.Q.; Pham, N.D.K. Leveraging artificial intelligence to enhance port operation efficiency. *Pol. Marit. Res.* **2024**, *30*, 140–155. [CrossRef]
65. Lee, J.M.; Sim, M.S.; Kim, Y.S.; Lim, H.R.; Lee, C.H. Strategizing Artificial Intelligence Transformation in Smart Ports: Lessons from Busan's Resilient AI Governance Model. *J. Mar. Sci. Eng.* **2025**, *13*, 1276. [CrossRef]
66. Kousteni, V.; Tsiamis, K.; Gervasini, E.; Zenetos, A.; Karachle, P.K.; Cardoso, A.C. Citizen scientists contributing to alien species detection: The case of fishes and mollusks in European marine waters. *Ecosphere* **2022**, *13*, e03875. [CrossRef]

environments

MDPI

Article

Climate Change-Related Salinity Fluctuations and Warming Induce Physiological Stress and Cellular Alterations in an Antarctic Intertidal Brown Alga

Francisca Morales [1], Pamela T. Muñoz [2], Agustina Undabarrena [3,4], Paula S. M. Celis-Plá [1,5,6], Consuelo Rámila [1], Polette Aguilar-Muñoz [1,6,7], Verónica Molina [1,6,7], Nelso P. Navarro [8,9,10], Claudio A. Sáez [1,5,6,11], Céline Lavergne [1,5,6,12], Beatriz Cámara [4,13], Catherine Tessini [14], Kerina González-Pino [1,15], Gabriela B. Pérez-Hernández [1,15] and Fernanda Rodríguez-Rojas [1,5,6,*]

1 HUB AMBIENTAL, Universidad de Playa Ancha, Subida Leopoldo Carvallo 207, Valparaíso 2360004, Chile; francisca.morales@upla.cl (F.M.); paulacelispla@upla.cl (P.S.M.C.-P.); consueloramila@gmail.com (C.R.); polette.aguilar@upla.cl (P.A.-M.); veronica.molina@upla.cl (V.M.); claudio.saez@ua.es (C.A.S.); lavergne@icm.csic.es (C.L.); kerina.gonzalez@alumnos.upla.cl (K.G.-P.); gabriela.perez@upla.cl (G.B.P.-H.)

2 Departamento de Ciencias Biológicas, Instituto de Ciencias Biomédicas, Facultad de Ciencias de la Salud, Universidad Autónoma de Chile, Talca 3460000, Chile; pamela.munoz@uautonoma.cl

3 Novo Nordisk Foundation Center for Biosustainability, Technical University of Denmark, Kongens Lyngby, 2800 Copenhagen, Denmark; a.undabarrena.canusso@gmail.com

4 Laboratorio de Microbiología Molecular y Biotecnología Ambiental, Centro de Biotecnología DAL, Universidad Técnica Federico Santa María, Valparaíso 2390123, Chile; beatriz.camara@usm.cl

5 Laboratorio de Investigación Ambiental Acuático, Universidad de Playa Ancha, Subida Leopoldo Carvallo 207, Hospital del Salvador, Valparaíso 2360004, Chile

6 Departamento de Ciencia y Geografía, Facultad de Ciencias Naturales y Exactas, Universidad de Playa Ancha, Subida Leopoldo Carvallo 270, Hospital del Salvador, Valparaíso 2360004, Chile

7 Centro de Investigación Oceanográfica COPAS COASTAL, Universidad de Concepción, Concepción 3349001, Chile

8 Laboratorio de Ecofisiología y Biotecnología de Algas (LEBA), Facultad de Ciencias, Universidad de Magallanes, Punta Arenas 6200000, Chile; nelso.navarro@umag.cl

9 Centro FONDAP de Investigaciones en Ecosistemas Marinos de Altas Latitudes (IDEAL), Punta Arenas 6200000, Chile

10 Network for Extreme Environments Research, NEXER-Universidad de Magallanes, Punta Arenas 6200000, Chile

11 Departamento de Ciencias del Mar y Biología Aplicada, Universidad de Alicante, Carretera de San Vicente del Raspeig s/n, 03690 Alicante, Spain

12 Departament de Biologia Marina i Oceanografia, Institut de Ciències del Mar, ICM-CSIC, 08003 Barcelona, Spain

13 Millennium Nucleus Bioproducts, Genomics and Environmental Microbiology (BioGEM) Avenida España, Valparaíso 2340000, Chile

14 Laboratorio de Electroquímica y Química Analítica, Departamento de Química, Universidad Técnica Federico Santa María, Valparaíso 2390123, Chile; catherine.tessini@usm.cl

15 Doctorado Interdisciplinario en Ciencias Ambientales, Facultad de Ciencias Naturales y Exactas, Universidad de Playa Ancha. Subida Leopoldo Carvallo 270, Valparaíso 2360004, Chile

* Correspondence: fernanda.rodriguez@upla.cl

Abstract: Antarctica is experiencing one of the fastest warming rates globally, profoundly impacting seawater temperature and salinity, with direct consequences for marine life. The present study examined the combined effects of salinity fluctuations at 20, 33 (control salinity), and 41 psu, and temperatures of 2 °C (control temperature) and 8 °C (thermal stress) for 3 days, on the health and physiology of the Antarctic intertidal macroalga *Adenocystis utricularis*. Photosynthetic activity, photoinhibition, and photoprotective processes were assessed alongside biomarkers of oxidative stress/damage (total ROS, lipid peroxidation, and protein carbonylation) and antioxidant/osmotic response (ascorbate, free amino acids, and proline). The results showed that maximum quantum yield (F_v/F_m) remained stable under both salinity and thermal stress. However, productivity (ETR_{max}), the photoprotection index (NPQ_{max}), and irradiance saturation (Ek_{ETR}) were significantly decreased at 8

 https://doi.org/10.3390/environments12100390

°C, remaining constant under salinity fluctuations. At 2 °C, oxidative stress and damage were significantly higher under hypo- and hypersalinity conditions. However, at 8 °C, oxidative stress indicators decreased, accompanied by increased ascorbate levels in both hypo- (20 psu) and hypersalinity (41 psu) treatments compared to the control salinity. While warming temperatures negatively altered the oxidative response of *A. utricularis* at a 33 psu, we report here an interactive effect between salinity and temperature, leading to an altered stress response to salinity fluctuations under thermal stress. This study provides key information to better understand the adaptation of Antarctic intertidal macroalgae to multifactor climate change consequences.

Keywords: climate change; antarctica; photosynthesis; macroalgae; biomarkers; thermal stress; salinity stress

1. Introduction

In recent decades, global temperatures have risen by approximately 1.25 °C, a trend largely driven by industrialization and the large-scale emissions of greenhouse gases into the atmosphere. According to projections from the Intergovernmental Panel on Climate Change (IPCC) Sixth Assessment Report (AR6), global mean sea surface temperatures are expected to increase by +1.5 to +2 °C by mid-century, with the worst-case scenario predicting a rise of up to +6 °C by the end of the century [1,2]. Such dramatic warming poses significant threats to ecosystems worldwide, and one of the most alarming regions experiencing rapid temperature increases is Antarctica [3].

Global warming is leading to the accelerated melting of glaciers and ice sheets, which in turn, contributes to rising global sea levels, threatening coastal communities and ecosystems worldwide [4]. Glacial melting and seawater evaporation, triggered by temperature increase, lead to significant shifts in salinity within Antarctic intertidal zones, creating highly dynamic environments [5]. These salinity changes, alongside variations in turbidity and water temperature, form complex gradients that can critically impact the distribution, physiology, and survival of marine species [6].

Intertidal ecosystems are among the most vulnerable to climate change impacts [7]. Within these ecosystems, intertidal macroalgae serve as foundational species, contributing to primary productivity, enhancing habitat complexity, and supporting local biodiversity [8,9]. However, their capacity to survive and sustain these critical ecological functions is increasingly threatened by stressors such as salinity fluctuations. For example, climate change-driven salinity declines in the Baltic Sea are expected to shift macroalgae distributions, reduce genetic diversity, and favor green algae over red and brown algae, ultimately impacting ecosystem complexity and higher trophic levels [5]. This underscores the critical need to understand the response of Antarctic intertidal macroalgae to salinity shifts, enabling better predictions of climate change impacts on macroalgae ecosystem services and the biodiversity of this region.

One effective approach to studying the potential effects of abiotic stressors on macroalgae fitness is the use of physiological indicators and stress biomarkers [10]. These tools provide insight into the organism's cellular response to specific environmental stressors. Abiotic factors, such as fluctuations in temperature and salinity, can trigger the production and accumulation of reactive oxygen species (ROS), leading to oxidative stress and physiological damage in primary producers [11]. ROS interact with a variety of cellular molecules, including membrane lipids (causing lipid peroxidation), proteins (leading to carbonylation), and DNA (resulting in mutations). However, the detrimental effects of ROS can be mitigated by antioxidant defense systems, which encompass both non-enzymatic and enzymatic mechanisms. Non-enzymatic antioxidants, such as glutathione and ascorbate, play a critical role in neutralizing ROS [12], while enzymatic antioxidants, including glutathione reductase (GR), ascorbate peroxidase (APX), and superoxide dismutase (SOD), are essential for scavenging ROS and reducing oxidative stress in many macroalgal species [13]. In addition to oxidative stress, fluctuations in salinity can also result in cell dehydration and a loss of turgor pressure, which further exacerbates stress conditions. To counteract these effects, primary producers, including macroalgae, employ osmoregulatory mechanisms, such as the synthesis of osmoprotectants (e.g., L-proline, L-arginine, and L-asparagine). These compounds help to maintain the osmotic balance and prevent excessive water loss, thus enhancing the organism's ability to survive under fluctuating salinity conditions [14].

Previous studies have demonstrated the utility of biomarkers in assessing cellular responses of Antarctic macroalgae species, including *Monostroma hariotii*, *Pyropia endiviifolia*, and *Adenocystis utricularis* under heat stress (8 °C vs. 2 °C) over a 5-day period [15]. This study indicated that, even under warmer conditions, intertidal algae exhibited sufficient physiological and metabolic tolerance, which may enhance their survival potential [15]. Similarly, in the Arctic, the macroalga *Palmaria palmata* displayed relative tolerance to daily irradiance fluctuations and varying salinities (18 and 28 psu), measured by ecophysiological indicators (i.e., photosynthesis performance) [16]. However, its physiological response deteriorated progressively as salinity decreased, likely due to the increased energy demands associated with acclimatization to hyposaline conditions [16]. While a few studies have investigated the effects of elevated temperatures on macroalgae cellular responses [15,17], our understanding of the combined impacts of temperature and salinity stressors on Antarctic macroalgae species remains limited. Thus, a critical knowledge gap remains; while most studies have examined single-factor stressors, the interactive effects of temperature and salinity on Antarctic intertidal macroalgae are still poorly understood.

A. utricularis is a brown, pear-shaped seaweed that dominates the mid- and low-intertidal zones along the coast of King George Island, Antarctica, forming extensive meadows [9]. However, the rapid melting of the Collins Glacier [18,19], situated near the coastal zone, could pose a significant threat to the resilience and ecological fitness of their populations in the region. In this study, we investigated the effect of salinity fluctuations—driven by rising temperatures (seawater evaporation) and glacial melting in intertidal pools—under current (2 °C) and projected future (8 °C) sea surface temperature scenarios on the cellular responses of this seaweed. To achieve this, we conducted a laboratory experiment under controlled conditions, utilizing two different temperatures, 2 °C (control) and 8 °C, and three salinity levels, 20 (hyposaline), 33 (control), and 41 (hypersaline) psu, for 3 days. To evaluate the health status of *A. utricularis*, our study involved multidisciplinary experimental approaches, including photosynthesis indicators along with diagnostic biomarkers, such as parameters for oxidative stress, damage, antioxidant response, and osmolyte production.

2. Materials and Methods

2.1. Samples Collection and Culture Conditions

Samples of *A. utricularis* were collected in plastic containers from the intertidal zone of Punta Artigas, located in Bahía Fildes, King George Island, Antarctica (62°11′06.2″ S 58°52′42.9″ W) during the Antarctic field campaign ECA-59 (austral summer of 2023) (Figure 1). The experiment was carried out in the facilities of the Professor Julio Escudero Scientific Base belonging to the Chilean Antarctic Institute (INACH), located in Las Estrellas Village on King George Island. Once collected, algae fronds were washed twice with filtered artificial seawater (0.22 μm) and stored in plastic containers for 48 h at 2 °C under dark conditions and constant aeration for acclimation to laboratory conditions. Afterward, approximately ~300 mg of fresh biomass was transferred to 1 L containers with filtered seawater, considering three biological replicates. Salinity treatments consisted of 20 psu (hyposaline condition), 33 psu (control salinity), and 41 psu (hypersaline condition) [16,20]. The hypersaline condition was obtained by adding artificial sea salts (InstantOcean) to filtered seawater until reaching the desired salinity, whereas the hyposaline condition was prepared by diluting filtered seawater with autoclaved ultrapure water. Salinity, temperature, and pH were monitored using a multiparameter probe (Hanna Instruments HI98194; Rhode Island, USA). The experiments were carried out in thermoregulated baths at 2 °C (control temperature) and 8 °C (thermal stress), as described previously [15], with a photoperiod of 20:4 (light: dark), 90 µmol m^2 s^{-1} photosynthetic active radiation (PAR), and constant aeration. After 3 days, the samples were collected and washed twice with filtered distilled water, frozen in liquid nitrogen, and preserved in a dryshipper container until their arrival at the HUB AMBIENTAL UPLA in Valparaíso, Chile, where they were stored at −80 °C until further analyses.

Figure 1. (**A**) Collection of *A. utricularis* samples from an intertidal meadow located at Punta Artigas, Bahía Fildes, King George Island, Antarctica. (**B**) *A. utricularis* fronds.

2.2. Photosynthesis and Energy Dissipation as In Vivo Chlorophyll a Fluorescence

The measurement of in vivo chlorophyll *a* fluorescence from Photosystem II was conducted using a portable pulse-amplitude modulated fluorometer (Junior PAM, Walz GmbH, Effeltrich, Germany). Macroalgae tissues were incubated in 10 mL dark chambers to generate rapid light response curves for each experimental condition. After 15 min of dark adaptation, F_o and F_m values were recorded to calculate the maximum quantum efficiency (F_v/F_m), where F_v is obtained as $F_m - F_o$. In this context, F_o represents the baseline fluorescence of the thalli after dark adaptation, while Fm corresponds to the peak fluorescence following a high-intensity light pulse exceeding 4000 µmol m^{-2} s^{-1} for a few seconds [21]. The electron transport rate (ETR) was evaluated by exposing the samples for 20 s to a series of twelve increasing intensities of actinic white light using the Junior PAM fluorometer (Walz, Germany) [22]. The ETR was then calculated following the formula presented in [21]:

$$ETR\ (\mu\text{mol electron m}^{-2}\ \text{s}^{-1}) = \Delta F/F'_m \times E \times A \times F_{II} \quad (1)$$

The effective quantum yield ($\Delta F/F'_m$) was defined as $\Delta F = F_m{}' - F_t$, where F_t represents the intrinsic fluorescence of the algae under light conditions, and $F_m{}'$ is the maximum fluorescence recorded after a saturation pulse in light-incubated algae. E corresponds to the incident PAR irradiance, expressed in µmol photons m^{-2} s^{-1}, while A denotes the thallus absorptance, representing the fraction of incident light absorbed by the algae [23]. The fraction of chlorophyll associated with Photosystem II (PSII) within the 400–700 nm range, denoted as F_{II}, is considered to be 0.8 in brown macroalgae [24]. The parameters related to electron transport rate (ETR), such as the maximum electron transport rate (ETR_{max}) and the initial slope of the ETR versus irradiance curve (α_{ETR}), which serves as an indicator of photosynthetic efficiency, were derived from the tangential function [25]. Additionally, the saturation irradiance for ETR (Ek_{ETR}) was determined by identifying the intersection point between ETR_{max} and α_{ETR}. Lastly, non-photochemical quenching (NPQ), used as a measure of photoprotection, was computed following Equation (2) [22]:

$$NPQ = (F_m - F_m')/F_m' \quad (2)$$

The maximum non-photochemical quenching (NPQ_{max}) and the initial rate of increase in NPQ as a function of irradiance (α_{NPQ}) were determined using the tangential function describing NPQ in relation to irradiance [25].

2.3. Total ROS Content

Total ROS levels were measured using the Intracellular ROS fluorometric kit (orange) provided by Sigma-Aldrich. Briefly, 30 mg of previously ground biomass was lysed with 300 µL of 0.5 M HCl, followed by vortexing for 10 min at room temperature. Subsequently, centrifugation was conducted for 5 min at 7500 rpm at 4 °C. Then, 100 µL of the supernatant was combined with 100 µL of 100 mM sodium phosphate buffer (pH 6.8) and 0.5 µL of the kit fluorophore. Additionally, a blank sample was prepared by mixing 100 µL of the supernatant with 100 µL of 100 mM sodium phosphate buffer (pH 6.8), together with an autofluorescence fluorophore blank prepared using 250 µL of 100 mM sodium phosphate buffer (pH 6.8) and 0.5 µL of the fluorophore. Relative fluorescence units (RFUs) were measured at 540 nm (excitation) and 570 nm (emission) in a fluorometric spectrophotometer, Cytation 5 (Agilent Biotek Instruments, Santa Clara, CA, USA). Results were expressed in relative fluorescence units (RFUs) per mg of fresh weight.

2.4. Quantification of Lipid Oxidative Damage

The quantification of thiobarbituric acid reactive substances (TBARSs) as an indicator of lipid peroxidation was conducted following the method described in [15]. Initially, 150 mg of biomass previously ground in liquid nitrogen was lysed with 200 µL of 10% TCA solution by adding four 3 mm glass beads and vortexed for 15 min at room temperature. After centrifugation at 14,000× *g* for 25 min at 4 °C, 100 µL of the supernatant was mixed with 100 µL of 0.5% TBA solution. The samples were incubated for 45 min at 95 °C in a thermoregulated bath, followed by cooling on ice for 5 min. The absorbance of the final mixture was measured at 532 nm using a SPECTROstarNano Microplate Reader (BMG-LABTECH). This measurement was conducted in a 96-well microplate, with 200 µL of the mixture per plate. The standard curve was constructed using a commercial malondialdehyde standard (MDA) (Sigma-Aldrich, Darmstadt, Germany).

2.5. Quantification of Protein Oxidative Damage

The quantification of carbonyl groups in proteins is a parameter to evaluate their oxidative damage by using the 2,4-dinitrophenylhydrazine (DNPH) method [26]. Briefly, 30 mg of fresh biomass was homogenized in liquid nitrogen and then mixed with 300 µL of FAPRB lysis buffer (Plant Total RNA Purification Mini Kit 50 preps FAVORGEN) and 300 µL of 0.5 M HCl, vortexed for 10 min at room temperature, and centrifuged at 10,000 g for 15 min at 4 °C. For total protein content, 5 µL of supernatant was mixed with 250 µL of Bradford reagent, incubated at 37 °C for 2 min for subsequent reading in the BMG-LABTECH SPECTROstar Nano spectrophotometer at 596 nm. For carbonylated proteins, 300 µL was rescued in a new tube, and 60 µL of 10 mM DNPH was added. After a 10 s vortex, the mixture was incubated in the dark for 30 min at room temperature. Then, 360 μL of 4 °C cooled 20% TCA was added and incubated for 15 min on ice. After vortexing for 10 s, the tube was centrifuged for 5 min at 10,000× *g* at 4 °C. The supernatant was carefully removed, and the pellet was washed with 1 mL of ethanol/ethyl acetate (1:1). After another round of vortexing for 10 s, the tube was centrifuged for 1 h at 10,000× *g* at 4 °C. Following removal of the supernatant, the pellet was inverted and air-dried for 20 min. Then, the pellet was resuspended in 500 μL of 6 M guanidine hydrochloride and incubated at 37 °C for 15 min. A 10 s vortex was performed halfway through the incubation period and again at the end. Finally, the sample was centrifuged at 10,000× *g* for 5 min at 4 °C. In a UV plate, 150 μL of the supernatant and 150 μL of 6 M guanidine (quantification of carbonylated proteins) were added. Measurements were performed in the SPECTROstarNano spectrophotometer (BMG Labtech, Ortenberg, Germany) at 366 nm. The values were replaced in the Lambert–Beer equation (cuvette length 0.8 cm and DNPH–carbonyl complex molar extinction coefficient 22,000 M^{-1} cm^{-1}). The results are expressed as nmoles of carbonylated proteins in mg of total proteins.

2.6. *Quantification of Free Amino Acids and Proline*

The quantification of ninhydrin-positive substances (NPSs) was used as an approximation for the quantification of free amino acids and proline related to the osmotic stress response (osmolytes compatible with water) [26]. Briefly, 100 mg of previously ground biomass was homogenized with liquid nitrogen and lysed with 400 μL of 0.1 M HCl by adding four 3 mm glass beads and vortexing for 10 min at room temperature. Then, samples were centrifuged at 7000 rpm for 15 min at 4 °C. A volume of 350 μL of the supernatant was mixed with 350 μL 100 mM sodium citrate buffer, pH 4.8, and 290 μL of the previous mixture, and this was mixed with 10 μL of 2% ninhydrin. Then, samples were incubated at 95 °C for 20 min and allowed to cool to room temperature for 10 min. Finally, 150 μL of incubated sample was diluted with 150 μL of 100 mM sodium citrate buffer, pH 4.8, and measured in a spectrophotometer, SPECTROstarNano, at 560 nm for free amino acid content and 403 nm for proline content. The concentrations were established by comparison with a standard curve constructed with a commercial solution of free amino acids obtained from Sigma-Aldrich, and free amino acid levels are expressed in μmoles of free amino acids per mg of fresh weight.

2.7. *Quantification of Ascorbate and Dehydroascorbate*

The quantification of ascorbate in its reduced (ASC) and oxidized (DHA) forms was evaluated through the reduction in the Fe (III)-2,4,6-tripyridyl-S-triazine (TPTZ) complex [15]. Briefly, 100 mg of previously ground biomass was homogenized with liquid nitrogen and lysed with 300 μL of 0.1 M HCl, 300 μL of FAPRB lysis buffer (Plant Total RNA Purification Mini Kit 50 preps FAVORGEN), and four 3 mm glass beads, with vortexing for 10 min at room temperature. The samples were centrifuged at 17,800× *g* for 15 min at 4 °C. To measure reduced ascorbate (ASC), 10 μL of supernatant was mixed with 290 μL of the FRAP solution (250 mM sodium acetate buffer, pH 3.6, 0.83 mM TPTZ, and 1.7 mM $FeCl_3$) and measured immediately at 593 nm. For total ascorbate (ASC + DHA), 250 μL of the lysed supernatant was mixed with 2.5 mL of 100 mM dithiothreitol (DTT) and incubated for one hour at room temperature. The reaction was stopped with 2.5 μL of 5% N-ethylmaleimide. Then, 10 μL of the reaction was mixed with 290 μL of the FRAP solution and measured at 593 nm. DHA was quantified by a simple subtraction of total ascorbate and reduced ascorbate. Concentrations were established by comparison with a calibration curve generated with commercial L-ascorbic acid obtained from Sigma-Aldrich as a standard. Results were expressed in μmoles of ascorbate per mg of fresh weight.

2.8. Statistical Analysis

Statistical analyses were performed in R version 4.3.0. Interactive effects between physiological variables (F_v/F_m, α_{ETR}, ETR_{max}, Ek_{ETR}, and NPQ_{max}) were analyzed using two-way ANOVA [27]. This test was performed for *A. utricularis*, including temperature (two levels) and salinity (three levels) as fixed factors for photosynthetic variables (n = 3), with a level of probability applied in the statistical analyses at $p < 0.05$. Homogeneity of variance was tested using Cochran tests and by visual inspection of the residuals. Student–Newman–Keuls tests (SNK) were performed after significant ANOVA interactions. For stress biomarkers, normality and homogeneity tests were evaluated using the Shapiro–Wilk and the Bartlett statistical tests, respectively. Two-way ANOVA with a confidence interval of 95%, together with a Tukey test, was carried out to establish specific statistical differences between treatments.

3. Results

3.1. Physiological Responses of A. utricularis to Salinity Fluctuations and Elevated Temperatures

Physiological responses in *A. utricularis* to temperature and salinity were assessed using photosynthetic parameters (Figure 2). The F_v/F_m ratio, as a photoinhibition index, showed no significant variation in response to changes in temperature or salinity (Figure 2A). The highest irradiance at the saturation point of the curve (Ek_{ETR}) was significantly affected by temperature and salinity fluctuations compared to the control, whereas the other treatments showed no statistically significant differences among them (Figure 2B). Photosynthetic efficiency (α_{ETR}) increased significantly when exposed to 8 °C and under 20 and 33 psu (Figure 2C). The maximal electron transport rate (ETR_{max}) was significantly higher under control temperature conditions, regardless of salinity levels (Figure 2D). Finally, maximal non-photochemical quenching (NPQ_{max}), as the photoprotection index, was significantly higher under control temperature conditions, irrespective of salinity (Figure 3).

3.2. Oxidative Stress and Damage in A. utricularis Under Salinity Fluctuations and Elevated Temperatures

The total ROS content in *A. utricularis* fronds varied significantly across different salinity conditions and between temperature scenarios (Figure 4). According to ANOVA results, salinity and temperature exhibited a highly significant interaction. At 2 °C, ROS content was significantly higher under hyposaline conditions compared to both the control and hypersaline conditions. In contrast, at 8 °C, ROS accumulation was significantly lower under hyposaline conditions than in the control and hypersaline conditions. Additionally, at 20 psu, ROS levels decreased at 8 °C compared to 2 °C, whereas at 33 psu and 41 psu, ROS levels increased. However, all salinity treatments at 8 °C showed higher ROS accumulation compared to the control (2 °C, 33 psu) (Figure 4).

Oxidative stress damage was assessed by measuring lipid peroxidation in cell membranes and protein oxidation levels; these two parameters responded differently to salinity and temperature (Figure 5). In the case of lipid peroxidation, salinity and temperature exhibited a highly significant interaction according to the two-way ANOVA analyses. At 2 °C, lipid peroxidation significantly increased under both hypo- and hypersaline conditions compared to the control salinity. Conversely, at 8 °C, hypo- and hypersaline conditions resulted in lower levels of lipid peroxidation compared to the control. Additionally, at 20 psu and 41 psu, TBARS levels decreased at 8 °C compared to 2 °C, whereas at 33 psu, these increased. However, all salinity treatments at 8 °C showed elevated TBARS levels compared to the control (2 °C, 33 psu), although this difference was not statistically significant in the 41 psu treatment (Figure 5A).

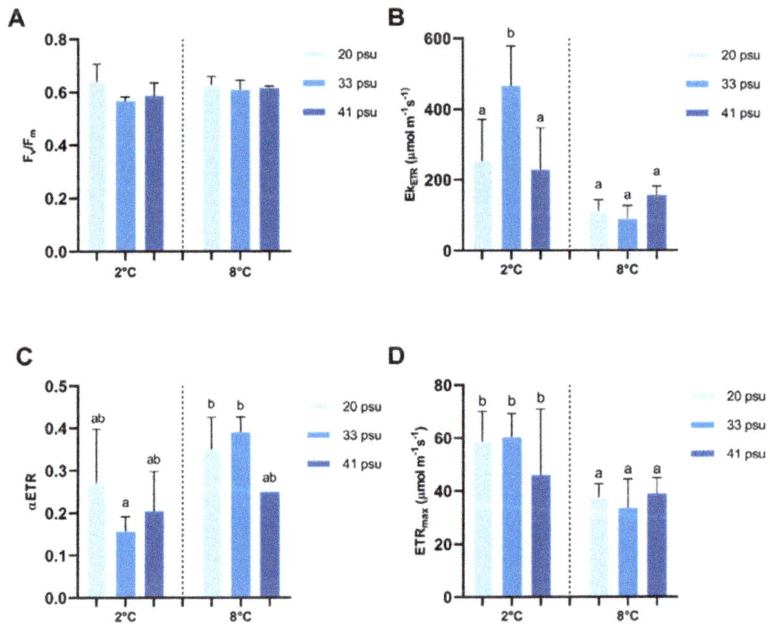

Figure 2. Effect of temperature and salinity on photosynthetic performance in *A. utricularis*. (**A**) Maximum quantum yield, F_v/F_m; (**B**) irradiance of saturation, Ek_{ETR}; (**C**) photosynthetic efficiency, α_{ETR}; and (**D**) maximal electron transport rate, ETR_{max}. The determination of photosynthetic parameters was conducted at 2 °C (control) and 8 °C, at three salinity levels: 20 (hyposaline), 33 (control), and 41 psu (hypersaline). The data are presented as mean ± SD, with $n = 3$. Lowercase letters indicate significant differences between treatments ($p < 0.00032$).

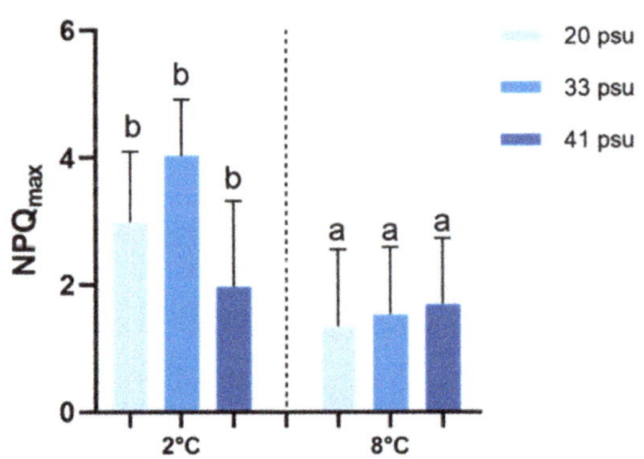

Figure 3. Effect of temperature and salinity on maximal non-photochemical quenching (NPQ_{max}) in *A. utricularis*. This parameter was determined at 2 °C and 8 °C, at three salinity levels: 20 (hyposaline), 33 (control), and 41 psu (hypersaline). The data are presented as mean ± SD, with $n = 3$. Lowercase letters indicate significant differences between treatments ($p < 0.00392$).

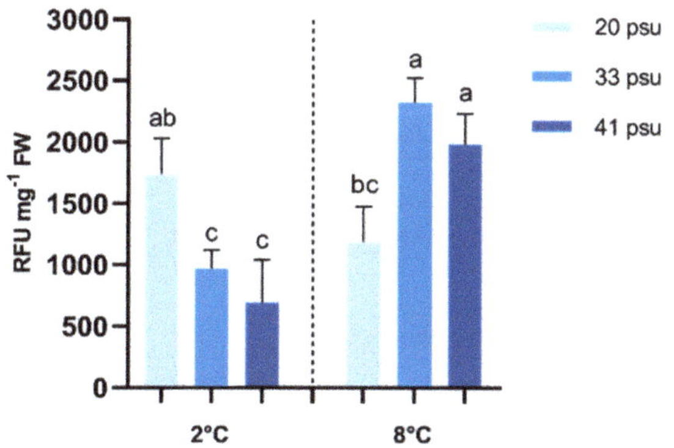

Figure 4. Effect of temperature and salinity on total ROS content in *A. utricularis*. The determination of ROS was conducted at 2 °C (control) and 8 °C, at three salinity levels: 20 (hyposaline), 33 (control), and 41 psu (hypersaline). The data are presented as mean ± SD, with $n = 3$. Lowercase letters represent significant differences between treatments ($p < 0.0456$).

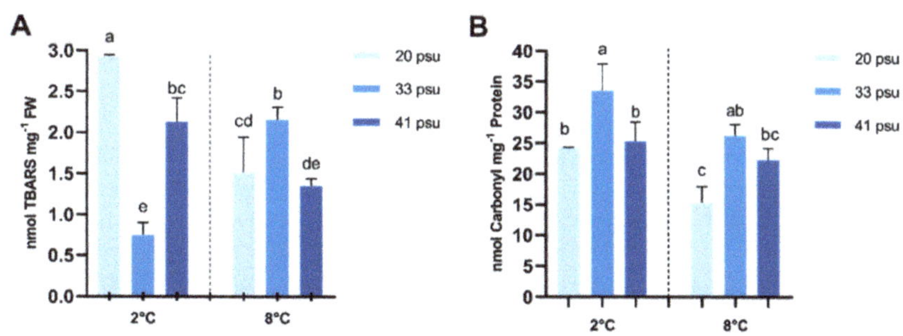

Figure 5. Effect of temperature and salinity on oxidative damage to macromolecules in A. utricularis. (**A**) Quantification of peroxidized lipids (TBARS). (**B**) Quantification of protein carbonylation. The determination of oxidative damage parameters was conducted at 2 °C (control) and 8 °C, at three salinity levels: 20 (hyposaline), 33 (control), and 41 psu (hypersaline). The data are presented as mean ± SD, with n = 3. Lowercase letters represent significant differences between treatments ($p < 0.00046$ and $p < 0.00089$).

Protein oxidation, measured as carbonylated proteins, followed a different pattern. Although both temperature and salinity showed a significant effect according to two-way ANOVA analyses, no significant interaction between salinity and temperature was observed. At 2 °C, lower levels of protein carbonylation were observed under hypo- and hypersaline conditions compared to the control. This trend persisted at 8 °C, where protein carbonylation remained elevated under control salinity conditions, although the difference was not significant with the 41 psu treatment. Additionally, all salinity treatments at 8 °C showed a tendency toward lower carbonylated protein levels compared to the control (2 °C, 33 psu), although not statistically significant in the 33 psu treatment (Figure 5B).

3.3. *Osmotic and Antioxidant Responses in A. utricularis Under Salinity Fluctuations and High Temperatures*

Osmolytes parameters, free amino acids (NPSs), and proline showed different responses to salinity and temperature fluctuations. According to the two-way ANOVA, salinity and temperature had no significant effect on NPS levels (Figure 6A). In the case of proline, however, salinity and temperature exhibited a highly significant interaction. At 2 °C, the hyposaline condition exhibited a tendency toward higher proline concentrations compared to both the control and hypersaline conditions, although not statistically significant. In contrast, at 8 °C, the hyposaline condition showed the lowest proline levels, but these differences were also not significant. Additionally, at 20 psu, proline levels significantly decreased at 8 °C compared to 2 °C, while at 33 psu and 41 psu, no significant differences in proline levels were observed across temperatures (Figure 6B).

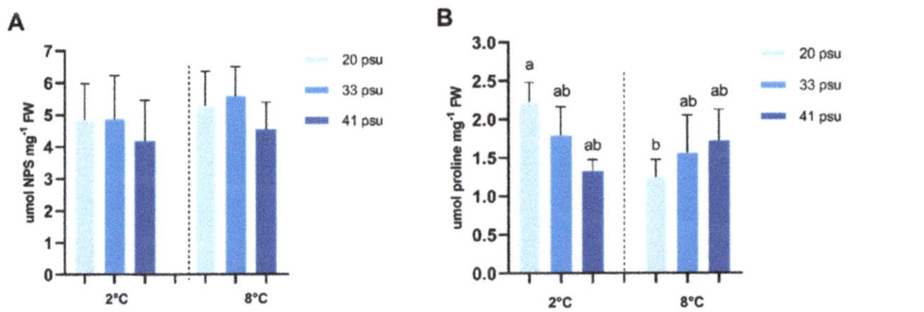

Figure 6. Effect of temperature and salinity on free amino acids and proline content in *A. utricularis*. (**A**) Quantification of ninhydrin-positive substances (NPSs) to determine osmolytes compatible with water. (**B**) Proline quantification to determine response to salinity stress. Both parameters were determined at 2 °C (control) and 8 °C, at three salinity levels: 20 (hyposaline), 33 (control), and 41 psu (hypersaline). The data are presented as mean ± SD, with $n = 3$. Lowercase letters represent significant differences between treatments ($p < 0.000879$ and $p < 0.000776$).

Antioxidant response was assessed by measuring ascorbate levels in both its reduced (ASC) and oxidized (DHA) forms (Figure 7). Total ascorbate levels (ASC + DHA) were significantly influenced by temperature, as indicated by two-way ANOVA, while no significant effects of salinity or interactions between salinity and temperature were observed. Total ascorbate levels were consistently lower at 8 °C compared to 2 °C within the same salinity treatments. Although these differences were not statistically significant, the only significant differences were observed between 20 psu at 2 °C and 33 psu at 8 °C. At both 2 °C and 8 °C, the highest total ascorbate levels were recorded under hyposaline conditions. DHA levels were also influenced by temperature, being consistently lower at 8 °C compared to 2 °C within the same salinity treatments. Although not statistically significant, a different trend was observed at 2 °C vs. 8 °C. At 2 °C, total ascorbate levels were higher under low salinity conditions (20 psu), whereas at 8 °C, higher values were obtained at 20 and 41 psu. Finally, the ASC/DHA ratio was higher at 8 °C, indicating an efficient reduction in ascorbate at this temperature.

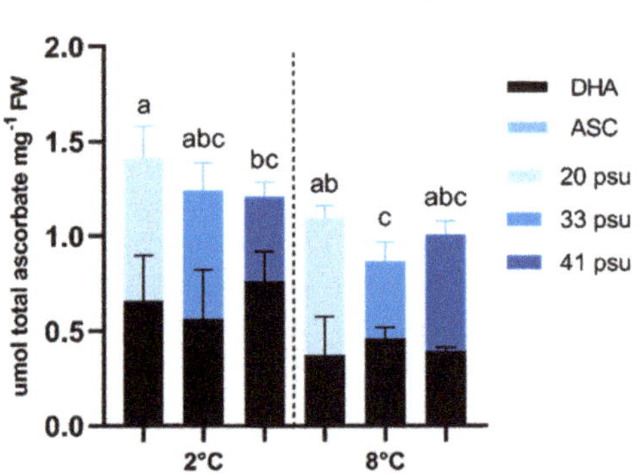

Figure 7. Effect of temperature and salinity on ascorbate (ASC in blue) and dehydroascorbate (DHA in black) in *A. utricularis*. Both parameters were determined at 2 °C (control) and 8 °C, at three salinity

levels: 20 (hyposaline), 33 (control), and 41 psu (hypersaline). The data are presented as mean ± SD, with $n = 3$. Lowercase letters represent significant differences between treatments ($p < 0.00073$).

4. Discussion

Despite the ecological significance of Antarctic intertidal macroalgae, cellular and molecular studies investigating their responses to salinity fluctuations and rising temperatures remain scarce. This knowledge gap limits our understanding of the mechanisms underlying their physiological and biochemical adaptations to the rapidly changing environmental conditions driven by climate change. In this study, we conducted a controlled short-term experiment to understand the cellular responses of an ecologically relevant macroalgae species of the intertidal zone to salinity and thermal stresses.

4.1. Photosynthetic Performance in A. utricularis Under Salinity Fluctuations and Elevated Temperatures

Photosynthetic efficiency in *A. utricularis* remained relatively high across all treatments, with F_v/F_m values around 0.6 [28], showing no significant differences. This indicates an ability to tolerate short-term temperature and salinity fluctuations. The lack of temperature effects aligns with findings of the authors of [15], who reported similar results for this species, exposed to 8 °C over the same period.

Regarding other photosynthetic parameters, the ecophysiological responses appear to be modulated only by temperature. Photosynthetic parameters did not show significant differences across the three salinity treatments. However, ETR_{max}, a proxy for productivity, and NPQ_{max}, an indicator of photoprotection, decreased under elevated temperature, in contrast to what was reported in [15]. This decline may be driven by various biological mechanisms, such as pigment content, and should be addressed in further research. The NPQ_{max} results suggest that heat dissipation could be an important mechanism in this species, alongside the use of the xanthophyll cycle as a photoprotection strategy [17]. In general, based on photosynthetic performance and ecophysiological parameters, the intertidal macroalgae *A. utricularis* exhibits biological viability and reduced vulnerability under increased temperature conditions, highlighting its potential adaptation to thermal and salinity stress.

4.2. Oxidative Stress and Damage in A. utricularis Under Salinity Fluctuations and Elevated Temperatures

Both temperature and salinity are well-known triggers of oxidative stress in marine photoautotrophs. Temperature increases beyond critical thresholds induce oxidative stress in these species, significantly affecting marine organisms in polar ecosystems [29]. Similarly, salinity fluctuations can cause oxidative stress in macroalgae [30,31]. Although the specific effects of temperature and salinity interactions on oxidative stress responses in macroalgae remain unexplored, evidence from Arctic macroalgae has shown that such interactions can influence their physiological and biochemical status [32], suggesting that oxidative stress responses could also be impacted.

Our study identified significant interactions between salinity and temperature across the responses of various biomarkers in *A. utricularis*, including ROS and TBARS, where temperature strongly influenced the species' response to salinity. At the control temperature (2 °C), the species exhibited higher stress under hyposaline conditions compared to the control salinity (33 psu), as evidenced by higher ROS and TBARS levels, whereas at 8 °C, the opposite trend was observed. This suggests that the studied species may have a greater capacity to withstand short-term hyposaline stress under future scenarios of elevated sea temperatures, while also highlighting its vulnerability to hyposaline conditions under current temperature regimes. Our findings also suggest that salinity could influence the species' ability to cope with thermal stress. At 20 psu, ROS and TBARS levels were lower under future climate scenario temperatures compared to current conditions, while at control salinity (33 psu), these levels significantly increased. Therefore, under hyposaline conditions, the species may exhibit a higher capacity to adapt to higher temperatures. Effects of salinity on modifying resilience to other abiotic stresses have been observed in other species like *Palmaria palmata*, which demonstrated significant tolerance to daily irradiance fluctuations, though this resilience diminished under hyposaline conditions [16]. Taken together, our findings and previous studies highlight the critical importance of investigating the combined effects of multiple stressors on marine primary producers.

Despite the interaction effects mentioned above, it is important to highlight that, after the short-term exposure period studied (3 days), the species consistently exhibited higher ROS and TBARS levels, along with lower total ascorbate and DHA levels, at elevated temperatures (20 psu, 33 psu, and 41 psu, at 8 °C) when compared to the control conditions (33 psu at 2 °C). This could suggest a potentially reduced fitness of *A. utricularis* under future climate change scenarios, regardless of salinity conditions. The observed short-term effects of temperature are consistent with previous studies on this species, which reported a slight increase in oxidative stress indicators, such as hydrogen peroxide (H_2O_2) and TBARS, during short-term exposure to elevated temperatures (3 days at 8 °C). However, these oxidative stress parameters were found to decrease or return to control levels during mid-term exposure (5 days), indicating the presence of adaptive responses to thermal stress [15].

4.3. Osmotic and Antioxidant Responses in A. utricularis Under Salinity Fluctuations and High Temperatures

Osmolytes, such as proline, despite their role related to the response to hypersaline stress, are also key indicators of antioxidant capacity in response to other abiotic stressors, enhancing the co-accumulation of antioxidant molecules, bolstering resistance to oxidative stress [33,34]. In *A. utricularis*, our results show an interactive effect of temperature and salinity on proline levels. These peaked under hyposaline conditions at 2 °C and also reached higher values when exposed to hypersaline conditions at 8 °C. These results suggest that, under current low-temperature conditions, proline may function primarily as an antioxidant molecule. However, in future warmer scenarios, it appears to shift its role to its canonical function as an osmolyte, helping to prevent dehydration at hypersaline conditions. Under low temperatures, the increase in proline coincided with elevated ROS and TBARS levels under hyposaline conditions, supporting its potential antioxidant role in redox buffering and membrane stabilization. Under warming, the strongest proline accumulation observed at hypersalinity is consistent with its canonical osmoprotective function in mitigating dehydration stress [33–35]. This context-dependent shift aligns with previous evidence indicating that proline can act both as a stress signal and a protective metabolite when organisms are exposed to combined abiotic stressors [35].

The ascorbate–glutathione cycle also plays a pivotal role in antioxidant defense mechanisms in macroalgae, mirroring processes observed in terrestrial plants [36]. Our results indicate that temperature and salinity have an interactive effect on ascorbate levels. ASC accumulated under both hypo- and hypersalinity conditions at 8 °C. In contrast, at 2 °C, ascorbate showed either no significant differences or was lower under hypo- and hypersalinity conditions compared to the control, respectively. Additionally, the ascorbate-to-dehydroascorbate ratio (ASC/DHA) showed a similar trend, suggesting efficient maintenance of reducing power within the cycle to withstand high temperatures. This indicates that the antioxidant response of *A.* antioxidant response to hypo- and hypersalinity would be better at higher temperatures. Therefore, *A. utricularis* may efficiently cope with salinity changes under a climate change scenario, highlighting the capacity of macroalgae to optimize antioxidant responses under multiple stressors. Data from [15] suggest that these mechanisms allow Antarctic macroalgae to maintain adequate metabolite levels to counteract ROS induced by temperature variations. Similarly, the Arctic marine macroalga *Polysiphonia arctica* demonstrated an enhanced enzymatic activity, particularly APX and catalase (CAT), and production of novel antioxidant compounds, which contributed to their resilience against environmental stressors [37]. Additionally, the tropical red alga *Hypnea musciformis* also exhibited elevated glutathione (GSH) and ASC synthesis at higher annual temperatures. This robust antioxidant system enabled the macroalgae to tolerate oxidative stress fluctuations characteristic of intertidal habitats during seasonal changes [38].

Despite the better adaptability to salinity shown at higher temperatures in *A. utricularis,* it is important to highlight that, after the short-term exposure period studied (3 days), the species presented lower total ascorbate levels, at elevated temperatures, when compared to the control conditions (33 psu at 2 °C). This could suggest a potentially reduced fitness of *A. utricularis* under future climate change scenarios, regardless of salinity conditions. However, under salinity fluctuations at 8 °C, total ascorbate levels were slightly higher compared to control salinity, suggesting that the combined stress conditions triggered ascorbate accumulation. Indeed, it has been evidenced that a combination of abiotic stresses, such as salinity, heavy metals, and desiccation, among others, could enhance the biosynthesis of metabolites and promote macroalgae adaptability against stressful conditions [11].

Although these results are valuable for understanding *A. utricularis's* short-term response to thermal and salinity stresses, it is important to mention that ecosystem resilience cannot be predicted solely by examining individual species' responses, as interspecies interactions may enhance the overall resilience of an ecological system. For example, Arctic mixed kelp communities have shown remarkable short-term tolerance to multiple climate stressors, including warming, freshening, and reduced light availability [7]. Although light availability remains the primary factor affecting net community productivity, the combined effects of these stressors did not significantly impact productivity or photosynthetic efficiency in short-term experiments [7]. Interestingly, these results contrast with previous studies on individual species, which reported significant sensitivity to environmental stressors [39–41]. Therefore, the antioxidant responses of macroalgae are diverse and dynamic, tailored to specific ecological niches and stressors, resulting in a complex spectrum of tolerance and adaptation among species.

Increasing the rate of exploration and study of Antarctic macroalgae is crucial, as these species play key ecological roles and may hold unique biochemical adaptations. Expanding our knowledge in this region will enhance understanding of ecosystem functioning under changing environmental conditions and provide valuable insights into biodiversity and resilience in polar marine systems.

5. Conclusions

The ability of macroalgae such as *A. utricularis* to activate stress response mechanisms under multiple environmental stressors highlights their potential for adaptation to fluctuating conditions. Under current temperature conditions, this species exhibits heightened oxidative stress and damage in hyposaline environments. However, this pattern reverses under higher temperature scenarios, where the species demonstrates higher tolerance to hyposaline conditions. Notably, this species experiences the least oxidative stress under control conditions (2 °C at 33 psu) compared to any other scenario studied. This observation suggests that *A. utricularis* may be less suited to the elevated sea temperatures expected with climate change, regardless of salinity levels. Nevertheless, within a warmer temperature scenario, this species appears better adapted to salinity fluctuations (i.e., hyposaline environment) than to control (33 psu), potentially driving shifts in its distribution and relative abundance toward lower-salinity regions. These findings indicate possible ecological implications, but this hypothesis requires further validation through long-term experimental studies. Understanding how these organisms adapt to interactive stressors, including warming, freshening, salinity changes, and UVR, will be essential for predicting the resilience of polar ecosystems. Additionally, integrating molecular, biochemical, and ecological approaches can provide a comprehensive understanding of the mechanisms driving adaptation and resilience in these vital ecosystems.

Author Contributions: F.M.: Methodology, formal analysis, data curation, visualization, writing—original draft preparation. P.T.M.: Methodology, investigation, formal analysis, writing—review and editing. A.U.: Methodology, visualization, formal analysis, writing—review and editing, funding acquisition. P.S.M.C.-P.: formal analysis, methodology, visualization, writing—review and editing. C.R.: formal analysis, validation, writing—review and editing. P.A.-M.: formal analysis, methodology, writing—review and editing. V.M.: Conceptualization, methodology, visualization, writing—review and editing. N.P.N.: Conceptualization, methodology, formal analysis, writing—review and editing. C.A.S.: writing—review and editing, resources. C.L.: conceptualization, validation, writing—review and editing. C.T.: Methodology, formal analysis. B.C.: Methodology, formal analysis, visualization, writing—review and editing. K.G.-P.: Methodology, formal analysis, visualization, writing—review and editing. G.B.P.-H.: Formal analysis, supervision, conceptualization. F.R.-R.: Conceptualization, resources, methodology, visualization, supervision, writing—review and editing, funding acquisition. All authors have read and agreed to the published version of the manuscript.

Funding: This study was financed by the INACH RT_30-21 grant from the Instituto Antártico Chileno, Chile. VM was funded by Fondecyt Regular N°1211977 and COPAS COASTAL ANID FB210021. CL was funded by the projects ANID InES I+D 2021, grant number INID210013, and the Marie Curie Postdoctoral Fellowship HORIZON-MSCA-2022-PF-01 project, grant number 101106387, and CL acknowledges the Severo Ochoa Centre of Excellence accreditation (CEX2019-000928-S), funded by AEI 10.13039/501100011033. BC acknowledges Fondecyt Regular N°1221264 and ANID-Milenio-NCN2023_054. KG was financed by the UPLA inner doctoral scholarship UPA22991. Data obtained for total ROS quantification were granted by Cytation 5 equipment from the Fondequip EQM160131 project of the University of Playa Ancha, Chile.

Data Availability Statement: The data that support the results of this study can be obtained from the corresponding author upon request.

Acknowledgments: During the preparation of this manuscript/study, the author(s) used ChatGPT4 to improve the readability and language of the text. The authors have reviewed and edited the output and take full responsibility for the content of this publication.

Conflicts of Interest: The authors declare no conflicts of interest.

References

1. Convey, P.; Peck, L.S. Antarctic environmental change and biological responses. *Sci. Adv.* **2019**, *5*, eaaz0888. [CrossRef] [PubMed]
2. IPCC. Climate Change 2023: Synthesis Report. In *Contribution of Working Groups I, II and III to the Sixth Assessment Report of the Intergovernmental Panel on Climate Change*; Core Writing Team, Lee, H., Romero, J., Eds.; IPCC: Geneva, Switzerland, 2023; pp. 35–115. [CrossRef]
3. Siegert, M.; Atkinson, A.; Banwell, A.; Brandon, M.; Convey, P.; Davies, B.; Downie, R.; Edwards, T.; Hubbard, B.; Marshall, G.; et al. The Antarctic Peninsula under a 1.5 °C global warming scenario. *Front. Environ. Sci.* **2019**, *7*, 102. [CrossRef]
4. Hill, E.A.; Gudmundsson, G.H.; Chandler, D.M. Ocean warming as a trigger for irreversible retreat of the Antarctic ice sheet. *Nat. Clim. Chang* **2024**, *14*, 1165–1171. [CrossRef]
5. Röthig, T.; Trevathan-Tackett, S.M.; Voolstra, C.R.; Ross, C.; Chaffron, S.; Durack, P.J.; Warmuth, L.M.; Sweet, M. Human-induced salinity changes impact marine organisms and ecosystems. *Glob. Change Biol.* **2023**, *29*, 4731–4749. [CrossRef]
6. Umbert, M.; De Andrés, E.; Sánchez, M.; Gabarró, C.; Hoareau, N.; González-Gambau, V.; García-Espriu, A.; Olmedo, E.; Raj, R.P.; Xie, J.; et al. Contribution of satellite sea surface salinity to the estimation of liquid freshwater content in the Beaufort Sea. *Ocean Sci.* **2024**, *20*, 279–291. [CrossRef]

7. Miller, C.A.; Gazeau, F.; Lebrun, A.; Gattuso, J.P.; Alliouane, S.; Urrutti, P.; Schlegel, R.W.; Comeau, S. Productivity of mixed kelp communities in an Arctic fjord exhibit tolerance to a future climate. *Sci. Total Environ.* **2024**, *930*, 172571. [CrossRef] [PubMed]
8. Watt, C.A.; Scrosati, R.A. Seaweeds as ecosystem engineers. *Bull. Ecol. Soc. Am.* **2014**, *95*, 250–251. [CrossRef]
9. Gómez, I.; Huovinen, P. *Antarctic Seaweeds: Diversity, Adaptation and Ecosystem Services, Adaptation and Ecosystem Services*; Springer International Publishing: Berlin/Heidelberg, Germany, 2020. [CrossRef]
10. Kaviraj, A.; Unlu, E.; Gupta, A.; El Nemr, A. Biomarkers of environmental pollutants. *BioMed Res. Int.* **2014**, *2014*, 806598. [CrossRef]
11. Kaur, M.; Saini, K.C.; Ojah, H.; Sahoo, R.; Gupta, K.; Kumar, A.; Bast, F. Abiotic stress in algae: Response, signaling and transgenic approaches. *J. Appl. Phycol.* **2022**, *34*, 1843–1869. [CrossRef]
12. Moenne, A.; González, A.; Sáez, C.A. Mechanisms of metal tolerance in marine macroalgae, with emphasis on copper tolerance in Chlorophyta and Rhodophyta. *Aquat. Toxicol.* **2016**, *176*, 30–37. [CrossRef]
13. Ren, C.G.; Liu, Z.Y.; Wang, X.L.; Qin, S. The seaweed holobiont: From microecology to biotechnological applications. *Microb. Biotechnol.* **2022**, *15*, 738–754. [CrossRef]
14. Liang, W.; Ma, X.; Wan, P.; Liu, L. Plant salt-tolerance mechanism: A review. *Biochem. Biophys. Res. Commun.* **2018**, *495*, 286–291. [CrossRef]
15. Celis-Plá, P.S.M.; Moenne, F.; Rodríguez-Rojas, F.; Pardo, D.; Lavergne, C.; Moenne, A.; Brown, M.T.; Huovinen, P.; Gómez, I.; Navarro, N.; et al. Antarctic intertidal macroalgae under predicted increased temperatures mediated by global climate change: Would they cope? *Sci. Total Environ.* **2020**, *740*, 140379. [CrossRef] [PubMed]
16. Marambio, J.; Rosenfeld, S.; Bischof, K. Hyposalinity affects diurnal photoacclimation patterns in the rhodophyte *Palmaria palmata* under mimicked Arctic summer conditions. *J. Photochem. Photobiol.* **2022**, *11*, 100124. [CrossRef]
17. Sáez, C.A.; Troncoso, M.; Navarrete, C.; Rodríguez-Rojas, F.; Navarro, N.; Trabal, A.; Lavergne, C.; Pardo, D.; Brown, M.T.; Gómez, I.; et al. Photoprotective responses of three intertidal Antarctic macroalgae to short-term temperature stress. *Front. Mar. Sci.* **2023**, *10*, 1223853. [CrossRef]
18. Piccini, C.; Bertoglio, F.; Sommaruga, R.; Martínez de La Escalera, G.; Pérez, L.; Bugoni, L.; Bergamino, L.; Evangelista, H.; García-Rodríguez, F. Prokaryotic richness and diversity increased during Holocene glacier retreat and onset of an Antarctic Lake. *Commun. Earth Environ.* **2024**, *5*, 94. [CrossRef]
19. Parodi, C.; Cerpa, L.; Zhu, Z.; Zhang, J.; Muniz, P.; Venturini, N. Tracking suspended particulate organic matter biochemistry from glacial meltwater runoff to coastal waters of an Antarctic fjord. *Mar. Chem.* **2024**, *267*, 104455. [CrossRef]
20. Kim, B.; Lee, Y.; Hwang, J.Y.; Kim, Y.K.; Kim, T.; Kim, I.N.; Kang, S.; Kim, J.H.; Rhee, J.S. Physiological and molecular responses of Antarctic harpacticoid copepod *Tigriopus kingsejogensis* to salinity fluctuations—A multiregional study. *Environ. Res.* **2022**, *204*, 112075. [CrossRef]
21. Schreiber, U.; Bilger, W.; Neubauer, C. *Chlorophyll Fluorescence As a Nonintrusive Indicator for Rapid Assessment of In Vivo Photosynthesis*; Springer Berlin Heidelberg: Berlin/Heidelberg, Germany, 1995; pp. 49–70. [CrossRef]
22. Celis-Plá, P.S.M.; Bouzon, Z.L.; Hall-Spencer, J.M.; Schmidt, E.C.; Korbee, N.; Figueroa, F.L. Seasonal biochemical and photophysiological responses in the intertidal macroalga *Cystoseira tamariscifolia* (Ochrophyta). *Mar. Environ. Res.* **2016**, *115*, 89–97. [CrossRef]
23. Figueroa, F.L.; Conde-Alvarez, R.; Gómez, I. Relations between electron transport rates determined by pulse amplitude modulated chlorophyll fluorescence and oxygen evolution in macroalgae under different light conditions. *Photosynth. Res.* **2003**, *75*, 259–275. [CrossRef]
24. Figueroa, F.L.; Domínguez-González, B.; Korbee, N. Vulnerability and acclimation to increased UVB radiation in three intertidal macroalgae of different morpho-functional groups. *Mar. Environ. Res.* **2014**, *97*, 30–38. [CrossRef]
25. Eilers, P.H.C.; Peeters, J.C.H. A model for the relationship between light. intensity and the rate of photosynthesis in phytoplankton. *Ecol. Model.* **1988**, *42*, 199–215. [CrossRef]
26. Pérez-Hernández, G.; Morales, D.; Pereira-Rojas, J.; Díaz, M.J.; Blanco-Murillo, F.; Sola, I.; Rámila, C.; González, C.; González, K.; Sánchez-Lizaso, J.L.; et al. The halotolerant white sea anemone *Anthothoe chilensis*, highly abundant in brine discharges zones, as a promising biomonitoring species for evaluating the impacts of desalination plants. *Desalination* **2024**, *581*, 117612. [CrossRef]
27. Underwood, A.J. *Experiments in Ecology: Their Logical Design and Interpretation Using Analysis of Variance*; Cambridge University Press: Cambridge, UK, 1996. [CrossRef]
28. Rothäusler, E.; Rugiu, L.; Jormalainen, V. Forecast climate change conditions sustain growth and physiology but hamper reproduction in range-margin populations of a foundation rockweed species. *Mar. Environ. Res.* **2018**, *141*, 205–213. [CrossRef]
29. Gordillo, F.J.; Carmona, R.; Jiménez, C. A Warmer Arctic Compromises Winter Survival of Habitat-Forming Seaweeds. *Front. Mar. Sci.* **2022**, *8*, 750209. [CrossRef]
30. Luo, M.B.; Liu, F. Salinity-induced oxidative stress and regulation of antioxidant defense system in the marine macroalga *Ulva prolifera*. *J. Exp. Mar. Biol. Ecol.* **2011**, *409*, 223–228. [CrossRef]

31. Rodríguez-Rojas, F.; López-Marras, A.; Celis-Plá, P.S.; Muñoz, P.; García-Bartolomei, E.; Valenzuela, F.; Sáez, C.A. Ecophysiological and cellular stress responses in the cosmopolitan brown macroalga *Ectocarpus* spp. as biomonitoring tools for assessing desalination brine impacts. *Desalination* **2020**, *489*, 114527. [CrossRef]
32. Diehl, N.; Karsten, U.; Bischof, K. Impacts of combined temperature and salinity stress on the endemic Arctic brown seaweed *Laminaria solidungula* J. *Agardh. Polar Biol.* **2020**, *43*, 647–656. [CrossRef]
33. Barera, S.; Forlani, G. The role of proline in the adaptation of eukaryotic microalgae to environmental stress: An underestimated tool for the optimization of algal growth. *J. Appl. Phycol.* **2023**, *35*, 1635–1648. [CrossRef]
34. Xing, H.; Li, Q.; Zhao, Y.; Gao, H.; Li, L.; Zhang, Y.; Yu, X. Exogenous proline boosts the co-accumulation of astaxanthin and biomass in stress-induced *Haematococcus pluvialis*. *Bioresour. Technol.* **2023**, *369*, 128488. [CrossRef] [PubMed]
35. Hayat, S.; Hayat, Q.; Alyemeni, M.N.; Wani, A.S.; Pichtel, J.; Ahmad, A. Role of proline under changing environments: A review. *Plant Signal. Behav.* **2012**, *7*, 1456–1466. [CrossRef]
36. Foyer, C.H.; Noctor, G. Ascorbate and glutathione: The heart of the redox hub. *Plant Physiol.* **2011**, *155*, 2–18. [CrossRef]
37. Dummermuth, A. Antioxidative properties of marine macroalgae from the Arctic Antioxidative Eigenschaften mariner Makroalgen der Arktis. *Ber. Polarforsch. Meeresforsch. Rep. Polar Mar. Res.* **2003**, *458*, 9–11.
38. Maharana, D.; Das, P.B.; Verlecar, X.N.; Pise, N.M.; Gauns, M. Oxidative stress tolerance in intertidal red seaweed *Hypnea musciformis* (Wulfen) in relation to environmental components. *Environ. Sci. Pollut. Res.* **2015**, *22*, 18741–18749. [CrossRef] [PubMed]
39. Filbee-Dexter, K.; Wernberg, T.; Grace, S.P.; Thormar, J.; Fredriksen, S.; Narvaez, C.N.; Feehan, C.J.; Norderhaug, K.M. Marine heatwaves and the collapse of marginal North Atlantic kelp forests. *Sci. Rep.* **2020**, *10*, 13388. [CrossRef] [PubMed]
40. Lebrun, A.; Comeau, S.; Gazeau, F.; Gattuso, J.P. Impact of climate change on Arctic macroalgal communities. *Glob. Planet. Change* **2022**, *219*, 103980. [CrossRef]
41. Niedzwiedz, S.; Bischof, K. Glacial retreat and rising temperatures are limiting the expansion of temperate kelp species in the future Arctic. *Limnol. Oceanogr.* **2023**, *68*, 816–830. [CrossRef]

MDPI AG
Grosspeteranlage 5
4052 Basel
Switzerland
Tel.: +41 61 683 77 34

Environments Editorial Office
E-mail: environments@mdpi.com
www.mdpi.com/journal/environments